DEVELOPMENTS
IN THE SCIENCE
AND TECHNOLOGY
OF COMPOSITE MATERIALS

EUROPEAN ASSOCIATION
FOR COMPOSITE MATERIALS

DEVELOPMENTS
IN THE SCIENCE
AND TECHNOLOGY
OF COMPOSITE MATERIALS

6TH EUROPEAN CONFERENCE ON COMPOSITE MATERIALS

SEPTEMBER 20-24, 1993
BORDEAUX - FRANCE

EDITORS : A.R. BUNSELL, A. KELLY, A. MASSIAH

Woodhead Publishing Limited

With the sponsorship of

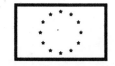

**Commission of the
European Communities**

Région Aquitaine

Copies of the publication may be obtained from :

EUROPEAN ASSOCIATION FOR COMPOSITE MATERIALS
2 Place de la Bourse - 33076 Bordeaux Cedex - France

WOODHEAD PUBLISHING LTD
Abington Hall - Abington Cambridge CB1 6AH - England

C EACM 1993

ISBN 1-85573-142-8

Printed in France

IV

SCIENTIFIC COMMITTEE ECCM-6

President :
Anthony KELLY
University of Surrey - Guildford
GU2 5XH Surrey - Great Britain

Vice Presidents :
Roger NASLAIN, LCTS (France)
Peter GARDINER, Smart Structures Research Institute (Great Britain)
Manfred NEITZEL, University of Kaiserslautern (Germany)
Anthony BUNSELL, Ecole des Mines de Paris (France)

Denmark
H. LILHOLT

France
C. BATHIAS
C. DROUET
J. JAMET
J.J. CHOURY

Germany
K. FRIEDRICH
G. GRÜNINGER
K. SCHULTE

Italy
G. D'ANGELO
A. CREDALI
A. SAVADORI
I. CRIVELLI VISCONTI

Latvia
Y.M. TARNOPOL'SKII

Russia
S.T. MILEIKO

Spain
A. GUEMES

Sweden
R. WARREN

Switzerland
R. PINZELLI

The Netherlands
A. BEUKERS

U.K.
M.G. BADER
B. HARRIS
F.L. MATTHEWS
D.C. PHILLIPS

COORDINATING AND ORGANISING COMMITTEE

General Secretary :
A. MASSIAH, EACM Bordeaux (France)

Director of ECCM Conferences :
D. DOUMEINGTS, EACM Bordeaux (France)

Denmark
H. LILHOLT

France
A.R. BUNSELL
A. MASSIAH
J.P. FAVRE
D. DOUMEINGTS

Germany
K. SCHULTE

Italy
I. CRIVELLI VISCONTI

U.K.
M.G. BADER

CONGRESS SECRETARIAT

C. MADUR, H. BENEDIC, N. TARDIEU
EACM - 2 Place de la Bourse
33076 Bordeaux Cedex - France
Tel +33 56 01 50 20, Fax +33 56 01 50 05

V

PREFACE

The ECCM conferences reflect the multidisciplinary nature of composites and the necessity of bringing together people with different interests to work together on these important subjects. The activities of the European Association for Composite Materials have generated both specialised meetings and the general conference series ECCM.

The Sixth conference in the ECCM series continues and enlarges the tradition set by the previous meetings which have encouraged a wide ranging debate on all aspects of composite materials.

ECCM-6 will be held for the fourth time in France following the success of both its predecessors. The meeting will take place in Bordeaux, the capital of AQUITAINE and a main centre for new technologies in France in which composite and advanced materials play a dominant role.

ECCM-6 will address the scientific, technical and industrial developments in composite materials. Colleagues from 24 countries proposed 180 papers, 125 of them are to be presented during the conference : 5 plenary papers, 90 oral papers and 30 posters, for which the importance of the poster sessions has been reinforced. ECCM-6 is set to be a most important scientific event attracting the majority of European specialists and many papers from the USA, China, Canada, Australia, Independent States Community and Japan. The papers come both from University and industrial research teams.

This conference has benefitted from generous support of the European Economic Community (EEC). The Directorate General 12 organises the first of a series of Brite Euram Workshops within the framework of ECCM-6 related to topics covered by the materials and industrial technologies programme.

Composites are finding increasing applications in an ever widening industrial context from bridges to gear wheels. There is an increase in papers by researchers applying their efforts to solve problems associated with the manufacture and use of composites. Non-destructive analysis, repair and design are subjects which are all attracting increasing attention. It is striking that the sessions dealing with fabrication and process modelling have attracted so many contributions of quality. Mechanical properties, ageing and damage tolerance continue to be major preoccupations of the composite research community and underline their importance.

In the name of all those who were involved in preparation and organisation of ECCM-6, we hope that ECCM-6 will show proof of the dynamism of this economic sector, and should show how to stimulate and orientate research along avenues even more closely related to industrial requirements.

ECCM has developed into a major scientific event which, although addressing first the interests of the European composite community, has found an equally important international role.

A. KELLY
President of ECCM-6

A.R. BUNSELL
President of EACM

TABLE OF CONTENTS

FIBRES, MATRIX, INTERFACE

FIBRE STRUCTURES

MECHANICAL PROPERTIES

XIII

ADVANCED COMPOSITES

FABRICATION TECHNOLOGIES, PROCESS MODELLING, NON DESTRUCTIVE TESTING

DESIGN AND NUMERICAL MODELLING

SESSION CHAIRMEN

- **FABRICATION TECHNOLOGIES AND MODELLING OF PROCESSES**
  A. KELLY, University of Surrey
  M. NEITZEL, University of Kaiserslautern
  S.T. MILEIKO, Russian Composites Society

- **FIBRES, MATRIX, INTERFACE**
  C. BATHIAS, CNAM
  K. SCHULTE, Technical University Hamburg-Harburg

- **DESIGN AND NUMERICAL MODELLING**
  E. EVENO, Du Pont de Nemours European Composites Development Center
  J.J. CHOURY, Sep

- **FIBRE STRUCTURES**
  I. CRIVELLI VISCONTI, University of Naples

- **MECHANICAL PROPERTIES**
  C. BAILLIE, University of Sydney
  D.C. PHILLIPS, Kobe Steel
  H. LILHOLT, Risø National Laboratory
  R. WARREN

- **AGEING AND DAMAGE TOLERANCE**
  P. GARDINER, Smart Structures Research Institute
  J. JAMET, Aérospatiale

- **MODELLING**
  A.R. BUNSELL, Ecole des Mines de Paris

- **APPLICATIONS**
  H. LILHOLT, Risø National Laboratory

- **ADVANCED COMPOSITES**
  K. FRIEDRICH, University of Kaiserslautern
  M.G. BADER, University of Surrey
  C. DROUET, Céramiques et Composites

- **FABRICATION TECHNOLOGIES, PROCESS MODELLING, NON DESTRUCTIVE TESTING**
  P.J. HOGG, Queen Mary & Westfield College

PLENARY PAPERS

- "Which materials for the future car"
  H. MATHIOLON

- "Simulation, detection and repair of defect in polymeric composite materials - a European research program"
  L.G. MERLETTI

- "The role of interface in polymer-based composites and test methods for evaluating its performance"
  C. GALIOTIS

- "Composite materials technology for civil and military aircraft in Spain : recent achievements"
  J.A. GUEMES, P. MUNOZ-ESQUER, F. LICEAGA

- "Composite materials in Russia : present position and prospectives"
  B.V. PEROV (not received to be published)

WHICH MATERIALS FOR THE FUTURE CAR

H. MATHIOLON

*RENAULT - Projet X08-Mosaic - 29 rue Chateaubriand
92500 Rueil Malmaison - France*

SUMMARY

The history of the motor car is inextricably tied up with the history of materials, since much progress is due to advances in materials. The motor car, which is a major market for materials, is changing rapidly from several viewpoints: customer needs with respect to comfort, spaciousness, safety and reliability, integration into the environment and environmental protection. The present paper is an attempt to determine possible developments in materials through the foreseeable development of the motor car.

Introduction

The present situation of automotive materials is still relatively stable. Metallic materials (sheet metals, steel, cast iron and aluminium alloys), although diminishing, still represent 70% of the weight of vehicles. Polymer materials now exceed 15%, while the remainder consists of inorganic materials, paints and liquids. As for ceramic materials, they are not yet present except in the form of wear rings or in the spark plugs or catalytic converter supports.

Starting from the present situation, we shall examine the projected evolution of the motor car, in order to deduce criteria for materials' selection. We shall then examine the possible trends for the main families of materials and their situation faced with the questions posed by environmental protection.

3

1. Present Situation

1.1. Knowledge of materials and processes

To build vehicle structures, the car makers chiefly use steel sheets, and their experience with this material and its processing has been refined over several decades.

The appearance of a few vehicles with a body made of polymer materials, such as the Renault Espace and the G.M. Saturn, is recent and limited.

The use of aluminium, although it has a long history (remember the shell body of Grégoire 1934 and the Panhard Dyna of the fifties), has not yet become widespread due to its cost, except on some top-of-the-range vehicles such as the Honda NSX and the Jaguar XJ 220.

For the mechanical parts, steel is used almost exclusively for machined parts, while for castings there is still fierce competition between cast iron and aluminium alloys, which represent only 4% to 5% of the weight of our vehicles.

1.2. The trends

Over the last decade, with the introduction of regulations concerning the environment (catalytic converter) and the improvement of safety, combined with the market demand for spaciousness, comfort and driving pleasure, the car makers have increased the weight of their vehicles by approximately 10 to 15%.

In the same period, we have seen a gradual reduction in the use of ferrous metals with increased use of all the other materials, especially sound insulation materials.

2. What Type of Motor Car for the Future?

2.1. The market demand

Customer demand will correspond to what the car makers are able to offer in the field of acoustic and vibratory comfort, spaciousness and the adaptability of the vehicle. The market will also evolve towards more customized cars, hence with more variants. The requirements with respect to reliability and long life will be more demanding and will need the implementation of Total Quality policies such as Renault has had for some years now.

2.2. Regulations

With the evolution of regulations relating to safety on the one hand and environmental protection on the other hand, and in particular the future standards on CO2, if they are applied, the car makers are likely to change all the parameters: reduction of vehicle weight, improvement of engine efficiency, reduction of friction, etc.

Moreover the non-polluting destruction of vehicles at the end of their life cycle implies that, as of now, materials should undergo stringent selection. They shall be either recyclable or destructible by combustion or any other process, to the exclusion of scrapheaps, which will become too costly, or even prohibited. Note that in this field the standards are becoming more stringent, with greater liability of the car makers according to the slogan "the polluter pays".

2.2. The consequences

The constraints outlined above show clearly that the vehicles of the future are very likely to become heavier to comply with the market demand and achieve improved safety, whereas their weight ought to be reduced to comply with the future measures on fuel consumption.

It therefore seems obvious that the solutions which can be employed will be more expensive, except if new technologies appear, which implies closer cooperation between the designers and the producers on the one hand, and also between the materials' suppliers and the car makers. We shall discuss further on an attempt along these lines under Renault's MOSAIC project.

3. Criteria for materials' selection

In fact, the material in itself is of no interest to the designer. The material is merely one of the components in a vast interactive system taking into account the production processes, the technologies already in place and thoroughly mastered, design tools (CAD), the potential for modelling and simulation, etc., leading to the required performance and quality of the new product at a minimum cost. It is the function defined by the technical specifications which is of prime importance.

The search for the optimum compromise between materials and manufacturing process therefore depends on the function to be fulfilled and as a consequence on the specific features of the motor car, but also technical and economic criteria specific to the material.

3.1. Criteria specific to the motor car

The motor car is a consumer good mass produced at a relatively low cost, of between FF70 and FF100 per kg depending on the equipment, which is to be compared with more sophisticated products such as aerospace equipment at FF4000 or even FF5000 per kg.

Mass production at a low cost implies very onerous investments in new vehicle research and development work and also in production facilities. This results in great inertia. Investments in production machinery are required to provide at least twenty years of service in most cases, and only those investments specific to a vehicle model are depreciated directly on that model.

The motor car also has a limited lifetime which is constantly tending to grow shorter; the first R5 was produced for twelve years and the Super 5 for seven years, with a face lift in mid-life.

This shorter production life is leading to shorter design times, which are gradually decreasing from five years to four years. However, given the desired improvement in product quality, only those solutions which have been tested from both the technical viewpoint and on the industrial production level can be selected in the design stage.

All these factors lead to complex decision-making structures and great inertia, which tend to keep the existing solutions in place and do not promote change, particularly in materials.

3.2. Materials criteria

These criteria are obviously technical, but also industrial and socio-economic.

Intrinsic technical criteria are related to mechanical properties (resistance, rigidity, fatigue limit, wear) and to specific properties (resistance to corrosion, aging, thermal or electrical conductivity, damping properties, etc.). The relevant quantities are generally well known.

Industrial criteria are related to the use and conversion of materials by machining, stamping, foundry work, injection, assembly by welding or bonding or mechanical fasteners, heat treatment and protective treatment, etc. These processes can influence the end properties of the part: the material cannot be considered separately from its manufacturing process. This is especially important for composite materials where very often the end material is obtained in the mould at the same time as the part is produced.

Economic criteria
These criteria concern on the one hand the price of the material itself due to its availability, the market trend and its speculative nature; an analysis, both static and dynamic, must therefore be made of the price trend for the raw material. The criteria also include the costs of use related to investments and the production costs which are influenced by volumes and production rates.

As the car maker has to know the cost trend at least two to three years before bringing out a new model and until the end of production of the model, the price variations must be known for a period of approximately ten years.

Social criteria
These criteria are related to social questions which will influence customer trends: energy savings, improved safety, environmental protection, etc. There are also criteria concerning the production personnel: safety, toxicity, noise, etc.

These important factors are not always easy to assess.

But with all these constraints, then, what materials should be selected?

4. What Materials?

Competition between the various families of materials is fierce and will become far more severe during the present decade, especially competition between ferrous metals and aluminium or magnesium alloys, and thermoplastic and thermosetting polymer materials.

Two types of development are considered, one not fundamentally bringing into question the existing production resources, i.e. continuous development, called Kaizen by the Japanese, and the other bringing everything into question, i.e. the scenario of radical technological change.

4.1. Scenario of continuous development

The most efficient material is that which corresponds to just what is needed to perform the function. In practice, the number of parameters to be allowed for leads to gradual optimization of the characteristics of the material itself and the manufacturing processes.

For **mechanical parts**, changes in the manufacturing processes for metallic materials, such as squeeze casting, high-pressure vacuum casting of light alloys, local heat treatments, etc. are leading to the development of new applications for well known materials or the creation of composite materials.

Some examples: suspension arms made from cast iron, forged steel or aluminium alloys, with some unsuccessful research on polymer composite materials.

After competition between cast iron, forged steel and aluminium, the connecting rods are now made from forged steel with surface treatment and optimized design, which represents the optimum compromise between weight reduction and cost. For the future, the development of aluminium metal-matrix composites and powder metallurgy seem very promising in terms of weight reduction but are too expensive at present.

As for **thermomechanical ceramics**, they aroused great interest for a decade but are now used only for friction and wearing parts such as rocker shoes, valve guides and piston pins. Their development is greatly limited by their cost, fragility and difficulty of use.

For **structural and body parts** which, for several decades now, have chiefly been made of steel sheet, there has been major development of this material and its corrosion protection properties. Examples are high-strength low-alloy steel sheet (HSLA), hot-dip galvanized and electrozinc coat steel sheet, etc.

Recent developments in these materials are moving in two directions, one towards **sandwich sheet metals** consisting of two thin metal sheets bonded together by a thin layer of polymer material, approximately 50 microns thick; these sheet metals have a

6

vibration damping factor which makes it possible to eliminate heavy sound insulation materials which create problems for recycling. The other avenue of development towards process improvements through the development of efficient software for **simulation of stamping,** so as to limit the material's thickness reduction during stamping, and hence allow thinner sheets to be used.

A major development, but which would not fundamentally compromise the production tool, could be a switch to aluminium sheet metal, which would require merely changes in thickness and assembly techniques but in which the tools for production, stamping and weld assembly could be adapted. Although the weight reduction achieved could be substantial, approximately 35 to 40%, the cost would probably be doubled; that is why few attempts have been made recently in this area, apart from the Honda NSX, and except for specific parts such as the engine bonnet.

4.2. Scenario of radical technological change

For **mechanical parts** a few examples may be mentioned, such as the Ford-Polymotor engine in which the oil pan, the cylinder block, the cylinder head and the cylinder head cover are made from polymer composite reinforced with glass fibre, while the central core of the engine block is still made from aluminium alloy and mechanical parts subject to heating are made from metal materials.

Another example is the suspension blades made from composite polymer materials reinforced with glass fibre or even carbon, with various levels of integration of the wheel steering function.

For **structural parts,** there could also be very great change, affecting not only materials but also the fabrication technologies.

As an example which will enable us to see the main avenues of change, I shall take the MOSAIC project (Matériaux Optimisés pour la Structure d'une Automobile Innovant dans la Conception - "Optimized materials for the structure of a motor car of innovative design"), which has obtained the EUREKA label, and which I have been managing for more than three years at Renault. The project aims at creating a vehicle structure of the Clio type, of lighter weight due to the use of new materials. Two paths have been followed, one according to the scenario of continuous development, and the other adopting radical technological change. The first path has given us a weight reduction of up to 10% at a very similar cost of production, using the products mentioned above, sandwich sheet metals and simulation of stamping as of the design stage. This interaction between design and production is very interesting and, in addition to weight reduction, allows shorter delivery times to be achieved, since the problems are known by the producers as of the design stage.

The other path represents a radical change, since it no longer uses steel, but merely light alloys and polymer composites reinforced with glass fibre. At the outset of the project, we determined the budget available for producing a structure approximately 20% lighter for the same cost price as the present structure of the Clio. We realized that the gross cost price of materials would have to be less than FF20/kg to succeed. This implies extensive integration of functions with a major reduction in the number of parts, by a factor of 3 or 4. For mass production, we have researched processes for achieving short cycle times, of approximately one minute. This approach led us towards materials in the form of semi-finished products bringing us closer to our objective, namely:

* aluminium alloys, preferably in the extruded form, for all parts in the shape of a longitudinal member, avoiding insofar as possible castings and sheet metals which are too expensive and include no major integration of functions;
* composite polymer materials reinforced with a high content of up to 50% glass fibre, converted by the SMC (Sheet Molding Compound) or SRIM (Structural Reaction Injection Molding) process for surface type parts, such as the floor or the wheel wells.

7

Insofar as concerns fabrication techniques, we have extensively developed bonding with our partners, because this solution is useful for assembling materials of different types, and in particular polymer composites with aluminium alloys. Major progress has been made in both adhesives and the surface treatments required, and in processes for accelerated polymerization.

Mechanical assembly techniques, which have also progressed, such as clinching, riveting, etc. and welding have been examined and are partly used alone or in association with another process.

Although I have no doubt not completely answered the question, I hope to have at least shed some light on it. In particular, I hope I have made you more aware of the difficult choices faced by the car makers.

Although the MOSAIC project is not completed, the results obtained at present suggest to us that our objectives were realistic and will be achieved.

For **body parts**, the same technologies are applicable with conventional skins, skins of aluminium sheet or of thermosetting polymer, as in the Renault Espace, or thermoplastic skins as in the G.M. Saturn. For this section of the vehicle the technical decision is greatly influenced by cost prices and production rates.

4.3. Applications

The progress achieved by continuous development is applicable with normal time limits compatible with the design of our new models, because, as we have seen, they do not require fundamental changes in our production tools.

The progress achieved according to the scenario of radical change, although much more promising, will be **slow to apply**, since it requires that the car makers re-examine almost all the specialist activities involved in the development of a new product, from design through to production and after sales.

5   The Environment

As you know, the French car makers have signed an undertaking with the government authorities. New models shall be 90% recyclable by the year 2002, while all models shall be 95% recyclable in 2010.

This commitment shows that the car makers want to conserve the environment; **car graveyards must disappear.** But this very laudable intention will inevitably pose many problems, especially for polymer materials with a high glass fibre content.

In fact, there is a lot of talk on recycling. However, environmental protection must not be confined to the mere recycling of materials, which, even if it allows a reduction in the number of scrapyards, which will become more and more expensive (FF1000/tonne or more) or will be prohibited, does not take into account the energy content of materials nor environmental protection during the production stages. In addition to recycling, the eco-balance must be looked at as a whole, i.e., environmental protection throughout the material's lifetime "from the cradle to the grave", associated with the energy savings and pollution reduction made possible by the material through a reduction in vehicle weight. It is generally considered that a 10% reduction in vehicle weight gives an average reduction in fuel consumption of approximately 5% for driving in the Paris region.

It is now, as of the design stage, that the car makers are facing up to these difficult questions, because they generally have economic repercussions. The decisions taken now will have effects over the next 20 to 25 years: five years for design, seven years for production of a model, and five to ten years of use.

In the present situation, steels are recycled, light alloys are completely recyclable and composite materials are progressing towards recyclability but still have a long way to go. When the question is looked at as a whole, as mentioned earlier, taking into account the energy savings obtained by reduced fuel consumption, the differences between material families are far smaller.

Insofar as concerns fabrication techniques, we have extensively developed bonding with our partners, because this solution is useful for assembling materials of different types, and in particular polymer composites with aluminium alloys. Major progress has been made in both adhesives and the surface treatments required, and in processes for accelerated polymerization.

Mechanical assembly techniques, which have also progressed, such as clinching, riveting, etc. and welding have been examined and are partly used alone or in association with another process.

Although I have no doubt not completely answered the question, I hope to have at least shed some light on it. In particular, I hope I have made you more aware of the difficult choices faced by the car makers.

Although the MOSAIC project is not completed, the results obtained at present suggest to us that our objectives were realistic and will be achieved.

For body parts, the same technologies are applicable with conventional skins, skins of aluminium sheet or of thermosetting polymer, as in the Renault Espace, or thermoplastic skins as in the G.M. Saturn. For this section of the vehicle the technical decision is greatly influenced by cost prices and production rates.

4.3. Applications

The progress achieved by continuous development is applicable with normal time limits compatible with the design of our new models, because, as we have seen, they do not require fundamental changes in our production tools.

The progress achieved according to the scenario of radical change, although much more promising, will be slow to apply, since it requires that the car makers re-examine almost all the specialist activities involved in the development of a new product, from design through to production and after sales.

5   The Environment

As you know, the French car makers have signed an undertaking with the government authorities. New models shall be 90% recyclable by the year 2002, while all models shall be 95% recyclable in 2010.

This commitment shows that the car makers want to conserve the environment; **car graveyards must disappear.** But this very laudable intention will inevitably pose many problems, especially for polymer materials with a high glass fibre content.

In fact, there is a lot of talk on recycling. However, environmental protection must not be confined to the mere recycling of materials, which, even if it allows a reduction in the number of scrapyards, which will become more and more expensive (FF1000/tonne or more) or will be prohibited, does not take into account the energy content of materials nor environmental protection during the production stages. In addition to recycling, the eco-balance must be looked at as a whole, i.e., environmental protection throughout the material's lifetime "from the cradle to the grave", associated with the energy savings and pollution reduction made possible by the material through a reduction in vehicle weight. It is generally considered that a 10% reduction in vehicle weight gives an average reduction in fuel consumption of approximately 5% for driving in the Paris region.

It is now, as of the design stage, that the car makers are facing up to these difficult questions, because they generally have economic repercussions. The decisions taken now will have effects over the next 20 to 25 years: five years for design, seven years for production of a model, and five to ten years of use.

In the present situation, steels are recycled, light alloys are completely recyclable and composite materials are progressing towards recyclability but still have a long way to go.

When the question is looked at as a whole, as mentioned earlier, taking into account the energy savings obtained by reduced fuel consumption, the differences between material families are far smaller.

There is no doubt that polymer composite materials will have a brilliant future in the automotive industry only to the extent that they can achieve progress in this area.

6. Conclusion

It is not easy to conclude, because forecasting in this area is very difficult and is tied up with changes not only in materials but also in legislative measures.

However, it can be asserted that the universal will to show greater respect for human health and our natural surroundings is leading all the car makers to work on improving safety and reducing pollution by a reduction in fuel consumption and car emissions, from production of the material through to vehicle destruction.

However, these applications will be introduced very gradually, since steel is holding its own and has not yet said its last word.

This is the best chance for developing so-called new materials for the motor car, such as light alloys and polymer composite materials.

SIMULATION, DETECTION AND REPAIR OF DEFECTS IN POLYMERIC COMPOSITE MATERIALS - A EUROPEAN RESEARCH PROGRAM -

L.G. MERLETTI

AGUSTA S.p.a. - Via giovanni Agusta 520 - 21017 Gascina Costa di
Samarate (VA) - Italy
EUROCOPTER DEUTSCHLAND GmbH - RISØ NATIONAL LABORATORY -
WESTLAND HELICOPTER Ltd

Abstract

This research project is supported by the European Communities under the BRITE EURAM program and it is addressed to the study of the behavior of CFRP in the form of solid laminates and of sandwich panels with Nomex honeycomb. As suggested by the title it develops into three main themes connected to each other which gives its acronym:**DeSiR** (**De**tection, **Si**mulation and **R**epair). The aim of the first theme is the definition of the role that defects and damage play in the behavior of composite materials. Porosity, resin content variation, delamination in solid laminates as well as impact damage both on solid laminates and on sandwiches have been evaluated. The aim of the second theme is the definition of the most appropriate Non Destructive Evaluation method to detect the presence of defect (or damage) and to measure its severity. The aim of the third theme is the definition of the most appropriate method to repair impact damage both in solid laminate and in sandwich structures.

Introduction

The use of polymeric composite materials in the manufacturing industry has given a solution to several typical problems connected with the use of other materials. One of these solutions is the opportunity to save weight by replacing metal parts and this strongly meets the sensitivity of the aeronautical design engineers; therefore the use of polymeric composite materials in the aeronautical field is continuously increasing even in the manufacture of critical parts. But while many problems have been solved, other problems have been created by the use of composite materials; for example we could mention the reduction of mechanical properties caused by voids or by the presence of delaminations or debonding, by the variation of the ratio of resin/fiber, the susceptibility to impact damage and therefore the repairability problems, the moisture absorption and so on. Consequently nowadays the aeronautical materials engineers are facing the problem of understanding the effects of these defects and damage firstly on the material itself and then on the behavior of critical parts and the influence that they could have on the aircraft certification.

In this program, the interest of the research has been devoted both to solid laminates and to sandwich structures. Carbon Fiber Reinforced Plastic (CFRP) was the selected material to manufacture them; more precisely the solid laminate structures have been manufactured with two different types of unidirectional pre-preg (indicated in the tests as material 1 curing at 125° C and material 2 curing at 175° C) and the sandwich structures have been manufactured with two different types of fabric pre-preg for the skin (again indicated with material 1 curing at 125° C and material 2 curing at 125° C) and nonmetallic honeycomb for the core. As suggested by the title of the program, it develops into three main themes; specifically: the Simulation of defects and damage, their Non Destructive Detection and the Repair of the damage.

Simulation

The aim of the first theme is the definition of the role that defects and damage play to affect the behavior of composite materials. The first approach to this theme was the definition of the type of defects and the properties affected by their presence, then the definition of the most suitable method to simulate them in specimens and thirdly the most appropriate method to evaluate them with a destructive test. Some defects expected during a manufacture process, such as voids or porosity, resin content variations and delaminations and some expected during service such as impact, were selected for this evaluation.

Porosity

The presence of voids or porosity in the resin matrix of a solid laminate is caused mainly by the lack of pressure during the cure cycle and themselves cause a reduced contact between two layers; consequently the interlaminar shear strength was selected as the mechanical property that would be affected by porosity and therefore 20 ply unidirectional 0° specimens were chosen for this test according the ASTM Standard Test Method D 2344 84. To simulate the presence of porosity in a solid laminate, two possible methods have been reported: pressure reduction of the cure and the lack of flow in the matrix. Of these two, it was decided that the method of reducing the pressure of the cure would be used with the aim to obtain 4 different levels of porosity (optimal laminate, 2%, 4% and 6%). Of course it was quite impossible to get plates with homogeneously distributed voids, but, starting from the Ultrasonic attenuation C-scan, homogeneous areas in the plates have been identified and Inter Laminar Shear Strength specimens have been obtained from them. Another point of discussion was about the method of measuring the void content in a specimen. Two methods are normally used for this purpose: nitric or sulphuric acid digestion and image analysis of polished cross sections. The main disadvantage of the former is that the resin and fiber densities can really only be estimated and a small change in either value causes a significant change in the obtained void content value, but on the other hand the latter method measures the void content only on a small section of the specimen and it relies greatly on operator interpretation of what actually is or is not a void. The logical consequence of these problems was the agreement to measure the void content using both methods and to compare the results. An example graph of Inter Laminar Shear Strength vs. percentage of voids evaluated with the two methods for the material 2, is reported in figure 1.

Resin Content Variation

Resin content variation was the second type of defect taken into account in solid laminate specimen. The variation of this parameter is mainly caused by an uncontrolled flow of the resin or by an inhomogeneous distribution of the pressure of the cure. Both flexure and tensile strength of 0° unidirectional solid laminate specimens are strongly effected by resin content variations but the former mechanical property was selected to evaluate its effect because the thickness of typical plates taken into account is 1.4-2.1 mm thick (13 plies) and it is difficult to find adhesive strong enough to bond tabs to so thick a 0° tensile specimen. The 3 point bending test according ASTM Standard Test Method D 790 was selected to evaluate the flexural strength. Two possible methods to create the variation of resin content have been investigated. In the first one, starting from the knowledge of the rehological behavior of the resin and therefore its viscosity curve during the cure cycle, some experiments have been tried applying the pressure at different times during the cure. This method didn't give us the expected results therefore another method has been investigated. By means of different combinations of bleeders, cure pressure and sealed or unsealed tool, we tried to promote or inhibit the squeeze out of the resin obtaining plates with four different levels of resin/fiber ratio which has been measured by means of the nitric acid digestion method. From these plates 25 specimens for flexural strength tests have been obtained. The graph flexural strength vs. resin content is reported in figure 2.

Delamination

Among the in manufacturing defects investigated in this program, the internal delamination (or debonding between layers of solid laminates) is the most critical one and its presence is not allowed in real parts because it affects several characteristics of composite materials; more specifically it strongly reduces the compression strength of a part. Therefore the compression strength test was chosen to evaluate its effects. The reduced thickness of typical plates of interest (approximately 2 mm of 17 ply in a quasi isotropic lay up) created the problem of buckling of the specimen during the test.A specific anti buckling device with 9 rails for each side of the plate has been developed for this purpose. To simulate the presence of internal delamination in a solid laminate, several methods have been reported but the most widely employed technique for this purpose is the placement of permanent inserts between adjacent pre-preg plies before curing. Moreover, in order to guarantee a perfect lack of adhesion and to reduce the influence of foreign material, a double layer of a very thin PTFE film was used to produce the permanent insert.The PTFE film was cut into rectangles with a 2:1 ratio of the sides and with different areas; then these rectangles were folded in half to produce double layered squares which were inserted between plies. In this manner 25 specimens in material 1 and 25 specimens in material 2 for compression tests have been manufactured.

Impact Damage

Impact damage is the typical damage expected in composite parts during handling and service; impact damage means delaminations and even fiber breakage in solid laminate structures and delaminations, debonding, crushed honeycomb, fiber breakage and even perforation of one or both face sheets in sandwich structures. This damage is caused by low energy impacts which may occur during aircraft manufacture and in operational service. Risk studies report that the impact energy seldom exceeds a level of 25-30J and that the most suitable technique to simulate the damage is the drop test using a hemispherical impactor. Therefore we selected to inflict the impact damage by means of a 25 mm diameter hemispherical impactor with the specimens clamped between two metallic plates with a concentric hole of 150 mm diameter and with the impactor dropping in the center of the free concentric area.

Three different states of damage were defined on the basis of preliminary tests both for sandwich and for solid laminate structures (in both cases defined as: barely visible impact damage or BVID, intermediate and maximum damage). On sandwich structures, manufactured both in material 1 and in material 2, it was agreed to define:

a) BVID as the damage obtained with a 3J impact and that appears as a 1-1.5 mm deep surface indentation,

b) intermediate damage as that caused by a 10J impact and that appears as one face sheet penetrated,

c) maximum damage as that caused by a 30J impact and that appears as complete perforation of the structure.

On solid laminate structures a different impact behavior between material 1 and material 2 was observed therefore the definition of BVID, intermediate and maximum damage was determined on the basis of the extent of the internal delamination caused by the impact as assessed by ultrasonic inspection. Specifically the BVID was defined as the damage corresponding to an internal symmetrical delamination of 150-200 mm^2 which was obtained with a 12J impact for material 1 and 4J impact for material 2, the intermediate damage (corresponding to an internal symmetrical delamination of 350-400 mm^2) was obtained with an impact of 20J for material 1 and 12J for material 2 and the maximum damage (corresponding to an internal symmetrical delamination of 800-1000 mm^2) was obtained with an impact of 30J for material 1 and 18J for material 2. The reduction in compression strength of both types of structures was chosen to evaluate the effect of impact damage; moreover, because the data obtained from this activity will be used as a reference basis for the evaluation of the restored characteristics from the repair activity, it was agreed to increase the evaluation of the behavior of the structures performing the compression tests under both static and dynamic load and in both undamaged and damaged condition.

Detection

The aim of the second theme is the definition of the most suitable Non Destructive Evaluation method and procedures to detect the presence of defects and to assess their severity. Moreover because impact damage could be inflicted in operational service, some activities were devoted to the in field application.

Porosity

The presence of porosity in solid laminates mainly causes the scattering of the ultrasonic wave, therefore the attenuation of the ultrasonic beam is the parameter to observe for the evaluation of the severity of void content. The inspection of the plates was performed in water immersion both in double through transmission and in pulse echo mode by means of precision scanning equipment with a focussed transducer. In both cases the amplitude of the echo was recorded; in the former from a mirror behind the plates, in the latter from the back wall. The best results have been obtained from the inspection in the double through transmission mode. Starting from the obtained C-scans, areas with the attenuation in a range of 2 dB were identified and considered as having a homogeneous void content.

Resin Content Variation

Starting from the fact that the short transverse elastic modulus of a composite plate increases with an increase in the fiber/resin volume ratio and from the fact that the sound propagates into the resin with a velocity which is very different from the propagation velocity of the sound into the fibers, we assumed that the resin/fiber ratio could be assessed by means of the evaluation of the velocity of the sound compression wave. Of course this assumption was made in first approximation because the real model of the sound propagation is more complex and it isn't within the scope of this program to study it. The short transverse elastic modulus strongly depends upon the void content of the specimen while in the specimens tested for the evaluation of the resin content variation effect, the void content was almost constant (in comparison with the resin content variations). However, as shown in the graph of figure 3, a very good correlation between the ultrasonic velocity and the resin content variation has been found.

Delamination

While there isn't any doubt that a double layer of PTFE really simulates a delamination in a laminate, it was found that the Non Destructive test method could not always detect the defect. In fact in the C-scan obtained from the ultrasonic inspection of some specimens, the central area of the simulated defect appears continuous; this was assumed to be due to the absence of a real separation between the resin and the PTFE film and between the two folded foils of PTFE that doesn't create a reflecting interface (on the other hand PTFE and CFRP have approximately the same acoustic impedance and therefore the ultrasound cannot detect the presence of the two different materials. Several attempts (e.g. to freeze the specimen or to drill a little hole in the center of the defect) are under investigation trying "to open the bad area" and to create a real reflecting interface, but at the moment no repeatable results have been obtained.

Impact Damage

Because impact damage could be caused to a composite part during both the aircraft manufacture and operational service, its detection was faced by means of two different approaches. In both cases the inspection was based on the ultrasonic method but in one case it was carried out in water immersion mode and in the second one by means of a portable manual scanner with the aim to make it applicable as an "in service inspection". In the water immersion inspection, the first precaution to take is to prevent the water penetrating into the delaminations produced by the impact because the water could improve the ultrasonic coupling reducing the sensitivity of the inspection; this was obtained by means of the application of a very thin adhesive film. Another opportunity to obtain a good resolution from the water immersion inspection is given by the use of high frequency focussing transducers, but the composite material filters the high frequency consequently reducing the resolution. The solution to this problem on sandwich structures lies in the through transmission inspection with the use of a very small hydrophone as a receiving transducer. In this case, taking care that the distance between the tip of the hydrophone and the surface of the specimen is less than 1 mm, the resolution was improved to 1 mm. An example of the obtained C-scan is reported in figure 4. For the inspection of solid laminate specimens a pulse echo technique was used. With this technique the time of flight of the echo with the highest amplitude within the gate is recorded; so, when the center line of the ultrasonic beam is crossing the border of a delamination, the echo from the delamination itself exceeds the back wall echo and therefore it will be recorded. Using a focussing transducer and an indexing of about 1mm during the inspection a C-scan with a good resolution was obtained. An example of C-scan is reported in figure 5.

Repair

The aim of the third theme is definition of the most appropriate method to repair the impact damage in both the structures of interest. From the mechanical tests and from the evaluation of the reduction of compression strength, it was inferred that the BVID could be tolerated and the maximum damage (which corresponds to the complete perforation of the structure) would cause the rejection of the part; therefore the intermediate damage was selected for repair activities. Three methods for repairing sandwich structures and three methods for solid laminate structures are under investigation for the selection of the most suitable one. These selections are based on the evaluation of the restored static mechanical properties as well as the complexity of the procedure in terms of products and tools, the quality of the repaired surface and so on. Then the selected methods will be subjected to characterization by means of the comparison of the restored mechanical properties of the repaired specimens. The characterization will be developed through conditioning before and after repairing and then testing under dynamic and static loads.

Interelationship Between Themes and Conclusions

Currently all data is loaded into a specific Data Base and will be used to validate a model that we are going to study.

From the Simulation theme activities, indications about the criticity threshold of defects and damage have been obtained. These indications governed the Detection activities in term of sensitivity of NDT methods to allow the detection and the assessment of defects. On the other hand these indications governed the choice of the damage to Repair. From the Detection activities and more specifically from the Non Destructive Evaluation associated with the Data Base and the Modelling activity, the possibility to oversee the influence that a damage could have on the mechanical behavior of a structure is aspected; moreover the Detection theme gave indications of the real extension of the damage to Repair despite of the external aspect of it. At last the Repair theme is aspected to give the possibility of a complete restore of the mechanical characteristics of a damaged structure.

During the present program, work was done on some typical configurations of composite materials; at this point and after the completion of the program, it appears fruitful to extend the investigation to other configurations such as big solid laminates, bond line between precured laminates, weak bonding and so on.

Aknowledgements

Many people have helped to make the DeSiR project a success. Particular credit should be given to R. Pezzoni and M. Rigamonti (Agusta), N. Davies and A. Lizza (CIRA), R. Schindler and K. Wolf (Eurocopter Deutschland), P. Brøndsted, S.I. Andersen, K.K. Borum and E. Gundtoft (RISØ) and A. Dew and N.G. Marks (Westland Helicopters).

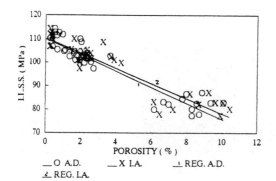

Figure 1
Interlaminar shear strenght vs. percentage of voids content in material 2 solid laminate specimens evaluated by menans of both acid digestion and image analysis.

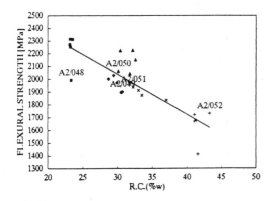

Figure 2
Flexural strength vs. resin content in material 2 solid laminate specimens.

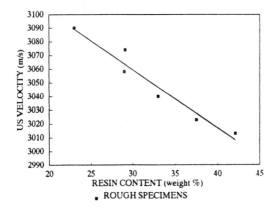

Figure 3
Ultrasound velocity vs. resin content variation (each point represents the average of resin content in 5 material 2 solid laminate specimens).

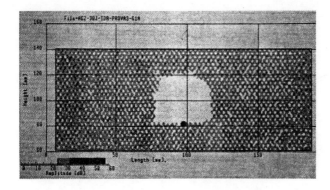

Figure 4
Amplitude C-scan
of a material 2
sandwich specimen
with a 30J impact.

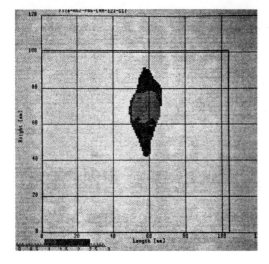

Figure 5
Time of flight (converted in
depth) C-scan of a material 2
solid laminate specimen with a
12J impact damage.

THE ROLE OF INTERFACE IN POLYMER-BASED COMPOSITES AND TEST METHODS FOR EVALUATING ITS PERFORMANCE

C. GALIOTIS

Queen Mary & Westfield College - Mile end Road - London E1 4NS - UK

ABSTRACT

This paper reviews the role of interface in polymer based composites and describes in some detail the most common state-of-the-art micromechanical methods for measuring fibre/ matrix adhesion. The creation of an interphase between fibre and matrix is discussed and its significance is briefly assessed. Finally, certain sets of data are presented to highlight the different modes of interfacial failure encoutered in model fibre/ matrix geometries.

1. INTRODUCTION

In the classical sense a surface of a material is usually defined with respect to vacuum whereas an interface develops whenever two surfaces come into close contact. For polymer composites, the interface between the fibre and the matrix plays an important role in (a) successfully transmitting stresses from the matrix to the fibre and (b) protecting the fibres from all forms of chemical attack. For analysis purposes it is normally considered to be of zero thickness /1/. In reality physical and chemical interactions, differential thermal effects as well as the presence of all sorts of additives (sizings, wetting agents, etc.) between the fibre and the matrix, can lead to the creation of an interphase of variable thickness and properties /2/.

Fundamental to the discussion of composite micromechanics is the representative volume element (RVE) /3/. It is the smallest region or piece of material over which the stresses and strains are macroscopically uniform. In **Figure 1** the two important dimensions which define the properties of the single lamina in this case are shown. These are: (a) the interfibre spacing, D, and (b) the lamina thickness, t. The third dimension is arbitrary. For perfect fibre/ matrix bond the RVE concept is of extreme importance for design purposes; most micromechanical models particularly the ones based on mechanics of materials approaches use the RVE to predict composite stiffnesses and hence Poisson's ratios. Microscopically, however, the stresses and strains in the RVE are non-uniform owing to the heterogeneity of the material. In polymer based composites under tension, fibre fracture(s) normally precede composite failure and therefore the exact stress distribution within the RVE should determine the mode of failure. This obvioudly leads to an enlargement of the RVE with applied stress to accommodate one or more broken fibres as shown in **Figure 2**. The axial stress in the fibre is zero at the broken fibre site and reaches the applied stress value at transfer length, l_t /4/. The interfacial shear stress is maximum at the fibre fracture and decays to zero at l_t. The ratio of the axial stress in a fibre adjacent to r broken fibres to the fibre stress at infinity is called the 'stress concentration factor', K_r /5/ and is directly related to the intefibre distance, D, the fibre and matrix properties, the number of broken fibres, as well as, the transfer length, l_t /6,7/ Since the latter evidently depends upon the strength of the fibre/ matrix bond, it follows that the fracture characteristics in composites can be modified by careful control of the interface. Other important parameters which are also affected by the strength of the interface are the off-axis, shear as well as the compressive strengths of continuous fibre composites.

2. THE ARCHITECTURE OF THE FIBRE/ MATRIX INTERFACE

The formation of a bond across the fibre/ matrix interface is the result of a number of of physico-chemical phenomena such as (a) van der Waals and acid-base interactions and (b) the formation of primary chemical bonds between chemically compatible groups of fibre and matrix /8/. Obviously, bond formation is influenced by adequate wetting of the fibre by the matrix. Fibre surface treatments tend to increase the wettability of the fibre and to promote the formation of chemical bonding. Mechanical interlocking between fibre and matrix can also lead to an interface which is very strong in shear but rather weak in transverse tension /8/.

The architecture of the fibre/ matrix interface is presented in **Figure 3**. At the macroscopic level the bond in the RVE is considered for design purposes to be perfect. However, at the microscopic level the picture is extremely complex due to the existence of an interphase of variable thickness comprising of fibre surface chemistry and topography, sizing, wetting and other coating agens, as well as, diffused matrix material. The properties of this intermediate phase is also affected by local thermal and mechanical environments. Finally, the integrity of this interphase and hence of the RVE/ composite, depends upon the events taking place at the molecular/ nanoscale level as a result of adequate contact between the two dissimilar surfaces. These include the physico-chemical interactions mentioned earlier and their effectiveness in the presence of oligomers, impurities, solvent molecules and other contaminants. Furthermore, molecular level events such as increases in surface functionality are a necessary precursor to increasing fibre-matrix adhesion through increases in fibre-matrix dispersion type interactions and increases in fibre surface energetics and wettability. The macroscopic, microscopic and molecular phenomena are strongly inter-related and therefore the overall physico-mechanical performance can be successfully tailored by suitable manipulation of the various critical variables at the molecular and/ or microscopic levels.

3. MICROMECHANICAL TEST METHODS

Interfacial Shear Strength (IFSS) measurements can be performed in two ways. Composite specimens can be made and standard short beam shear, four point shear or flexural strength tests can be conducted. All these tests can yield values of IFSS, however, the state of stress in these specimens is rarely in-plane shear therefore the results obtained may have no relevance to the true IFSS between fibre and matrix. Another much simpler approach is to measure the IFSS using a single fibre model composite. This has its limitations in that it is conducted on an isolated single fibre so it cannot be considered a truly composite test. In spite of that, comparative assessments between fibres with different treatments and the detection of true interfacial phenomena, can be achieved in many cases, hence the ever increasing popularity of these type of tests.

A number of single fibre test methods have been employed over the past 30 years and a large volume of -sometimes- conflicting data has been produced. The most important of these tests are:

the **pull-out tests** (**Figure 4**); a fibre is partly embedded in a volume (plate, sheet, drop etc.) of resin. The interface strength is derived from the force needed to pull out the fibre from the resin /9-11/

the **fragmentation test** (**Figure 5**); a single fibre is embedded in a high-strain -to-failure matrix. By loading the coupon the fibre will fracture. The interface strength is derived from the fragment length l_c at the saturation point /12/.

the **microindentation test** (**Figure 6**); a single fibre is pushed into the matrix and the load at which interface failure occurs is measured. The interface strength is derived by means of the force for interface failure via a finite element analysis package /13/.

In spite of the popularity of these tests large discrepancies exist between results of different tests on the same fibre matrix system and between results of different laboratories using the same type of micromechanical tests for the same material syste. This has been confirmed by a European round-robin study co-ordinated by DRA (UK) and ONERA (France) /14/. This lack of reliability of the micromechanical test methods can be attributed to problems related to the **experiment** and problems related to the **data reduction schemes** employed to derive values of interfacial shear strength in each case.

More specifically, in the **pull-out methods** the analysis employed to derive a value of interfacial shear strength constant shear at the interface. However, as it will be shown in this presentation this assumption is not valid and in many cases yields erroneous results of ISS. As far as, the **fragmentation test** is concerned, a value of the "interfacial shear strength" is obtained which is related to the average critical fragmentation length, the fibre diameter and the tensile strength of the filament. The latter is not easily quantifiable at the critical length level. Moreover, the analysis employed assumes again constant shear at the interface which for polymeric matrices it is only valid if stress transfer occurs by friction. Tests conducted on a series of carbon fibres of different moduli have indicated that this assumption leads to an underestimation of the interfacial shear strength by a factor of at least two /15/

Regarding the **microindentation test**, todate only a limited volume of data have been produced. Although it has been argued that the strength of this test lies in its application to full composites /13/, the selective loading of single fibres is not, actually encountered when stressing a practical composite. Other problems associated with this test are (a) the shape and size of the indenter (b) the various finite element packages employed to reduce a value of interfacial shear strength from the force required to debond the fibre.

A recent review of all the above tests has been presented by Herrera-Franco and Drzal /16/.

4. INTERFACIAL FAILURE MECHANISMS IN SIMPLE GEOMETRIES

The major drawback of all micromechanical tests is their inability to determine the stress field around fibre at the microscopic level /4/. This has led to a number of very simplistic assumptions about the state of stress at the interface and, quite often, the 'interfacial shear strength' values that these techniques provide can be the product of the analysis employed in each case. More recently, the technique of laser Raman spectroscopy (LRS) has been developed in our lab and elsewhere, to measure the stress or strain in individual fibres at a resolution of 1-2 μm /4/. This technique can be applied to almost all fibre/ matrix systems provided that the matrix is reasonably transparent. Thus, reinforcing fibres such as carbon or aramid can be used as internal strain gauges in the composites they are incorporated in, without the need of external strain gauges or other devices. It is worth mentioning here that this technique can be applied to single-fibre geometries, as well as, to full composites, and, therefore, provides the necessary link between single-fibre and multi-fibre composites. A detailed description of the LRS technique and its principles, can be found in previous publications /4, 15-17/.

4.1. Continuous Fibre Geometries

The fibre strain distributions of representative fragments of similar lengths for three continuous **unsized carbon fibre/ epoxy resin** systems, are shown in **Figure 7**. In all cases the maximum strain supported by the fibre is 1%. As can be seen, the strain profile for the untreated fibre (HMU) is virtually linear indicating a frictional type of reinforcement. For the treated fibre systems (HMS and IMD), however, the strain take up is more-or-less in accordance with the elastic stress transfer models /19,20/. These fibre strain distributions are converted to interfacial shear stress (ISS) distributions via a balance of forces equation /4/ and the resulting curves are also shown in **Figure 7**. The following observations can be made at this point: (a) the interfacial shear stress is nearly constant along the HMU/MY-750 fragment (b) the surface treated fibre/resin systems, HMS and IMD/ MY-750, exhibit distributions which reach a maximum value near the fibre end and decay to zero towards the middle of each fragment and, finally, (c) the higher the maximum ISS value for the HMS and IMD/ MY-750 systems, the shorter the distance from the fibre end where this maximum appears.

Since a value of maximum interfacial shear stress cna be derived at each level of applied strain then an interfacial or interphasial shear strength (IFSS) of a carbon fibre/ resin system can be defined as the maximum value of ISS developed throughout the fragmentation test. It should be stressed, however, that there is a statistical distribution of the ISS maxima at each level of applied strain, as large number of fragments are sampled with the laser Raman probe. It is, therefore, more appropriate to derive, an **average maximum ISS** value at each level of applied strain, along with the standard deviation of the mean. Such plot of **average maximum ISS** as a function of applied strain for all three systems, is shown in **Figure 8**. As can be seen, the average max. ISS increases with applied strain for both systems and reaches an upper limit of 36±6 MPa and 66±15 MPa for the HMS and IMD/ MY-750 systems, respectively. These values are good estimates of the interfacial shear strength (IFSS) of the two systems. The average max. ISS for the untreated HMU/ MY-750 system appears to be insensitive to applied strain and is approximately six times lower than that of the treated HMS/ MY-750 system.

The ISS distribution in unsized carbon fibre/ epoxy systems before and after interfacial failure is presented schematically in **Figure 9**. At low applied strains, the τ(x) function does not intersect the IFSS line (interfacial shear stress limit) and, therefore, the maximum ISS develops at or very close fibre tip of the fibre fracture. At higher applied strains, the $\tau(x)$ function intersects the IFSS line and, therefore, the region B between the fibre tip and the intersection point, is subjected to extremely high shear stresses, that exceed momentarily the interfacial shear strength of the system. This is the region where failure of the interfacial bonds or, at even higher strains, complete detachment of the fibre from the matrix is expected to occur (**Figure 10**).

4.2. Discontinuous Fibre Geometries

The variation of fibre strain along the whole length of a short **sized Kevlar 49/ epoxy** system at 1.3%, 1.9% and 2.8% levels of applied strain, is shown in **Figure 11**. At 1.3% applied strain, the strain builds up parabolically from the fibre ends and reaches a maximum value at the middle of the fibre. At 1.9% and 2.8% applied strain, however, "trapezoid" strain-transfer profiles are obtained, which are indicative of a marked deviation from the purely elastic stress-transfer characteristics. It is worth noting, that the first sign of fibre fracture is clearly observed at 2.8% applied strain **Figure 11**. The corresponding ISS distributions are also shown in **Figure 12**. As can be seen at 1.3% applied strain, the ISS is maximum at the fibre ends and decays to zero at the middle of fibre. In contrast, at 1.9% applied strain, the maximum ISS values shifts away from the fibre ends exactly indicating the onset of interfacial failure. From then onwards, the ISS distribution fluctuates around a constant value of approximately 45MPa over a distance of about 150μm, before it decays to zero at the middle of the fibre. Indeed, the maximum ISS for all strain levels shown in **Figure 13**, increases linearly up to 1.3% applied strain and then forms a plateau at around 45 MPa.

The results obtained from the sized Kevlar 49/ epoxy composites are presented schematically in **Figure 14**. Up to 1.3% of applied strain, the stress transfer characteristics of this system correspond to the elastic stress transfer models described earlier. Above 1.3% of applied strain, the interfacial shear stress fluctuates around a maximum plateau value of approximately 45MPa **Figure 13** over a certain distance x=B away from the fibre end. This behaviour is quite similar to "shear slip" in metal composites /20/ and is due to local yielding of the fibre/ matrix interphase. Therefore the interfacial shear strength (IFSS) in this system clearly represents the *interphasial yield stress* of the sized Kevlar 49/ epoxy composite.

ACKNOWLEDGEMENTS

The author would like to thank Drs. N. Melanitis, P. L. Tetlow and Mr. C. Vlattas for performing the experiments and assisting with the presentation of this paper.

REFERENCES

1. Chamis, C.C., Composite Materials, **6** (1974), Academic Press, New York, 31-37.
2. Drzal, L. T., Rich, M. J., and Lloyd, P.F., J. Adhesion, **16** (1982), 1-30
3. Jones, R. M. "Mechanics of Composite Materials", Hemisphere Publ. (1975), N.Y.
4. Galiotis C, Comp. Sci. & Techn., **42** (1991), 125-150
5. Rosen, B.W. "Tensile failure of fibrous composites", AIAAJ., **2** (1964), 1985
6   Fukuda, H. and Kawata, K., Fibre Sci. & Techn, **9**, (1976), 189
7   Eitan, A. and Wagner, H.D., Appl. Phys. Lett. **58** (1991), 1033

8. Hull, D., "An Introduction to Composite Materials", Cambridge University Press, (1988)
9. Broutman, L. J., **ASTM STP 327**, Amer. Soc. for Testing Mats.,(1963),133
10. Miller et al, Comp. Sci. and Techn., **28** (1987), p.17
11. Piggott, M. R., Comp. Sci. & Techn., **42** (1991), 57-76
12. Kelly A. and Tyson W.R.D., Mech. Phys. Solids, **13** (1965), 329
13. Mandell, J.F., Chen, J.H. and McGarry, F.J., Int. J. Adhesion & Adhesives, **1** (1980), 40
14. Pitkethly, M. J.,Favre, J. P., "Round robin on interfacial test methods", Composites, in press
15. Melanitis N., Galiotis C., Tetlow P.L., Davies C.K.L., J. Composite Materials, **26** (1992), 574-610
16. Herrera-Franco, P. J. and Drzal, L.T., Composites, **23** (1992), 2
17. Jahankhani H., Galiotis C., J. Comp. Materials, **25** (1991), 609-631
18. Melanitis N., Galiotis C., Proc. Royal Soc. Lond., **440** (1993), 379-398
19. Cox H.L., The British J. of Applied Phys., **3**,(1952) pp 72-79
20. Piggott M.R., Load Bearing Fibre Composites (1980), Pergamom Press, Oxford.

FIGURES AND FIGURE CAPTIONS

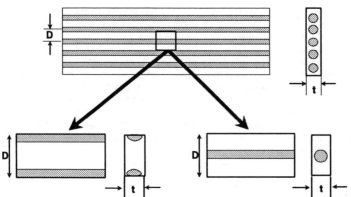

Figure 1 Representative volume element (RVE)

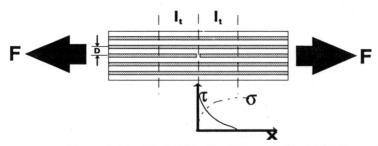

Figure 2: Modified RVE with stress and ISS distributions

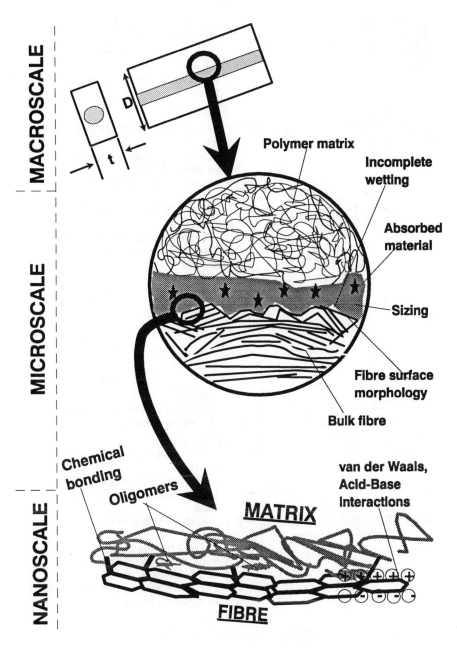

Figure 3 The Architecture of the Interface

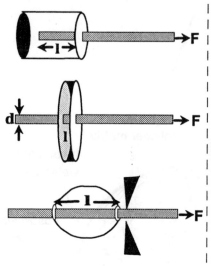

Formula: $\tau = \dfrac{F}{\pi l d}$

Method:

Direct measurement of 'debonding force'

Problems:

-Assumes constant shear along interface

-Clamping/ loading the matrix can affect results

-Resin meniscus alters stress state

-Point of failure initiation difficult to detect

-Loading configuration induces a transverse tensile force at interface

Figure 4 Pull-out tests

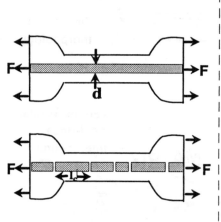

Formula: $\tau = \dfrac{\sigma_f\, d}{2\, l_c}$

Method:

Direct measurement of 'critical length' at saturation

Problems:

-Assumes constant shear along interface

-Requires matrix strain to failure at least 3 times higher than that of fibre

-Requires fibre strength-length data

-The interface at point of saturation may be affected by matrix yielding, etc.

-The critical length may be difficult to measure/ define

Figure 5: Fragmentation tests

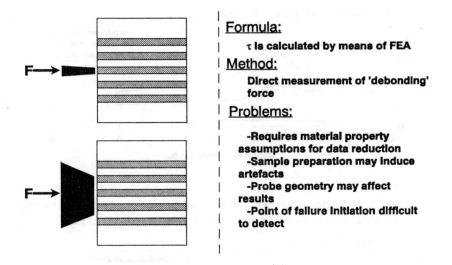

Formula:

τ **is calculated by means of FEA**

Method:

Direct measurement of 'debonding' force

Problems:

-**Requires material property assumptions for data reduction**
-**Sample preparation may induce artefacts**
-**Probe geometry may affect results**
-**Point of failure initiation difficult to detect**

Figure 6: Micro-indentation tests

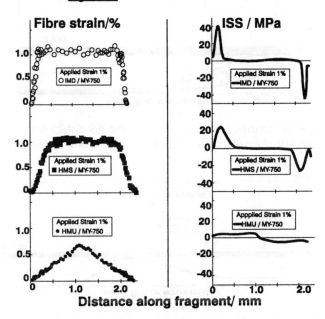

Distance along fragment/ mm

Figure 7: The fibre strain (left) and ISS (right) distributions or representative fragments obtained from 3 different unsized carbon fibre/ epoxy systems

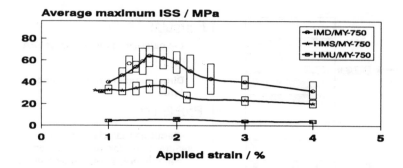

Figure 8: The average maximum ISS as a function of applied strain for the three different unsized carbon fibre/ epoxy systems of Fig.7

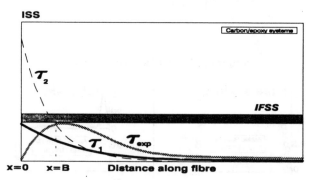

Figure 9: A schematic representation of the ISS distribution for unsized carbon fibre/ epoxy systems before and after interfacial failure

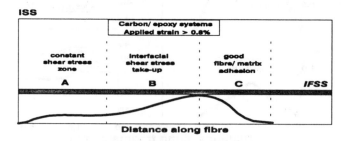

Figure 10: A schematic representation of the ISS distribution for unsized carbon fibre/ epoxy systems at high applied strains

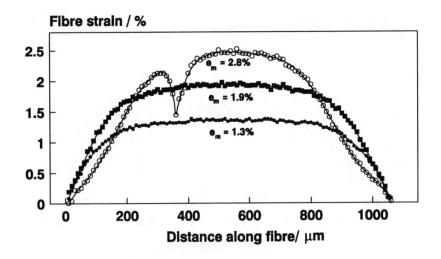

Figure 11: Fibre strain profiles for a short Kevlar 49/epoxy resin system

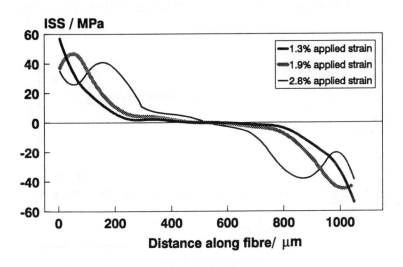

Figure 12: ISS profiles for the short Kevlar 49/epoxy resin system of Fig.11

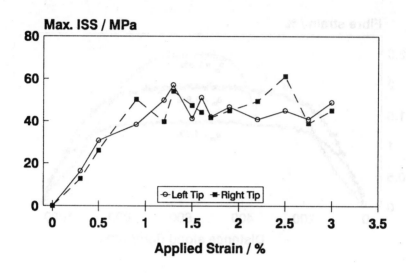

Figure 13: The maximum ISS as a function of applied strain for the two fibre ends of the short Kevlar 49/ epoxy system

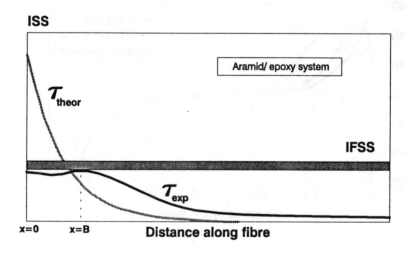

Figure 14: A schematic representation of the ISS distribution for the sized Kevlar 49/ epoxy system

COMPOSITE MATERIALS TECHNOLOGY FOR CIVIL AND MILITARY AIRCRAFT IN SPAIN : RECENT ACHIEVEMENTS

J.A. GUEMES, P. MUNOZ-ESQUER*, F. LICEAGA**

ETSI Aeronauticos - Universidad Politecnica de Madrid
- 28040 Madrid - Spain
*CASA - Getafe Madrid - Spain
** Inasmet - San Sebastian - Spain

ABSTRACT

CASA leads a team of Spanish companies involved in manufacturing advanced composite structures using various technologies such as automatic lay-up, filament winding and RTM. Details about the manufacturing abilities of these companies will be given in this paper.

Most of the recent CASA programs are done in cooperation with other European and International partnership; the most significant success being the design and manufacturing of the horizontal tail plane of the Airbus A320, the first primary structure of a civil aircraft fabricated in composite materials.

A brief look of the materials and equipment suppliers and on the activities of the Research Centers will also be given.

THE BEGINNINGS

Construcciones Aeronauticas S.A. (CASA) began its activities into the field of nonmetallic structures at the mid-sixties, with facilities for structural adhesive bonding and for manufacturing sandwich structures.

The whole spectrum of technologies needed to design and manufactures composite materials were developed during the 70's, and minor applications of these technologies were done on CASA products at those times; the jet trainer C-101 (first flight 1979) incorporated an air brake in metallic sandwich structure; the CASA C-212 is a turbopropeller for military transport, and had several secondary elements, such as fairing, in GRP.

At the beginnings of the 80's, the basic technologies for design and stress analysis, quality assurance, production and inspection were mastered. The two most representative applications from this time were:

The CN-235, a pressurized medium range transport aircraft. The development began in 1980, and first flight was in 1983. It incorporates a significant number of secondary structural parts in Composites, basically glass or aramid skins and Nomex sandwich structures.

The Airbus A300 and A310 began development in 1978, first flight 1982. CASA had the responsibility for the wheel doors, made out of CFRP, in addition to the horizontal tail plane, in conventional metallic structure.

THE AIRBUS FAMILY

The involvement into high responsibility structures began in 1984, with the start of the development of the Airbus A320 (First Flight 1987). CASA receives the responsibility for design, manufacturing and certification of the horizontal tail plane. At that time this was the first primary structure in composite to be put in service into a civil aircraft (primary structure means that its failure in service implies the loss of the airplane).

The tail plane is composed of the following elements: Torque box, leading and trailing edges, elevators and fittings. The torque box is formed by integrally stiffened skin panels, mechanically joined to front and rear spars and to thirteen ribs. The skin panels are made with a modular tooling technique. The controlled thermal expansion of the tool parts give the right pressure for consolidating the stiffeners. Fittings (five to join the elevator to the torque box, one for the trimming of the whole tail), are made in CFRP fabric, by a Net Laminate Matched Die technique.

With the successful completion of this program, demands for higher production capabilities promoted the birth of new factories, working under CASA transferred technology. These industries will have a large share of its work from CASA projects, but also they have a growing activity as independent manufacturers, within and outside the aerospace sector.

Airbus A330/340 is an evolution of the A320 technologies and concepts; advanced procedures, such as automatic tape lay-up, are used in the serial production of the components. One of the features of A330 is that the tail plane is an integral tank, and then problems related to leakage and composite-fluid compatibility were also addressed.

OTHER AEROSPACE PROGRAMS

The European Fighter Aircraft (FE) was a major booster for the development of technologies. CASA, jointly with BAE, had the responsibility for the starboard wing. First aircraft flew at the end of 1992.

Activities in composites also includes the manufacturing of the outer flaps for the Boeing B-757, and the horizontal tail plane for the McDonnell Douglas MD-11, among the civil aircrafts. For the F/A-18, in service in the Spanish Air Forces, several structural elements are manufactured in Spain. Advanced technologies for bonding and superplastic forming are used for the wing of the Saab-2000, of which CASA is the main contractor.

For space applications, CASA has designed and manufactured in CFRP sandwich various antennas and reflectors used in satellites of the European Space Agency. CASA manufactures the electrical equipment box for the Ariane launcher; The RTM technology has been successfully tested to manufacture this large and complex structure.

COMPANIES AND PROCESSES

Table 1 shows the companies qualified in Spain for manufacturing composites according to aerospace standards. In Table 2, the quoted area for the autoclave represents the maximum size of a part that can be manufactured. The surface of the lay-up room is an indirect measure of the maximum volume of work that the company can undertake.

In all the above mentioned cases, the facilities include automatic ultrasonic C-Scan, automatic cutting of prepregs and Quality Control laboratory; only CASA and INTA have, at the present time, the necessary equipment for qualification of new materials and processes.

It is worth to mention that Hercules installed a prepreg factory in Spain, which started its operation in 1989, with a capability of 680.000 Kg/year. Among the equipment suppliers, there are local manufacturers for Numerically Controlled Prepreg Cutting machines (INVESTRONICA), N.C. machines for drilling and milling laminates and for machining nuclei (DYE), autoclaves (TEICE), automatic ultrasonic C-Scan (TECAL), ovens (ISAM) and portable equipment for repairing composites (CASA).

The Universities of Madrid and Zaragoza, and Research Institutes, such as INTA, CENIM and INASMET organize courses on specific topics of Composites, and they have a participation into the European research programs.

TABLE 1

COMPANIES MANUFACTURING ADVANCED COMPOSITES IN SPAIN

	LAY-UP	RTM	FILAMENT WINDING	BONDING
CASA (1966)	X	X		X
FIBERTECNIC (1987)	X	X		
ARIES COMPLEX(87)	X			
HTC (1990)	X			X
ICSA (1990)	X		X	

FIRM	CITY	LAY UP ROOM	AUTOCLAVE Diam X Length (m)	PRESSURE	TEMPERATURE
ARIES COMPLEX	TRES CANTOS (MADRID)	1500 M2	3.6 X 12.0 3.0 X 14.0	10.5 16.0	340°C 400°C
CASA	GETAFE (MADRID)	3650 M2 948 M2 (ADH)	3.6 X 12.2 2.1 X 6.0 1.4 X 1.5 2.7 X 10.0 3.6 X 12.2 1.1 X 2.0	10.5 3.5 14.0 14.0 10.5 20.4	260 °C 260 °C 430 °C 350 °C 250 °C 350 °C
CASA	ILLESCAS (TOLEDO)	3604 M2	4.5 X 13	17.2	400 °C
CASA	BARAJAS (MADRID)	750 M2	3.0 X 4.0 0.8 X 1.0 1.0 X 2.0	16.0 10.5 10.5	250 °C 250 °C 250 °C
FIBERTECNIC	VITORIA (ALAVA)	1360 M2	3.0 X 9.0 3.0 X 9.0	16.0 18.3	287 °C 420 °C
H.T.C.	SEVILLA	1152 M2 525 M2 (ADH)	3.6 X 12 3.6 X 14	17.5 17.5	400 °C 400 °C
ICSA	TOLEDO	1400 M2	3.6 X 12	17.0	400 °C
INTA	MADRID	60 M2	1.2 X 2.5	20.4	375 °C

TABLE 2.- AUTOCLAVES FOR COMPOSITES IN SPAIN

FABRICATION TECHNOLOGIES
AND MODELLING OF PROCESSES

THERMAL AND MECHANICAL ANALYSIS OF OVERINJECTED COMPOSITE COMPONENTS

A.M. HARTE, J.F. McNAMARA

University College - School of Engineering - Galway - Ireland

ABSTRACT

A technology for the formation of complex components by means of injecting short fibre reinforced thermoplastic items onto long fibre reinforced thermoplastic bases is currently being developed. Extensive numerical modelling and laboratory studies are being carried out to determine the effects of variations in the processing parameters on the quality and strength of the finished part. This paper focuses on the finite element modelling of the non-linear transient heat transfer and the thermomechanical response of the part during the production cycle. Results of the analysis work are compared with laboratory thermocouple readings and warpage measurements.

INTRODUCTION

This paper describes some aspects of computational work being conducted into the overinjection of reinforced thermoplastics. The study forms part of an ongoing research project aimed at developing viable design and processing techniques for the commercial applications of this process. The overinjection process consists of two steps. The first step is the compression forming of a continuous fibre reinforced thermoplastic part, with a relatively simple geometry, to serve as a structural frame and to carry the main load. This part is known as the base item. In the second step, this base is overinjected with a short fibre reinforced thermoplastic in order to produce the required functional detail. The overinjected part is known as the overitem. Knowledge of the processing conditions needed to ensure an adequate bond between the base and overitem is essential. In addition, the design process must take into account the effect of the processing parameters on the quality of the finished part in terms of such factors as surface finish and part distortion.

Essential to the study of the behaviour of an overinjected part is the determination of the thermal and stress history of the part during the production process and this is the purpose of the present study. Numerical modelling of the process is carried out in the main using the ABAQUS(1) finite element package. The analysis of the mould filling part of the process is carried out elsewhere. Side by side with the computational work reported here, a programme of experimental testing is taking place. The results of the finite element work are compared with experimental measurements for a particular part.

The conduct and results of the numerical study are presented hereafter as follows. Firstly, the details of the part analysed and the physics of the injection process are briefly described. Then the heat transfer analysis is discussed and results presented. Next the thermal stress analysis procedure is described and selected results given. The paper concludes with a summary discussion, acknowledgements and references.

PART DEFINITION AND PROCESS DESCRIPTION

The part selected for this study is shown in Figure 1 and is intended to represent a typical rib stiffener. The base item material is a polyamide 6 reinforced with a woven glass fibre fabric. This material is supplied by Bayer in prepreg form under the trade name "Polystal". The base lay-up is shown in Figure 2. Layers of carbon fibre reinforced PEI are placed at the top and bottom of the stack. The top layer forms a heating element for the base heating phase of the process while the bottom layer in included to form a balanced laminate and therefore reduce warpage during cooling. The overitem material is a short glass fibre reinforced polyamide 6 with a fibre content of 30% by volume. This is also supplied by Bayer and bears the trade name "Durethan BKV30H". Future studies will investigate other base and overitem materials.

The processing cycle can be summarized as follows. Initially the mould of the injection moulding machine is preheated to 100°C by circulating oil at a specified temperature through a system of thermal regulation channels. The base item is then placed in the mould and the base preheating starts. This preheating is carried out using an electrical resistance method. This envolves passing a specified current through the heating element for a prescribed time until the desired temperature is achieved on the face of the base which will be in contact with the overinjected material. Presently a range of base temperatures from 180°C to 235°C are being considered. In the application considered in this paper, a current of 19 amps is passed through the heating element, shown in Figure 2, for 100 seconds to achieve a temperature of 235°C at the top of the base at the centre point. Having reached the desired base temperature, the injection moulding stage begins. While a range of injection temperatures from 260°C to 280°C are being considered, the results presented here are for an injection temperature of 260°C. The injection stage itself lasts for about 2 seconds.This is followed by a packing stage lasting 30 seconds and by a period of cooling in the mould lasting about 300 seconds. At this point the finished part is ejected from the mould and is allowed to cool to room temperature.

HEAT TRANSFER ANALYSIS

A heat transfer analysis is required for three distinct sections of the work(2). These are (i) the design of the thermal regulation system for the mould, (ii) the selection of a suitable power input for the base heating, and (iii) the prediction of the thermal history of the part during the production cycle in order to be able to predict the mechanical response. Finite element meshes for each of the tasks were generated using two-dimensional 8-noded isoparametric heat transfer elements. Heat losses due to convection and radiation from the exterior surfaces of the model are included. Non-linear temperature dependent material properties are used throughout.

Thermal Regulation System

A mould temperature of between 80°C and 100°C is specified by the manufacturers of the overitem material. In order to achieve this, a system of thermal regulation circuits are embedded in the mould during manufacture. Finite element models of a number of different circuits are considered. The design chosen is considered to be the most suitable because it achieves a constant temperature in the region of the wall cavity and the heat-up time is reasonable. The mould is instrumented with a number of thermocouples and pressure transducers. These are a valuable means of checking on the accuracy of the analytical work and especially in determining what modelling assumptions are valid. The modelling of the thermal regulation channels is carried out using a fixed temperature boundary condition for the releant nodes. The elements corresponding to the channels are given the properties of oil with adjustments made to account for convection effects. This straightforward approach to modelling the heating system is very effective. Heat-up/cool-down experiments carried out on the mould show excellent agreement between the thermocouple readings and model predictions. Results for one of the thermocouple locations are shown in Figure 3.

Base Heating

Studies are carried out to determine the effect of varying the input power supply on the heatup rates for the base. A 2-dimensional model of the base, mould and insulation is developed. During this stage of the process, a gap is present between the bottom surface of the base and the mould wall. In the numerical model, this gap is modelled using gap elements with specified gap radiation and conductance. The high mould pressure during the subsequent injection phase closes this gap. Heat losses due to convection and radiation from the exterior surfaces of the mould are to a sink temperature of 20°C while those from the top surface of the base to the mould cavity are to a sink temperature of 100°C. The resistence heating is modelled as a distributed flux over the top surface of the base. A large number of bases are instrumented by embedding thermocouples directly under the heating element. Curves showing the predicted and measured temperature rise in different locations for various power settings are available and the heat-up curve for a 19 amp current is shown in Figure 4. Excellent agreement is found between the experimental and numerical results. These heat-up curves are used to determine the required time to achieve any desired base temperature for any specified current setting.

Thermal Loading During Processing

A sequential analysis approach is adopted in order to investigate the thermomechanical behaviour of the part during processing. In the first stage, a transient heat transfer analysis is conducted following each phase of the production cycle as closely as possible. This analysis provides the thermal loading for the stress analysis stage described below. The analysis phases are as follows :

Phase 1	- Preheating of Mould
Phase 2	- Base Preheating
	- (Injection Phase)
Phase 3	- Post Injection Cooling in Mould
Phase 4	- Post Ejection Cooling in Air

The finite element mesh changes for each phase as indicated in Figure 5. Phase 1 includes the mould and its insulation only. For Phase 2, the model is extended to include the base item. For Phase 3, the overitem elements are added. Finally, for Phase 4, the mould and its insulation are removed to leave the base and overitem elements only. The individual phases are linked by using the temperature distribution at the end of one phase as initial conditions for the following phase.

In Phase 1, the mould preheating phase, the mould is initially at a uniform temperature of 20°C. The nodes corresponding to the thermal regulation channels are fixed at 100°C throughout. The analysis continues until a steady state is reached.

In Phase 2, the base preheating phase, the base nodes are initially set at 20°C while the initial temperatures for the mould nodes are the temperatures at the end of Phase 1. The power input to the base is modelled as a distributed flux as described above. The duration of this analysis phase is selected from the appropriate base heatup curves to give the desired base temperature. At the end of Phase 2, the injection of the overitem material onto the preheated base takes place. The mould filling analysis is carried out elsewhere(4) for the appropriate injection temperature, pressure and mould wall temperatures. This analysis provides the temperature distribution in the overitem at the end of the mould filling phase.

In Phase 3, the post injection cooling phase, the mesh now includes the entire part and mould. Initial conditions for the mould and base are nodal temperatures at the end of Phase 2 while those for the overitem are taken from the filling analysis. The analysis time is set at 330 seconds which is the approximate time required to cool the part to the recommended ejection temperature of 100°C. At this stage of the analysis work, it is assumed that the shrinkage of the part away from the mould during cooling does not affect the heat transfer regime. Future work on a fully coupled thermal stress analysis will take into account the effects of intermittent contact/gaps on the heat flow. Figue 6 shows the predicted versus measured temperatures in this phase for a thermocouple located below the base. Very good agreement is found for this mould location. The predicted peak temperatures for those thermocouples which are located near the top at the part in the region of the overitem are higher than the experimental measurements. The model predicts a faster rate of heat flow out of the overitem than that found experimentally. This is consistent with the uncoupled heat transfer/deformation assumption. The preliminary results do, however, give a very good initial estimate of the thermal history.

In Phase 4, the post ejection cooling of the part in air, the initial temperatures of the part are the temperatures at the end of Phase 3. The analysis continues until the temperature distribution in the part reaches a steady state.

THERMAL STRESS ANALYSIS

Stage 2 incorporates a two-dimensional non-linear stress analysis to determine the mechanical response of the part during the cooling Phases 3 and 4. The finite element mesh of the base and overitem in this stage of the analysis uses 8-noded isoparametric plane strain elements with nodal points coincident with those used in the thermal analysis. The analysis is carried out in two steps. Step 1 represents the part cooling within the mould. The initial temperatures in the part are identical to those used in Phase 3 of the thermal analysis and the temperature loads from this phase are applied. The mould deformations are assumed to be so small in comparison with the part deformations that the mould may be modelled as a set of rigid surfaces enclosing the part. The interaction between the mould and the part is modelled using quadratic rigid body interface elements. Step 2 of the analysis detemines the part deformations when cooling in air after it is released from the mould. Mould release at the end of Phase 3 is simulated by specifying rigid body displacements for each rigid surface away from the part. Temperature loads from the Phase 4 thermal analysis are applied.

Figure 7 shows the displaced mesh at the end of the mould cooling step. The shrinkage of the part away from the mould can be clearly seen. The displaced mesh after the part has cooled in air is shown in Figure 8. The deformation pattern is consistent with that observed in the laboratory. Table 1 compares computed and measured values of part deformations indicated in Figure 8. The values shown are very satisfactory preliminary results for a simplified uncoupled analysis approach. In addition, it should also be noted that some of the mechanical properties of the materials at elevated tempertures are not yet available. When these are included in the model, a further inprovement in the displacement results is anticipated.

CONCLUSIONS

An approach to the finite element modelling of the thermomechanical response of a thermoplastic part during processing is presented. The technique of varying the finite element mesh at different stages of the process appears to work satisfactorily when results are compared with experimental measurements. An uncoupled displacement analysis, which models the mould as a set of rigid surfaces, gives a very good preliminary estimate of the deformation fields. The model can easily be adjusted for different processing conditions making it a very useful design tool.

Work on a fully coupled approach to the problem is currently underway. In addition, experiments using different materials are in hand which will provide further means of verifying the analytical modelling approach.

ACKNOWLEDGEMENTS

Work on this project has been carried out under EC Brite-Euram Contract BE3532. The authors would like to acknowledge the contribution of all of the research partners to the work presented. In addition, analysis work carried out by G. O'Donnell, MCS and P. O'Brien and experimental work carried out by P. Murphy, UCG is gratefully acknowledged.

REFERENCES

(1) ABAQUS, Users Manual, Hibbitt, Karlsson & Sorenson, Inc., Providence, RI 02906, 1987
(2) Harte, A.M. & McNamara, J.F., "Numerical Modelling of Heat Transfer During Overinjection of Thermoplastic Composite Materials", Proceedings of Irish Materials Forum No. 8, Dublin, Sept. 1992
(3) O'Brien, P., "Development of Design and Processing Techniques for the Overinjection of Thermoplastic Laminates", M. Eng. Sc. Thesis, Dept of Mechanical Engineering, University College, Galway, 1992.
(4) Vincent, M., CEMEF, Valbonne, France, Private Communication.

	Shrinkage in x (%)	Warpage y (mm)
ABAQUS	0.8	0.72
Experimental	0.5	0.94

Table 1. Numerical v Experimental Part Distortions

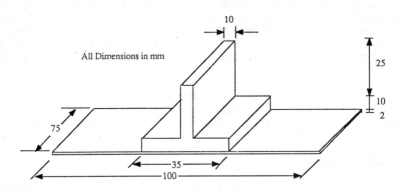

Figure 1   Dimensions of Overinjected Part

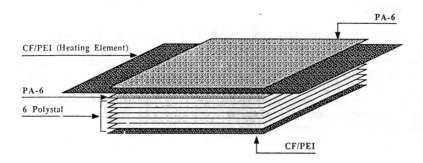

Figure 2   Lay-Up of Base

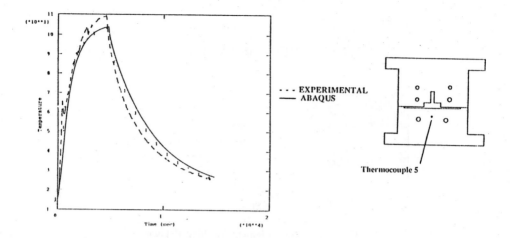

Figure 3   Thermal Regulation of Tool. Heat Up/Cool Down Curves

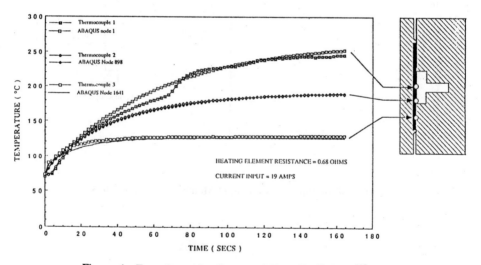

Figure 4   Experimental v Computed Heat-Up Curves [3]

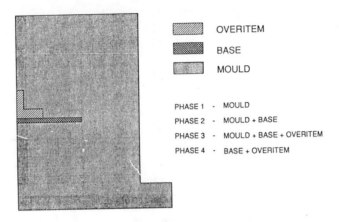

Figure 5 Finite Element Meshing Regions

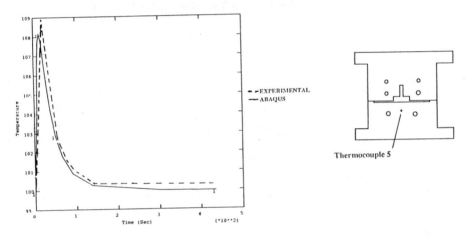

Figure 6 Post Injection Cooling in Mould - Thermocouple 5

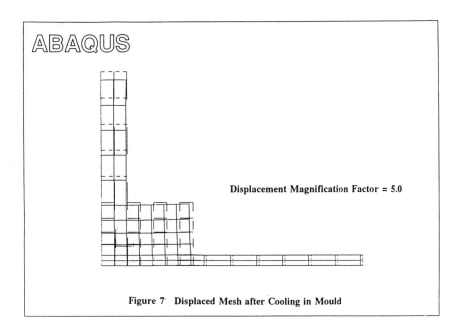

Figure 7 Displaced Mesh after Cooling in Mould

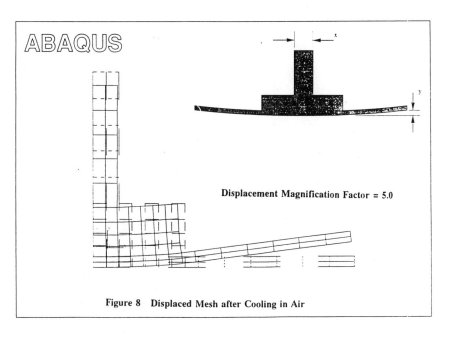

Figure 8 Displaced Mesh after Cooling in Air

FACTORS INFLUENCING IMPREGNATION AND TEST METHODS IN COMPOSITE MATERIALS

L. MADARIAGA, J.R. DIOS

Gaiker - Centro de Transferencia Tecnologica - Parque Tecnologico de Zamudio - Edificio 202 - 48016 Zamudio (Bizkaia) - Spain

ABSTRACT

A brief summary of impregnation and adhesion test methods is reviewed in this paper. First an introduction to glass fibre - matrix interface, focused on organosilane coupling agents. Following this general introduction, data of filament wound fibreglass-thermoset composites under different process conditions and formulations clearly show that fibre volume fraction is not the only factor that influence the mechanical properties and that impregnation helps adhesion.

INTRODUCTION

Fibre reinforced thermoset composite materials are able to withstand a broad range of service conditions, depending on the selection of the reinforcement and matrix. But even if the selection of proper materials has been done, there is a key factor in the performance of the composite: fibre-matrix adhesion. Adhesion is enhanced by using proper coupling agents but also by asuring a good wettability of fibre by the resin during the process.

Anyway, adhesion of final parts is not easily controlled. The main reasons are:

-      Interface adhesion mechanism is not usually a matter of knowledge by Porcess engineers.

-      The adhesion improvement is not sensitive to many test procedures.

There is a great number of factors that influence adhesion.

I. INTERFACE AND COUPLING AGENTS

There is a region around the fibre reinforcement responsible of providing good adhesion between the fibres and the matrix. This region is called the interface, and the strength, water absorption and reduction of mechanical properties due to chemical and environmental attack are directly related to the performance of this thin layer.

Non-treated glass fibre filaments and organic matrix are non-bonding surfaces. In order to obtain durable bonding between both materials, the glass fibre is pretreated with a coupling agent.

Silanes are the most common and useful coupling agents in glass fibre reinforced composites. They are a family of organosilicon monomers defined by the following formula:

$$X_3 Si (CH_2)_n Y \qquad (1)$$

where

       X: hydrolizable group.
       Y: organofunctional group hydrolytically stable.

The aliphatic chain $(CH_2)_n$ is usually a propylene group (n=3). The hydrolizable group (X) is usually methoxy /1/ or ethoxi /2/.

The coupling agent must first be converted to a reactive silanol form by hydrolysis:

$$R\text{-}SiX_3 + 3 H_2O \text{ ------ > } R\text{-}Si (OH)_3 + 3HX \qquad (2)$$

where HX is usually an alcohol.

This reaction may be done during the sizing of the glass fibre or brought about directly on the surface at the time the resin and fibre are put in contact.

The silanol formed can react with other silanol groups of the glass surface:

$$R\text{-}Si(OH)_3 + (OH)\text{-glass} \text{ -----------> } R\text{-}Si (OH)_2\text{-}O\text{- glass} + H_2O \qquad (3)$$

It must be pointed out that the above reaction is reversible. A dynamic equilibrium between bonding sites in the glass surface is formed, thus allowing relief of internal stresses built up during the shrinkage in the curing process or by external loads applied to the composite. A theory of hydrogen bridges has also been proposed instead of covalent bonds between the glass surface and the coupling agent.

The efficiency of silane coupling agent adhesion to glass fibre depends on the number of functional groups present in the glass surface and steric limitations due to size and shape of the organosilane molecule /3/. That´s the reason why flexible molecule structures are preferred.

II. SELECTION OF SILANE COUPLING AGENT

The coupling agent organic side reacts with the polymeric matrix. Proper selection of Silane type becomes a matter of matching the functional group of the silane with that of the organic phase.

Generally speaking the following conditions are necessary to satisfy good adhesion /4/:

-        Silane functional groups must be matched to those on polymer matrix, in such a way that a reaction may occur and so covalent bonds may be created.

-        The rate of reactivity of silane functional groups must be the same than the one of the functional groups reacting in the organic matrix. Selecting silanes in this way will satisfy that the polymer reactive bonds will not be completely consumed in their own reactive process.

A self-condensation reaction between silanol forms creates siloxane polymers, with properties similar to those of silicone resins. This side reaction is important in some pH ranges and when an aqueous solution is used in order to treat the inorganic surface.

The improvements obtained with Silane coupling agents can be summarized as follows:

-        Improved tensile, flexural and compressive strength.

-        Improved retention of properties after exposure to the enviromental conditions and moisture attack.

III. ADHESION TEST METHODS

Due to the fact that adhesion is the most important parameter related to final properties, several methods are used to quantify the obtained degree of adhesion.

These methods are commonly applied to a single filament, not considering adhesion effects generated by process conditions.

3.1. Fibre Pull-out /5/

In this test method, a single filament is embebbed in a matrix system. After curing the sample, a free end of the fibre is pulled out. The adhesion is quantified by measuring the force applied to pull out the fibre and divided by the contact surface between fibre and matrix.

The main disadvantages of this method are due to the need of a sophisticated equipment and difficult sample manufacturing process. The results may not be in accordance with the fibre performance in the final composite where multiple adjacent fibres are laid together.

3.2. Fibre fragmentation

The sample used is a single filament completely embebbed in the matrix. The shape of the sample is similar to a traction test specimen. Applying a traction force to the specimen, the fibre cracks at specific points following a Weibull distribution.

If the sample is clear, it is possible to follow the chemical interactions in the interface using Laser-Raman spectroscopy. The interface bonds vibrational frequencies are directly related to the actual stresses in the interface.

This method usually gives better results (10%-20% improved) than the Fibre Pull-out technique and a better understanding of the interface performance.

3.3. Micro compression /6/

The method consist of applying a compressive force to single fibres in a composite sample until a slippage of the fibre occurs. It has the advantage of using test samples similar to those obtained in composites, and so getting more realistic results.

3.4. Dynamical Mechanical Analysis (D.M.A.) /7/

Measuring the loss factor at the glass transition temperature of the composite is a direct measure of the elastic energy dissipated by the sample. When the material becomes more elastic, the loss factor decreases and then when the adhesion is improved the loss factor lowers.

3.5. Mechanical tests after boiling water immersion /8/

This is one of the most common used test methods to quantify coupling agents ability to retain properties after environmental and moisture attack. The obtained results are directly related to the adhesion degree.

The samples are tested after being immersed in boiling water. S.S. Oleesky and J.G. Mohr in /9 / showed that 2 hours in boling water are equivalent to 30 days immersed in water at room temperature.

IV. ADHESION AND IMPREGNATION

Even though the right coupling agent were used in the proper amount, the final performance of the composite structure might not be as good as expected due to the lack of a good impregnation procedure. This fact is more frequently encountered when high glass fibre content is used in the process.

Figure 1 shows a comparison results of flexural tests made over filament-winding samples manufactured at ⊕ 54° winding angle using different process conditions. It can be seen that even if the glass fibre content in sample A is greater than in B, properties obtained with B are better because impregnation conditions   are favoured.

Consider now two samples of almost the same glass fibre content as it´s shown in Figure 2. The use of a wetting agent and good impregnation conditions, improve flexural properties before and after boiling water immersion.

Two main benefits are obtained with impregnation improvement:

-    Imrpoved wettability of the fibre and stronger fibre-resin adhesion effect.

-    Reduced void content of final composite.

V. CONCLUSSIONS

The impregnation process and methods to improve it are key factor in order to obtain the best properties of composite materials.

Adhesion tests for single filaments are a good way to study coupling agents performance. But if we really want to measure the actual interface performance in a composite structure we must consider impregnation as a real important parameter influencing in adhesion.

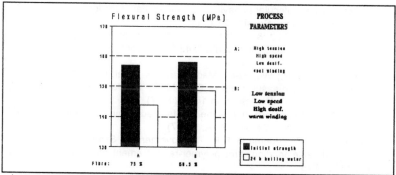

Figure 1: Effect of process conditions in flexural strength and glass fiber content by weight

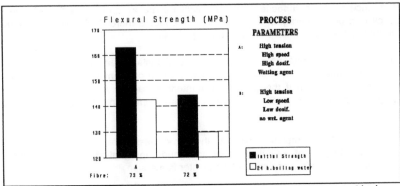

Figure 2: Effect of wetting agent and process conditions in flexural strength

ACKNOWLEDGEMENTS

This research was supported by the Province Council of Bizkaia. We thank the financial support of the Economic Promotion Department. Thanks also to the GAIKER support in the redaction of this paper.

REFERENCES

1.   Dow - Corning " Silane Coupling agents ". Technical bulletin.
2.   Union Carbide " Organo Functional Silanes ". Technical bulletin.
3.   Arkles,B.; Steinmetz,J. & Hogan, J. " Polymeric Silanes: An evolution in coupling agents ", 42nd Ann. Conf. SPI Inc., Session 21-C, (1987).
4.   Jang, J.; Ishida, H. & Pleuddemann, E.P. " FT-IR Study of the Solvent Effects on the molecular behaviour of methacrylate functional silanes ", 42nd Ann. Conf. SPI Inc., Session 21-F, (1987).
5.   Drzal, L.T., Rich, M.J. & Ragland, W. " Adhesion between fiber and matrix - its effect on composite test results ". 42nd Ann. Conf. SPI Inc., Session 7-A, (1987).
6.   Herrera-Franco, P.; Wu, W-L ; Madhukar,K. & Drzal, L.T. " Contemporary methods for the measurement of Fiber-Matrix interfacial shear strength ", 46th Ann. Conf. SPI Inc., Session 14-B, (1987).
7.   Chuk, P.S. " Characterization of the interface adhesion using tan delta " 42nd Ann. Conf. SPI Inc., Session 21-A, (1987).
8.   Pape, P.G. & Plueddemann, E.P. " Improvements in Silane coupling agents for more durable bonding at the polymer-reinforcement interface " ANTEC´91 pp. 1870-1875.
9.   Oleesky, S.S. & Mohr, J.G. " Handbook of reinforced plastics " ed. Van Nostrand Reinhold. New York. (1964).

FINITE ELEMENT SIMULATION OF THE RESIN TRANSFER MOLDING PROCESS

R. GAUVIN, F. TROCHU, J.-F. BOUDREAULT, P. CARREAU

Mechanical Engineering - Ecole Polytechnique - C.P. 6079 - Succ. "A" - Montréal - Québec H3C 3A7 - Canada

ABSTRACT

The computer simulation of the injection phase in resin transfer molding (RTM) can help the mold designer to properly position the injection ports and the air vents, to select an adequate injection pressure and to optimize the cycle time. The purpose of this article is to present RTMFLOT, an integrated software environment especially designed for the numerical simulation of the resin transfer molding (RTM) process. An application to a subway seat is described to illustrate the various stages of the simulation process.

INTRODUCTION

In resin transfer molding (RTM), a stack of fiber mats or woven rovings is first laid in the mold cavity. Then the mold is sealed and resin is injected at low pressure. As the RTM process is being used more frequently for the manufacturing of composite parts, the need of computer simulations has arisen. An adequate positioning of injection ports and air vents is critical to ensure a proper filling, to minimize the resin loss and to optimize the cycle time. The pressure inside the mold must be monitored also during the filling process in order to prevent any significant deflection of the mold walls. The software environment RTMFLOT was designed to address these needs.

The impregnation of a fibrous preform is governed by Darcy's law, which states that the flow rate through a unit area is proportional to the pressure gradient. The factor of proportionality is called the permeability of the reinforcement. In the software environment

RTMFLOT (see Fig. 1), it can be evaluated by DATAFLOT using micro-mechanical or empirical models. The second step consists in subdividing the mold into a set of triangular slices (program MESHFLOT). Triangles were chosen because arbitrary shapes can be better approximated by triangles. Darcy's equation is solved at each time step by non-conforming finite elements (program FLOT). The resin front positions and the pressure field can be computed at any time and visualized by the program interface VISUFLOT. Finally, the heat transfer phenomenon can be analyzed by HEATFLOT if the mold or the resin is heated.

Several numerical methods have been proposed for the numerical simulation of mold filling in resin transfer molding. The first family of models were based on finite differences on a moving mesh [1-2]. This approach gives a good reproduction of the resin front profiles, but becomes quite difficult to apply in the case of dividing or merging fronts /3/. More recently, the finite element method was applied by Trochu and Gauvin /4/ using a non-conforming approximation which respects integrally the important physical condition of resin mass conservation across the inter-element boundaries. The experimental and calculated resin front profiles were found to be in good agreement with experiments for two-dimensional molds /5/. In this paper, the simulation is carried out for a three-dimensional mold of complex shape, a subway seat, in order to illustrate the capabilities of the program.

The subway seat shown in Fig. 2 is reinforced by one layer of OCF 8610 fiber mat of surface density 300 g/m^2. For this part of uniform thickness 1/8", the porosity is 0.89. The isotropic permeability of the mat was measured in the laboratory using a one-dimensional mold. An accurate evaluation of the permeability is a critical factor which governs the accuracy of any subsequent mold filling simulations.

1.MESHING OF THE MOLD (MESHFLOT)

The first step consists in creating a geometric model of the part on the computer. The triangular mesh of uniform thickness of a subway seat is shown in Fig. 2. Note that a special refinement is created automatically in the vicinity of the injection port, where it is important to use smaller elements in order to follow accurately the progression of the resin flow.

2.MOLD FILLING SIMULATION (FLOT)

The mold filling simulation performed by the program FLOT is based on a non-conforming finite element approximation of the pressure field on triangles /4-5/. Figure 6 shows an example of resin front profiles. Three types of special elements have been added to address the molder's practical needs: the injection port, the air vent and the channel. The **injection port** is represented by a tube with a manometer or a piston depending on the type of injection selected: constant

pressure or constant flow rate. The injection port can be posit
anywhere on the mold boundary. Note that the pressure read on
installation is not the pressure that must be used by the compu
program. It should be corrected by the head loss in the tube. If
air vent is introduced in the mold to evacuate the residual air, i
will be associated on the mesh to a specific node J. At this location
the boundary condition is $P = P_J$, where P_J is the hydrostatic pressure
depending on the cumulative resin loss. Finally, a **channel** is a
preferential path for the resin flow that can be created in the mold.
It is a practical solution used sometimes by molders to facilitate the
filling of the mold.

3.VISUALIZATION OF THE NUMERICAL RESULTS (VISUFLOT)

An interactive visualization of the numerical results is of paramount
importance to simplify the interpretation of the large amount of
information created by the finite element analysis. In RTMFLOT this
function is performed by VISUFLOT, a menu-driven interface which
simulates the filling of the mold in real time. The pressure field
may be displayed at any requested time.

4.HEAT TRANSFER ANALYSIS (HEATFLOT)

The mold and the resin can be heated in order to accelerate the
filling process by diminishing the viscosity of the resin. So it is
necessary to analyze the heat transfer phenomenon associated with the
resin flow.

5.CONCLUSION

The integrated software environment RTMFLOT was used to simulate the
mold filling of a complex three-dimensional part. Three-dimensional
meshes of thin shell molds are generated by MESHFLOT. Then program
FLOT performs the numerical calculations. Finally, simulation results
are displayed by VISUFLOT. Several particular features have been
added to adapt the numerical model to the practical procedures
followed by mold designers, the main objective being to speed up the
design process and to reduce the time required for prototype testing.

Several aspects must be considered when evaluating a numerical model.
The first one is undoubtedly the reliability, which can be verified by
comparing calculated results with experimental observations. Another
important feature concerns the program's flexibility, namely its
ability to provide users with a variety of physically meaningful
boundary and initial conditions. Without such a set of well-
documented numerical tools, it is difficult to produce consistent and
sound results. Finally, the simplicity of the tool itself and a
correct integration of its various components into an efficient
programming environment is of paramount importance for its acceptance
by end-users.

REFERENCES

1. COULTER, J. P. & GÜÇERI, S. I., J. Reinf. Plast. & Comp., 1988, pp 7(3):200-219.
2. LI, S. & GAUVIN, R., J. of Reinf. Plast. and Comp., No 10-3, 1990, pp 314.
3. TROCHU, F. & GAUVIN, R., J. of Reinf. Plast. and Comp., July 1992.
4. TROCHU, F. & GAUVIN, R., Adv. Comp. Let., Vol. 1, No. 1, 1992, pp 41-43.
5. TROCHU, F., GAUVIN, R. & GAO, D.-M., "Numerical Analysis of the Resin Transfer Molding Process by the Finite Element Method", accepted in Advances in Polymer Technology, May 1993.

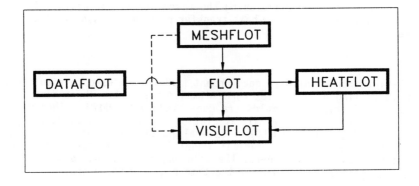

Fig 1   Schematics of **RTMFLOT** program module

Fig 2   Triangular mesh of the subway seat

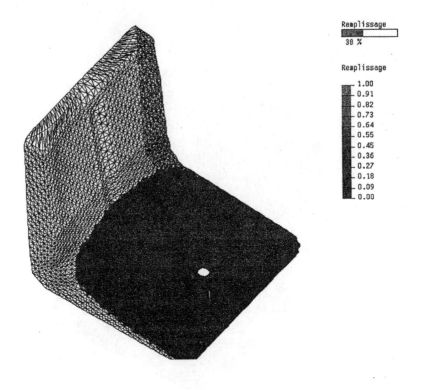

Remplissage
30 %

Remplissage

1.00
0.91
0.82
0.73
0.64
0.55
0.45
0.36
0.27
0.18
0.09
0.00

Fig 3   Example of position of the resin front

DIELECTRIC MONITORING OF THE CHEMORHEOLOGICAL BEHAVIOR OF EPOXY BASED COMPOSITES

A. MAFFEZZOLI, J.M. KENNY*, A. TRIVISANO**, L. NICOLAIS**

Dipartimento Scienza dei Materiali - Universita' di Lecce - Via per Arnesano - 73100 Lecce - Italy
*Istituto Chimico - Facolta' di Ingegneria - Universita' di Catania Vle. A. Doria - Catania - Italy
**Dipartimento Ingegneria Materiali e Produzione - Universita' di Napli P. Tecchio 8125 Naples - Italy

ABSTRACT

Dielectric properties are sensitive to the chemical-physical changes occurring in reacting systems and can be related to processing parameters. The availability of an "in-situ" measurement of a structure sensitive parameter quantitatively related to degree of reaction and viscosity represents a big potential for on line cure monitoring. A commercial TGDDM-DDS based epoxy system is studied by calorimetric, rheological and dielectric analysis under isothermal and non-isothermal conditions. A correlation between the main cure parameters, the degree of reaction and viscosity, and the dielectric properties of the reactive system is proposed.

1. INTRODUCTION

Dielectric sensors have been proposed in the last years as in-situ sensors capable to provide informations that can be correlated to the fundamental process variables during cure [1-2]. The dielectric behavior of thermosets during cure was deeply analyzed by Senturia and Sheppard [1]. However, intrinsic difficulties and unresolved issues related to the generation and interpretation of the results of dielectric measurements are still being discussed [2]. In this paper a commercial high temperature epoxy system for aereonautical applications is studied by calorimetric, rheological and dielectric analysis under isothermal and non-isothermal conditions. Then a relationship between a dielectric property (ionic resistivity) and the degree of reaction and viscosity during cure of an epoxy matrix for advanced composites is proposed.

2. RESULTS AND DISCUSSION

Ionic conductivity (σ) or the ionic resistivity (ρ) is measured from dielectric loss (ϵ "), considering that the contribution of the dipole orientation to the dielectric loss response becomes negligible and therefore:

$$\sigma = 1/\rho = \omega \epsilon_0 \epsilon " \qquad (1)$$

where ω is the angular frequency and ϵ_0 is the permittivity of the free space equal to 8.854×10^{-12} Farads/m. The effects of dipole orientation are frequency dependent and the ionic resistivity can be evaluated with Eq. 1 properly chosing the frequencies during polymerization, according with the procedures indicated elsewhere [3]. The advancement of the polymerization reaction is measured by DSC experiments. The heat of reaction developed during an isothermal (Q_{it}) test is usually lower than the maximum amount of heat of reaction obtained during dynamic scans (Q_T) indicating that the polymerization process can not be completed in this conditions. Then, a maximum degree of reaction achieved during isothermal tests can be calculated as $\alpha_m = Q_{it}/Q_T$ [4]. The behavior of a_m as a function of the isothermal test temperature shows a linear dependence:

$$\alpha_m = a + b\, T \tag{2}$$

It is clear that Eq. 2 holds for temperatures lower than the maximum glass transition temperature characteristic of the reacting system ($T_{g\infty}$). Parameters a and b are given in Table 1. The evolution of resistivity is modelled using an empirical relationship between resistivity and degree of reaction (α) accounting also for the effects of temperature:

$$\alpha = \alpha_m \frac{\log(\rho)-\log(\rho_0)}{\log(\rho_{max})-\log(\rho_0)} \left[\frac{\log(\rho_{max})}{\log(\rho)}\right]^p \tag{3}$$

Where ρ_0 is the ionic resistivity of the unreacted resin, ρ_{max} is the maximum value reached at the end of an isothermal cure and p is a parameter temperature independent. In Fig. 1 degree of reaction curves obtained from calorimetric analysis are well compared with resistivity data processed according with Eq. 3, at four different temperatures with an average value of p = 1.75.

In order to calculate the degree of reaction also in non-isothermal conditions, the temperature dependence of ρ_0 and of ρ_{max} is needed. The ionic resistivity ρ_0 is modelled with the classic WLF equation [1]:

$$\rho = \rho_{g0} \exp \left[\frac{C_1(T-T_g(\alpha))}{C_2+T-T_g(\alpha)} \right] \tag{4}$$

where C_1 and C_2 are constants, and ρ_{g0} represents the resistivity of the unreacted resin at the glass transition temperature T_{g0}. Figure 2 shows the good correspondence between a WLF expression and the experimental data obtained at constant heating rate. Parameters of Eq. 4 are listed in Table 1. In isothermal experiments the reaction end is mainly determined by vitrification and the maximum resistivity measured for a glassy system. In dynamic experiments the actual temperature is always higher than the glass transition of the sample and the maximum resistivity measured at the rubbery state. In fact the values of ρ_{max} obtained in isothermal conditions corresponding to lower mobility conditions are higher than those obtained at temperatures higher than $T_{g\infty}$. This behavior is shown in Fig. 3 where the resistivity ρ_{max} of glassy samples and of rubbery samples measured at temperatures higher than $T_{g\infty}$, are represented. The linear behavior of the logarithm of the resistivity as a function of $1/T$ indicates that the temperature dependence of O_{max} can be well represented by an Arrhenius equation:

$$\rho_{max} = K \exp (E/RT) \qquad (5)$$

The parameters of Eq. 5 are given in Table 1 for each of the two temperature ranges. Calorimetric results and predictions of the complete model (Eqs. 5-8) for a typical autoclave cycle are shown in Fig. 4. The same value of the parameter p in Eq. 4 (p=1.75) is used for isothermal and non-isothermal conditions.

The viscosity of studied system system is correlatde with ionic resistivity using a model previously developed by Kenny and Opalicki [5]:

$$\eta = \eta_{go} \exp \frac{C_{1\eta}(T-T_g(\alpha))}{C_{2\eta}+T-T_g(\alpha)} \left| \alpha_g/(\alpha_g-\alpha) \right|^n \qquad (6)$$

where η is the viscosity, η_{go} is the viscosity of the unreacted resin at its glass transition temperature, α_g is the extent of reaction at the gel point and $C_{1\eta}$, $C_{2\eta}$ and n are constants. The dependence of T_g from the degree of reaction is well represented by a linear relationship:

$$T_g = q + s \, \alpha \qquad (7)$$

The parameters of the rheological model given are summarized in Table 1. Degree of reaction values obtained applying the complete model (Eq. 3,5) are for viscosity calculation by Eq. 6,7. Experimental data and model results are compared for isothermal and non isothermal conditions in Figs. 5 and 6.

3. CONCLUSIONS

The experimental results obtained by thermal, dielectric and rheological measurements have been used for the development of a relationship between the ionic resistivity and the degree of reaction during isothermal and non-isothermal cure of an epoxy matrix for advanced composites. Degree of reaction data calculated from dielectric measurements have been used in a chemorheological model for viscosity calculation, obtaining very good correspondence between measured and predicted values in isothermal and non-isothermal conditions.

4. REFERENCES

1. S.D. Senturia and N.F. Sheppard, Adv.Polym.Sci., 80, (1986) p.1.
2. J. Mijovic, J.M. Kenny, A. Maffezzoli, A. Trivisano, F. Bellucci and L. Nicolais, Compos. Sci Tech., in press.
3. A. Maffezzoli, A. Trivisano, M. Opalicki, J. Mijovic and J.M. Kenny, J. Mater. Sci., in press.
4. J.M. Kenny and A. Trivisano Polym. Eng. Sci. 31 (1991) p. 19.
5. M. Opalicki and J.M. Kenny, Macromolecular Symposia, in press.

Table 1. Parameters of Eqs. 5-10.

$T_{go} = 267$ K	$T_g = 500$ K a $= -2.211$	b $= 0.006715$ K^{-1}	p $= 1.75$
$C_1^o = 25.1$	$C_2^o = 37.4$ K	$\rho_{go} = 7 \, 10^{14}$ ohm cm	
ln(k) $= 1.372$ ohm cm	E/R $= 11076$ K^{-1}, for T $< T_{g\infty}$		
ln(k) $= -31.43$ ohm cm	E/R $= 26959$ K^{-1}, for T $< T_{g\infty}$		
$C_{1\eta} = 32.7$	$C_{2\eta} = 37.4$ K	$\eta_{go} = 10^{12}$ Pa s	
n $= 1.91$	$\alpha_g = 0.271$	q $= 267$ K	s $= 232$ K

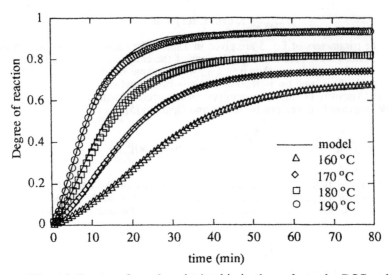

Figure 1. Degree of reaction obtained in isothermal cure by DSC and by application of the proposed model to ionic resistivity data

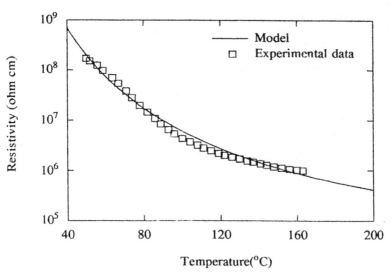

Figure 2. WLF behaviour of the resistivity of unreacted resin

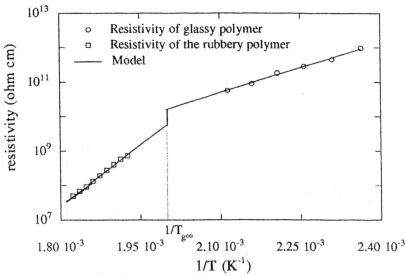

Figure 3. Resistivity as function of temperature
for cured resin below and above $T_{g\infty}$

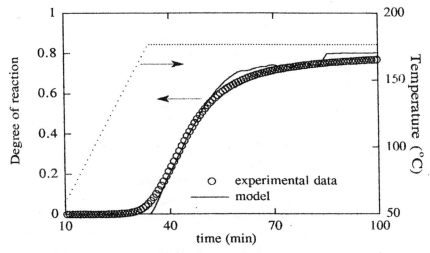

Figure 4. Comparison between degree of reaction obtained
in an autoclave cycle simulation (ramp 5 °C/min)
by DSC and calculated using ionic resistvity data

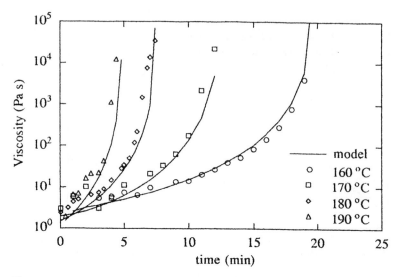

Figure 5. Comparison between viscosity values obtained by the proposed
model and experimental data in isothermal conditions

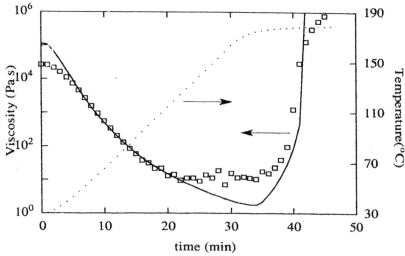

Figure 6. Comparison between viscosity values obtained
by the proposed model and experimental data
in an autoclave cycle simulation (ramp 5 °C/min)

PROCESSING STRATEGY FOR FILAMENT WINDING OF THERMOSET MATRIX COMPOSITES BASED ON A MATHEMATICAL PROCESS DESCRIPTION

J. KENNY, M. MARCHETTI*, C. CANEVA*
T. PHILIPPIDIS**, A. FRANCIOSA***

*Dept of Materials and Prod. Eng. - University of Naples
P. Tecchio - 81020 Naples - Italy*
University of Rome - Dept Chem. Eng. - V. Eudossiana 18 - Rome - Italy
** *University of Patras - Dept Mech. Eng. - Patras - Greece*
***Sistema Compositi - Cast. Pagliano - Colleferro - Italy*

ABSTRACT

This research, is a part of a Brite-Euram project and its main objective is the development of rational procedures for the optimisation and control of the manufacturing of filament-wound components. The project includes the application of reported manufacturing models to the development of a control system able to correlate numerical predictions, real variables and final properties of the processed part. In fact, at the present the main parameters of the winding and curing processes are empirically defined, limiting a major application of the filament winding technology for production of advanced composite parts where high manufacturing reliability and high quality standards are required. In the first part of the project a master model of the filament winding process has been developed and is presented here.

INTRODUCTION

General software packages have been developed for the numerical simulation of geodesic and non-geodesic winding paths with regard to axi-symmetric and non axi-symmetric bodies /1-6/. However, the fabrication of wound composites is still based mainly on a trial and error procedure, because numerical simulation has not yet been applied to the development of control systems which can correlate the main parameters values of the winding and curing processes with the processing behaviour and the final properties of the components.

Understanding of this processing-structure-properties relationship and its implementation in industrial manufacturing procedures, is the starting point to guarantee the high quality and reliability required for mass production, allowing the production of filament-wound composites at low cost, in order to confer them a more important role in the advanced composites market. These objectives have been addressed in this work through the development of mathematical models that simulate the phenomena that occur during the winding and curing processes. In particular the following activities have been programmed and partially performed:

1.-Development of a mathematical master model for the simulation of the winding and curing processes. The model provides the wound composite temperature, degree of cure, viscosity, fibre position and fibre tension as functions of position and time during filament winding and subsequent curing, and the residual stresses and strains within the wound composite during and after the cure.

2.-Development, manufacture and implementation of sensor systems able to monitor the main variables during winding and curing processes.

3.-Development of a Data Acquisition and Experiment Supervisor System to work as an interface between sensors and control system.

4.-Development of a complete software package to combine and correlate the results of the mathematical with the sensor system output.

At the end of the first year of this three years project, the first step has been completely finished including material characterisation and experimental validation.

1. MODELLING OF THE FILAMENT WINDING PROCESS

Processing of thermosetting matrix composite materials by filament winding is achieved by winding a fibre tape impregnated with resin around a mandrel. Fibre impregnation can occur immediately before the wound process, conveying the fibre through a resin bath (wet winding). Alternatively, a commercial pre-impregnated fibre tape can be used. The second possibility offers a cleaner and less toxic, but more expensive, process than the first one. From the point of view of the curing model the same equations can describe both processes. During winding, the impregnated fibre band is guided by an automatic winding arm which controls the fibre tension (Fo) and the winding angle (ϕ). As a result of the winding process a cylinder with a continuously growing thickness is formed till the final product is obtained. Generally, the process is performed at room temperature. However, the mandrel can be heated or cooled. Once the composite is formed the system is heated to promote and complete the matrix polymerisation reactions. The application during post-cure of an hydrostatic pressure will favour the composite consolidation avoiding the development of gas bubbles during resin reaction.

For simplicity, a circular section of the composite will be considered in this analysis. Moreover, the thickness of the composite cylinder is assumed to be much smaller than the length. Then, thermal and mechanical effects associated with composite ends are neglected and only transfer phenomena through the thickness are considered.

Modelling a technological process means describing, through mathematical equations, the phenomena that occur during the process. Usually, it requires mass, force and energy balance equations. The strongly non-linear character of the resulting system has been reduced by many authors by de coupling the phenomena and dependent variables into sub-models /1-3/. Our case objective is the development of relationships between the thermochemical properties of the resin-fibre system and the controlled processing parameters (fibre angle and tension, winding speed, heating rate, cure temperature and time). In particular, the following variables will be described as a function of time and radial position: mandrel and composite temperature, degree of cure of the resin, resin viscosity, fibre position and tension, stress and strain on the mandrel-composite cylinder, mechanical properties of the final composite part. In order to achieve the proposed goal, different sub models which consider the balance equations of the different complex physic and chemical phenomena occurring during the process must be developed. This analysis subdivides the process into four sub-models:

- fibre motion sub-model
- thermal sub-model
- kinetic and rheological sub-model
- stress-strain sub-model

As shown in Fig. 1 integration of these sub-models constitutes the master model of the filament winding process. Input and output parameters and variables and the interrelation between the sub-models, that are briefly described in the following, are also shown in Fig. 1 /6/.

1.1 Fibre motion sub-model

Considering the balance of forces acting on the fibres at each layer, this sub-model computes the position, displacement and instantaneous tension of the fibres, the thickness of each layer and of the whole composite, and the fibres volume fraction. Input data for this sub model are the physical and mechanical properties of the raw materials, the geometrical dimensions of the mandrel and the composite, and processing parameters such as winding angle and tension. Other input data including actual temperature, degree of polymerisation and viscosity of the resin are provided by other sub-models.

Two main factors which govern the displacement of the fibres should be considered. The first one is associated with the radial component of the fibre tension which produces a radial displacement $w_f(j)$, where the index j indicates the j-layer of the composite. After the resin gel point the fibres are blocked, no more displacements are allowed and the final fibres position is reached. Temperature changes in the cylinder, and the consequent variation in the degree of polymerisation of the resin, produce expansions and contractions of the system, with a radial component that is computed in the stress-strain sub-model.

The radial displacement $w_f(j)$ depends on the balance between the radial component of the fibre tension and the resistance that the viscous resin presents to the fibre movement /4/. Radial tension is a function of the fibre modulus, E_f, while the viscous resistance, following Darcy's law of flow in porous media, is a function of the permeability of the fibre fraction, S. Considering these two contributions, the displacement of a fibre layer during a short time interval, dt, is given by the following expression:

$$\Delta wf(j) = - S\ \sigma_f(j)\ \Delta t\ sin^2\ \phi(j)\ /\ \mu(j)\ r_f(j) \qquad (1)$$

In Eq. 1 μ is the resin viscosity, r_f is the radial coordinate indicating the position of the considered j-layer and σ_f is the fibre tension given by the ratio between the applied force (F) and the transversal area of the fibre tape (A). Evidently, the effective force on the fibre is continuously changing from its initial value Fo as the winding proceeds /6/. Assuming an elastic behaviour the expression of the fibre tension, to be solved coupled with the displacement equation, is given by:

$$\sigma_f(j)\ (t+\Delta t) = \sigma_f(j)\ (t)\ [1 - E_f\ S\ \Delta t\ sin^4\ \phi(j)\ /\ \mu(j)\ r_f(j)^2] \qquad (2)$$

1.2 Thermal sub-model

This sub model, based on the solution of the energy balance, provides the distribution of temperature through the cylinder radius as a function of time. Input data are the raw material thermal properties and thermal boundary conditions. The geometry and the instantaneous position of the fibres are given by the fibre motion sub-model, while the reaction kinetics, responsible for the rate of heat developed by the polymerisation reaction, is given by the kinetic sub-model.

The main contributions to the heat transfer is given by the heat developed by the polymerisation reaction, H_r, and by the diffusion of heat in the radial direction. This quantities are reflected in the energy balance equation, that for the cylindrical geometry adopted, is written in the following form:

$$\rho C\ \partial T/\partial t = (1/r)\ \partial(r\ Kr\ T/r)/\partial r + Hr\ d\alpha/dt \qquad (3)$$

71

where p,C and K_r are, respectively, density, specific heat and thermal conductivity of the composite; Hr is total reaction heat and α is the degree of cure, defined as a function of the partial heat developed since the reaction starts:

$$\alpha(t) = H(t) / Hr \tag{4}$$

The development of appropriate models for the reaction rate ($d\alpha/dt$) will be discussed in the next section. Moreover an appropriate model of the composite properties (p, C, Kr) as a function of fibre volume fraction, temperature and degree of polymerisation must be also provided /7/. Finally, initial and boundary conditions must be also defined. Finite differences have been applied to solve the parabolic differential equation (Eq. 3).

1.3 Kinetic and rheological sub-model

This model provides polymerisation rate, degree of cure, heat of reaction and viscosity of the resin as a function of position and time. Input data are the parameters of the kinetic and rheological equations obtained with an appropriate experimental program. The composite geometry and the instantaneous position of the fibres are given by the fibre motion sub-model, while the temperature distribution is given by the thermal sub-model. Reactive epoxy resins will be considered as they constitutes the most used matrices for advanced composites. Several kinetic models have been developed to describe the kinetics of polymerisation of these matrices /2,3,7-9/, however, for commercial systems only empirical models can be used. The characterisation method, which is described in detail elsewhere /8-10/, included several isothermal tests performed at different temperatures in the range of the normal processing temperature, and dynamic tests performed at different heating rates. Test results, expressed in terms of degree of cure according with Eq. 4 were processed by multiple non-linear regression analysis. The following kinetic model was found to describe the kinetic behaviour of these systems:

$$d\alpha/dt = (K_1 + K_2\alpha^m)(\alpha_m - \alpha) \tag{5}$$

where α_m is the maximum degree of cure obtained at lower temperatures and is a linear function of T /8/. In Eq. 6, m and n are reaction orders and K_1 and K_2 are the kinetic constants depending on temperature through an Arrhenius type expression:

$$Ki = .Ai \exp(-E/RT). \tag{6}$$

For the rheological behaviour of this reactive systems two main contributions must be considered. The first one is associated with the molecular mobility changes as a consequence of temperature variations (physical effect), the second one is related to the molecular structural changes induced by the cure reaction (chemical effect). Several theoretical and empirical models have been reported in the scientific literature /7/. Again, only empirical models were considered in this work as useful tools to describe the rheological behaviour of commercial systems with unknown composition. The characterisation of the epoxy systems studied was performed using a parallel plates rotational rheometer with the same test conditions used in the kinetic characterisation. Mobility changes due to temperature effect were modelled following the well-known Williams, Landel and Ferry (WLF) equation, while chemical effects due to the growing molecular structure were modelled using an empirical term which predicts infinite viscosity at the gel point (α_g) /7,9/. The full rheological model is represented by the following expression:

$$\mu(T,\alpha) = \mu_g \exp[C_1 (T-T_g(\alpha)) / (C_2+T-T_g(\alpha)] (\alpha_g/(\alpha_g - \alpha))^n \quad (7)$$

where μ_g is the viscosity at the glassy state, C_1 and C_2 are the constants of the WLF equation, n is an empirical exponent and $T_g(\alpha)$ is the continuously changing glass transition temperature of the reactive system.

1.4 Stress-strain sub-model

The stress sub-model gives the distribution of stress and strains at the layer interfaces as a function of time. Input data required for this computation are the mechanical properties, the thermal expansion coefficients and the chemical shrinkage coefficients of the materials. This sub-model uses the geometrical distribution of the layers and the temperature and degree of polymerisation distribution through the composite thickness provided respectively by the fibre motion, thermal and kinetic sub models.
Evidently, the practical utilisation of the four sub-models requires the solution of a complex system of coupled differential equation describing the different phenomena applying numerical techniques. The principal results are described in the mid-term report of the corresponding BRITE-EURAM project /10/.

2. CONCLUSIONS

The development of a master model for the filament winding technology and its implementation in a simulation code has been presented. The practical utilisation of the four described sub-models has required the numerical solution of a system of coupled differential equations describing the different phenomena. Some relevant results giving an idea of the potential of the simulation code are part of the achievements of our project that will continue with the experimental validation of the models.

3. ACKNOWLEDGEMENT

This work was performed in the frame of the Brite/Euram Project BREU-CT91-0451

4. REFERENCES

1. B. Spencer in "Proceedings of the 34th International SAMPE Symposium", 1989, 1556-1570
2. S. Lee and G. Springer, J. Composite Materials, 24 (1990) 1270-1298
3. E. Calius and G. Springer, Int. J. Solid Structures, 26 (1990) 3
4. G. Di Vita, P. Perugini and M. Marchetti in " Proceedings of ICCM-VIII", Honolulu, USA, 1991
5. G. Di Vita, P. Perugini, P. Moroni and M. Marchetti, Composites Manufacturing, 3 (1992) 53-58.
6. G. Di Vita, PhD Thesis, University of Rome, 1993
7. J.M. Kenny, A.Apicellla and L. Nicolais, Polym. Eng. Sci., 29 (1989) 973-983
8. J.M. Kenny and A. Trivisano, Polym. Eng. Sci., 31 (1991) 19
9. J. Kenny, M. Opalicki, in "First Annual Report, Project Brite/Euram n. BREU-CT91-0451", 1992
10. A. Franciosa et al. in "Mid-Term Report, Project Brite/Euram n. BREU-CT91-0451", 1993

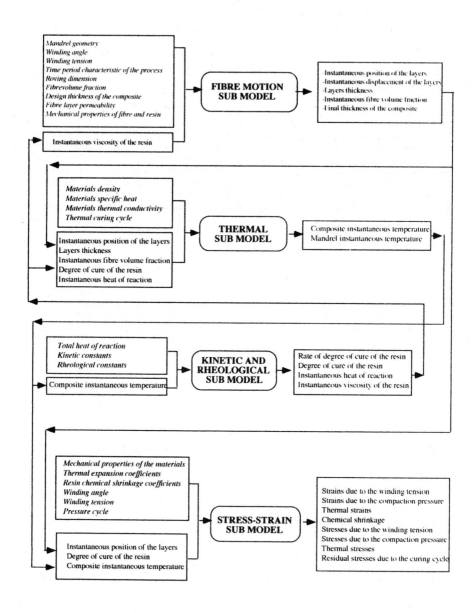

Fig. 1: Scheme of the master model of the Filament Winding Process

74

PRODUCTION SYSTEMS FOR OPTIMUM PULTRUSION PROCESS

F. SANZ

BMO Composites - St Julien d'Empare - BP 55 - Capdenac - France

Abstract : The most important parameters of the pultrusion process
are the die heating and the pulling speed of the profile.
Through the presentation of a fully automatic pultrusion line, this
paper presents how a Computerized CNC Control and a specific
ready-to-use program allow pultruders to well perform their quality
production criteria.

Filament winding units as usefull ancillary equipment for the
production of pultruded tubes by the "pull winding" process, are
also described.

1 - INTRODUCTION

Whatever the manufacturing process, quality assurance and optimum
processing is achieved by knowledge of several parameters,
experience on the subject, particular knack and adequate machine
use. This paper gives an illustration about the pultrusion process.

2 - PULTRUSION PROCESS : General review

/FIG. 1/ : Principle of composite material production
/FIG. 2/ : Principle of pultrusion process
Pultrusion is the process with higher productivity, when the subject
is about producing profiles with a constant cross-section. As a
matter of fact, this continuous process, technically perfected and
well tuned, delivers 24 h a day, a finished product with a high
repeatibility, relative low material and manufacturing costs and of
course a high optimum production rate.

Length of profiles are unlimited, fiber can be used in the form of rovings mats, spun-rovings, wovens with variable reinforcement rates and with a wide choice for the type of resin.
Complex profiles can be achieved with a very high reinforcement in longitudinal direction. In addition, a transversal reinforcement can be supplied for some profiles, by means of a "pull-winding" system.

3 - PROCESS PARAMETERS

Some factors have an overwhelming influence on the properties of pultruded profiles and of course on the related production phase. The design of pultrudable geometries and the choice of right components (resin, fiber, fillers...) as well as laminate construction are obviously primordial for a correct pultruded parts production.

The heart of the pultrusion process is the die. /FIG 3/
The main parameters are the following :
- die heating temperature,
- pulling speed,
- reactivity of resin formulation,
- fiber fraction,
- temperature of resin formulation (i.e viscosity)

Three of these parameters are usually constant and have to be held at their predefined values during the whole time of the production. The reactivity of the resin formulation depends on type and amount of all ingredients; its viscosity is linked to the temperature that is very often the ambiant one (preheating for epoxies); the fiber weight fraction is determined by the number of rovings and mats.

The other 3 parameters are not only design parameters but production parameters with direct consequences on the production rate, on the respect of the characteristics expected for the finished product.

The internal pressure in the die, which is an image of the chemical reaction can be measured undirectly by the pulling force value. The die temperature is measured directly and controlled on-line as the pulling speed.

4 - PULTRUSION LINE

/FIG. 4/ : machine picture.
Four weldment modules are dedicated to specific functions.

4-1 Wet out module

This unit has 2 functions :
Fiber feeding : conforming plates guide the fibers from the creel stand to the impregnation tank.
Fiber impregnation : to ensure complete wet-out of the fiber by the resin.

As option, additional tools can be mounted as mandrel support system, forming plates for wovens and mats...

4-2 Die/cooling module

This unit ensures :
Profile curing : the die is located very accurately and gives a precise adjustement with the machine centerline. The heating system and its control is described later on this paper.
Profile cooling : the weldment is extended beyond the die bed to allow profile cooling at room temperature before the puller entrance, even at high speeds of production.

4-3 Pulling/clamping module

The product is pulled through the die by two pullers working in alternative sequence. Each puller is equipped with a gripper unit.

4-4 Saw module

This unit is a flying cut-off saw.

5 - PROCESS CONTROL

5-1 General review

The whole machine is fully automatically controlled by a CNC device. A specific pultrusion program has been developped and is delivered with the machine, ready to use.
The user has only to input his specific parameters.
A PLC device is also incorporated.

5-2 Die heating

/FIG 5/ : controllers and connectors.
Four heating zones are possible with the basic machine and as option number can be increased up to 12 zones.
Each heating zone benefit of :
- 4 electrical connectors supplying 8 KW
- 1 thermocouple sensor
- 1 PID temperature controller
The temperatures are regulated within a +/- 1° C accuracy, according to the use of low response-time relays and performing PID feed-back loop. Temperature values, orders and measures are displayed permanently on the main console. Two threshold temperatures can be programmed with alarm feasabilities. At any moment the user can define changes in the temperature values.

5-3 Internal pressure in the die

The pulling force is permanently measured and displayed. This is done alternatively when puller A or puller B is active, or when both are in work during the "tandem cycle". An alarm is trigerred if 90% of the nominal pulling force is exceed.

5-4 Pulling speed

The skill and know-how of BMO in the design and manufacture of machine-tools since several years, induce the designer team of BMO Composites to set this postulate for the pultrusion line:
"As soon as production cycle is activated, stops or even pulling speed variations are not permitted".

The pulling speed remains constant :
- during and after modification of any part program parameter
- when tandem cycle is launched and released,
- when cutting cycle is launched or released (manually or automatically).

An exception is of course valid : at any moment, if user wants to change the pulling speed, by an input of a new value, by means of the keyboard of the CNC.

6 - OPERATING PROGRAM - SCREEN MENUS

The pultrusion system is extremely simple to use by means of an user-friendly man/machine dialogue.

7 - PULL-WINDING

/FIG 6/ : filament winding unit pictures.
The pultrusion principle is intended to give fiber reinforcement particularly in the longitudinal way. Wovens and mats can be added in a relative low proportion.
The high tensile and flexural resistance obtained can be reinforced with fiber helicallly wound to achieve also a high compression resistance. For this, one (or more) filament winding unit is (are) integrated in the pultrusion line, combining the advantages of both techniques : pultrusion and filament winding.
The filament winding unit is connected to the pultrusion line and becomes a new axis fully synchronised with the former ones, by the CNC. The value of the desired pitches are taken into account by the program, resulting in the direct drive of the filament winding unit, according to the actual pulling speed.

8 - CONCLUSION

For a pultruded composite part production, several factors and their interaction must be considered, to tail the finished product to fit the requirements of the specific applications.

The design of the profile, the definition of the raw materials and the process parameters have to find in the production equipment all the means necessary to accomplish the production process in optimum conditions.

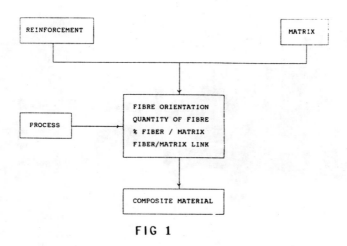

FIG 1

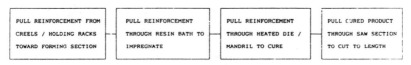

FIG 2

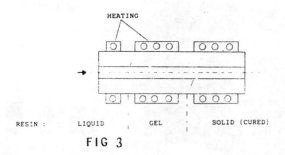

FIG 3

FIG 4

FIG 5

FIG 6

LONGITUDINAL AND TRANSVERSAL PERMEABILITY MEASUREMENT OF GLASS FIBER REINFORCEMENT FOR THE MODELLING OF FLOW IN RTM

G. GOULLEY, P. PITARD, J. PABIOT

Dept Technologie des Polymères et Composites
Ecole des Mines de Douai - 941 rue Charles Bourseul
BP 838 - 59508 Douai Cedex - France

ABSTRACT

As a first step, a bibliographic study enabled to know the methods used to determine permeability values. As a second step, devices were developed to determine longitudinal and transversal permeabilities values from Darcy's law. We compared permeabilities of several glass fiber structures and we calculated the ratios of longitudinal to transversal permeabilities to explain the difficulties of propagation of the resin encountered in some cases.

INTRODUCTION

The Resin Transfer Molding (RTM) makes it possible to make structural parts by injecting a liquid resin through a fibrous reinforcement which has been pre-placed in a mold. This technology is growing and nowadays, the industrial purpose is the modelization of the process. So it is necessary to know how to determine the permeabilities values of the reinforcements.

Therefore, the Ecole des Mines de Douai in the context of a contract with the Research Ministry, RENAULT, PSA Peugeot Citroën, SNPE and with the collaboration of CISI INGENIERIE and the CEMEF and LMA laboratories has developed devices to determine the permeability values in the plane and in the thickness of the reinforcements.

I. GENERALITIES

1.1. Definition of the permeability

The permeability is that property of a porous material which characterizes the ease with which a fluid flows through the material under the driving force of a pressure gradient. It can be defined as the fluid conductivity of a porous material.

1.2. Determination of the permeability

Mostly, permeability values are obtained from Darcy's law - Equation 1 -, considering a Newtonian fluid in laminar permanent flow :

$$Q = K.\frac{A}{\mu}.\frac{\Delta P}{\Delta L}$$

(1)

with Q : flow rate (m³.s⁻¹) ; K : permeability (m²) ; A : cross section of the cavity (m²) ; μ : viscosity (Pa.s) ; DP : pressure drop (Pa) on the distance DL (m).

1.3. Necessity of transversal permeability measurement

Preliminary works have shown that a transversal flow is superposed to the main longitudinal flow when the injection is perpendicular to the plane of the reinforcement - figure 1. So, transversal permeabilities have to be determined.

II. METHODS FOR THE DETERMINATION OF THE PERMEABILITIES

A bibliographic study has been done to know the methods used to determine the permeability values.

2.1. Determination of the longitudinal permeability

About twenty research teams have already developped technics to determine the longitudinal permeability in fibrous reinforcements. These technics can be divided into two main groups.

2.1.1. First method /1;2/

From flow rate and pressure values obtained with a stationary flow of a fluid in a parallelepipedic mold whose inlet is at one of its extremities - figure 2 -, it is possible to obtain the permeability value in the main direction of the flow by simple application of Darcy's law.

With that kind of method some problems can occur : slide of the fabric (washing effect) - figure 3 - and a flow of the fluid near the wall of the mold (boundary effects) - figure 4.

2.1.2 Second method /3;4/

We can obtain permeabilities in several directions in the plane of a reinforcement with a plane transparent square mold - figure 5 - with a central inlet perpendicular to the plane of the mold and from the evolutions of the pressure at the inlet and of the radii of the front's shape. Moreover, central inlet avoids boundary and washing effects. Meanwhile, a packing effect facing the inlet can compress the structure and create a flow between the wall and the first ply of the fabric. Furthermore, this technic is limited by a compromise between the size of the mold and the flow rate. When the flow rate increases, the mold has to be small enough to avoid a deformation of the transparent wall which is maintained at its periphery. Otherwise, the size of the mold must be large enough to observe the propagation of the front.

2.2. Determination of the transversal permeability

Few researchers have worked on the determination of the transversal permeability of fibrous fabrics. The principle of the determination is the same as that used for the determination of the longitudinal permeability : by measuring the flow rate and the pressure drop the permeability deduced from Darcy's law. Generally, the reinforcement is placed between two porous plates which are themselves positioned between two cylinders - figure 6.

2.3. Injection technics
In order to measure the longitudinal or transversal permeabilities, two methods can be used to inject the fluid.

2.3.1. Constant flow rate injection /1;2/
Constant flow rate is usually obtained with a jack set on a dynamometer - figure 7.

2.3.2. Constant pressure injection
A tank which contains the resin is maintained at an adjustable pressure - figure 8.

2.4. Materials
2.4.1. Reinforcements
In most cases glass fibers are used in the studies on the permeability, meanwhile, some teams have worked on carbon or graphite fibers. Moreover, all sorts of structures are being studied : often mat but also woven fabrics.

2.4.2. Fluids
Often, the permeants are not resin : water, glycerin mixed with water, DOP /3/, silicon oil, motor oil, corn syrup,...Polymers can also be used : polyester /2/ or epoxy /4/.

3.MATERIALS DEVELOPED BY THE ECOLE DES MINES DE DOUAI
3.1. Device for the determination of longitudinal permeabilities
A pallelepipedic mold -figure 9 - with a cavity measuring 1500x100x4 was used. The feed at one of the mold's extremities occupies all the width. The other extremity is open in order to create a stationary flow.
Three pressure transducers were mounted among the six locations on the mold axis. The flow rate was measured by weighing at the outlet. A PMMA sheet enabled to verify that boundary effects were small. The tightness was realized by a peripheral seal. The reinforcement is cramped upstream of the first transducer to avoid the washing effect.
The injection system -figure 10 - is composed of a transfer cylinder driven by a proportional distributor in order to have a regular flow rate.

3.2. Device for the determination of transversal permeability
A hollow punch is used to cut the sample and the whole is positioned in the measure chamber of the device - figure 11. The gap for the sample is controled with a set of adjusting washers between the plug and the body. A pressure transducer is positioned upstream of the sample. The difference between the pressure drop with and without the sample gives the pressure gradient due to the sample. A transfer cylinder driven by a dynamometer - figure 8 - was used to obtain a constant flow rate. The flow rate is calculated with the movement of the crosshead and the sizes of the cvlinder.

3.3. Materials
3.3.1. Fluid
The fluid used for the experiments is a plasticizer, its viscosity is 0,1 Pa.s at 20°C. Measures were made in an air-conditioned room.

3.3.2. Reinforcements
We used the following fiber glass fabrics :
- mat Unifilo (MA) : 450 g.m²
- unidirictional woven fabric (UD) : 420 g.m²
  - warp direction (UDwa)
  - weft direction (UDwe)
- bidirectional woven fabric (BD) : 1740 g.m²
- mixed fabrics (MI) : 1UD/6MA/1UC
2UD/4MA/2UD
3UD/2MA/3UD

Glass rates were 20%. 35% and 55% (by volume).

4. RESULTS
4.1. Methodology
From pressure drop and flow rate measured values we plotted Q (flow rate) versus $(A/\mu.\Delta P/\Delta L)$. According to Darcy's law, the slope at the origin of the graph gives the permeability - figure 12.

4.2. Results
See also table 1.

We verified that permeability values decrease when the glass ratio increases -figure 13. For example. the longitudinal permeability of the mat is divided by three when the glass ratio increases from 20% to 35% (by volume).

Otherwise. we showed - figure 14 - that the ratio of the longitudinal permeability to the transversal permeability is very high in the case of the woven fabrics. Such an anisotropy explains the bad propagation of the resin in the thickness which could happen in that kind of reinforcement.

CONCLUSIONS
During this work. devices have been designed for the determination of the longitudinal and transversal permeabilities of the reinforcements used in RTM. These devices have enabled to compare the permeabilities of several reinforcements. The ratio of the longitudinal permeability to the transversal permeability can be calculated to forecast the flow direction of a fluid in a reinforcement..

Moreover, the permeability values could be used in a tridimensionnal flow software for RTM which could take into account the longitudinal and the transversal permeabilities.

REFERENCES
1. R. Gauvin and M. Chibani, Intern. Polymer Processing, (1986) 42-46
2. L. J. Lee, J. A. Molnar and L. Trevino, Modern Plastics International, (1989) 64-69
3. K. L. Adams, B. Miller and L. Rebenfeld, Polymer Engineering and Science, (1986) 1434-1441
4. W. I. Lee and M. K. Um, Polymer Engineering and Science, (1991) 765-771

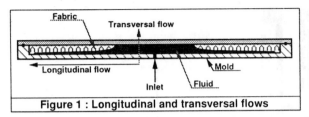

Figure 1 : Longitudinal and transversal flows

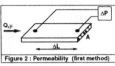

Figure 2 : Permeability (first method)

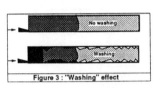

Figure 3 : "Washing" effect

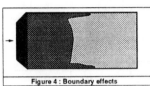

Figure 4 : Boundary effects

Figure 5 : Permeability (second method)

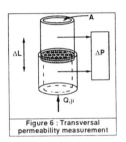

Figure 6 : Transversal permeability measurement

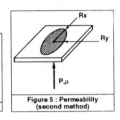

Figure 7 : Constant flow rate injection

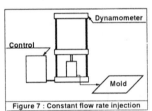

Figure 8 : Constant pressure injection

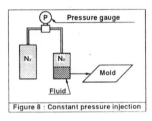

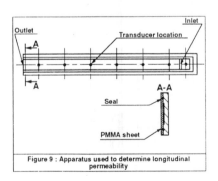

Figure 9 : Apparatus used to determine longitudinal permeability

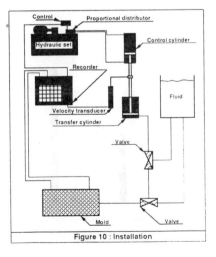

Figure 10 : Installation

87

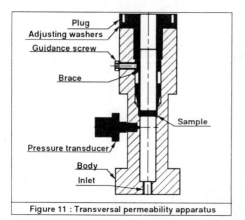

Plug
Adjusting washers
Guidance screw
Brace
Sample
Pressure transducer
Body
Inlet

Figure 11 : Transversal permeability apparatus

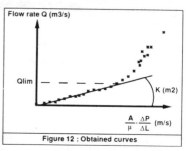

Flow rate Q (m3/s)

Qlim

K (m2)

$\frac{A}{\mu} \cdot \frac{\Delta P}{\Delta L}$ (m/s)

Figure 12 : Obtained curves

Table 1 : Results	PERMEABILITIES ($x\ 10^{-12}\ m^2$)						RATIO		
	Transversal			Longitudinal			Long./Trans.		
FABRIC	20%	35%	55%	20%	35%	55%	20%	35%	55%
BD		47	10		5000	240		106	24
UDwa UDwe		12 12	2 2		1000 1000	30 12		83 83	15 6
MA	1344	619		2600	800		2	1	
MI1/6/1 MI2/4/2 MI3/2/3		165 155 32			760 1300 3500			5 8 109	

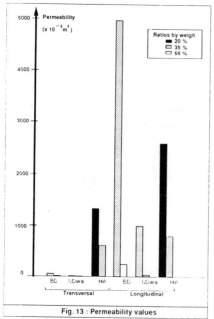

5000 ─ Permeability
($x\ 10^{-12} m^2$)

Ratios by weigh
■ 20 %
▨ 35 %
☐ 55 %

4000 ─

3000 ─

2000 ─

1000 ─

0 ─

BD UDwa MA   BD UDwa MA
└─ Transversal ─┘ └─ Longitudinal ─┘

Fig. 13 : Permeability values

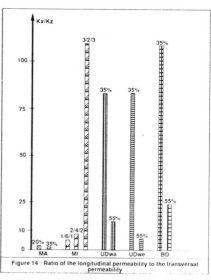

Kx/Kz

100

3/2/3

35%

35% 35%

75

50

25

2/4/2   35%

1/6/1   55%   55%

10

20% 35%   55%   55%

0
MA   MI   UDwa   UDwe   BD

Figure 14 : Ratio of the longitudinal permeability to the transversal permeability

88

FRICTION AND DAMAGE ACCUMULATION, FORMATION, MORPHOLOGY AND TRANSPORT OF DUST FROM FIBRE HANDLING

B. CHRISTENSSON, C-H. ANDERSSON*, H. ANDERSSON**
O. MANSSON***, S. KRANTZ, O. LASTOW****, K.R. AKSELSSON****

National Institute of Occupational Health - 17184 Solna - Sweden
*Swedish Institute of Textile Research - TEFO - PO Box 5402
40229 Göteborg - Sweden
**Lund Institute of Technology - Dept of Production and
Materials Eng. - PO Box 118 - 22100 Lund - Sweden
***Lund Institute of Technology - Dept of Working Environment
PO Box 118 - 22100 Lund - Sweden

ABSTRACT

The mechanical stress build up in fibre handling, the dynamical frictional phenomena between fibres and curves surfaces and formation of dust have been studied. Typical are stick slip behaviour and threshold stress levels for formation of dust. The emissions of dust can be limited by electrostatic charging.The measured coefficients of friction of the fibres depend on the mechanical properties, micro structure, morphology and sizing of the fibres, the friction materials, the deformation rates and applied loads.

1. BACKGROUND AND PURPOSE

Accumulation of fibre damage in the handling and processing of fibres is a crucial feature limiting the technical possibilities of economical use of the potential of mechanical properties of high performance load bearing fibres. The subsequent formation and transport of fibrous dust is not only a potential work environment hazard, but also a problem regarding the reliability in the processing, i.e. electrical functions, wear of equipment etc.

The aim of this work is a mapping relating the emission of dust to the processes of handling and the mechanical properties, microstructure, morphology, geometrical arrangement and sizing of fibres.

2 THEORY

The mechanisms of frictional damage, fracture and formation of fragments are related to the load-strain behaviour and micro structure of the fibres:

- Linear elastic   i) isotropic. Examples: glasses and SiC.

                               ii) fibrillar fully oriented. Example: UHM-C.

- Non linear elastic, strain hardening, fibrillar incompletely oriented. Examples: HT-C, aramides.

- Strain softening, viscous behaviour. Example: Al-wire.

- Complex, non linear elastic/strain softening/strain hardening, incompletely oriented fibrillar microstructure with viscous behaviour. Examples: most textile fibres, metal wires.

Fibrillar structures are highly anisotropic regarding mechanical properties and can be modelled as composite materials, with a second grain boundary phase acting as matrix . The load bearing capacity in axial compression is limited by fibrillar buckling, /5-7/.

Polymeric surface coatings, sizings, are extensively used in order to improve handleability and provide :
- Ductile surface behaviour
- Lubricated friction
- Compatibility and bonding to matrix materials

The handling properties and the interfacial behaviour in composites are however physically closely related, as to be seen from the apparently contradictory requirements for boundary lubrication : i) good bonding i.e. low surface energy or reactive bonding, ii) high transverse compression strength and iii) very low shear strength.

Dust is usually classified with respect to its aerosol properties and the risks of lung damage as:
- Respirable dust
- Inhalable dust
- Non inhalable fragments

There are also practical and fundamental reasons to distinguish between the dust formed in the processing and the dust existing among the virgin fibres. The dust delivered with the fibres may be more or less mobile. A reasonable classification of the dust related to the fibres and the handling processes is thus :
- Mobile dust
- Sessile dust
- Dust formed in the handling processes

The loss of material in the formation of dust from fibres by applied stresses is associated with some kind of damage. Damage is also likely to occur for example due to local high shear or indentation stresses during the dust mobilization processes, not nessessarily after the mobilization of the dust.

In the bending of a fibre the difference between the outer and inner radius gives a strain difference (3). The law of friction is given by. equations (4) for plane and (5) for curved surfaces of elastic materials /2,3/. Assuming linear elastic behaviour, expressions for the contact stresses (6) and (7) are derived :

$$\sigma = \varepsilon\, E \qquad\qquad (1)$$

                                                   σ stress

                                                   ε strain

$$\sigma = \sigma_{pl} \qquad\qquad (2)$$

                                                   E   elastic modulus

$$\varepsilon = \frac{d}{R+d} \qquad (3)$$

$$F_1 = F_0\,\mu \qquad (4)$$

$$F_1 = F_0\exp(\mu\,\theta) \qquad (5)$$

$$\sigma_1 = \frac{F_0\exp(\mu\,\theta)}{R\,b} \qquad (6)$$

$$\tau_1 = \frac{\mu\,F_0\exp(\mu\,\theta)}{R\,b} \qquad (7)$$

R   bending radius
d   diameter of fibre

μ coefficient of friction

F_0 applied preload force
F_1, F_2 tensile forces due to friction
θ contact angle
σ_1 frictional contact stress
b width of contact zone
τ_1 frictional shear stress

An analysis of the Herzian contact stress distribution, i.e. the tensile stresses at the edge of the contact zone, at compression of crossing fibres is given in Ref.4.

3. EXPERIMENTAL WORK

The emission of dust from tensile testing and friction over small radius bent surfaces was studied for some typical multifilament fibres with and without sizing:
- Aramides: Kevlar, Twaron and Technora types
- Glass: E and EC
- Carbon: HT, HM and IM types
- SiC: Nicalon and Tyrrano types
- Textile: PET, PP

Most fibres are studied as rowings, i.e. as bundles of multifilament fibres without twist. Kevlar 29 is also studied in some different twisted yarn configurations.

For the studies of dust emission from handling of continuous fibres i.e. filaments, rowings and yarns, a dust sampling chamber working with clean gas, nitrogen AGA SR, has been developed, Fig.1.

For the frictional stress build-up studies with modified ASTM method /1/, pins of diameter $\phi = 3.0$ mm of two typical industrial fibre guide materials also suitable as model materials were used and compared with a relatively coarse abrasive of the kind used to texturize PET-fibres:
- Normalized polished high carbon low alloy steel
- Fine grain polished alumina
- Alumina abrasive, 400 grit

The analysis of the friction is based on equation (5). The coefficient of friction is thus given from :

$$\ln F_1 - \ln F_2 = 2\,\mu\,\theta \qquad (8)$$

F_1 denotes the tensile force in downward motion and F_2 in upward.

Direct counting instruments and filter take up were used parallel for the analysis of the emitted dust. The morphology of the dust and changes in fibre morphology were characterised in SEM.

All samples were conditioned and all measurements were performed at 20°C and 60%RH standard atmosphere.

4. RESULTS AND DISCUSSION

The dependance of bending radii and contact angles on the stress build-up is obvious from equations (5)-(7). The frictional build-up force exhibits stick slip behaviour. The force maxima provide applied contact stress maxima and thus a risk of fragmentation of fibres. Friction arises primarily from the shear of the junctions between the contacting materials, by adhesion or by asperity interlocking. The risk of fibre damage is thus strongly depending on the physics behind these peak loads.The coefficients of friction are functions of the surface characteristics, speed, pre-load and earlier handling history, as shown for Kevlar 49 on raw alumina pins in Fig.2.

The emission of dust was found very low below threshold pre-loads for sized aramides, glass, HT-C and textile multifilaments, Fig.3. Fine particle concentrations below the background values were found for some aramide samples at low pre-loads on alumina pins. Small but positive emissions were however found when the fibres were discharged.

The fragments of glass, Nicalon and carbon fibres exhibit brittle fracture types, Fig.4. In some cases fracture and debonding of the sizings of glass fibres gave fragments similar to respirable sharp ended fibrous dust when analyzed in optical microscope.

Typical for aramide and PET-fibres are remaining transverse deformation from the contact stresses, instead of brittle fracture . The aramide and PET-fibre fragments exhibit fibrillar types of morphology of the kind illustrated in Fig.5.

It is from statistical point of view impossible to completely exclude any kind of emission of dust from fibres subjected to mechanical stresses in handling. The existance of threshold stresses for emission of dust however indicates possible conditions for technically reasonable handling.

Non conducting fibres can become electrostatically charged on non conducting frictional surfaces and thus act as particle traps. The imobilization of the fine dust by electrostatocal charging can thus act as the rate limiting step for the emission.

The dust from aramide fibres exhibit morphologies much resembling the morphology of fibrillar asbestos fragments.The mechanical properties of asbestos and aramides are however quite different. Asbestos is almost linear elastic and brittle. Aramides however posess ductility with very low stiffness in transverse compression and undergo plastic deformation at small strains in any compression mode /5-7,12/. It is thus for example very likely that aramide fragments buckle instead of penetrate human tissue.

The fragments of glass fibres etc exhibit blunt ends of the brittle 45-90° fracture types instead of fibrillar morphology, as predicted from the stress state and the isotropic micro structure of the fibres /4/.

.ACKNOWLEDGEMENT

The financial support from the Swedish Work Environment Fund is greatfully acknowledged.

REFERENCES:

1. ASTM D 3108-83,"Standard test method for coefficient of friction , Yarn to metal"
2. Kowalski K, Melliand Textilberichte (1991) 171-175 (in German)
3. Andersson C-H, TEFO Rpt TTT 889005 FR 1989 (in Swedish)
4. Andersson C-H, Månsson O.and Karlsson M., ECCM 6, Bordeaux 1993
5. Allen S.R., J. Mater. Sci. 22 (1987) 853-859
6. Knoff W.F., J. Mater. Sci.Lett. 6 (1987) 1392-1394
7. déTeresa S.J., Porter R.S. and Ferris R.J. J. Mater. Sci. 23 (1988) 1886-1894
8. Latzke P.M., "Textile Fasern, Rastermikroskopie der Chemie und Naturfasern"
   (1988) Deutscher Fachverlag, Frankfurt (M) (in German)
9. Oudet Ch. and Bunsell A., J. Mater. Sci. 22 (1987) 4292-4298
10. Wagner H.D., J. Mater. Sci. Lett. 5 (1986) 439-440
11. Calil S.F., Clark I.E. and Hearle J.W.S., J. Mater. Sci. 24 (1989) 736-748
12. Kawabata S., J. Text. Inst. 81(1990) 432-447

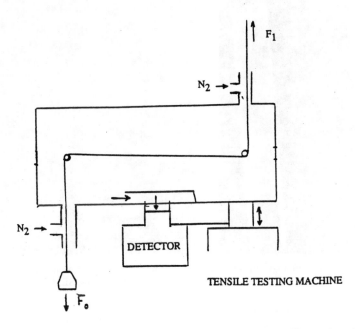

Fig.1. The principle of the dust sampling chamber in use for handling testing of fibres, as used for frictional stress build up studies in a tensile testing machine. i) Frictional pins, $\phi = 3.0$ mm ii) Gas inlets, a flow of pure gas, nitrogen AGA SR-quality is used in order to avoid contamination from the surrounding laboratory background.

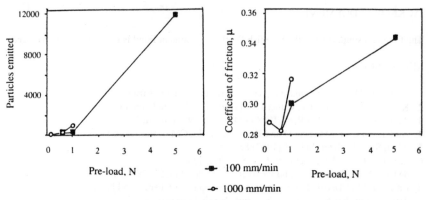

Fig.2. Emission of particles from Kevlar 49 on Al$_2$O$_3$ pins as a function of pre-load.

Fig.3. Coefficient of friction of Kevlar49 on Al$_2$O$_3$ pins as a function of pre-load.

Fig.4. Brittle fracture morphology of E-glass fibre fragments, SEM 750X

Fig.5. Fibrillar fragment morphology from Kevlar 49 on Al$_2$O$_3$ pins, SEM 750X

FABRICATION TECHNOLOGIES INDUCED POROSITY AND ITS INFLUENCE ON THE MECHANICAL BEHAVIOUR OF THERMOSET MATRIX COMPOSITES

P. KRAWCZAK, J. PABIOT

Dept Technologie des Polymères et Composites
Ecole des Mines de Douai - 941 Rue Charles Bourseul - BP 838
59508 Douai Cedex - France

ABSTRACT

After a presentation of some examples of porosity in industrial made structures, the main technological parameters inducing this type of defect are given. Finally the influence of the void volume content on the mechanical behaviour of industrial composites is investigated and an experimental modelisation of porosity is undertaken in order to take into account all the parameters characterizing porosity.

INTRODUCTION

Composites are light and anisotrop bi-phasic materials made up of a fibre reinforced organic matrix. They have very attractive mechanical and physical assets given by a combination of the properties specific to each phase. Their main advantages come from the great set of available materials - according to reinforcement structure or kind and resin type - as well as from the processing easiness by the means of different technologies. The development of composites as structural materials in aeronautic, railway, automobile ... industries however still meet uncertainties relating to the quality and the reproductibility of manufactured parts: The fabrication of such structures is likely to induce defects of porosity, which are the result of a defective control of the processing technologies and of the associated materials. The first aim of this paper is to position the problem by showing the qualitative state of the porosity in some industrial parts. Then the influence of the porosity on the mechanical behaviour of industrial composites is investigated. Finally the first results of a parametric study of the incidence of voids are presented.

1. INDUSTRIAL REALITY OF POROSITY LINKED PROBLEMS

Practical experience shows that in spite of firm efforts of converters to restrict the problem, it remains difficult at the present time to avoid the occurence of voids in composite materials. Whatever the choice of the resin (phenolic, epoxy or unsaturated polyester) or the processing technology (contact moulding, filament winding, compression, resin transfer moulding, centrifugation moulding) may be, composite structures present characteristic defects of variable shape (spherical, ellipsoïdal, random, ...), size (from µm up to mm), distribution and location (volume or interface) with void contents from 1% (for autoclave moulding for instance) up to 15% in certain cases for SMC or BMC (figure 1). The causes of these structural defects are linked mainly to material manufacturing operations (impregnation of the fibre or laminating) but also to the selection of technological parameters (pressure level, temperature steps, vacuum assistance) according to the rheological characteristics of the resin.

2. INFLUENCE OF VOID VOLUME CONTENT ON THE MECHANICAL BEHAVIOUR OF INDUSTRIAL COMPOSITES

Glass/epoxy woven composite materials showing void contents varying from a few percent to several ten percents (evaluated by image analysis) were fabricated by autoclave moulding thanks to a modification of the technological parameters, and their mechanical properties were measured. The results show that such a defect could effectively have a detrimental effect and justify all the efforts made to optimise moulding processing in order to avoid the porosity induced by fabrication.

Decreases in strength of several ten percents appear as a matter of fact at void contents of about ten percent, whether it be for interlaminar shear strength with a decrease of nearly 50% (figure 2), for flexural strength (figure 3) or tensile strength (figure 4) (with respectively 25% and 20% decrease). The great spreading of the results, confirmed by a synthesis of bibliographical data carbone reinforced composites (figure 5) /1 to 6/, suggests nevertheless the deficiency of total void volume content to characterize porosity. The previous reason incited us to undertake an experimental modelisation of porosity, taking into account not only the total void volume content, but also the shape, the size and the distribution of porosities, which are potential parameters of influence, as the computations of certain authors show /7/.

3. EXPERIMENTAL MODELISATION OF POROSITY

In order to take into account these different parameters characterizing porosity, the interlaminar shear strength has been measured on composites containing model unit porosities (voids), which were artificially created during the processing of the material.

The first experiments were carried out under a 3 point bending sollicitation (short beam shear test). Porosity was simulated by the means of flat or tubular defects, which were inserted in the mid-plane of the laminate just before autoclave curing. The flat defects of rectangular cross section were simulated thanks PTFE films of different thickness (50 to 500 µm) and width (1 to 6 mm). Silicone rubber tubes of different diameters (0,3x0,64 mm to 1x2,2 mm) represent the tubular defects.

The influence of the width and thickness of flat defect on interlaminar shear strength was studied. It appears that the most influential parameter is the defect width, and hence the void surface content in the plane where maximum shearing occurs. The flaw thickness does not seam to play a role. The influence of the size of the tubular defect on the interlaminar shear strength was investigated for different cases (presence (insert) or absence (void) of the silicone tube). Noteworthy is the fact that the presence of a low rigidity insert (silicone tube) does not modify the value of the measured shear strength at a given defect size (fig.6). Figure 7 shows a synthesis of the results obtained for the two kinds of defect geometries (flat and tubular) in terms of equivalent void surface contents in the plane of shearing. It appears that both shape and size of defects affect the interlaminar shear strength. A tubular defect seems to be more detrimental than a plane one.

CONCLUSION

Porosity in composite materials remains a badly mastered but highly detrimental phenomenon whether it be from an industrial or from a scientific point of view. These first results showed the importance of the problem as well as the necessity to improve our knowledge of the incidence of the parameters characterizing porosity, such as shape or distribution.

ACKNOLEDGEMENT

The presented work was carried out in collaboration with the following companies : CARLIER PLASTIQUES, COMPREFORME, HERMEX, LERC, PLASTIREMO-AERAZUR, SOTIRA 59, STRATIFORME, all members of the "Club Composites + 20" , BYK CHEMIE, EUROPEAN OWENS CORNING and VETROTEX.

REFERENCES

1/ YOSHIDA H., OGASA T., HAYASHI R., Statistical approach to the relationship between ILSS and void content of CRFP, Composite Science and Technology, Vol.25, p.3-18, 1986
/2/ YOKOTA M.J., In-Process controlled curing of resin matrix composites, SAMPE J., july/august 1978, pp. 11-17

/3/ CLARKE B., Nondestructive evaluation of composite materials, Inspection and Testing, march 1990, pp. 135-139

/4/ TANG J.M., LEE W.I., SPRINGER G.S., Effects of cure pressure on resin flow, voids and mechanical properties, J. of Composite Materials, Vol.21, may 1987, pp.421-440

/5/ HANCOX N.L., The compression strength of unidirectional carbon fibre reinforced plastics, J. of Mat. Science, 10(1975), pp. 234-242

/6/ GUNYAEV G.M., The effect of structure processing defects on mechanical properties of polymeric composites (Ch. VII) in Handbook of Composites, Vol.3 - Failure Mechanics of Composites, Ed. G.C. Sih and A.M. Skruda, Elsevier Science Publishers, pp. 375-391, 1985

/7/ GRESZCZUK L.B., Micromechanics failure criteria for composites subjected to transverse normal loading, AIAA 12th Struct. Dynam. Mater. Confer., California, AIAA/ASME, Paper n°71-355 (1971) pp.1-9

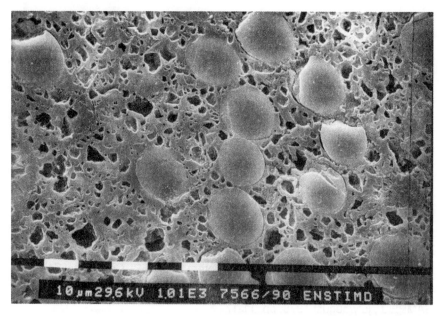

Fig.1: Example of porosity in a BMC industrial composite structure

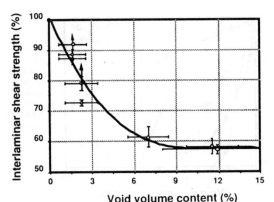

Fig.2: Influence of void volume content on interlaminar shear strength

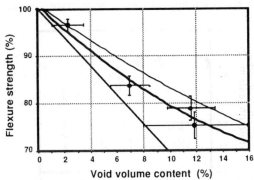

Fig.3: Influence of void volume content on flexural strength

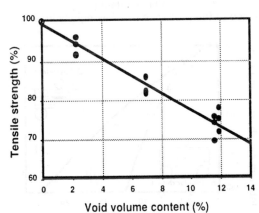

Fig.4: Influence of void volume content on tensile strength

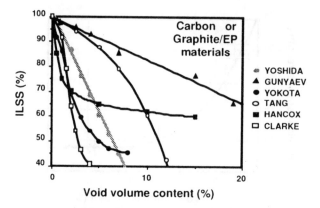

Fig.5: Synthesis of bibliographical data - Influence of void volume content on the interlaminar shear strength of carbon or graphite / epoxy composites

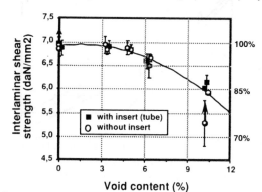

Fig. 6: Influence of the presence of a low rigidity insert on the ILSS

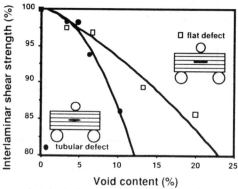

Fig.7: Influence of defect shape on interlaminar shear strength (ILSS)

SOLIDIFICATION AND PROPERTIES OF 2024 ALUMINIUM Al2O3 PARTICULATE COMPOSITES

A.N. ABD EL-AZIM, S.F. MUSTAFA, Y.M. SHASH*, A.A. YOUNAN

Central Metalurgical Research and Development Institute (CMRDI)
PO Box 87 - Helwan - Cairo - Egypt
*Cairo University - Cairo University - Mech. Dept - Cairo - Egypt

ABSTRACT.

Aluminium alloy (2024)-Al_2O_3 particle composites containing up to 30 vol.%. Al_2O_3 have been prepared by direct casting technique. Surface treatment to Al_2O_3 particles and addition of small amounts of magnesium found to aid wettability and bonding between matrix and reinforcement.
Metallographic examinations showed that the presence of alumina particles refined the structure of the matrix. Also, the mechanical properties of the produced composites and matrix alloy were studied in the heat treated conditions.
It was found that the addition of Al_2O_3 particles to the matrix alloy improves the wear resistance and the yield strength whereas the ultimate tensile strength and ductility are decreased in both natural (T_4) and artificial aging (T_6) at 170°C.

1. INTRODUCTION.

Particle reinforced aluminium-based composites prepared via a liquid metallurgy route are relatively low cost materials, in comparison with other discontinously reinforced aluminium composites /1-3/. By selecting appropriate reinforcement materials, the properties of particle reinforced aluminium composites can be tailored /4/. It is commonly agreed that alumina is an ideal reinforcing material for aluminium since the two are physically compatible at the projected service temperature of the resultant composites /5/.
Liquid metal technique to produce Al-base alloy particulate composite is considered to be simple and economical. The properties of these MMCs are controlled by many variables such as reinforcement distribution, wetting of reinforcement by matrix alloy, reactivity at reinforcement matrix interface etc. Surface treatment or surface coating is made to promote wetting and to suppress excessive reaction between the reinforcement and matrix /6-9/.

Currently, moderate knowledge exists regarding the mechanical properties of particulate-reinforced aluminium-matrix composites /10,11/.

The present work, details the process for fabrication of aluminium alloy composites containing up to 30 vol.% alumina particles and their properties. Also, the effects of addition of particulates on the mechanical properties of composites are discussed.

2. EXPERIMENTAL.

Commercial 2024 Al-alloy was used as matrix and different vol.% (10-30) of Al_2O_3 particles were introduced into the melt through vortex method. The particle sizes of alumina lay in the range of 50-150 μm. The details of this method were discussed in a previous investigation /12/. For the purpose of comparison separate ingot of the unreinforced alloy was also produced. The starting composition of the matrix alloy was 4.12 wt. pct. Cu, 1.94 pct. Mg, 0.252 pct. Si, 0.686 pct. Mn, 0.181 pct Fe, 0.118 pct. Zn, balance Al.

The structure of the produced composites and matrix alloy were examined microscopically. Also specimens from these alloys were machined and used for wear and tensile tests after being subjected to various heat treating procedures. These treatments consisted in solution treatment at 500°C for 2 hours followed by water quenching. Both tests were carried out in the cast condition and after aging in case of natural (T_4) and artificial aging (T_6) at 170°C. Wear tests were performed on a testing machine of the pin and ring type. Specimens used were in the form of a cylinder 8 mm diameter & 12 mm height pressed against a steel wheel (SAE 1045 ring) of 73 mm diameter.

3. RESULTS AND DISCUSSION.

3.1 Fabrication of Composites.

In the present study it was possible to introduce up to 30 vol.% Al_2O_3 in molten 2024 alloy and produce composite castings. Pretreatment of the alumina particles prior to addition into the melt by impregnation in Na^+ ions-containing solution found to decrease the wetting angle between the molten al and Al_2O_3 particles. The turblance generally involved by the vortex method results in a high gas content and oxide inclusions level in the melt. Degassing by dry nitrogen is the most effective way of removing gases from MMC melts. Also, the use of N_2 atmosphere has further aided in improving wettability between molten Al and alumina particles. In addition, Magnesium was added in small amounts (< 1wt.%) into the melt in order to improve wettability.

3.2. Structural Features.

The unreinforced and reinforced alloys structures are shown in Figs 1 and 2. It is obvious from Fig 1, that the columnar grains which nucleated from the chilled zone disappeared completely as Al_2O_3

of alumina from 10 to 30 % resulted in the formation of agglomerated particles, (Fig 2). The matrix structure of composites show much smaller grain sizes. These results may be due to the presence of Al_2O_3 particles which act as sites of nucleation during solidification of the melt.
The boundary layer between Al_2O_3 particles and matrix (Fig. 2b) is formed due to the reaction of Mg with Al_2O_3 particles at the interface to form spinel, $MgO.Al_2O_3$ /5/.

3.3 Mechanical Properties.

3.3.1 Strength

The results of tensile tests at room temperature of naturally and artifically aged at 170°C on 2024-Alloy with varying Al_2O_3 content are shown in Table 1. It can be noticed that the introduction of Al_2O_3 particles into the matrix improve the yield strength whereas the ultimate tensile strength and ductility are reduced. This improvements in yield strength is obviously due to stiff Al_2O_3 particles. However, the lower values of tensile strengths seem to be due to excessive reaction between Al_2O_3 particles and Magnesium during the fabrication of the composites to form spinel. These reaction product will indirectly affect the strength of the composites as a result of Mg loss of the matrix and degradation of the reinforcement.

3.3.2 Wear

Numerical results of the wear tests performed in naturally and artificially aged at 170°C on 2024- Al alloy with different percent of Al_2O_3 particles under a constant load of 50 N are shown in Fig. 3. It can be noticed that the addition of Al_2O_3 particles to the matrix decreases the volume losses in all cases. The higher the Al_2O_3 particles content of composite, the lower the volume losses and the higher the wear resistance, Fig 3. This is due to the presence of hard Al_2O_3 particles, and its inoculation effect which results in finer grain structure.

4. CONCLUSIONS

1. Aluminium alloy (2024)-Al_2O_3 particle composites containing up to 30 vol.% Al_2O_3 p have been prepared by direct casting technique.
2. The presence of Al_2O_3 particles refine the structure of the matrix. The higher the percentage of Al_2O_3 particles, the smaller is the grain size of the matrix.
3. The addition of Al_2O_3 particles to the matrix improve the yield strength, whereas tensile strength and ductility are decreased.
4. The presence of hard reinforcing Al_2O_3 particles in 2024-Al alloy matrix increases wear resistance.

REFERENCES

1. P.K. Rohatgi, Mod. Casting, 78 (1988), 47-50.
2. T.Z. Kattamis and J.A. Cornie, Cast Reinforced Metal Composites, Ed. S.G. Fishman and A.K. Dhingra (ASM International, 1988), 47-51.
3. D.M. Schuster, M. Skibo and F. Yep, J. Metals, 39, (1987), 60-61.
4. H.J. Heine, Foundry Management & Technology, 7, (1988), 25-30.
5. A.P. Levitt, Whisker Technology, 1st ed. Wiley Interscience, N.Y. (1970), P. 245-250.
6. A. Banerjee, P. K. Rohatgi W. Reif, Metall., 38; (1984), 656-660.
7. A.G. Kulkarni et al., J. Mat. Sci., 14, (1979), 592-599.
8. F.Delanney, L. Rozen and A. Deryttere, J. Mat., Sci., 22, (1987), 1-5.
9. L. Salvo, M. Suery, J.G. Legoux and G.L. Esperance, Mat. Sci. and Eng., A135 (1991), 129-133.
10. S. Yajima, et al., J. Mater. Sci. 16 (1981), 3033-3036.
11. R.L. Trumper, Metals and Mater., 3 (1987), 662-667.
12. A.N. ABDEL-Azim, et al., 2nd. Inter. Conf. of Automated Composites, Netherlands, (1988), 4/1 - 4/9.

Table 1. Tensile properties of 2024-Al alloy with different prcent. of Al_2O_3 in case of natural (T4) and artificial aging(T6).

Material	0.5% Proof Stress(MPa)	U.T.S (MPa)	Elongation (%)
2024-T6	60	220	6
2024-10 Vol% Al_2O_3-T6	67	132	3.2
2024-20 Vol% Al_2O_3-T6	75	114	2.8
2024-30 Vol% Al_2O_3-T6	79	92	2.2
2024-T4	57	214	7.4
2024-10 Vol% Al_2O_3-T4	63	97	3.8
2024-20 Vol% Al_2O_3-T4	96	86	2.3
2024-30 Vol% Al_2O_3-T4	72	82	2.4

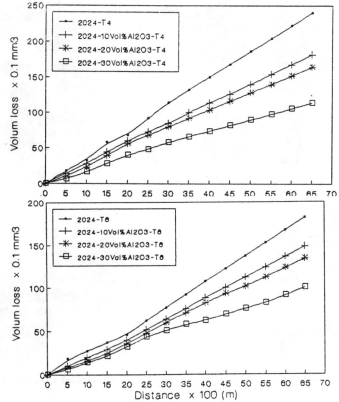

Fig.3 Volume loss vs. travelling distance of investigated alloys in case of T_4 and T_6 condition.

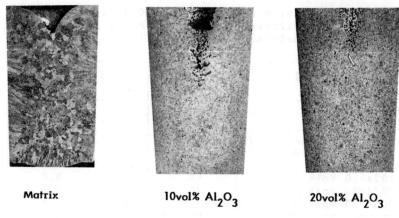

Matrix   10vol% Al_2O_3 20vol% Al_2O_3

Fig.1 Macrosturctures of unreinforced and reinforced alloys.

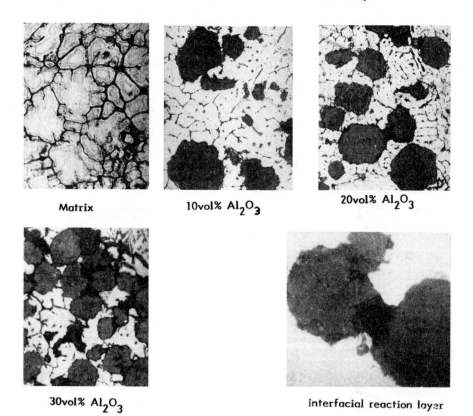

Matrix   10vol% Al_2O_3 20vol% Al_2O_3

30vol% Al_2O_3 interfacial reaction layer

Fig.2 Microstructures of unreinforced and reinforced alloys.

SQUEEZE INFILTRATION CASTING OF ALPHA-ALUMINA FIBRE CONTAINING Al 4Cu1Mg COMPOSITE

A.N. ABD EL-AZIM, S.F. MUSTAFA, S.A. ABDEL-HADY*, S.A. BADRI

*Central Metallurgical Research and Development Institute (CMRDI)
Non Ferrous Alloys Lab. - PO Box 87 - Helwan - Cairo - Egypt
Helwan University - Egypt

ABSTRACT.

In this investigation 2024 Al alloy was infiltrated into a preform of unactivated and/activated α-alumina fibre. The perform, die and ram were heated to a temperature of 250°C before applying the pressure, up to 100 MPa to the molten alloy. The fibre loading under control which was equivalent to a fibre loading in the composite of 0.28 volume fraction. Macrostructural examination revealed very fine equiaxed grains and microstructural studies showed a good fibre/matrix bonding without any chemical reaction at the interface. Both modified and unmodified composites showed higher tensile properties and excellent wear resistance compared with their matrices in both T4 and T6 conditions.

1. INTRODUCTION

Production of metal matrix composites (MMCs) by casting technique is one of the low-cost manufacturing materials for a variety of engineering applications /1-3/. The major difficulty of this process is poor wettability of most of the ceramic particles by the molten aluminium and its alloys. Also, sever chemical attack of the reinforcements can occur during fabrication which has a strong influence on the properties of the products /4/. Surface treatment or surface coating is usually made to promote wetting and control the reaction between the fibres and the matrix/5-8/. However, the surface treatment requires complex facilities and hence increases the fabrication cost. Using squeeze casting, which is being used to effectively fabricate fibre reinforced metal (FRM) /9,10/, the surface treatment may be omitted. It is possible to fabricate FRM in a short time using high pressure and therefore to shorten the time during which fibres are exposed to high temperature.

In this work, MMCs have been produced by infiltrating alumina fibre performs with molten 2024 Al alloy using a squeeze casting technique. The infiltration process, structure and composite properties were studied.

2. EXPERIMENTAL

The reinforcement used was FP-alumina fibre with a diameter of 0.02 mm and 2024 Al alloy was used as matrix. Activation of fibres was carried out by impregnation in Na^+ ions-containing solution. Unactivated and activated fibres were used in this investigation. The alumina fibres were previously pressed in a preform having a diameter of 100 mm, 10-20 mm height and 0.28 volume fraction of fibre (vf) using a plunger/sleeve die. The perform was then set into the preheated metal die, and heated to 250°C. The molten alloy of 730°C pouring temperature was poured into the die cavity and squeezed into the perform at a low plunger speed (9 m/s) at predetermined pressure (50-100 MPa) . The full size of the fabricated samples, which was 100 mm in diameter and 40 mm in height, were heat treated, and then the tensile and wear specimens were machined from both composite and matrix alloy. The tensile tests were carried out at room temperature and 200°C. Wear tests were performed on a pin and ring wear machine under dry sliding condition after aging at room temperature and 170°C. Optical and scanning electron microscopy were used to examine the morphology and structure of the product phases.

3. RESULTS AND DISCUSSION

3.1. Metallographic Examination

3.1.1. Macrostructure

Macrostructure of matrix alloy (upper half), and the composite (lower half) are shown in Fig. 1. It is clear that the structure of unreinforced matrix shows fine equiaxed grains. This is apparently due to the high pressure applied during fabrication by squeeze casting process. Also, the macrostructure of composite shows finer structure compared with the matrix alloy. Eventually in addition to the high pressure applied, the very rapid heat exchange resulting when the matrix encounters the cold fibres, leads to the formation of very fine grain characteristic of the reinforced alloy /10,11/. Similar results were obtained in case of reinforced and unreinforced alloys with activated preform solidified under identical conditions.

3.1.2. Microstructure

Metallographic examination of as-cast ingots showed a fairly uniform distribution of randomly oriented fibres in the unactivated composite, Fig. 2. However, the fibre distribution in the activated preform is more uniform than the unactivated one, Fig. 2. This could be attributed to the improved wettability between fibre and molten alloy due to activation, resulting in a reduction in friction forces between fibre and melt during infiltration process. It was noticed that the grain structure within all the composites was finer than else-where and that the grain boundaries were linked to the fibres. This may be partly as a result of the effect of fibres which are associated with the cast regions to solidify. No reaction layer was observed at the fibre/matrix interface of both modified and unmodified composites, Fig. 2. It can be concluded that, the molten metal infiltrate the preform and solidifies in a very short time and therefore get no chance to react with alumina fibre. Meanwhile the presence of a certain degree of fibre breakage could be related to the high pressure applied during squeeze forming. The composites showed, relative fibre motion during squeeze forming. This fibre movement seems to be very limited and within the elastic range, in case of activated fibre. This may be attributed to better wettability. On the other hand, poor wettability observed between matrix and the fibre in case of unactivated composite is due to the presence of oxide film on the melt surface or adsorbed contaminants on the fibre structure, Fig. 2. Metallographic investigation of the heat treated composites showed an apparent interaction zone, of limited thickness, at the matrix-reinforcement interface, Fig. 3. Experimental observations indicate that $MgAl_2O_4$ and possibly $CuAl_2O_4$ coexist in this interaction zone /12/ due to long-time heating during both solution treatment and aging.

3.2. Mechanical Properties.

3.2.1. Tensile Properties.

The tensile properties of the fibre composites and matrix alloy in the heat treated conditions, (T4 and T6) are shown in Figs 4 and 5. In general, the strength levels of the composites were higher than those of the unreinforced matrix measured at peak hardness, Fig 4. The magnitude of this increase ranging from 20 to 40 pct. On the other hand, the tensile ductilities of the composites, were significantly reduced, Fig. 5. The high pressure used has also contributed to the improvement in the strength of both matrix and composite alloys. This increase in strength could be attributed to the higher hardness of the fibres, and the presence of local dislocation concentrations around the fibres usually considered as a barrier for dislocation motion. Further improvements in the tensile properties are achieved in the modified composites, the magnitude of this increase ranging from 20 to 50 pct., Fig. 4. This may be due to better wettability and limited chemical reactions at the interface in the modified composites.

Tensile properties of matrix and composite at 200°C temperature are shown in Fig. 4. It is clear that the strength for both reinforced and unreinforced alloys decreases significantly at 200°C. Decrease in strength of both materials, at an identical temperature, implies that the high temperature strength of the composites are governed by the properties of the matrix which appears to be caused by over aging of the alloy matrix. Also, it can be seen that the strength of the composites at 200°C equals to the strength of the 2024 Al alloy at room temperature. The main reasons for this may be attributed to the strength of alumina fibre which is almost unaffected at 200°C. In addition the Al/Al_2O_3 interface bonding strength is high, even at this temperature.

Fig 5 shows that the ductility of the composites is much lower than that of the matrix alloy. This is mainly related to the presence of the fibre which strongly prevented the plastic deformation of the composite. Increasing the temperature, results in an increase in the fibres during plastic deformation of the composite, and therefore the tensile elongation of the composite is increased with the increase of temperature.

3.2.2. Wear Properties

Numerical results of the wear tests performed at T4 and T6 of the matrix and fibre composites solidified under pressure of either 50 or 100 MPa under a constant load of 50 N are shown in Fig. 6. In all cases, the weight loss due to wear of the matrix is the highest and the modified one is the least. The decrease in wear rates due to the presence of α-Al_2O_3 fibres at different aging and pressure condition is obviously due to the hard alumina fibres and its inoculation effect which result in finer structure. High wear resistance of the modified composite was achieved due to the presence of strong bonding between the fibres and the matrix alloy. It is now possible to use special surface treatment of the fibres in order to improve the wear resistance of the composites. It can be concluded from these results that optimum wear resistance can be achieved with the modified composite solidified under a pressure of 100 MPa in the T6 condition.

4. CONCLUSIONS

1. A metal matrix composite was manufactured with the squeeze casting technique. The matrix was based on 2024 Al alloy and fibre preform contained 0.28 (vf) alumina fibre.
2. The macrostructures of the modified and unmodified composites show finer structure compared with the matrix alloy.
3. Fibre distribution in the activated preform is more uniform than the unactivated one. Also, No reaction layer was observed at the fibre/matrix interface of both composites in the cast condition.
4. The tensile strength of the composites were higher than those of the matrix measured at the peak hardness. The magnitude of this increase ranging from 20 to 50 pct. On the other hand, the tensile ductilities of the composites, were significantly reduced.
5. High wear resistance of the composites was achieved due to the presence of hard alumina fibres and its inoculation effect which results in finer structure.

REFERENCES

1. K.K. Chawla, Composite Materials Science and Engineering, Springer. Verlag, New York, (1987).
2. M. Skibo, P.L. Morris, and D.J. Lioyd, Proceedings of the World's Material Congress, Chicago, IL, ASM, 1, (1988), 257-259.
3. N. Eusthopoulos, et al., J. Mater. Sci., 9, (1974). 1233-1238.
4. Y. Kimura., J. Mater. Sci., 19, (1984) 3107-3110.
5. A.G. Kulkami, B.C. Pai, and . N. Balasubramanian, J. Mater. Sci., 14, (1979) 592-596.
6. F. Amateau, J. Composite Materials, 10 1976), 279-285.
7. F.A. Badia and P.K. Rohatgi : Trans. Am. Foundrymen's Soc., 79 (1969), 402.
8. A.R. Champion, et al., : 2nd Int. Conf. on Composite Materials, Toronoto, Canada, (1978).
9. H. Fukunaga and T. Ohde: Progress in Science and Engineering of Composites, ICCM IV, Tokyo, (1982), 1443.
10. T.W. Clyne et al., J. of Mater. Sci., 20, (1985), 85-96.
11. T.W. Clyne, Met. Trans., 18A (1987), 1519-1530.
12. C.G. Levi, G.J.Abbaschian, and R. Mehrabian, Met. Trans., SA, (1978), 697-711.

Fig.1 Macrostructures ofboth matrix(upper half)and
alumina fibre composite(lower half), X1

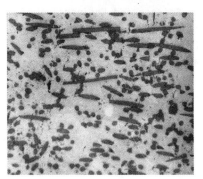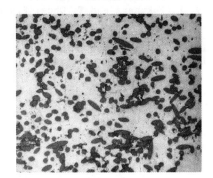

unmodified   modified
Fig.2 Micro distribution of alumina fibre composite
in the cast condition , X100

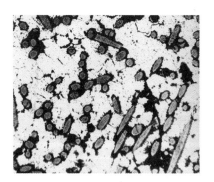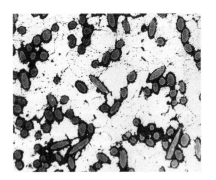

unmodified   modified
Fig.3 Micro distribution of alumina fibre composite
after heat treatment, X200

112

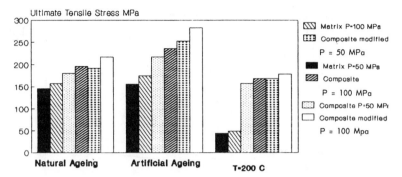

Fig. 4 Unltimate tensile strength of the investigated material.

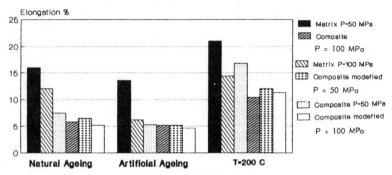

Fig. 5 Elongation percent of the investigated materials.

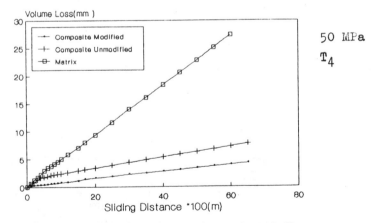

Fig. 6 Volume loss VS. Travelling distance of investigated alloys

113

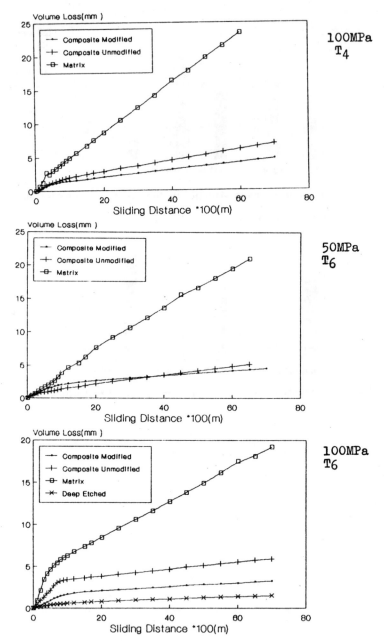

'Fig.6 Volume loss vs. travelling distance of investigated alloys, continued.

MATHEMATIC MODELS FOR SYSTEMS OF COMPUTER-AIDED DESIGN OF THIN-WALLED POSITIVE PRODUCTS MANUFACTURING

V. BOGOLYUBOV, A. BRATUKHIN, G. LVOV, O. SIROTKIN

*Research Center of Aviation Industry - 24 Petrovka Str.
103051 Moscow - Russia*

ABSTRACT

The report presents the basic principles of CAPP system generation for production of composite structures in aircraft industry. It considers the mathematic models describing conditions in preforms, separate composite components, operation of manufacturing equipment and tooling.

INTRODUCTION

Problems of engineering development for composite products manufacturing are classified as difficult-to-formalize problems. A wide variety of composites and their manufacturing processes along with discrepancy of technical and technological requirements cause to generate CAPP system not as a problem program oriented to the definite set of problems, but as a structured system of programmes and means intended for specific manufacturing planning. In principle, internal structure of CAPP system for composite products manufacturing and organization of interfaces with the external programmes approach the schemes of operational system generation.

BASIC PRINCIPLES OF CAPP SYSTEM GENERATION

Basic objects of development of computer-aided system
include reduction of CAPP time and labour input, increase of productibility and quality of composite structures and expansion of their range.
Development stages are varied for different composite
structures, hence it can be highlighted the certain general stages of production planning.
Analysis of structure service conditions and original
technical requirements are the basis for a choice of
type and specific structural features of composites. The
following step is the choice of manufacturing process
type and design (or selection) of equipment and tooling.
CAPP system for composite products manufacturing is
generated as the single structured system of mathematic
models bases, databases of information and inquiry types,
packages of application programmes and systems means.
This system provides a possibility of inventions in field
of mathematic modeling of manufacturing processes as well
as gaining and summarizing the technological experience
for efficient use of computer art in solving the scientific and applied problems in the field of design of composite products used in aircraft industry.
Theoretical base of production planning is mathematic
modeling. For use it in engineering practice every class
of mathematic models must be carried out to the level of
program means in the composition of the structured system.
Results of systematic and combined developments of mathematic models for manufacturing process of production of
polymeric material forming tooling are presented in
monograph /I/.
Mathematic model of forming of three-layer metal-polymer
thin-walled panel has been developed /2/ as inverse problem.
Operating experience for tooling made of polymeric composites shows that accuracy characteristics of shaping surfaces varies in time. Therefore, development of mathematic models of polymeric composite products ageing with
consideration for variations of their physical-mechanical
properties is of high priority /3/.
During production of products made of polymeric composites propagation of the residual manufacturing stresses
takes place in polymeric matrice curing. This event has
to be taken into account during production planning.

STATEMENT OF PROBLEM OF POLYMERIC COMPOSITES CURING

In the process of heat treatment of original semifinished material, including stages of heating, exposure and

cooling, matrice transfers sequentially to gel (polymerization) from liquid state and then to glassy state
(glass transition) in cooling. In this case its shrinkage results from density increase. Variation of physical-
-mechanical propert ies of matrice taking place during
curing cycle can be given by the equation:

$$\Phi = \Phi_h + \eta_p (\Phi_p - \Phi_h) + \eta_g (\Phi_g - \Phi_p) h_0 (t - t_0) \qquad (I),$$

where $\eta_p \in$ /0,I/-relative degree of polymerization
(conversion depth), $\eta_g \in$ /0,I/ - similar characteristic for glass transition, Φ_h , Φ_p , Φ_g correspond
to three states of matrice, t - starting time of cooling
stage and h -Heaviside step function.
Variation of temperature field within preform volume and
matrice shrinkage are the causes of growth in stress-
-strained state in the interior of composite material.
Hazard of defect initiation is complicated by inhomoge-
neity of composite, reinforcing fiders of which have the
characteristics significantly different from those of
matrice. During investigation of interaction between polymeric composite structural components the simplest model
of unidirectional reinforced composite is the long
axially symmetric compound cylinder, the internal part of
which is reinforcing fiber I and the external part is
polymeric matrice II (Fig.I).
Calculation of technological stresses and strains at being
investigated instant of time includes two steps:
I.definition of distribution of temperature T and conversion degree within the material volume by solution of
equations of non-stationary heat conduction and curing
kinetics.
2.definition of strain and stress components in matrice
and reinforcing elements with consideration for their
joint deforming.
Heat exchange process for the compound cylinder under
consideration is given by two equations of non-stationary
heat conduction /4/ :

$$\frac{1}{a^1} T^1 = \frac{1}{r} \frac{\partial}{\partial r} (r \frac{\partial T^1}{\partial r}), \quad \forall r \in [0, R_1];$$

$$\frac{1}{a^{II}} T^{II} = \frac{1}{\lambda^{II} r} \frac{\partial}{\partial r} (\lambda^{II} r \frac{\partial T^{II}}{\partial r}) + \frac{q}{\lambda^{II}} \eta_p h_0 \times \qquad (2)$$

$$\times (T^{II} - T_1), \quad \forall r \in [0, R_2]$$

that are supplemented by the initial condition at t=0,
boundary condition on the surface $r=R_2$ and conditions of
temperature fields conjugation on the surface $r=R_I$. In

117

equation (2) indexes I and II correspond to components of compound cylinder (see Fig.2), a and λ are temperature - and heat conductivity coefficients that are defined for matrice according to (I), q is thermal effect of polymerization reaction and T_T is temperature of the onset of gelling. Polymerization rate is defined by the kinetic equation /5/

$$\eta_p = k_0 (1 - \eta_p) \exp(-U/RT) \qquad (3)$$

with the initial condition $\eta_p(t_1) = 0,$ where k_0 is reaction rate constant, U is activation energy at polymerization and R is Boltzmann's constant.
Solution of represented initial-boundary problem is carried out numerically on the basis of unexplicit finiteness-difference scheme. Defined temperature-conversion fields make it possible to calculate the technological stresses and strains in composite at any instant of time. Compound cylinder being studied is under conditions of flat deformation $(\mathcal{E}_z = 0)$; at initial instant of time t=0 stresses in it are absent. In the process of heat treatment progressive stress accumulation occurs. For small intervals of time $t_k - t_{k-1} = \Delta t$ at issue increments of stress-strained state components can be described by the system of equillibrium equation:

$$d/dr (\Delta \sigma_r) + (\Delta \sigma_r - \Delta \sigma_\Theta)/r = 0, \qquad (4)$$

simultaneous equation:

$$d/dr (\Delta \mathcal{E}_\Theta) - (\Delta \mathcal{E}_r - \Delta \mathcal{E}_\Theta)/r = 0, \qquad (5)$$

and geometric relationships:

$$\Delta \mathcal{E}_r = d/dr (\Delta u), \qquad \Delta \mathcal{E}_\Theta = \Delta u/r \qquad (6)$$

with the appropriate boundary conditions and the physical dependences considering the thermal strains of components and shrinkage of matrice; in this case fiber characteristics don't depend on temperature and for matrice they are defined according to (I).
Cylinder partitioning into elementary rings with piecewise-constant approximation of temperature allows one to receive the set of coaxial cylinders with constant characteristics. Substitution of physical relationships into (5) in view of (4) transforms each of them to resolving equation relative to radial stress increment $\Delta \sigma_r$. Its solution can be received easily and constants found from conjugation conditions and boundary condition on the surface face r=R_2. Complete components of stress-strained state at the end of K-interval of time are calculated according

to principle of superposition. Application of complete space-time discretization makes it possible to conjugate solutions of heat conduction and mechanics problems. During autoclave curing of polymeric composite products temperature conditions are of great importance. Calculation of heat conduction processes for products and tooling is necessary to provide for the predetermined properties of composite with minimum expenditure of energy and time for expensive equipment utilization.

STATEMENT OF HEAT CONDUCTION PROBLEM

In order to investigate non-stationary heat conduction in the thin-walled elements of tooling of frame type we consider the problem of temperature field definition in the gentle sloping shell h thick inder convective heat transfer from the aurface z= -h/2 to autodave gas medium (coordinate z is measured from the middle surface). Operating surface z= +h/2 is assumed to be heat-insulating due to low heat conduction of part. Diagram of heat exchange is shown in Fig. 2.
Heat conduction is given by the equation:

$$\partial^2 T / \partial z^2 = T/a \qquad (7)$$

in conjunction with the initial condition

$$T(z, 0) = T_0 = \text{Const.} \qquad (8)$$

and boundary conditions on the surfaces $z = \pm h/2$

$$\partial T / \partial z = 0, \qquad z = h/2;$$
$$\partial T / \partial z - a / \lambda (T - \Theta) = 0, \quad z = -h/2, \qquad (9)$$

where Θ is temperature of gas medium and a is coefficient of conductive heat transfer on the surface $z = -h/2$.
To solve the problems (7)-(9) we applied the methods of operational calculus based on Laplace integral transform /6/. As a result, for condition of medium temperature variation shown in Fig.3. we took the closed solution in the form of relation between the shell temperature, dimensionless coordinates $\zeta = z/h$ and time $\tau = at/h^2$. Form of solution is defined by the interval of time under investigation, for example for τ $[\tau_4, \tau_5]$

$$T = \Theta_2 + \beta_2 (\tau - \tau_4 - \frac{2 + Bi}{2Bi}) + \frac{\beta_2}{2}(\zeta - 1/2)^2 +$$

119

$$+ 2\text{Bi} \sum_{n=1}^{\infty} \frac{\cos \mu_n (\zeta - 1/2)}{\mu_n^2 (\mu_n^2 + \text{Bi}^2 + \text{Bi}) \cos \mu_n} e^{-\mu_n^2 \tau} \times$$

$$\times [\beta_0 (1 - e^{\mu_n^2 \tau_1}) + \beta_1 (e^{\mu_n^2 \tau_2} - e^{\mu_n^2 \tau_3}) + \beta_2 e^{\mu_n^2 \tau_4}],$$

(10)

where $\text{Bi} = a h / \lambda$ is Bio criterion,

$\beta_0 = (\Theta_1 - \Theta_0) / \tau_1$, $\beta_1 = (\Theta_2 - \Theta_1) / (\tau_3 - \tau_2)$,

$\beta_2 = (\Theta_3 - \Theta_2) / (\tau_5 - \tau_4)$, μ_n $(n = 1, 2, \ldots)$

are roots of transcendental equation $\operatorname{tg} \mu_n = \text{Bi} / \mu_n$.
The calculation results performed at $h = 0,0001\,\text{m}$,

$\lambda = 0,4\,\text{W} \cdot \text{m}^{-1} \cdot \text{K}^{-1}$, $a = 60\,\text{W} \cdot \text{m}^{-2} \cdot \text{K}^{-1}$, $a = 0,0001\,\text{m}^2 \cdot \text{K}^{-1}$,

$\Theta_0 = \Theta_3 = 293\,\text{K}$, $\Theta_1 = 353\,\text{K}$, $\Theta_2 = 413\,\text{K}$, $t_1 = 0,64\text{h}$, $t_2 = 1,04\text{h}$,

$t_3 = 1,68\text{h}$, $t_4 = 3,22\text{h}$, $t_5 = 4,0\text{h}$ are shoun in Fig. 4

Plots I-5 conform to the instants of time $t_1 - t_5$.
In large-scaled products moulding irregularity of temperature distribution in thickness of the shell causes the
operating surface distortion. Estimation of tooling accuracy parameters is performed on the basis of the mathematic model of thermoelasticity for the thin-walled shell.

CONCLUSIONS

Mathematic modeling of manufacturing processes for composite products results in the statement of the complex of
elasticity, plasticity, viscoelasticity and the other
boundary problems of elliptic type. The process of non-
-stationary heat conduction are also enough complicated.
Therefore, during the development of CAPP system we selected the numerical methods as the main means of mathematic problems solution /7-IO/. Programmes that realize
the numerical methods of boundary and optimized problems
solution can be introduced included in the structure of
CAPP system as universal operating programmes and special
programmes oriented to the definite class of problems
alike.

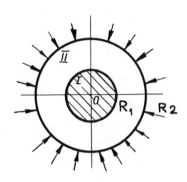
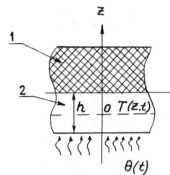

Fig I Estimated
composite model

Fig 2 Diagram of heat exchange between tool
and medium
I - product; 2 - tool

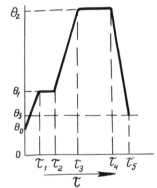
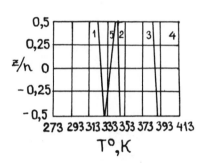
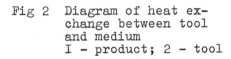

Fig 3 Cycle of tempera-
ture conditions of
acrylic fiber-epoxy
composite curing

Fig 4 Distribution of temperature in forming tool
thickness

REFERENCES

I.Bogolyubov V.S. "Polymer material forming tooling"
Moscow. Mashinostrojenije. (I979)
2.Bogolyubov V.S., Lvov G.I. & Kostromitskaja O.A. "Inverse problems of three-laver shell forming" Moscow.
USSR's Academy of Sciences Publishing House. MTT. N5.
(1989)
3.Christenson R. "Introduction to composites mechanics"
Moscow. Mir. (1982)
4.Gus A.N. et al."Technological stresses and strains in
composites" Kiev. Vyschaja Shkola. (1988)

5. Tomashevsky V.T., Jakovlev V.S. Technological Problems of Composite Mechanics. Applied Mechanics. 20(11) (1984)
6. Lykov A.V. "Theory of heat conduction" Moscow. Vyschaja Shcola. (1967)
7. Marchuk G.I. "Methods of computing mathematics" Moscow. Nauka. (1977)
8. Samarsky A.A., Andreev V.V. "Difference methods for elliptic equations" Moscow. Nauka. (1976)
9. Pobedrya B.E. "Computing methods in theory of elasticity and plasticity" Moscow. State's University Publishing House. (1981)
10. Moisejev N.N., Ivanilov Y.P. & Stolyarova E.M. "Optimization methods" Moscow. Nauka. (1978)

FIBRES, MATRIX, INTERFACE

INFLUENCE OF MALEIC ANHYDRIDE MODIFIED POLYPROPYLENE ON THE PERFORMANCE OF CONTINUOUS GLASS FIBRE REINFORCED POLYPROPYLENE COMPOSITES

A.A.J.M. PEIJS, H.A. RIJSDIJK, M. CONTANT

Centre for Polymers and Composites - Eindhoven University of Technology - PO Box 513 - 5600 MB Eindhoven - The Netherlands

ABSTRACT

This study investigates the influence of maleic anhydride modified polypropylene (m-PP) on static mechanical properties of continuous glass fibre reinforced polypropylene (PP) composites. Maleic anhydride modified polypropylene was added to the PP-homopolymer to improve the adhesion between the matrix and the glass fibre. Mechanical tests on unidirectional glass/PP composites with 0 wt.% and 10 wt.% m-PP, showed only a small increase in fibre dominated properties such as longitudinal tensile strength and strain, whereas, composite properties that are governed by the interphase such as transverse, shear and compressive strength, showed a significant increase due to an enhanced interaction between the glass fibres and the PP matrix.

INTRODUCTION

The economic potential of continuous fibre reinforced composites based on glass fibres and commodity plastics like poly(ethylene terephthalate) (PET), polyamide (PA) and polypropylene (PP) may attract, in combination with the potential advantage of high production rates, large volume markets such as automotive industries. An important aspect with regard to the performance and durability of such fibre reinforced thermoplastics is the adhesion between the fibre and the matrix.

In case of glass fibres, organofunctional silanes may be used as adhesion promoters between the organic matrix and the glass fibre [1,2]. Apolar polymers such as polypropylene (PP) and polyethylene (PE) have limited interaction with a sizing specific for these polymers. To resolve this problem, the polymer should be modified to increase the interaction between the matrix and the sized fibre. Fibre/matrix systems, that can be used to receive optimum adhesion between the fibres and the matrix, are acrylic acid grafted PP [3] and maleic anhydride modified PP in combination with glass fibre with compatible sizings. Such a sizing, containing amide groups, can increase the interaction between the PP matrix and the glass fibre by chemical reaction [4].

In a previous paper [5] it was shown that for both transverse and longitudinal flexural testing an optimum in strength was observed for composites based on a PP matrix with 10 wt.% m-PP. In this paper we will consider some further engineering properties of PP/glass composites focusing on two composite systems reflecting the extremes in interfacial bond strength; first, the reference composite based on PP-homopolymer and secondly, a composite with 10 wt.% maleic anhydride modified PP (m-PP) added to the PP-homopolymer.

1. EXPERIMENTAL

1.1 Materials

In this study we used a continuous glass fibre of PPG Industries Fiber Glass BV (PPG 854) and a polypropylene matrix (homopolymer VM6100, MFI: 20) of Shell Chemicals. To increase the interfacial bond strength between the glass fibre and the PP matrix, 10 wt.% maleic anhydride modified polypropylene of BP Chemicals Ltd. (Polybond 3002) was added to the PP matrix.

Unidirectional laminates were manufactured by a film stacking method, where fibres are wound on a rectangular mandrel with alternating layers of polymer films which were prepared using film blowing equipment. A typical laminate with a thickness of 2 mm consisted of approximately 10 layers of fibres with alternating layers of polymer film. This lay-up was melted at 200°C and pressed for 45 minutes at 25 bar. After hot-pressing, the composite was slowly cooled in the hot-press resulting in laminates with a fibre volume fraction of 58%, showing good overall impregnation and wet out of fibres.

1.2 Testing

Tensile testing: Longitudinal tensile tests were performed on a universal testing machine at a crosshead speed of 1 mm/min. The crosshead speed used for transverse tensile tests was 0.5 mm/min. The cross-sections of the test specimens for longitudinal and transverse testing were 0.3x12 mm and 2x25 mm respectively. Specimens were provided with adhesively bonded aluminium tabs.

Shear testing: The inplane shear strength (τ_{12}) was determined from two types of shear tests viz. $10°$ off-axis and $\pm 45°$ tensile tests, at a crosshead speed of 1 mm/min. Both specimens had a cross-sectional area of 2x12 mm. The interlaminar shear strength (ILSS) was determined using the short beam shear (SBS) test. Tests were performed according to ASTM D-2344 specifications on UD-composites with a cross-section of 3x10 mm, using a span-to-depth ratio of 4 and a test speed of 1 mm/min.

Compression testing: Compression tests were performed on a universal testing machine, at a crosshead speed of 0.5 mm/min. Test specimens with a thickness of 4.5 mm and a width of 20 mm were in accordance with ASTM D-3410, with the exception that the Celanese type of test fixture used in this study also allowed for specimen widths other than the standard 6 mm.

2. RESULTS AND DISCUSSION

2.1 Longitudinal tension

The results of the longitudinal tensile tests of the specimens with 0 wt.% and 10 wt.% m-PP are summarized in Table 1. A relatively small increase in longitudinal strength and strain is reported, which can be explained by greater effectiveness of the reinforcing fibres, due to an increase in fibre-matrix interaction in the composite system with 10 wt.% m-PP. Based on a rule of mixture relationship for the tensile strength of fibre reinforced composites, including an effectivity parameter k the influence of m-PP on the fibre efficiency can be demonstrated:

$$\sigma_c = k [\, \sigma_f \; V_f \; + \; E_m \, \varepsilon_f \, (\, 1\text{-}V_f \,) \,] \qquad (1)$$

where σ_c is the composite tensile strength, σ_f is the fibre tensile strength, E_m is the stiffness of the matrix, ε_f is the failure strain of the fibre and V_f is the fibre volume fraction. A typical value for the tensile strength of E-glass fibres as obtained from impregnated strand tests is approximately 2400 MPa. Using a matrix modulus of 1.5 GPa, a fibre failure strain of 0.033 and a fibre volume fraction of 0.58, the effectivity parameter k yields values of 0.51 and 0.63 for composites with 0 wt.% and 10 wt.% m-PP respectively. However, even in the case of a 10 wt.% m-PP composite system, fibre efficiency is rather low in comparison with that of glass fibre reinforced systems based on epoxy resins, yielding values for k in the order of 0.85-0.95.

2.2. Transverse tension

Due to the addition of m-PP the transverse strength and strain increased with 130% and 180%. Since off-axis composite properties such as transverse strength and strain are strongly dominated by the interface, an increase in interaction between matrix and fibre due to the addition of m-PP resulted in a remarkable increase in transverse properties. Scanning electron micrographs of the fracture surfaces of failed specimens (Fig. 1a-b) showed a change in failure mode from interface failure for the reference composite based on the PP-homopolymer to matrix fracture for composite based on a blend of 90 wt.% PP-homopolymer and 10 wt.% m-PP.

2.3. Shear

As shown in Table 1 also an increase in inplane shear strength as calculated from the 10° off-axis and ±45° tensile tests was observed. Addition of 10 wt.% m-PP resulted in an increase in inplane (1-2) shear strength of 49% and 42% for the 10° off-axis and ±45° tensile tests, respectively. The [±45]$_{4s}$ cross-ply laminates were manufactured from 0.3 mm thick prepregs which were moulded in a similar manner as the UD-composites. Short beam shear tests yielded the highest values for the composite shear strength, being 16 MPa and 21 MPa for 0 wt.% and 10 wt.% m-PP composites respectively. However, addition of 10 wt.% m-PP only resulted in a 30% increase in interlaminar (1-3) shear strength, which is significantly lower than the increase in inplane shear strength as obtained from the off-axis tensile tests.

Typical stress-strain curves for both types of ±45 laminates are shown in Fig. 2 revealing the non-linear response in tension. Especially in the case of 0 wt.% m-PP composites, this non-linearity leads to high failure strains in the order of 25-30%. The failure strain of 0 wt.% m-PP composites possessing poor levels of adhesion was much higher than that of the 10 wt.% m-PP composite, due to a more brittle failure process of the latter (Fig. 3).

Although the initial modulus is the same for both types of laminate, there is an increase in stress at which irreversible deformation starts to occur. Remarkably, the maximum failure stress for the 0 wt.% m-PP composites was higher than that for composites with m-PP. Because of the poor adhesion in the 0 wt.% m-PP composite, delamination between the individual plies manifests itself at low loads, resulting in a rotation of the fibres in the loading direction from ±45° to ±34°. In composites with improved fibre/matrix adhesion, only little interply failure occurs, preventing extensive fibre rotation (±43°) and causing a more local failure process. Since these 'scissoring' effects lead to false readings, the inplane shear strength as listed in Table 1 was calculated using the stress at the first knee in the stress-strain curve.

2.4. Compression

The compression test results show an approximate 60 % increase in longitudinal compressive strength for the 10 wt.% m-PP composite. A combination of longitudinal splitting and fibre buckling as the predominant failure modes was observed for all specimens tested. The occurrence of matrix and interface dominated failure modes rather than fibre dominated failure modes yielded relatively low values for the compression strength of PP/glass composite in comparison with data for glass fibre reinforced epoxies. Since the resistance to fibre buckling is provided by the shear modulus of the matrix this reduction can be related to the relatively low shear moduli and yield stresses of PP in comparison with highly crosslinked systems.

3. CONCLUSIONS

Tensile, shear and compression tests on UD-glass/PP composites with 0 wt.% and 10 wt.% m-PP showed that improved adhesion due to the addition of m-PP did not have much influence on the composite moduli. A small increase in longitudinal tensile strength and strain was observed for 10 wt.% m-PP composites, whereas off-axis composite properties that are governed by the interphase, such as transverse, shear and compression strength showed a relatively large increase.

REFERENCES

1. H. Ishida, 'A review of recent progress in studies of molecular and microstructure of coupling agents and their functions in composites, coatings and adhesive joints' *Polymer Composites*, 5 (1984) p 101.
2. E.P. Plueddemann *'Silane Coupling Agent'* Plenum Press, New York (1982).
3. J.M.H. Daemen and J. den Besten 'The influence of glass fibre sizing on the properties of GF/PP' *Engineering Plastics*, 4 No2 (1991) p 82.
4. A.R.C. Constable and J.A. Humenik, *SPI 44th Annual reinforced plastics/composites institute conference*, Anaheim 1989, p 11A.
5. H.A. Rijsdijk, M. Contant and A.A.J.M. Peijs 'The use of maleic anhydride modified polypropylene for performance enhancement in continuous glass fibre reinforced polypropylene composites' *in: Proc. ICCM-9* (A. Miravete, ed.) (1993) Madrid.

(a)   (b)

Fig. 1.   Scanning electron micrographs of transverse fracture surfaces showing: (a) interface failure in reference composite (PP-homopolymer) and (b) matrix failure in modified composite (10 wt.% m-PP).

Table 1. *Tensile shear and compressive properties of UD-composites with 0 wt.% and 10 wt.% m-PP.*

Test	0 wt.% m-PP	10 wt.% m-PP
Tension		
[0] E_{11} (GPa)	43.4 ± 1.4	43.6 ± 2.0
[0] σ_{11} (MPa)	720 ± 50	890 ± 60
[0] ε_{11} (%)	1.8 ± 0.02	2.0 ± 0.01
[90] E_{22} (GPa)	7.2 ±.0	5.8 ± 0.6
[90] σ_{22} (MPa)	4.2 ± 2.0	9.7 ± 1.5
[90] ε_{22} (%)	0.05 ± 0.02	0.14 ± 0.02
Shear		
[10] τ_{12} (MPa)	11.4 ± 3.0	17.0 ± 1.0
[±45] τ_{12} (MPa)	9.6 ± 0.8	13.6 ± 1.0
SBS τ_{13} (MPa)	16.0 ± 0.3	21.0 ± 0.4
Compression		
[0] σ_{11c} (MPa)	250 ± 70	396 ± 40
[90] σ_{22c} (MPa)	35.7 ± 1.5	37.5 ± 1.5

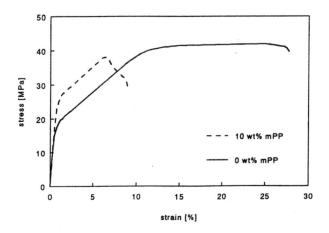

Fig. 2. Stress-strain curves for [±45]₄ₛ PP/glass laminates in tension.

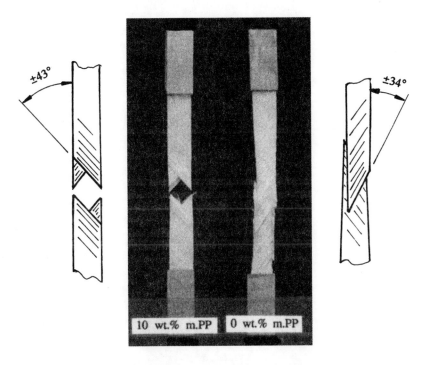

Fig. 3. Failed ±45 tension specimens showing: (a) extensive matrix cracking and delamination throughout the entire 0 wt.% m-PP composite and (b) more local failure in the 10 wt.% m-PP composite.

DEFORMATION MICROMECHANICS OF ARAMID FIBRES IN COMPOSITES

M.C. ANDREWS, R.J. YOUNG, R.J. DAY

Manchester Materials Science Centre - UMIST
Grosvenor Street Manchester M1 7HS - UK

ABSTRACT

Raman spectroscopy has been used to study the deformation micromechanics of aramid fibres in a model single-fibre composite with an epoxy resin matrix. The strain induced shift of the 1610 cm^{-1} aramid Raman band can be used to map the distribution of stress or strain along a short Kevlar 49 fibre inside the resin from which the interfacial shear stress can be calculated. The technique is used to study the efficiency of stress transfer from the matrix to the fibre by varying the mechanical properties of the epoxy resin. It is demonstrated that it is possible to determine the interfacial shear stress which is found to reach a maximum value close to the shear yield stress of the matrix.

INTRODUCTION

It has been shown that Raman spectroscopy can be used to study the deformation micromechanics of fibres in a model single-fibre composite with an epoxy resin matrix /1-6/. The use of Raman spectroscopy to follow composite deformation is based upon the finding that certain bands in the Raman spectrum of the fibre shift in wavenumber upon the application of stress or strain. This allows accurate determination of the deformation of the fibres within a composite matrix. The strain-induced band shifts can be used to map the distribution of stress or strain along individual fibres in composite matrices from which the interfacial shear stress can be calculated. This is of fundamental importance in the study of stress transfer from the matrix to a fibre.

In recent years Raman spectroscopy has been applied to several aspects of composite deformation. These include the effect of fibre orientation /7/, fibre fragmentation of both aramid /2/ and carbon fibre /8,9/ systems, the effect of stress concentrations associated with neighbouring fibres /10/, and more recently the effect of surface treatments applied during processing to both carbon fibre /11/ and aramid fibre /2/ systems. This present study uses Raman spectroscopy to study the effect of the matrix properties on the transfer of stress from the matrix to the fibre. This is of fundamental importance for the study of plastic deformation of the matrix /12/ which is not accounted for by classical shear-lag theory /13/.

EXPERIMENTAL

A commercial grade of sized aramid fibre, Kevlar 49 was used in this study. The fibres were 12 μm in diameter with a tensile modulus, E_f, of 125 GPa and a failure strain of 2.8%. Single, short-fibre composites were prepared in which short Kevlar 49 fibres, 3-4 mm long were placed in a two-part, cold-curing epoxy resin matrix with different levels (parts by weight, pbw) of curing agent as described elsewhere /1,2/.

The single-fibre composite samples were deformed using a Miniature Materials straining rig, (Polymer Laboratories Ltd., UK), as shown schematically in Figure 1. The rig was placed on the stage of a modified Olympus BH-2 optical microscope connected to a SPEX 1000M single monochromator. Raman spectra were obtained as described elsewhere /1,2/. The peak position of the 1610 cm^{-1} aramid Raman band was recorded at intervals of 10 μm close to the end of an aramid fibre in the epoxy resin matrix at matrix strain levels of 0-2.4% in intervals of 0.4%.

Tensile test pieces were cut and shaped into dumb-bell specimens, from a flat sheet of epoxy resin made for each level of added hardener and deformed in tension with the strain measured using an extensometer. A block of epoxy resin was also cut and polished to form small rectangular blocks which were subsequently compressed between two steel plates attached to the base and cross-head of an Instron. For both the tensile and compression experiments the yield stress and the Young's modulus in both tension and compression were determined.

RESULTS

Figures 2(a) shows the variation of fibre strain, e, with distance, x, along the left-hand end of a Kevlar 49 fibre in an epoxy resin matrix, with 30 parts of hardener, for matrix strains ranging from 0% to 2.4%. The solid lines are a fit of the experimental data to an asymmetric sigmoid function with a correlation coefficient greater than 98%. At low matrix strains, ($e_m < 1.2$%), the fibre strain increases from the fibre end, ($x=0$), up to a plateau value along the middle of the fibre. This is in qualitative agreement with that predicted by classical shear-lag theory /13/ and is similar to previous results /1,2/ for an aramid/epoxy model composite.

Figures 2(b) shows the variation of interfacial shear stress, τ, with distance, x, along the fibre calculated assuming /1,2,14/ that the interfacial shear stress is proportional to (de/dx) which is the slope of the lines in Figure 2(a). At matrix strains up to 1.2% the data are in qualitative agreement with that predicted by the classical shear-lag theory /13/ where the interfacial shear stress is a maximum at the fibre ends, ($x=0$), decreasing to zero at a distance, x, along the fibre. The interfacial shear stress reaches a maximum value at an applied matrix strain level, e_m, between 0.8% and 1.2%. This is shown in Figure 3 in which the maximum interfacial shear stress, τ_{max}, is plotted for each level of applied matrix strain for a series of aramid/epoxy systems made with different levels of hardener. At matrix strain levels greater than 1.2 % the maximum interfacial shear stress values decrease and the maximum is no longer at the fibre end where the matrix has yielded.

It is shown in Figure 3 that the maximum interfacial shear stress, τ_{max}, is dependent upon the mechanical properties of the resin. This can be seen more clearly by comparing the interfacial shear stress data in Figure 3 with the compressive stress/strain data in Figure 4 for all the epoxy resin systems used. As the amount of hardener is increased both the compressive modulus and the compressive yield stress increase. The yield stress reaches a maximum value of 100 MPa whereas the compressive modulus appears to reach an optimum value of 3.4 GPa for the 30 parts of hardener system and then decrease to 2.8 GPa for the 38 parts of hardener system. This is consistent with previous studies /15/ which have shown that the mechanical properties of epoxy resins depend upon the level of hardener.

The shear yield stress, σ_y, can be calculated from the tensile and compressive yield stresses /16/. Figure 5 shows the variation of maximum interfacial shear stress, for an applied matrix strain level of 1.2%, with the calculated shear yield stress for each of the resin systems employed. It is found that the values of maximum interfacial shear stress correlate extremely well with the values of shear yield stress of the epoxy resin matrix. The equality between the two measurements implies very strongly that shear yielding takes place at the end of the fibres when the Cox-type behaviour /13/ breaks down. It must, however, be noted the variability of data obtained for the resin systems with only 20 parts of hardener is greater than the data obtained for the other resin systems. This could be due to a poor dispersion of the low level of hardener within the resin resulting in the possibility of localised differences in the state of resin cure.

CONCLUSIONS

It has been demonstrated that Raman spectroscopy is an extremely powerful technique for the study of composite micromechanics and the factors which affect the transfer of stress from the matrix to the fibre. It has been shown that there is a transition in behaviour at high strains due to plastic deformation of the matrix when the interfacial shear stress exceeds the shear yield stress of the matrix.

ACKNOWLEDGEMENTS

This work was supported by a Research Grant from the Science and Engineering Research Council and the Ministry of Defence. The authors are grateful to E.I. du Pont de Nemours for supplying the aramid fibres used in this study. One of the authors (RJY) is grateful to the Royal Society for support in the form of the Wolfson Research Professorship in Materials Science.

REFERENCES

1. M.C. Andrews, R.J. Day and R.J. Young, *Comp. Sci. and Tech.*, in press.
2. M.C. Andrews and R.J. Young, *J. Raman Spec.*, in press.
3. R.J. Young, C. Galiotis, I.M. Robinson and D.N. Batchelder, *J. Mater. Sci.*, **22** (1987) 3642.
4. R.J. Young and P.P. Ang, 'Interfacial Phenomena in Composite Materials '91', Edited by I Verpoest and F.R. Jones, Butterworth-Heinemann Ltd., Oxford, 1991, pp 45-52.
5. H. Jahankhani and C. Galiotis, *J. Comp. Mat.*, **25** (1991) 609.
6. C. Vlattas and C. Galiotis, 'Developments in the Science and Technology of Composite Materials- ECCM5', Edited by A.R Bunsell, J.F. Jamet and A. Massiah, European Association for Composite Materials, Bordeaux, France, 1992, pp 415-420.
7. M.C. Andrews, R.J. Day, X. Hu and R.J. Young, *J. Mater. Sci. Lett.*, **11** (1992) 1344.
8. Y. Huang and R.J. Young, 2nd International Conference on 'Deformation and Fracture of Composites', UMIST, Manchester, P22.
9. N. Melanitis, C. Galiotis, P.L. Tetlow and C.K.L. Davies, *J. Comp. Mat.*, **26** (1992) 574.
10. K.M. Atallah and C. Galiotis, *Composites*, in press.
11. C. Galiotis, N. Melanitis, C. Vlattas and A. Wall, 2nd International Conference on 'Deformation and Fracture of Composites', UMIST, Manchester, 1993, p14.
12. A. Kelly and N.H. Macmillan, 'Strong Solids', 3rd Edition, Clarendon Press, Oxford, 1986, p264.
13. H.L. Cox, *Brit. J. Appl. Phys.*, **3** (1952) 72.
14. G.S. Holister and C. Thomas, 'Fibre Reinforced Materials', Elsevier Publishing Co. Ltd., London, 1966, p18.
15. L.T. Drzal, in 'Advances in Polymers Science', Vol.75, Springer Verlag, Berlin, 1986, p.7.
16. R.J. Young, 'Introduction to Polymers', Chapman and Hall, London, 1981, p266.

Single-Fibre Model Composite

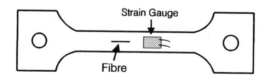

Figure 1. Schematic representation of the single-fibre model aramid epoxy composite.

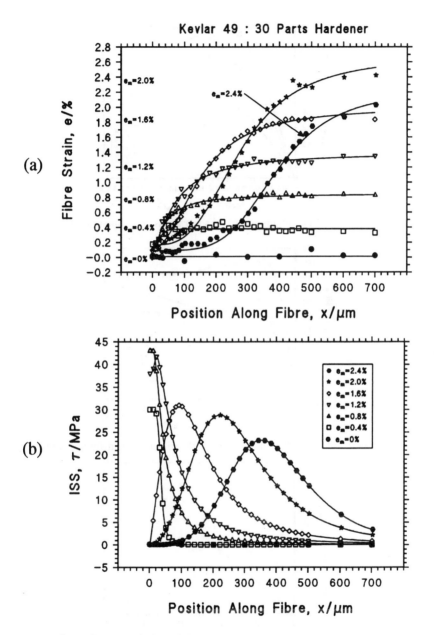

Figure 2. (a) Variation of fibre strain, e, with position along the fibre.
(b) Derived variation of interfacial shear stress, τ, with x.

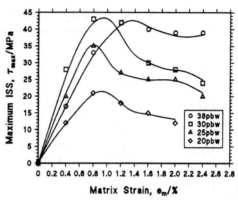

Figure 3. Dependence of maximum interfacial shear stress, τ_{max}, upon the matrix strain.

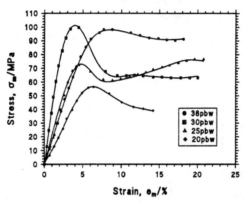

Figure 4. Compressive stress/strain curves for the different resin formulations.

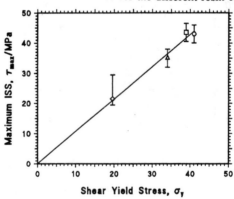

Figure 5. Dependence of τ_{max} upon the shear yield stress, σ_y, of the resin.

EVIDENCE OF CHEMICAL BOND FORMATION BETWEEN SURFACE TREATED CARBON FIBRES AND HIGH TEMPERATURE THERMOPLASTICS

K.J. HÜTTINGER, G. KREKEL, U. ZIELKE

Universität Karlsruhe - Institut für Chemische Technik
Kaiserstrasse 12 - 7500 Karlsruhe - Germany

Abstract

The paper concerns the adhesion of HT-thermoplastics to carbon fibre surfaces. For the solution of the problem carbon fibres of different surface treatments were used and the surface chemistry was analysed by (1) Temperature Programmed Desorption of functional groups (TPD) and (2) determination of the work of adhesion with aqueous solutions of pH values from 1 to 14. The surface chemistry of the HT-thermoplastics was studied by determination of the work of adhesion only. The adhesion was determined by measuring the Interlaminar Shear Strength (ILSS) of UD-composites. This property was shown to be proportional to the maximum work of adhesion as found in the wetting studies. It was also found that an increase of the processing temperature leads to an increase of the ILSS, but only with surface treated fibres. From this result and the desorption temperatures of the most reactive functional groups the formation of chemical bonds at the interface was concluded. Maximum ILSS values are even higher than those found for an epoxy resin.

Introduction

Adhesion studies in relation to carbon fibre reinforced polymers are mainly focussed on reactive polymers such as epoxy resins, whereas little is known about the adhesion of thermoplastics and thus on high temperature thermoplastic polymers [1-3]. Such polymers only exhibit functional groups. It may be assumed that the adhesion forces at the polymer/carbon fibre interface are limited to acid/base interactions. As compared to chemical bonds such interactions are weak [4], and consequently, the adhesion of the polymers to the carbon fibre surface should be weak. This paper summarizes main results of a fundamental study on this problem. Within the study evidence for chemical bond formation in the carbon fibre/high temperature thermoplastic interface was found [5].

I. EXPERIMENTAL
1.1 Materials

For all studies Celion G30-500 carbon fibres (BASF AG) were used. These high tenacity fibres were available without and with commercial surface treatment. Unoxidized fibres were treated with 0,75% ozone in oxygen at 100°C for 60 seconds. The high temperature thermoplastics polyethersulphone (Ultrason E1010) and polycarbonate (Makrolon 2400) were obtained from BASF AG and from Bayer AG, respectively.

1. 2 Fabrication of Composites

For the fabrication of unidirectionally reinforced composites with thermoplastic matrices five plies of carbon fibres were wrapped on a frame and subsequently impregnated with a solution of 13 wt% of the polymer in methylene chloride. In order to evaporate residual solvent, the impregnated fibres were dried in air and treated in a vacuum at 110°C for six hours. The prepregs were moulded in a die at temperatures of 330°C (polyethersulphone) and between 260°C and 320°C (polycarbonate) applying a pressure of 10 MPa. The fibre volume fractions of the composites varied between 57 and 62 %.

II. Analytical Methods
2.1 Temperature Programmed Desorption

The temperature programmed desorption (TPD) experiments were performed by heating carbon fibre samples with a rate of 10 K/min to 1050°C in pure argon (oxygen content < 0.1 ppm) applying a flow rate of 8 l/h. Carbon dioxide and carbon monoxide desorbed from the fibre surfaces were continuously analysed by NDIR-analysers (Leybold Heraeus, Binos 1.2).

2.2 Contact Angle Measurements

Wetting studies of the fibres were performed according to the Wilhelmy-technique with the aid of a Sartorius microbalance (sensitivity: 0.1 μg) [6]. The measurement of the wetting force F_W and the tear-off force F_{TO} directly yields the contact angle Θ:

$$F_W/F_{TO} = \cos \Theta \qquad (1)$$

With the value of $\cos \Theta$ the work of adhesion W_{Sl} can be calculated if the surface tension of the test liquid γ_l is known [7,8]:

$$W_{SL} = \gamma_l * (1 + \cos \Theta) \qquad (2)$$

Contact angle measurements on plane polymer surfaces were conducted using the sessile-drop technique [8]. The contact angle was measured every minute for a period of ten minutes. A decrease of the contact angles was found until a plateau was reached after six to ten minutes. By extrapolating the curve back to zero time, the advancing contact angle was obtained. The work of adhesion W_{Sl} was calculated according to eqn. (2). The feature of the wetting studies performed represents the test liquid. Due to the application of aqueous solutions with pH values from 1 to 14, well defined acid-base complexes (Brønstedt type) are formed with

140

functional groups at the surface of the solid. Such acid-base complexes are reflected by a decrease of the contact angle ϴ or an increase of the work of adhesion W_{sl}. A decisive prerequisite for comparing the work of adhesion obtained with solutions of different pH values represents the independence of the surface tension g_L of the pH value. All test liquids exhibit a value of 72.8 mN/m according to the value of pure water [9,10].

For measuring the interlaminar shear strength (ILSS) the short beam test according to ASTM 2344 was applied (span to depth ratio: 4).

III. RESULTS AND DISCUSSION
3.1 Functional Groups at the Carbon Fibre Surfaces

Oxygen containing functional groups at the carbon fibre surfaces exhibit different thermal stabilities. They are decomposed by formation of carbon dioxide and/or carbon monoxide [11,12]. For this purpose, TPD is an appropriate method for a quantitative analysis of the functional groups. Most important is the desorption of carbon dioxide below 250°C, because it can exclusively be attributed to the decomposition of the most reactive strongly acidic carboxylic groups [11]. Desorption rates of carbon dioxide with unoxidized, commercially oxidized, and ozone treated fibres are presented in Figure 1. From the surface of the unoxidized fibre, only a small amount of carbon dioxide is desorbed below 250°C. A characteristic peak at 600°C is due to the desorption of peroxidic groups [13]. The commercially oxidized fibre shows enhanced desorption rates, but compared to the unoxidized fibre the structure of the desorption spectrum is similar. A significantly different desorption spectrum is found with the ozone treated fibre. It is characterized by a pronounced peak at about 200°C. Only little carbon-dioxide is split off at higher temperatures.

3.2 Work of Adhesion at the Carbon Fibre Surfaces
The work of adhesion according to eqn. (2) is a direct measure of the interactions between the surface of a solid and an adjacent liquid. On a molecular basis dispersion and non-dispersion forces determine the extent of the interaction [8]. Thus, the work of adhesion changes if the chemical nature of the solid surface or the liquid is altered. The work of adhesion of unoxidized, commercially oxidized, and ozone treated carbon fibres as a function of the pH value of the applied aqueous test liquids is presented in Figure 2. An increase of the work of adhesion with increasing pH value indicates the formation of acid-base complexes due to acidic surface groups of the fibres [9,10]. In accordance with the low concentration of surface groups found by the TPD studies, the unoxidized fibre exhibits only low values of the work of adhesion over the total pH range. On the other hand, the highest values up to 140 mJ/m^2 were found with the ozone treated fibre. These results are in accordance with the amount of carbon dioxide measured in the TPD experiments.

3.3 Work of Adhesion at the Polymer Surfaces
As follows from Figure 3, the work of adhesion of both high temperature thermoplastics, polyethersulphone and polycarbonate, increases with decreasing pH value. Thus it can be concluded that the surface chemistry of the polymers is determined by basic functional groups. The work of adhesion of

polyethersulphone is about 10 mJ/m^2 higher than that of polycarbonate. The basic character of the polymer surfaces should favour the formation of acid-base complexes in the interface with the acidic carbon fibre surface. The meaning of these complexes for the adhesion will be discussed below.

3.4 Influence of the Surface Groups on the Work of Adhesion

In principle, the TPD results should be sufficient to correlate the functionality of carbon fibre surfaces with the adhesion of the basic polymers polyethersulphone and polycarbonate. However, the maximum work of adhesion as found by increasing the pH value is at least of similar relevance for the acidic functionality of the carbon fibre surfaces. This will be shown later. On the other hand, the maximum work of adhesion of differently treated fibres plotted as a function of the amount of carbon dioxide desorbed up to 250°C gives a fairly linear relationship [5]. This correlation indicates that the increase of the maximum work of adhesion of the fibres mainly results from carboxylic groups.

3.5 Relevance of the Surface Chemistry of the Fibres in the Adhesion of High Temperature Thermoplastics

Figure 4 shows the ILSS values of the composites with polyethersulphone and polycarbonate matrix, respectively, as a function of the maximum work of adhesion of the fibres. Linear relationships can be observed in all cases. In other words, there exists a direct relationship between the acidic functionality of the carbon fibre surfaces and the adhesion of the polymers. More remarkable is the steep increase of the ILSS with increasing work of adhesion in the case of polyethersulphone. For polycarbonate, two straight lines are shown which are obtained for composites with two different processing temperatures (see Fig.4). The steep increase of the ILSS in the case of the polyethersulphone matrix and the strong influence of the processing temperature in the case of the polycarbonate matrix suggest that chemical bonds are formed in the interface. For that two steps should be responsible:
(1) An oriented adsorption of the polymer at the carbon fibre surface during the impregnation process due to acid-base interactions between acidic groups at the fibresurface and basic functional groups of the polymer.
(2) Chemical reactions between carboxylic groups at the fibre surface and sulphone or carbonate groups of the polymers, initiated by the decomposition of the thermally instable carboxylic groups during the processing of the composites.
For polycarbonate, Figure 5 shows a model of acid-base interactions which should be formed in the first step. However, acid-base interactions cannot directly be considered to be responsible for the enhanced adhesion as follows from the small increase of the ILSS for a processing temperature of 260°C, but a strong increase for 320°C (polycarbonate, see Fig.4). On the other hand, acid-base interactions should be most important for the formation of chemical bonds, namely in an indirect manner during the second step. This statement is explained in a model presented for polyethersulphone in Figure 6. It shows both steps, the interaction of an acidic carboxylic group with a basic sulphone group due to a preferred adsorption of the polymer and the reaction between the same groups. During processing of the composites at 330°C, the strongly acidic carboxylic groups are decomposed (see Fig.1) and acid-base interactions can be transformed into

chemical bonds. The mechanism shown in Figure 6 only represents an example, because many other types of reactions are conceivable. These considerations explain the above statement that acid-base interactions are indirectly decisive for the adhesion.

The formation of chemical bonds is strongly supported by the results with polycarbonate processed at a higher temperature of $320\,^{\circ}C$ as compared to $260\,^{\circ}C$. Nevertheless, the ILSS values of polyethersulphone are not reached. This result is understandable as only in the case of polyethersulphone charges and radicals can easily be transferred to an oxygen atom of the sulphone group since they can be delocalized in the neighbouring aromatic systems of the polymer.

IV. CONCLUSIONS

The results obtained give a new view on the adhesion of high temperature thermoplastics to carbon fibre surfaces. The contribution of carboxylic surface groups for an enhanced fibre/matrix adhesion can mainly be attributed to reactions in the interface. These interface reactions are heterogeneous solid/liquid reactions, for which the adsorption as the primary step is decisive. The adsorption is favoured by acid-base interactions which necessitate the orientation of the interacting groups in a mutual steric favourable position. Chemical bond formation in the fibre/matrix interface is initiated by the decompostion of carboxyl groups at the fibre surface.

ACKNOWLEDGEMENT

Financial support by the German Research Foundation (DFG) is gratefully acknowledged.

REFERENCES

[1]   R.Weiß, Proc. Verbundwerk , Wiesbaden, Demat, 12.0 (1988)

[2]   H.Keller Proc.2.Symposium Materialforschung , Dresden, BMFT, 1, 48 (1991)

[3]   F.Orth, G.W.Ehrenstein Proc.2.Symposium Materialforschung , Dresden, BMFT, 2, 1307 (1991)

[4]   P.W.Atkins "Physikalische Chemie" , 1.Aufl., Verlag Chemie, Weinheim (1987), pp. 598-602

[5]   G.Krekel, K.J.Hüttinger, W.P.Hoffman, D.S.Silver, U.J.Zielke J.Mat.Sci 28 (1993), paper submitted
        G. Krekel, Einfluß von Oberflächenstruktur und Oberflächenchemie von Kohlenstoffasern auf die Verstärkung von Hochtemperaturthermoplasten, VDI Verlag GmbH ISBN 3-18-149505-0, Düsseldorf 1993

[6]   L.A.Wilhelmy Ann.Physik 119, 177 (1863)

[7]   W.A.Zisman in "Contact Angle, Wettability, and Adhesion" R.F.Could (Ed.), American Chemical Society, Washington D.C., ACS Adv.Chem.Series 43, 1 (1964)

[8]   K.J.Hüttinger in "Carbon Fibres, Filaments and Composites" , J.L.Figueiredo et al. (Eds.), KluwerAcademic Publishers, Dordrecht (1990), p. 245

[9]   K.J.Hüttinger, S.Höhmann-Wien, G.Krekel Carbon 29, 1281 (1991)

[10]   K.J.Hüttinger, S.Höhmann-Wien, G.Krekel J.Adhesion Sci.Tech. 6, 317 (1992)

[11]   H.P.Boehm, E.Diehl, W.Heck, R.Sappock Angew.Chem. 76, 742 (1964)

[12]   G.Tremblay, F.J.Vastola, P.L.Walker jr. Carbon 16, 35 (1978)

[13]   K.Kinoshita in "Electrochemical and Physicochemical Properties of Carbon" , J.Wiley & Sons, New York,(1988), p. 88

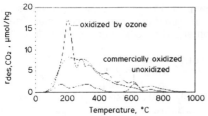

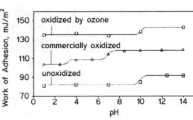

Fig. 1: Desorption rates of carbon dioxide of fibres with different surface treatments as a function of the desorption temperature

Fig. 2: Work of adhesion of fibres with different surface treatments as a function of the pH value of the test liquid

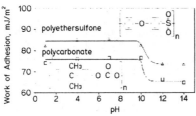

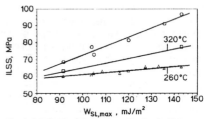

Fig. 3: Work of adhesion of the high temperature thermoplastics polycarbonate and polyethersulfone as a function of the pH value of the test liquid

Fig. 4: Interlaminar shear strength of polyethersulfone (○) and polycarbonate (△,□) composites with differently treated fibres as a function of the maximum work of adhesion of the fibres

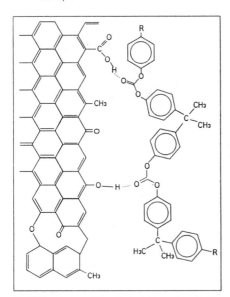

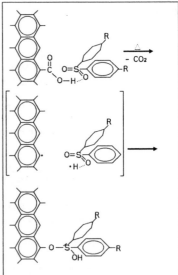

Fig. 5: Model of possible acid-base interactions at the fibre/matrix interface in prepregs (i.e. for polycarbonate as matrix)

Fig. 6: Model for postulated chemical reactions at the fibre/matrix interface which should occur during the fabrication of the composites (i.e. for polyethersulfone as matrix)

144

MICROINDENTATION TECHNIQUE FOR CHARACTERIZING FIBRE/MATRIX INTERFACE OF FIBRE REINFORCED POLYMERS

B. LARGE-TOUMI, M. SALVIA, L. CARPENTIER, O. VEAUVILLE
P. KAPSA, L. VINCENT

Ecole Centrale de Lyon - Département MMP - 36 Av. Guy de Collongue - BP 163 - 69131 Ecully Cedex - France

ABSTRACT

This paper presents the results obtained on glass/epoxy and polypropylene composites by a microindentation method. The existence of a debonding force -above which no fibre/matrix debonding occurs- is established. The influence of both the matrix shear modulus and the fibre tested neighbourhood is pointed out. The modelling (of a shear-lag analysis type) takes into account these two parameters as well as the local geometry of each experiment. This model provides a consistent evaluation of the interfacial shear strength.

INTRODUCTION

Fatigue of UD(0°) glass fibre reinforced polymers has been discussed as the resistance of the composite in the vicinity of the first nucleated defects. This resistance mainly depends on interfacial properties. Thus, specific tests like micromechanical tests have been set up to characterize such properties. The "pull-out" test /1/ and the "microdebond" test /2/, which both consist in pulling out a single fibre embedded into a matrix, provide a measure of the interfacial debonding strength. Their drawback lies in the kind of samples required, which are more or less representative of a real composite. Other techniques based on macromechanical testing are also used but the quantification of the interfacial contribution is often difficult.

The microindentation method also evaluates the debonding strength; its major advantage is to allow tests on real composites. This test is commonly used on ceramic-ceramic composites for which modelling of the interfacial behaviour has already been established /3/, /4/. In this study, experiments have been performed on glass reinforced epoxies and polypropylene (with the same glass fibres). These materials require a specific modelling which we will detail here.

145

I-EXPERIMENTAL METHOD

The microindentation test consists in pushing out, using an indentor, a fibre which is perpendicular to a polished surface of a real composite sample. During a loading-unloading cycle of indentation, both normal load and penetration depth are recorded (Fig.1) thanks to data acquisition and computer treatment; the apparatus is shown schematically in Fig.2.

Before a test, the sighting system permits the choice of the area to be indented and when the indentation is done, it permits a visual control of the damage which has occurred during the test (fibre fracture, debonding) and of the indentation localisation itself. The sighting system, coupled with a monitor, allows a maximum magnification of 5600. For this study, we use a Vickers diamond indentor; the indentation speed is 0,1 or 0,2µm/s, the accuracy on the displacement measurement is 0,05µm and on the load measurement 1mN.

II-MATERIALS

The materials characterized in this study are three E glass reinforced polymers:
- the sample called "PP" is a glass reinforced polypropylene processed by hot compression moulding, with a fibre volume fraction of 50%;
- the samples called "MP" and "EP" are two glass reinforced epoxy (Epon 828) processed by filament winding with a volume fraction of 50%; the specimens only differ by the nature of the glass sizing: the EP sample has an epoxy specific sizing and the MP one has a multipurpose one.

The fibre diametre range is from 12 to 21 µm -which is rather small. Moreover, the fibres are fragile and during indentation fractures often occur; one way to reduce this phenomenon is to have a "perfect" polishing of the sample surface. This is also necessary to allow an accurate indentation. Therefore the samples are carefully polished, using classical abrasive grits and diamond pastes (up to 1µm).

III-RESULTS

Fig.1 shows a typical curve obtained on the MP sample; the Fig. A revealed that, for this experiment, debonding occurred but this did not appear on the curve. That means that no debonding criterion can be established from the study of an indentation curve on FRP; the permanent fibre displacement due to debonding can only be evaluated using a Scanning Electron Microscope (Fig.B). This displacement is generally low (< 100 nm).

Experiments performed at various loads revealed the existence of a debonding force, F_d, on each material. This is representative of the chemical nature of adhesion between glass fibres and matrix (if the applied load remains inferior to F_d, no debonding occurs).

The optical and SEM views pointed out, on the one hand, the effect of a low modulus matrix and, on the other hand, the influence of the indented fibre neighbourhood. Indeed, until the applied load reaches F_d, the surrounding matrix is sheared -more precisely, the matrix is sheared between the fibre tested and its nearest neighbours. This effect is increased when the matrix shear modulus is low-; when debonding has occurred, the matrix probably comes up to its initial position as shown in Fig.3. Some of our experiments resulted in partially debonded fibres, where debonding only occurred when the distance between the fibre and its neighbours was small enough. Debonding appears easier whenever the fibres are touching. The influence of a low modulus matrix and of the fibre packing has already been discussed with great attention /5/, /6/.

IV-MODELLING

Modelling already described by Piggott /7/ has been used; it was modified to agree with our boundary conditions.We thus consider a perfect hexagonal fibre arrangement (Fig.4a) and we introduce the notion of an "equivalent radius", R_{eq}, to reduce the initial geometry to an equivalent one (Fig.4b and 4c); R_{eq} is defined by:

$$\pi R_{eq}^2 - \pi r^2 = A$$

where r is the radius of the fibre and A is the area of matrix contained in the circle of radius R (Fig.4a).

Then we assume that the longitudinal displacement is zero at the "equivalent" fibre/matrix interface (at a distance R_{eq}) -indeed, it appeared from our tests that the bordering fibres did not move and that the interfaces were not damaged. Using Piggott's approach, this leads to the differential equation (for further details see /7/):

$$\frac{d^2 \sigma_f}{dx^2} = \frac{n^2}{r^2} \sigma_f \qquad \text{where } n^2 = \frac{2G_m}{E_f Ln\left(\frac{R_{eq}}{r}\right)}$$

σ_f is the fibre longitudinal stress, E_f the fibre Young's modulus and G_m the matrix shear modulus. The solution is: $\qquad \sigma_f = B\, sh(nx/r) + D\, ch(nx/r)$

To write the boundary conditions, we assume first that σ_f is homogeneous on a section of fibre (even on the upper surface) and second that L >>R, where L is the thickness of the sample; we obtain:

at x=0, $\sigma_f = F/\pi\, r^2 = \sigma_0$ and, at x= L, $\sigma_f = 0$. Thus:

$$\sigma_f = \sigma_0 \left[ch\left(\frac{nx}{r}\right) - \coth\left(\frac{nL}{r}\right).sh\left(\frac{nx}{r}\right) \right]$$

If τ_e is the interfacial shear stress, then the force equilibrium written on a section of fibre

leads to: $\tau_e = -\frac{r}{2}\frac{d\sigma}{dx}$ This gives: $\tau_e = -\frac{n\sigma_0}{2}\left[sh\left(\frac{nx}{r}\right) - \coth\left(\frac{nL}{r}\right).ch\left(\frac{nx}{r}\right) \right]$.

τ_e is maximum at x=0 and $\tau_{e\,max} = \frac{n\sigma_0}{2}\coth\left(\frac{nL}{r}\right)$ and as $L/r \to \infty$: $\tau_{e\,max} = \frac{n\sigma_0}{2}$

(indeed,experimentally $L \approx 1cm$ and $r \approx 10\mu m$).

Then, for $F=F_d$, $\tau_{emax} = \tau_i$, where τ_i is the interfacial shear strength; thus:

$$\tau_i = \frac{F_d}{2\pi r^2}\sqrt{\frac{2G_m}{E_f Ln\left(\frac{R_{eq}}{r}\right)}}$$

Determination of R_{eq}:

The real neighbourhood of a fibre is different from the idealized case: the nearest fibres are positioned at various distances and generally they do not have the same diameter. The tests are performed to induce only a partial debonding of each indented fibre; this permits to apply the upper-presented modelling with the parameters of the local geometry where debonding occurred (Fig.5). The equivalent radius, R_{eq}, is then defined: $\qquad \theta R_{eq}^2/2 = \theta r^2/2 + A_m$ ou $\qquad \theta R_{eq}^2/2 = \theta r^2/2 - A_f$

A_m(resp.f) is the area of matrix (resp.the nearest fibre -of radius c) included in the sector of angle θ and radius R. A_f is approximated by:

$$A_f = \frac{(\pi - \theta)c^2}{2} + cR\left(1 - \cos\left(\frac{\theta}{2}\right)\right)$$

So: $\qquad R_{eq}^{\ 2} = R^2 + c^2 - \frac{\pi c^2}{\theta} - \frac{2cR}{\theta}\left(1 - \cos\left(\frac{\theta}{2}\right)\right)$ where $\theta = 2Arc\sin\left(\frac{c}{R}\right)$

CONCLUSION

The values of τ_i calculated with our model are:

- for the EP material: $\tau_i = 120MPa \pm 20$ ($G_m = 1GPa$, $E_f = 73GPa$)
- for the MP material: $\tau_i = 105MPa \pm 18$ ($G_m = 1GPa$, $E_f = 73GPa$)
- for the PP material: $\tau_i = 53MPa \pm 14$ ($G_m = 0,5GPa$, $E_f = 73GPa$)

For example, the results obtained for the PV sample were plotted in a $Ln(R_{eq}/r) = f(\sigma_0^2)$ diagram (Fig.6).

First, we should point out that the result scattering is rather low for this kind of experiment. In the case of PP, the largest scattering can be due to a "structural" variation of τ_i in the specimen. Indeed, the fibre impregnation for this material is different if the fibre is in an inner or an outer part of a bundle, owing to the viscosity of the matrix during the process. We should also note that glass/polypropylene composites are known to have a poor fibre/matrix adhesion. As regards the glass/epoxy samples, the EP sizing gives a better adhesion, from a chemical point of view -these two points are in agreement with our own results.

So, the model used seems to provide a coherent relative evaluation of τ_i for the three materials tested. Moreover, for the interlaminar shear strength (ILSS), three-point shear tests performed on the EP (resp.MP) material gave a value of 80MPa (resp.75MPa). The ILSS value is generally considered to be lower than the fibre/matrix interfacial shear strength for this kind of material -which would be verified in our case. However, the values obtained for τ_i appear slightly high when compared with results of pull-out tests, for example ($\tau_i = 50MPa$, /1/, for an EP material kind). The question is: are these two sets of results (microindentation and pull-out ones) to be compared directly and how can we connect them to macromechanical test results?

ACKNOWLEDGEMENTS

The authors are grateful to the DRET for financial support, to Koninklijke/Shell-Laboratorium, Amsterdam and to Vetrotex-St Gobain, Gallet and Duflot for sample furnishing and to Mr N.Chavent (Soretrib, Lyon) for his help.

REFERENCES

/1/ M.R.PIGGOTT, A.SANADI, P.S. CHUA, D.ANDISON "Mechanical interactions in the interphasial region of fibre reinforced thermosets" Composite Interfaces, H.Ishida and J.L. Koenig Editors, Elsevier Science Publishing Co. Inc., 1986.
/2/ B.MILLER, P.MURI, L.REBENFELD "A microbond method for determination of the shear strength of a fiber/resin interface". Composites Science and Technology 28 (1987) 17-32.
/3/ D.ROUBY, G.NAVARRE "Role of interfaces on mechanical properties of ceramic-ceramic fibre composites". Proceedings of the 11th Riso International Symposium on Metallurgy and Materials Science: Structural Ceramics; ed.:J.J.BENTZEN and all, Roskilde, Denmark, 1990
/4/ D.K.SHETTY "Shear-lag analysis of fibre push-out (indentation) tests for estimating interfacial friction stress in ceramic-matrix composites". J.Am.Ceram.Soc.,71 [2] C-107 C-109 (1988)

/5/ **J.F.MANDELL, D.H.GRANDE, T.-H. TSIANG, F.J.McGARRY**
"Modified microdebonding test for direct in situ fiber/matrix bond strength determination in fiber composites". Composites Materials:Testing and Design (Seventh Conference), ASTM STP 893, J.M.Whitney,Ed., American Society for Testing and Materials, Philadelphia, 1986, pp.87-108.
/6/ **L.B.GRESZCZUK** "Theoretical studies of the mechanics of the fiber-matrix interface in composites". Interfaces in composites, ASTM STP 452, American Society for Testing and Materials, 1969, pp.42-58.
/7/ **M.R.PIGGOTT** Load-bearing fibre composites, Pergamon Press,1980, pp.83-87.

FIGURES

Fig.A: optical micrographs of a fibre before and after an indentation test (MP material)

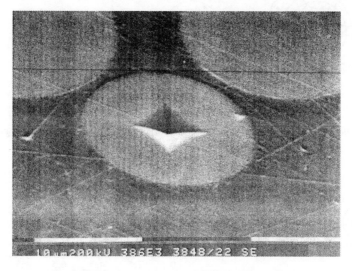

Fig.B: SEM view of an indented fibre (MP material)

149

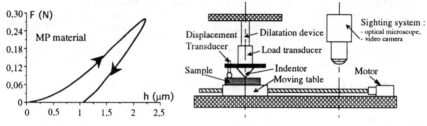

Fig.1: indentation curve

Fig.2: indentation system

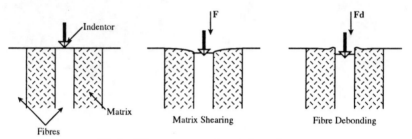

Fig.3: different stages of an indentation test

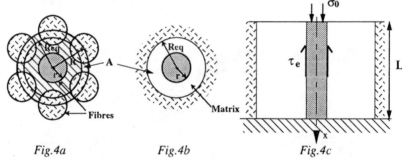

Fig.4a   Fig.4b   Fig.4c

Fig.4a: hexagonal fibre arrangement, 4b and 4c: definition of the equivalent geometry

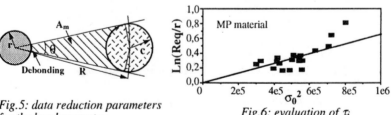

Fig.5: data reduction parameters
for the local geometry

Fig.6: evaluation of τ_i

INFLUENCE OF THE FIBRE/MATRIX INTERFACE ON THE BEHAVIOUR OF E-GLASS REINFORCED THERMOPLASTICS : AN ADHESION EFFICIENCY RATING

S. BONFIELD, M. GARET, A. BUNSELL

Ecole Nationale Supérieure des Mines de Paris - BP 87
Centre des Matériaux - 91003 Evry - France

ABSTRACT
A qualitative rating of the "interfacial adhesion efficiency" has been established for the different methods of fibre/matrix adhesion enhancement available for E-glass reinforced thermoplastics. This was achieved by comparing an experimental 0.5% secant modulus with a theoretical value, assuming perfect bonding. The adhesion improvement was further quantified with the microdrop test, which measures the interfacial shear strength of the system.

1. INTRODUCTION

The mechanical properties of an E-glass reinforced thermoplastic are dependent on the nature of the fibre/matrix interphase, and its efficiency in transferring shear stresses created in the matrix, around the fibre ends. Interfacial adhesion, and hence the composite behaviour, may be enhanced by either coating the fibre with a size specific to the matrix, or by modifying the matrix polymer itself. Both modifying systems have been examined. The former of these techniques requires the tailoring of the size to produce adequate adhesion with a given matrix. An E-glass reinforced polyamide 6.6. (PA66) has been produced with a variety of model coatings applied to the reinforcement to improve interfacial bonding. The second method by which interfacial adhesion can be enhanced is by matrix modification. A size is still a necessary element at the interface, however its chemistry can be simplified and a unique size is suitable for all matrices types. A modified polypropylene matrix composite was examined. This particular polymer is highly unreactive, due to its non-polar nature, and hence requires activation by modification to induce good fibre/matrix adhesion. Modification was achieved using a novel reactive processing technique (1).

The present paper is concerned with the "adhesion-efficiency" of each system. A qualitative interfacial adhesion rating was evaluated by comparing the experimental 0.5% secant modulus with a theoretical value, assuming perfect bonding or 100% efficiency. For the calculation of the theoretical modulus the formula suggested by Berlin et al. (2) was applied, this equation is considered as a reasonable predictor of short fibre composite properties and is of practical use. As all the composites were in an injection moulded form the fibre volume fraction, fibre length and fibre orientation distribution experimental data were exploited in these calculations, hence allowing all mechanical differences to be attributed to a particular modification of the fibre/matrix interfacial adhesion.

2. EXPERIMENTAL

2.1 Materials

a) "Size-modified" composites
 The interfacial adhesion improvement produced by the application of two model sizes to the fibres, before injection moulding, was examined. The sizes were based on a matrix reactive polymer (polyurethane) and a silane coupling agent/polyurethane mixture. The code names and composition of the sizes are given in Table 1.
 Size 2 is the basis of industrial sizes used for PA66/glass fibre composites. The silane coupling agent, γ-aminopropyltriethoxysilane, bonds the polymer to the glass, with stable siloxane (Si-O-Si) bonds forming at the glass/size interface. Adhesion is further improved by the addition of the polyurethane which acts as a bridging molecule, forming strong bonds with both the A1100 and the PA66.

b) "Matrix-modified" composites
 Unstabilized isotactic polypropylene powder was used as the basis for the modified matrix composites. Modification was produced during an in-situ reactive processing technique (1), in the presence of various combinations of three chemical additions: an initiator, I, (peroxide), a modifying agent, MA, (maleic anhydride) and a functional monomer, FM.
 Many commercial modified polypropylene composites, such as ICI grade HW60, are based on a combination of MA and a peroxide. The ICI sample was compared to an improved system (1) which contains the peroxide, MA, and an additional functional monomer, which should further enhance fibre/matrix adhesion. The code names and compositions of the specimens are given in Table 1.

2.2 Methods

The mechanical behaviour of thermoplastic matrix composites is determined not only by the intrinsic properties of the matrix and the reinforcement but also by the interface between them. Additionally the mechanical properties of injection moulded samples are dependent on the fibre volume fraction, the fibre length and the fibre orientation distribution.

2.2.1 Structural characterization

a) Percentage Weight of Fibres
   The percentage weight of fibres was measured by pyrolysing the central section of a tensile test specimen to remove the polymer. Each specimen was weighed before and after complete removal of the polymer to evaluate the relative proportions of the reinforcement and the matrix.

Calculation
   From the percentage weight of fibres results, the volume fraction, V_f, of the reinforcement can be calculated with:-

$$V_f = \rho_m / (\rho_m - \rho_f + (\rho_f / W_f)) \quad [1]$$

where ρ_m = matrix density $\rho_{(PP)}$ = 0.90Kg m^{-3}
$\rho_{(PA66)}$ = 1.14Kg m^{-3}
ρ_f = fibre density = 2.55Kg m^{-3}
W_f = weight fraction of fibres

   Hence, assuming there is no porosity, the matrix volume fraction, V_m, is obtained as:-

$$V_m = 1 - V_f \quad [2]$$

b) Fibre Length Distribution
   Fibre length is a primary controlling factor of the mechanical behaviour of the composites. The fibre length distribution was measured by image analysis of the filaments obtained from the pyrolysis of the tensile test specimen. Approximately 5000 fibres were measured for each distribution. Histograms of fibre length as a function of frequency were produced with the Tracor Northern "FIBRE ANALYSIS" program.

c) Fibre Orientation Distribution
   Injection moulded samples exhibit a random fibre orientation in the plane of the specimen as described by MacNally (3), with the reinforcement at 0° to the loading direction providing the maximum load bearing capabilities. The fibre orientation distribution was evaluated with image analysis of a longitudinal cross-section of the injection moulded sample. Histograms of fibre orientation as a function of frequency were obtained with the Tracor Northern "SIZING" program.

Calculation
   The Krenchel (4) or orientation factor, η_0, was calculated from these results with the following equation:-

$$\eta_0 = \Sigma\, a_k \cos^4 \Theta_k \quad [3]$$

where a_k=fibre fraction oriented in an angular increment away from the applied load Θ_k.

d) Fibre diameter

The diameters of the fibres were assessed with a Watson Shearing Eye-piece mounted on an optical microscope. An average of 30 fibres were measured for each value.

2.2.2 Micromechanical characterization - Microdrop test

Although there are numerous methods for the measurement of the interfacial shear strength the nature of the thermoplastic matrices renders the preparation of many of the required samples particularly difficult. For the microdrop test the matrix is deposited on the fibre as a droplet, which is relatively easy to produce. Drops of an average diameter of 350μm were formed on single fibres. The specimens were tested with an Universal fibre tensile test machine at a cross-head speed of 1mm/min.

Calculation

The interfacial shear strength, τ_i, can be calculated with:-

$$\tau_i = F/dL \quad [4]$$

where F= fibre pull-out force
    d= diameter of the fibre
    L= length of the microdrop

2.2.3 Macromechanical Characterization - Tensile test

The mechanical properties of the injection moulded matrix polymer and the composite were evaluated by tensile testing at cross-head speeds of 5mm and 1mm per minute. Composite deformation was measured with an extensometer with an accuracy of 3.10^{-6}m and a deformation limit of 8.1%. The cross-head displacement was used to evaluate the deformation of the polymer as the elongation of the matrices rendered all other available measuring techniques unsuitable. A 0.5% secant modulus was calculated for each composite type.

3. RESULTS AND DISCUSSION

A theoretical modulus, E_{th}, for the composite at 0.5% strain, was calculated with:-

$$E_{th} = \eta_0 E_f V_f \left[\int_{L_{min}}^{L_c} L/2L_c Q(L)dL + \int_{L_c}^{L_{max}} (1-L_c/2L)Q(L)dL \right] + E_m V_m \quad [5]$$

where η_0 = fibre orientation factor (see equation 3)
    E_f = modulus of the fibres = 73GPa
    V_f = fibre volume fraction (see equation 1)

Critical fibre length, Lc

The Cox model was assumed to hold for all the composites, as at low strains both the matrix and the fibre behave elastically. Additionally this formula assumes perfect bonding and is therefore appropriate for the theoretical modulus calculations. The Cox model is given by:-

154

$$\sigma_f = E_f \varepsilon_c [1 - \frac{\cosh(\beta(L/2 - x))}{\cosh(\beta L/2)}] \qquad [6]$$

where
$$\beta = [2G_m / E_f r^2 \ln(V_f^{-1/2})]^{1/2} \qquad [7]$$

and ε_c = composite strain = 0.005
L = fibre length
x = distance from the end of the fibre
G_m = shear modulus of the matrix
r = radius of the fibre

Equation 7 was adapted by Bader (5) from those published by Cox, by using the fibre volume fraction in place of length and interfibre spacing. The critical length, L_c, was defined by the point at which the stress of the fibre reaches $0.9 E_f \varepsilon_c$.

Fibre length distribution at 0.5% strain

At 0.5% strain, the stress in the fibres, σ_f, is 0.37GPa. The probability of fibre failure is therefore minimal as shown by Figure 1. The fibre length distribution measured by image analysis for the unstrained composite hence can be considered as analogue to that of a composite at 0.5% strain.

$Q(L)$ = normalized fibre length distribution

The normalized fibre distribution, $Q(L)$, was calculated using a polynomial fit of the histogram of the fibre length as a function of frequency.

$$E_m = \text{modulus of the matrix} = E_{PA66} = 3.2\text{GPa}$$
$$= E_{PP} = 1.5\text{GPa}$$
V_m = matrix volume fraction (see equation 2)

The above parameters were determined experimentally, except for those where a numerical value is given.

3.1 "Size-modified" composites

The results of the structural and micromechanical characterization for the "size-modified" composites are summarized in Table 2. The fibre volume fraction, of 0.3, and the orientation factor, of 0.80, were constant for both samples, indicating that these parameters were not influenced by the nature of the size.

The mechanical properties of the PA66 composites show a relatively small variation in modulus, however a significant difference in failure stress was measured between the Size 1 and Size 2 specimens. This is in good agreement with the results obtained by Bader and Collins (6). A comparison of the failure strengths and of the adhesion efficiency values showed the improved performance, hence enhanced interfacial bonding of Size 2. This difference was further quantified by the microdrop test. PA66 proved unsuitable for this method as the matrix polymer underwent severe degradation when heated in air. A modified polypropylene was therefore employed. The addition of the silane coupling agent to Size 2 produces an improved interfacial shear strength as shown by the values given in Table 2.

3.2 "Matrix-modified" composites

The structural and micromechanical results are tabulated in Table 2. A constant fibre volume fraction, of approximately 0.12 and an orientation factor of 0.64 were measured for both the commercial and the improved-system composites. As for the size-modified composites these parameters appear to be independent of the type of modification. The orientation factor, however, has been shown to be proportional to the fibre volume fraction, by comparing the results for both series of composites.

Both the modulus and the failure stress of the ICI samples were less than those of the MAF specimens. The addition of the functional monomer results in a 12% improvement of the mechanical characteristics. The enhanced interfacial adhesion can be further quantified by considering the interfacial shear stress measured with the microdrop test. An interfacial shear stress of 13.2MPa was measured for the ICI sample. Values for the MAF matrix polymer were not obtained as fibre failure preceeded interfacial failure. The interfacial strength of this particular modification hence exceeds 18MPa. Hill (7) measured an interfacial shear strength of 19.3MPa for a coupled, non-modified polypropylene composite, whereas Folkes et al. (8) measured a value of 30MPa in a similar composite. Discrepancies between these results are most likely to be associated with differences in the type of modification (either size-modified (Hill and Folkes) or matrix-modified for the measured samples) and variations in the specimen preparation. The microdrop test involves heating the matrix polymer to form the droplet. No mechanical shear is therefore involved in the production of the specimens. Shear is considered to play an important role in the fibre/matrix adhesion improvement produced by reactive processing. This, with any slight degredation induced by heating, may be responsible for the relatively low measured value for the ICI specimen.

The theoretical analysis gives an underestimation of the mechanical properties of the real composite. However as this value was employed as an index, representing 100% or perfect bonding, the relative bonding efficiencies are valid.

4. CONCLUSIONS

Adhesion enhancement can be achieved by either size or matrix modification. From the interfacial adhesion efficiency values it can be seen that the A1100 coupling agent improves the performance of the size-modified composites by 28% and that a matrix-modified composite offers the strongest interfacial bonding when in the presence of a functional monomer. The proposed model allows the examined systems to be ranked as follows:- Size 1, ICI, Size 2, and MAF (from the weakest to the strongest bonding). The interfacial adhesion efficiency rating is confirmed by the interfacial shear strengths measured with the microdrop test. This model therefore is an effective method of comparing interfacial bonding in composites with not only different adhesion enhancement systems, but also different matrices, fibre volume fractions, and fibre length and orientation distributions.

5. ACKNOWLEDGEMENTS

The manufacture of the samples by Vetrotex and Aston University, and the support of Vetrotex and the Commission of the European Communities in providing the funding for these projects, is greatfully acknowledged.

6. REFERENCES

1. S. Al Malaika and H. Sheena, *Private communication*, Aston University
2. A. Berlin, S. Volfson, N. Enikolopian and S. Negmatov *"Principles of Polymer Composites"*, Springer-Verlag, Berlin (1986)
3. D. MacNully, *Polym. Plast. Tech. Eng.*, 8 (1977) p.101
4. H. Krenchel "Fibre Reinforcement", Akademisk Forlag, Copenhagen (1964)
5. M. Bader and A.Hill, to be published
6. M. Bader and J. Collins in *"Progress in Science and Engineering of Composites"*, Ed. T. Hayashi, K. Kawata and S. Umekawa, ICCM-IV, Tokyo, (1982) p.1067
7. A. Hill, PhD thesis, University of Surrey, (1991)
8. M. Folkes and D. Kells, *Plastics and Rubber Proceedings*, 5 (1985) p.125

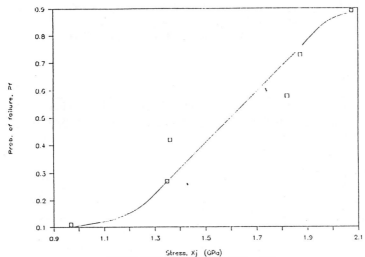

Stress, Xj (GPa)

FIGURE 1: Probability of fibre failure

CODE NAME	COMPOSITION	NOMINAL WEIGHT OF FIBRES (%)
"SIZE-MODIFIED" COMPOSITES (1)		
SIZE 1	Polyurethane (PU)	50
SIZE 2	Polyurethane + A1100 (2)	50
"MATRIX-MODIFIED" COMPOSITES (3)		
ICI	I + MA	30
MAF	I +MA + FM	30

TABLE 1: Specimen code names and compositions
(1) Samples produced by Vetrotex
(2) The silane coupling agent is marketed by Union Carbide as A1100
(3) Samples manufactured by Aston University

157

CODE	W_f (%)	V_f	V_m	L (μm)	Lmax (μm)	η_0	d (μm)	L_c	τ MPa
"SIZE-MODIFIED" COMPOSITES									
SIZE 1	47.7	0.29	0.7	126 (78)	650	0.80	17.6 (1.6)	146	12.0 (0.8)
SIZE 2	51.3	0.31	0.7	158 (109)	475	0.80	17.6 (1.6)	146	14.0 (1.6)
"MATRIX-MODIFIED"COMPOSITES									
ICI	28.4	0.12	0.88	196 (162)	900	0.64	15.0 (0.7)	251	13.2 (3.2)
MAF	29.1	0.13	0.87	274 (217)	1225	0.64	15.0 (0.7)	251	>18

TABLE 2: Structural and micromechanical characteristics (standard deviation)
Definitions of symbols are those given for equation
L = average fibre length measured by image analysis.

CODE	MEASURED VALUES		THEORETICAL VALUES	ADHESION EFFICIENCY (%)
	σ_f (MPa)	E (GPa)	E_{th} (GPa)	E
"SIZE-MODIFIED" COMPOSITES				
SIZE 1	67.7	11.1	14.8	75
SIZE 2	109.4	12.8	13.8	93
"MATRIX-MODIFIED" COMPOSITES				
ICI	65.6	5.6	6.5 ·	86
MAF	82.0	6.0	6.2	98

TABLE 3: Adhesion efficiency calculation

INFLUENCE OF FIBRE COATING ON THE TRANSVERSE PROPERTIES OF CARBON FIBRE REINFORCED COMPOSITES

J. DE KOK, P. VAN DEN HEUVEL, H. MEIJER

Eindhoven University of Technology Centre for Polymers and Composites - PO Box 513 - 5600 MB Eindhoven - The Netherlands

ABSTRACT

The influence of thin elastomer fibre coatings on the transverse properties of unidirectional carbon fibre reinforced composites has been investigated using a combination of experimental and numerical techniques. For the experiments a solution technique is used to apply coatings on the carbon fibres, varying in coating thickness and modulus. In the finite element micromechanical analyses a third constituent is introduced at the fibre interface, which represents the coating applied. Both experiments and numerical analyses show that with a brittle matrix an increase in coating thickness and/or a decrease in coating stiffness significantly reduces the composite's transverse stress and strain to failure. However, the numerical analyses indicate that with matrices with higher failure strains a rubbery coating will lead to an increase of the transverse failure strain.

INTRODUCTION

The low transverse strain to failure of fibre reinforced composites composes a major drawback which limits their application. This investigation is part of a detailed study focusing on the transverse properties of unidirectional composites. The general aim is to find routes to improve the transverse failure strain. In this study, material parameters governed by either fibre, matrix or interphase are systematically varied and analyzed using a combination of experimental and numerical-micromechanical analyses. Previous research on the influence of fibre surface treatment on the transverse properties of carbon/epoxy composites has shown that the interfacial region affected by the surface treatment is very thin and has mechanical characteristics close to those of the matrix /1,2/. Consequently, the interfaces influence neither the overall composite transverse modulus nor the stress concentrations /1,2,3/. However, the micromechanical analyses showed that fairly thicker interfacial regions than those investigated via fibre surface treatment, with a low elastic modulus, will result in a strong decrease of the transverse modulus combined with a considerable change in the stress concentrations /1,4,5/. Especially because the stress situation within a composite is strongly influenced by the presence of such a interphase, also the transverse strength and failure strain are expected to be affected. In this paper the influence of these thick interfacial regions are investigated in more detail, for carbon fibre reinforced epoxy.

EXPERIMENTAL

In order to exclude the influence of fibre-matrix debonding, surface treated high strength carbon fibres (Courtaulds Grafil XA-S) have been used /1,2/. The fibres were coated using a solution technique with thermoset elastomers based on DGEBA (diglycidyl ether of bisphenol-A, LY556 of Ciby Geigy) and Jeffamine (poly oxipropylene diamine) hardeners. Coatings were applied in varying thicknesses up to approximately 0.06 μm (2% of the fibre radius). With thicker coatings the fibres coalescent, which prevents impregnation. A variation in rubber-coating properties was obtained by the use of several Jeffamines (D2000 and D4000 of Texaco) with different molecular weights, leading to different crosslink densities. Homogeneous rubber materials were obtained varying in modulus from 0.8 MPa to 4 MPa (Table 1). As matrix material a standard quite brittle epoxy system of Ciba Geigy (LY556/HY917/DY070) has been used. 50 Vol.% carbon fibre reinforced composites were manufactured by filament winding.

Transverse properties were determined by three-point bending. Figure 1 shows the transverse failure stress of different composites as a function of the relative coating thickness. With increasing coating thickness a lower the transverse stress (and strain) to failure of the composites is obtained. Coatings with a lower modulus (based on D4000) show a reduction in transverse stress (and strain) to failure is much stronger and is not that dependent on the coating thickness.

NUMERICAL ANALYSES

To analyze the composites with coated fibres with the aid of a finite element micromechanical model a third constituent is introduced at the fibre interface representing the coating applied (figure 2a). The numerical analyses in transverse tensile loading are based on a generalized plane strain model of a quarter fibre in a block of matrix. For the interphase being, elements are used for incompressible Mooney-Rivlin material behaviour to describe elastomer behaviour with a high strain capability. The mechanical properties of the thermosets and the transversely isotropic carbon fibres are listed in table 1. The matrix is considered to be linear elastic until failure.

The use of a Von Mises yield criterion as a failure criterion for this type of epoxy matrix has turned out to give good predictions of the transverse strength. By using this failure criterion for common carbon fibre reinforced epoxy a transverse strength is predicted higher than the matrix strength (compare table 1 with figure 3). The maximum stress between the fibres (at the pole) is higher than the transverse stress applied due to stress concentrations. However, because the contraction of the matrix is constrained by the presence of the stiff fibres, reasonably high tensile stresses perpendicular to the direction of the applied stress are found, yielding a maximum Von Mises stress lower than the stress applied. Because the Von Mises stress in the composites is lower than the transverse stress applied, the composites are stronger than the matrix itself.

When a thin elastomer coating is applied on the fibres, the contraction of the matrix will not be completely constrained by the fibres, because the coating can deform considerably as a result of the shear stresses. This is illustrated in figure 2c. Consequently, the stresses perpendicular to the stress applied are reduced or even disappear and the Von Mises stress will increase. Finally it will be higher than the stress applied due to stress concentrations. Therefore, the presence of elastomer fibre coating reduces the overall composite transverse strength (see figure 3). Obviously, this effect is stronger when the coating is thicker or when it has a lower modulus.

For the brittle epoxy matrix investigated, the introduction of an elastomer coating appears to have only a negative effect on the transverse stress (and strain) to failure. However, with a more ductile matrix the introduction of a rubbery coating proves to have a positive effect on the transverse failure strain, see figure 4. The predicted failure strain is shown as a function of the coating thickness, based on D2000, with as parameter the maximum matrix plastic strain. This is an interesting result, since without the application of a rubbery coating around the fibres a substantial increase in matrix ductility hardly results in an increase of the composites overall transverse failure strain /6/.

CONCLUSION

The application of a flexible elastomer fibre coating in carbon fibre reinforced epoxy yields a decrease of the transverse stress and strain to failure with increasing coating thickness. With coatings with a lower modulus the reduction of the transverse strength is stronger.

Micromechanical analyses showed that the matrix deformation in the composites is less constrained when fibre coatings are applied. This leads to a reduction of the transverse failure strain with brittle matrices. However, for matrices with higher failure strains a rubbery fibre coating will eventually lead to an increased efficiency of the matrix ductility and an increased composite transverse failure strain. Consequently, combining rubber coated carbon fibres with ductile matrices is interesting for developing unidirectional composites with a high transverse tensile failure strain.

REFERENCES

1. J.M.M. de Kok, "The influence of the interface on the transverse properties of carbon fibre reinforced composites", Int. Report Eindhoven Univ. of Tech, CIP-DATA Kon. Bibliotheek, Den Haag, (1992)
2. J.M.M. de Kok, E.J. van Klinken and A.A.J.M. Peijs, "Influence of Fibre Surface Treatment on the Transverse Properties of Carbon Fibre Reinforced Composites", Proc. Advanced Composites '93, Wollongong (Australia), (1993)
3. J. de Kok, E.J. van Klinken and T. Peijs, "Micromechanical Modelling of the Fibre-Matrix Interface in Transversely Loaded Carbon Fibre Reinforced Epoxy", Proc. ECCM-5, (1992) 385-390
4. T. Riccò, A. Pavan and F. Danusso, Polymer Eng. and Sci., " Micromechanical Analyses of a Model for Particulate Composite Materials with Composite Particles; Survey of Craze Initiation", 18, 10 (1978) 774-780
5. H. Albertsen and P. Peters, Proc. "The Influence of Fibre Surface Treatment on Interphase Properties in CFRP", Proc. ECCM-5, (1992) 391-396
6. J.M.M. de Kok, H.E.H. Meijer and A.A.J.M Peijs, "The Influence of Matrix Plasticity on the Failure Strain of Transversely Loaded Composite Materials", Proc. ICCM-9, (1993)

Table 1. Mechanical properties of the materials.

Material		Epoxy matrix	Coating D2000	Coating D4000	Carbon fibre
Young's Modulus	E	2.7 GPa	4.0 MPa	0.8 MPa	-
Poisson ratio	ν	0.41	≈ 0.50	≈ 0.50	-
Yield stress	σ	94 MPa	-	-	-
Mooney-Rivlin	C_{10}	-	0.619 MPa	0.094 MPa	-
parameters	C_{01}	-	0.119 MPa	0.063 MPa	-
Long. modulus	E_L	-	-	-	235 GPa
Transv. modulus	E_T	-	-	-	20 GPa
Long. Poisson ratio	ν_{LT}	-	-	-	0.013
Transv. Pois. ratio	ν_{TT}	-	-	-	0.25
Shear modulus	G_{LT}	-	-	-	18 GPa

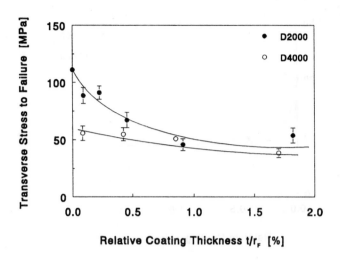

Figure 1. Influence of coating thickness on the transverse failure stress.

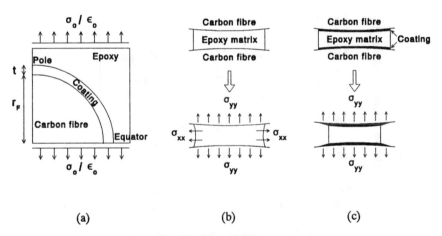

Figure 2. Micromechanical model (a) and illustration of matrix deformation in common composite (b) and composite with fibre coating (c).

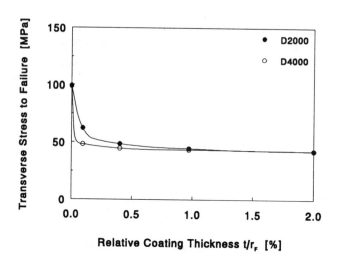

Figure 3. Influence of coating thickness on the predicted transverse strength.

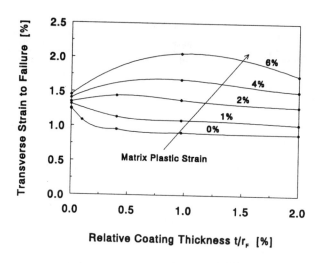

Figure 4. Influence of coating thickness on the predicted transverse failure strain, with variation in the matrix failure strain.

THE EFFECT OF ABSORBED WATER ON THERMOSET RESINS, THE FIBRE RESIN INTERFACE AND MECHANICAL PROPERTIES OF GRP'S

P. G. VANDOR, M.G. PHILLIPS

University of Bath - School of Materials Science - Claverton Down BA2 7AY Bath - UK

ABSTRACT.

The water content of four commercial glass fibre/thermoset resin laminates has been varied in a progressive manner by exposure at 70 degrees centigrade to vacuum or to water immersion, thereby covering a range from the dry to the saturated state. The effect of water content on the quality of the laminates has been assessed by measurement of flexural strength, and the effect on fibre matrix adhesion by measurement of interlaminar shear strength. In the saturated condition all four systems: epoxy, polyester, vinyl ester and phenolic showed comparable properties, despite wide variations in absolute water content and differences in mechanical properties in the dry condition.

1. INTRODUCTION.

Thermoset polymers used as matrices in glass reinforced composites are prone to absorb water. The matrix is usually plasticised /1/ and those mechanical properties which are not wholly fibre dependent are generally degraded. Cold cure phenolic resins have recently entered the field of large scale laminating where their main advantages lie in low cost and favourable combustion performance. The resins currently on offer and under developement are finding application in the fields of mass transit, and in offshore structures. These are both areas where limited room for access, and therefore escape, means that fire resistance, in all its many guises, is required for safety reasons. The use of cold cure phenolics in offshore applications means that these materials are likely to be exposed to wet environments and would therefore be prone to any degradation associated with absorbed water. This paper examines the effect of absorbed water on the flexural and interlaminar shear strengths of several thermoset resins which were reinforced with chopped strand mat (CSM) glass. The resins were a cold cure phenolic A, a vinyl ester B, an epoxy C and a polyester D.

2. MATERIALS.

The resins used in this work were all commercially available examples used for hand laminating. Panels of composite were reinforced with 3 layers of CSM, weight 450gsm. After hand laminating they were all cured following the various manufacturers recommended schedules. Panels were approximately 0.5m square and the average thicknesses were: roughly 3.5mm for A, C and D, and 2.5mm for B the vinyl ester. This corresponds to a fibre volume fraction of approximately 0.16 for A, C and D , and 0.22 for B. After curing the panels were cut into test coupons using a high speed diamond impregnated saw. Since water content was the main variable to be studied, the cutting was done dry.

3. EXPERIMENTAL.

3.1. Exposure programme.

Batches of samples were exposed at 70 degrees centigrade in one of two environments, either placed in a vacuum oven or immersed in water. Samples were exposed at 70 degrees so as to accelerate sorption or desorption. Sorption curves were constructed by removing specimens from their respective environments and weighing. The data was plotted as % weight change versus square root of time. Vacuum exposure continued until no further reduction in weight could be detected. This is refered to as the "dry" condition.

3.2. Mechanical testing..

Batches of mechanical test specimens were removed from the exposure environments after the requisite number of days. Tests were performed at ambient temperature in air, the elevated temperature exposure being used only to speed up the sorption process. Two types of mechanical test were carried out to determine flexural strength and interlaminar shear strength (ILSS). The flexural strength test was based on ISO 178 where a span to depth ratio of 16:1 is specified. The span used for the flexural testing was 56mm for the phenolic, epoxy and polyester samples while a span of 40mm was used for the thinner vinyl ester specimens. A cross head speed of 2mm/minute was used for all flexural strength tests. ILSS tests were carried out at a crosshead speed of 1mm/minute and a span to depth ratio of 5:1 was used. The span used for the ILSS test was 18mm for all materials except the vinyl ester where a span of 13mm was used.

4. RESULTS.

4.1. Exposure of composites

The results of the composite exposure programme are shown in Table 1 the scatter in water content for any group of specimens was typically +/- 5%. The weight changes are assumed to be due mainly to water sorption or desorption. The desorption curves showed typical Fickian behaviour , and weight changes tended to stabilise after 40 or 50 days exposure. Similar Fickian behaviour was seen for the absorption plots except for the polyester which showed some weight loss after 40 days or so. This weight loss was attributed to leaching out of soluble components in the resin. Water contents for the polyester after prolonged exposure, shown in Table 1, were calculated by extrapolating the initial part of the absorption curve and neglecting the losses due to leaching.

Desorption in vacuum showed that in the as received state the composites all contained some water, ranging from around 0.5wt% for the vinyl ester to 8.4wt% for the phenolic. The immersed specimens absorbed water. In this case the vinyl ester absorbed 1wt% while the phenolic absorbed 25wt%. Based on dry weight the vinyl ester, polyester, epoxy and phenolic absorbed 1.48, 3.58, 5.63 and 37wt% respectively. The absorption of water into polymers has been studied extensively [1] , however in the case of phenolics the absorption process is greatly affected by the presence of a large volume fraction of microvoids [2] which is of the order of 20% or so. The phenolic panel used in this work contained a great deal of coarse porosity and absorbed more than the normal 25wt% water.

4.2. Mechanical testing.

The flexural strength specimens all exhibited failure initiation at the tensile face of the three point bend specimen, normally just below the central loading point. The ILSS specimens all failed by shear, normally single shear, in the vicinity of the mid plane of the specimen. The flexural strength and ILSS results are shown graphically in Figures 1 and 2 respectively where error bars represent +/_ one standard deviation. The mechanical properties are plotted as a function of percentage of saturation with water, a parameter defined such that "dry" samples have the value zero and samples "saturated" after 3 months exposure have the value 100%. Presenting the results in this way offers a method of normalisation which will allow a direct comparison to be made between composites of widely different absolute water content.

Figure 1 shows that in the dry state the flexural strengths of the different composites varied considerably. The vinyl ester was strongest whilst the phenolic was weakest, the mean strengths being 234 and 140MPa respectively. The dry strengths of the epoxy and polyester were intermediate between the two at roughly 180MPa. In the saturated state the mean flexural strengths of the vinyl ester and phenolic dropped to 89 and 100MPa respectively whilst for the epoxy and polyester the strengths were reduced to 113 and 82 MPa.. In the saturated condition the epoxy and and phenolic retain 66% and 71% of the dry strength whilst the vinyl ester and polyester retain only 38% and 45% of the dry strength.

The interlaminar shear strengths of the phenolic and vinyl ester laminates are shown in Figure 2a and those of the epoxy and polyester in Figure 2b. The pattern of behaviour for ILSS is similar to that for the flexural strength. Again in the dry condition the phenolic and vinyl ester are at the extremes the former having an ILSS of 12.8MPa while the latter had a value of 28MPa. The epoxy and polyester again had intermediate values at around 21MPa. In the saturated state the vinyl ester, phenolic, epoxy and polyester had ILSS values of 9.1, 7, 12.1 and 6.4MPa respectively. Compared to the dry condition these figures represent retentions of 33%, 55%, 58% and 30% respectively

5. DISCUSSION.

Composite samples dried in vacuum showed that the vinyl ester had the greatest flexural and interlaminar shear strengths. It should be noted however that this laminate was considerably thinner than the others, giving it a voulme fraction of reinforcement of roughly 0.22 compared with roughly 0.16 for the others. Applying the simple mixtures rule, the greater volume fraction of fibre in the vinyl ester would give a 38% greater flexural strength. In the dry state the vinyl ester is actually 34% stronger than the epoxy and polyester composites. The ILSS of the vinyl ester is also 33% greater than for the epoxy and polyester materials.

Both flexural strength and ILSS of the dry phenolic are lower than for the epoxy and polyester by about 30 to 40 %. The presence of large voids in the phenolic laminate, which also resulted in exceptionally high water absorption, is at least partially responsible for the low values of strength and ILSS. This has been reported to be the case for other composite systems/3/. Flexural strength tests on a less porous phenolic laminate, using the same reinforcement have given strength values in the range of 170 to 180MPa /4/.

Exposure to water at 70 degrees centigrade results in saturation of all of the composites within a three month period. For the phenolic, epoxy and polyester the change in both flexural and interlaminar shear strengths was gradual over the whole range of water contents. This contrasts with the behaviour of the vinyl ester which showed a sudden reduction in both properties in the vicinity of 60% saturation which would correspond to exposure to water at 70 degrees for roughly 1 week. The polyester also differs from the other materials in that it lost some weight due to the leaching out of material under the harsh conditions quoted here.

The changes in flexural strength with increases in water content are followed almost exactly by changes in the ILSS. This applies to all of the laminates tested but is most noticeable for the vinyl ester which shows the sudden drop in strength around 60% of saturation. This suggests that in both tests the failure is affected by similar factors.

6. CONCLUSIONS.

1. The flexural strength and ILSS of a series of laminates have been measured in the dry state. The properties of the phenolic were rather lower than the other 3 laminates which was at least partially attributable to the presence of large voids in the laminate. The properties of the vinyl ester were apparently superior to the two 'standard' materials, an epoxy and a polyester but this was attributable almost entirely to the greater volume fraction of fibre in the vinyl ester laminate.

2. When exposed to water at 70 degrees the phenolic composite absorbed much more water than the other 3 composites, up to 37wt%, compared with 5.63, 3.58 and 1.48 wt% for the epoxy, polyester and vinyl ester respectively.

3. The reduction in mechanical properties caused by saturation with water was smallest for the phenolic and largest for the vinyl ester such that in the saturated state all the materials showed similar strengths.

ACKNOWLEDGEMENTS.

The authors would like to thank BP Chemicals Ltd. for funding the work.

REFERENCES.

1. G. Springer (ed). "Environmental Effects on Composite Materials" Technomic Publishing Co., Westport, CT, USA, 1981.
2. S. M. Tavakoli, R. A. Palfrey and M. G. Phillips. Composites, 20, (1989) 159-165
3. C. C. Chamis, M. P. Hanson, T. T. Serafini, in "Criteria for Selecting Resin Matrices for Improved Composite Strength". Proc. 28th Tech. Conf. on Reinforced Plastics. SPI. Paper no. 12c 1973.
4. BP Physical Test Data for CSM.

TABLE 1

**WATER ABSORPTION AND DESORPTION (WEIGHT%) IN COMPOSITES
AFTER EXPOSURE FOR VARIOUS TIMES AT 70 DEGREES CENTIGRADE**

EXPOSURE TIME DAYS	PHENOLIC	EPOXY	POLYESTER	VINYL ESTER
90 WATER	+25.0	+4.70	+2.72	+1.0
28 WATER	+24.7	+3.35	+2.33	+0.58
4 WATER	+12.4	+2.46	+0.67	+0.42
1 WATER	+6.7	+1.47	+0.38	+0.37
0	0	0	0	0
1 VACUUM	-2.7	-0.57	-0.37	-0.21
4 VACUUM	-5.1	-0.66	-0.49	-0.37
28 VACUUM	-8.2	-0.78	-0.66	-0.42
90 VACUUM	-8.4	-0.88	-0.83	-0.47

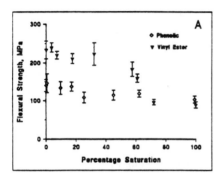
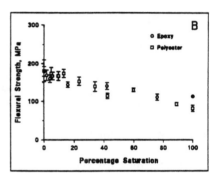

Figure 1. Flexural strength as a function of water content for four commercial thermoset resins with a common glass-fibre (CSM) reinforcement.

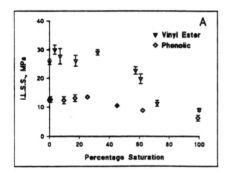
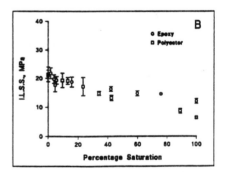

Figure 2. Interlaminar shear strength as a function of water content for four commercial thermoset resins with a common reinforcement.

EVALUATION OF THE TRANSVERSE STRENGTH IN A (0°, 90°, 0°) FIBROUS MULTILAYER WITH CRACKS

R. EL ABDI

Laboratoire Génie de Production - Ecole Nationale d'Ingénieurs de Tarbes - Chemin d'Azereix - BP 1629 - 65016 Tarbes - France

ABSTRACT

Four possible cases of crack extension in the neighbourhood of a fiber embedded in an elastic matrix and subjected to a transverse load are analytically studied. The interaction of several cracks in a (0°, 90°, 0°) multilayer with long fibers, is evaluated numerically by a finite element program.
A detailed study of the effect of the presence of fibrous reinforcements on transverse strength, compared to an unreinforced matrix is considered. This study also quantifies the effect of crack length on the global behaviour of the composite material, and the role of the geometrical and material parameters of the fiber/matrix interface.

1. INTRODUCTION

The relatively weak transverse strength of fibrous composites is a major drawback in the design of advanced composite materials in particular of oriented long fiber composites.
In practice, even if this kind of material is supposed to work along the fiber direction, in fact, transverse stress is unavoidable when flaws exist and may then cause a premature failure in the composite.
In this context several questions have to be answered, and three frequently reappear :
- Whether the presence of fibrous reinforcements increases or decreases the transverse strength compared to an unreinforced cracked resin ?
- What is the influence of crack dimensions on the global behaviour of the composite ?
- What is the contribution of the fiber/matrix interface and the influence of its geometrical and mechanical characteristics on the transverse strength ?

In order to plan a thorough study of a fibrous composite in a field of microcracks, some authors first attacked the analysis of a single fiber in an infinite elastic body containing one crack. Tests conducted by Maron and Arridge/1/ of photo-elastic simulation on a single fiber, were focused on a general failure of resin plates with a pre-vulcanized inserted inclusion. These tests were further developed by Knauss and Muller/2/ in order to analyse the evolution of transverse strength. Tirosh/3/ has numerically studied the interaction of several cracks present in an elastic multilayer with multi-parallel fibers.

Recently, Hine/4/ based on an experimental study which used an unidirectional carbon fiber/poly (ether-ether-ketone) (PEEK) cracked composite which was loaded along the fiber direction, was particularly concentrated on the treatment of the reinforcement surfaces so as to increase the strength of the composite when intralaminar or interlaminar fractures occur.

Several techniques are used when improving interface fiber/matrix properties. In some cases these interfaces can be the site of the initiation and propagation of flaws.

For this reason, we are studying a (0°,90°,0°) cracked laminate with long fibers under a load perpendicular to the fiber direction in order to analyse the interaction between the various layers. In particular, the numerical study will expand on the role of the fiber/matrix interface relative to the transverse strength; of crack lengths compared with fiber diameter and of the fiber volume proportion.

We shall limit the study to stress intensity factors sufficient for the case in question. In ref./5/ the emphasis is given to other rupture criteria and their use in a finite element analysis (see Barsom/5/).

2. FRACTURE MECHANICS - THE CASE OF A SINGLE FIBER

Consider a well-bonded circular fiber in an elastic matrix (Poisson's ratio ν), subjected to a transverse load σ_∞ sufficiently far away from a crack (Fig.1). The stress expressions in polar coordinates are given by /6/:

$$\sigma_{rr} = \frac{\sigma_\infty}{2}\left\{1 - \gamma\left(\frac{R}{r}\right)^2 - \left[1 - 2\beta\left(\frac{R}{r}\right)^2 - 3\delta\left(\frac{R}{r}\right)^4\right]\cos(2\theta)\right\}$$

$$\sigma_{\theta\theta} = \frac{\sigma_\infty}{2}\left\{1 + \gamma\left(\frac{R}{r}\right)^2 + \left[1 - 3\delta\left(\frac{R}{r}\right)^4\right]\cos(2\theta)\right\}$$

$$\sigma_{r\theta} = \frac{\sigma_\infty}{2}\left[1 + \beta\left(\frac{R}{r}\right)^2 + 3\delta\left(\frac{R}{r}\right)^4\right]\sin(2\theta) \qquad (1)$$

$$\text{with } \gamma = -\frac{(\kappa - 1)}{2} \; ; \; \beta = -\frac{2}{\kappa} \; ; \; \delta = \frac{1}{\kappa} \; ; \; \kappa = 3 - 4\nu$$

(in plane strain)

The analytical study shows the the tangential stress $\sigma_{\theta\theta}$ is always lower than the radial stress σ_{rr} and the location at which this stress reaches maximum is not on the interface border r=R (R: fiber radius) but at a distance r* such that:

$$\left(r^*/R\right) = \left\{6/[\kappa(\kappa - 0.5) + 4]\right\}^{\frac{1}{2}} \qquad (2)$$

For an epoxy matrix with $\nu = 0.35$, this distance is equal to r*=1.2R .

The four possible cases of cracks in the neighbourhood of a fiber (assumed more rigid than the matrix) are shown in figure 2.

If σ_0 represents the tensile strength of the matrix, and σ_c the critical stress of the fiber/matrix body, the expressions for this critical breaking stress are (for the four cases in figure 2) :

$$\left(\sigma_c/\sigma_0\right)_a = \left[1 + \gamma/2(1 + \hat{l}/2)^{-3/2} - 3/2\delta(1 + \hat{l})^{-7/2}(1 + \hat{l}/2 + \hat{l}^2/8)\right]^{-1} \qquad (3)$$

$$\left(\sigma_c/\sigma_0\right)_b = \left[\gamma/2(1+\hat{l}/2)^{-3/2} + 3/2\delta(1+\hat{l})^{-7/2}(1+\hat{l}/2+\hat{l}^2/8)\right]^{-1} \qquad (4)$$

$$\left(\sigma_c/\sigma_0\right)_c = \left[1-(\gamma+2\beta+3\delta)/2 - (1-2\beta-2\delta)m/\pi\right]^{-1} \qquad (5)$$

$$\left(\sigma_c/\sigma_0\right)_d = \left[(-\gamma+2\beta+3\delta)/2 + (1-2\beta-3\delta)m/\pi\right]^{-1} \qquad (6)$$

with $m=\alpha\sin^{-1}\sqrt{(2\alpha)}+\sqrt{[2\alpha(1-2\alpha)][-\alpha^2+(4/3)\alpha^3-(32/3)\alpha^4]}$ and

where γ, δ, and β have been defined previously (in (1)).

(lo is the crack length, $\hat{l}$ = lo/R, and α =lo/2r)

Figure 3 gives the evolution of the stress σ_c/σ_0 versus the fibers radius R/lo for different configurations of figure 2. If Vm, Vf, Gm, Gf respectively represent the Poisson's coefficients of the matrix and of the fibers, as well as their shear moduli, we see that case c (circumferential crack under an opening load) minimizes the critical stress. Weakening of the matrix is shown to be nearly 30% as soon as R/lo > 5. Therefore, to improve the matrix strength against crack propagation, it would seem to be preferable to use **small radius fibers.**

2. STUDY OF A LAMINATE WITH SEVERAL CRACKS

3. 1 Finite element analysis - Presentation and result analysis.
The Ansys code used in this study has a subroutine with automatic meshing when some principal data are given, and permits having a mesh as refined as possible around sensitive zones.

2.1.1 Choice of the crack-tip
The first question needing answer is whether it is necessary to conduct two parallel studies for the two crack tips indicated in figure 4. Let b be the fiber/matrix interface thickness (circular and uniform), K1- K10 respectively, the stress intensity factors in the open mode of the laminate consisting of three components (fibers-matrix-interface) and of the same medium consisting only of the matrix.
Note that the material used here consists of glass fibers and of epoxy-resin. The result curves of the K1/K10 ratio for different volume of fibers respectively calculated near the *high* crack tip and near the *low* crack tip show that the results near the two crack-tips are very similar. In the following we focus around the *low* crack tip neighborhood.

On the other hand, for long cracks, the presence of reinforcements and of fiber/matrix interfaces does not increase the composite strength because said composite behaves as a resin (K1/K10 ratio tending towards 1).
This is the opposite for small cracks, where high reinforcement content percentages increase the global composite strength.

2.1.2 Choice of fibers.
If the fibers represent the principal constituent concerning the mechanical strength in a tensile test for example, it is not the same for the tranverse strength. In figures 5a,5b the evolution of the K1/K10 ratio is shown for two composites, one contains glass fiber (Gglass=30 Gpa), the other contains carbon fibers (Gcarbon=20 Gpa) and both use the same resin (Epoxy). To simplify the analysis of the fiber choice, we consider only two composite constituents (reinforcement and matrix). In the figures quoted earlier, the resistance of the two composites to a transverse loading is practically identical. The type of reinforcement used has only slight influence on this property. Our further study will use a glass/epoxy .

3.1.3 Influence of crack length

The composite behaviour is tied not only to the different component properties of the composite, but also to the geometric flaws it contains. When the crack length reaches a non-negligible value, even before reaching the critical propagation length, the decrease of the transverse strength of the material can happen rapidly. Figure 6 in which the fiber/matrix interface has a thickness b/R equal to 0.05 and a fiber percentage of 35%, shows a fast decrease of the strength as soon as the crack moves, and rapidly tends towards an asymptote ($K_1/K_{10} = 1$) as soon as the crack length is equal or superior to the fiber diameter ($l_0/R > 2$) no matter what the fiber/matrix interface shear modulus (G_b). The same conclusion is imposed with a change in the fiber volume percentage.

3.1.4 The influence of mechanical fiber/matrix interface characteristics

The fiber/matrix interface is a transition medium. If its charateristics are well e selected, it leads to the establishment of a zone of "relative stress continuity" by minimizing the fiber/matrix shear. In figure 7 related to fibers percentage equal to 50% (surrounded by an interface of thickness b such as $b/R = 0.1$), for different shear moduli of the fiber/matrix interface (divided by the matrix modulus) (G_b/G_m), we note that the more the interface is "soft", the more the ratio K_1/K_{10} is small; apparently the increase of the interfacesuppleness is a " positive" factor for the transverse load behaviour. In some cases ($G_b/G_m=0.01$) it can decrease the stress intensity factor K_1 of the composite with respect to that of the unreinforced matrix ($K_1/K_{10} < 1$).

It is worth noting that a fiber percentage equal to 20% associated with a shear module interface equal to a tenth of that of the matrix, does not improve the resistance of the composite to the crack propagation, because the composite will behave as if consisting only of matrix (K_1/K_{10} almost constant, and equal to 1). This behaviour ties in with the conclusions of some authors (see /4/,/7/,/8/). Hine/4/ uses a chemical treatment of the carbon fiber surface in order to give an increased suppleness to the fiber/matrix interface. He found (on debonded fibers) that the prescribed treatment had resulted in a "scaling" of fiber before fiber rupture thereby absorbing part of the energy which otherwise would have been destined for crack propagation.

3.1.5 The influence of geometrical fiber/matrix interface characteristics

In addition to its mechanical characteristics, the fiber/matrix interface dimensions also influence the transverse strength of the composite. For different interface thicknesses, figure 8 shows that a non-negligible interface thickness (compared to the fiber radius ($b/R = 0.1$)) permits a decrease of the stress intensity factor, which can become less than that of the unreinforced matrix ($K_1/K_{10} < 1$).
In fact, at the time of utilization, the composite is never submitted to a pure opening mode. Combined actions resulting in interlaminar shear, induce among other things, a debonding of fibers if the fiber/matrix interface is "too" soft or "too" thick.

3.3.6 The influence of fiber volume

In the case of small cracks, an increase of fiber volume leads to an increased value of K_1 .
A rather low percentage of fibers hardly has any effect on the ratio K_1/K_{10} ; it must reach at least 35% to a significantly increase K_1/K_{10} for this composite type (fig.9), and this is only valid in the case of small cracks. In fact as soon as the crack length reaches the fiber diameter ($l_0/R=2$) there is a small variation of K_1 .

CONCLUSION

The theoretical analysis of a fiber embedded in an elastic matrix has shown that the crack moves at a distance r* (depending on the matrix caracteristics) which can be *greater* than the fiber radius, therefore avoiding a debonding of the fiber.
Contrary to a loading parallel to the fiber direction, in the case of transverse loading, the choice of fiber type *hardly* influences the transverse strength, as opposed to its *volume percentage*. In the case of *small cracks*, a fiber volume increase leads to an increase in the stress intensity factor K_1.
This is not the case when the cracks reach a non-negligible length compared to the fiber radius (here $l_0/R=2$). In fact the ratio K_1/K_{10} becames almost *insensitive* to the crack length even for fiber volumes of more than 35% whatever the interface characteristics.
An interface layer of shear modulus *smaller* than that of the matrix surrounding each reinforcement, is shown to improve the strength of the composite compared with an unreinforced matrix. The same tendency is true for a non-negligible thickness of fiber-matrix interface (fig.8).

REFERENCES

1. G.Marom and R.G.C. Arridge "Stress Concentration and Transverse Mode of Failure Composites with Soft Fibre Interlayer". Mat.Sci. Engng, Vol.23, pp.23-32   1976
2. W.G. Knauss and H.K.Muller "Polymer Reinforcement from View of Fracture Mechanics". Caltech Rep.4 1976
3. J. Tirosh, E. Katz and G.Lifschuetz "The Role of Fibrous Reinforcements Well Bonded or Partially Debonded on the Transverse Strength of Composite Materials". Engng. Fract. Mech., Vol.12, pp.267-277 1977
4. P.J. Hine, B. Brew, R.A. Duckett and I.M. Ward "Failure Mechanisms in Carbon Fibre Reinforced Poly(ether ether ketone).II: Material Variables". Composite Science and Technology, Vol.40, pp.47-67 1991
5. R. Barsoum "Finite Element Application in The Fracture Analysis of Composite Materials - Delamination". Key Engineering Materials, Vol.37, pp.35-58. Trans Tech Publications. P.O. Box 10-Hardstr.1". CH-4714 Aedermannsdorf Switzerland 1989
6. N.I. Muskhelishvili "Some Basic Problems in the Mathematical Theory of Elasticity". Noordhoff, Groningen, The Netherlands 1953
7. W.D.Bascom et al. "The Fracture of Epoxy and Elastomer-Modified Epoxy Polymers in Bulk and as Adhesives". Journal of Applied Polymer Science, Vol.19, pp.2545-2562 1975
8. H.Chai "Bond Thickness Effects in Adhesive Joints and its Significance for Mode I Interlaminar Fracture Composites". Composite Materials: Testing and Design (seventh Conference), ASTM STP 893, J.M. Whitney ED., American Society for Testing and Materials, Philadelphia, pp.209-231 1986

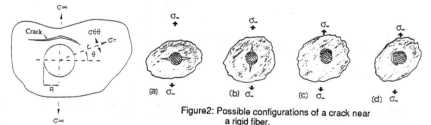

Figure1: Circular fiber in an elastic medium.

Figure2: Possible configurations of a crack near a rigid fiber.

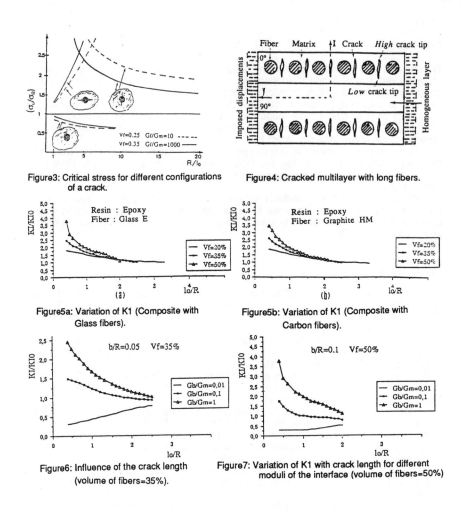

Figure3: Critical stress for different configurations of a crack.

Figure4: Cracked multilayer with long fibers.

Figure5a: Variation of K1 (Composite with Glass fibers).

Figure5b: Variation of K1 (Composite with Carbon fibers).

Figure6: Influence of the crack length (volume of fibers=35%).

Figure7: Variation of K1 with crack length for different moduli of the interface (volume of fibers=50%)

Figure8: Variation of K1 for different thicknesses of the interface.

Figure9: Variation of K1 with crack length for different volumes of fiber.

176

EVALUATION OF THERMO-ELASTIC CONSTANTS OF MATERIALS USED AS INTERPHASES IN TITANIUM MATRIX COMPOSITES

K. DEBRAY, E. MARTIN, B. COUTAND, J.M. QUENISSET, R. HILLEL*

Laboratoire de Génie Mécanique - IUT "A" Domaine Universitaire
33405 Talence Cedex - France
*IMP - 66860 Perpignan - France

ABSTRACT

Titanium matrix composites reinforced by SiC filaments are good candidates for aeronautic applications. Decreasing their sensitivity to chemical incompatibility between fibre and matrix implies the interposition of intermetallic or semi-metallic compounds. Thermo-elastic constants of these interphases must be evaluated to optimize the conditions of load transfer between fibre and matrix. In the present work, an experimental procedure has been developped to estimate Young's moduli and thermal expansion coefficients of thin coatings processed by CVD on graphite substrates.

INTRODUCTION

Significant benefits in term of specific strength, stiffness and creep resistance can be expected by designers from titanium alloy matrix composites reinforced by SiC filaments. Their fields of application can be mainly found in aircraft and aerospace industries (structure, turbine engine). However the extent of these domains strongly depend on the reliability of this kind of composites after high temperature exposure and thermomechanical fatigue [1].

Nevertheless, the chemical incompatibility between SiC filaments and titanium alloy matrices gives rise to the formation of interfacial zones consisting of brittle phases like titanium carbide (TiC) and titanium silicides (TiSi$_2$, TiSi, Ti$_5$Si$_4$ Ti$_5$Si$_3$) [2]. As a consequence, the reinforcement is partly damaged by chemical interaction, the conditions of load transfer are modified and early fractures of the brittle interfacial zone lead to drastic drops in mechanical performance. These phenomena are diffusion dependent and therefore increase with rising exposure temperature and time.

Coating the filament surface is able to prevent extensive chemical reactions between filament and matrix. Oxide coatings (MgO, Y$_2$O$_3$) act as diffusion barriers but their efficiency is insufficient and their brittleness induces a reduction in the mechanical properties of the composites. An adequate sequence of interphases located between filament and matrix in order to promote the thermodynamical stability of the composite system may be a preferable choice. Titanium silicides like intermetallic compounds $TiSi_2$ and Ti_5Si_3 have been shown to be good candidates for this purpose /3/.

In addition to the required thermodynamical stability of the filament/matrix system, optimization of the thermomecanical coupling of the fibre/matrix interface must be achieved to control the conditions of damage and failure of the composites. Thermomechanical models are now available to predict the influence of the interphase nature on the mechanical behaviour on the interfacial zone /4,5/. Use of these models implies the knowledge of the thermo-elastic constants and fracture energies of the interfacial zone components in order to select the appropriate combination and thickness that will minimize the effects of thermal expansion mismatch and catastrophic microcrack propagations.

This paper presents an experimental method that allows the evaluation of Young's moduli and coefficient of thermal expansion (CTE) of coatings like intermetallic compounds to be used as interphases in a SiC/Ti composites. The coatings are processed by chemical vapour deposition on a graphite substrate with a thickness in the range of 150 - 200 µm.

I.   EXPERIMENTAL METHODS

Due to the interest of the mechanical behaviour of layered structures in the use of micromechanical devices, many different test methods have been suggested and evaluated for the thermomechanical characterization of thin coatings /6/. Experimental techniques for measuring elastic moduli of thin films on substrate or free standing films include microtensile tests, the bulge test, the microindentation test, the resonnance methods and tests based on the measurement of elastic wave speeds. Evaluation of radius of curvature of a layered substrate can provide the thermal stresses or the coefficient of thermal expansion of the coating film /7/.

The experimental method proposed here allows the evaluation of the coating Young's moduli with the help of a three points bending test. Furthemore, a layered beam tested with a cantilever configuration and submitted to a thermal loading is used to estimate the CTE of the coating.

1.1.   Three points bending test

The testing fixture is schematically shown in Fig.1. The tests are performed with a testing machine operating with a crosshead speed of 0.1 mm/min. The resolution of the load cell and the capacitive displacement transducer are respectively of 0.01 N and 0.25 µm. Great care is taken to ensure alignement of the whole testing system.

The slope of the load-deflection curve is used to evaluate the compliance of the composite beam. Knowing the dimensions (L, b, h) and the Young's modulus E_S of the substrate and the thickness of the coating, the Young's modulus E_c of the coating can be easily computed in the case of a beam symmetrically coated on both sides (Fig. 3a). In the case of specimens only coated on one side (Fig. 3b), deriving E_c becomes more complex and requires the use of an iterative procedure. To take into account the geometrical irregularities of the coating related to the elaboration process, a numerical model was developped by discretizing the composite beam as an assembly of coated beams with linearly interpolated thickness (Fig. 3c). The influence of shear stresses is also included in the analytical relations used to evaluate the beam deflection.

Fig. 4 shows the predicted change in the beam compliance as a function of the coating Young's modulus for various thickness values of the films. The plotted values represent the compliance deviations between uncoated and coated graphite substrates whose dimensions are b= 4.5 mm, L= 22 mm, h= 1.3 mm with a graphite Young's modulus of 9.8 GPa. In fact, these deviations are indicative of the specimen compliance sensitivity to the stiff coatings.

1.2. Cantilever test

Fig.2 schematically illustrates the experimental set-up used for measuring the bending curvature radius. A one side coated specimen is mounted on a clamp and submitted to successive temperature increases ΔT. The deflection δ at the free end of the specimen which is induced by thermal stresses is then measured continuously. A thermocouple located on the test fixture is used to monitor the temperature. Knowing the dimensions (length L, width b, thickness h), the Young's modulus E_s and the CTE α_s of the substrate, the average thickness e and the Young's modulus E_c of the coating and assuming that the thermo-elastic constants of the beam components are independent of temperature, the CTE α_c of the deposit is given by :

$$\alpha_s - \alpha_c = \frac{\dfrac{E_s}{E_c}.e^4 + \dfrac{E_c}{E_s}.h^4 + 2eh.(2e^2 + 2h^2 + 3eh)}{3L^2 eh.(h+e)\dfrac{(\Delta T)}{\delta}}$$

1.3. Validation of the experimental methods

Specimens made of a 25 μm or a 100 μm thick copper sheet glued on one side of a graphite substrate were used to validate the experimental methods. Care was taken to perform at the copper/graphite interface an epoxy layer with a thickness lower than 1 μm. During the three points bending test, typical loads of 3 N were applied and typical specimen deflections of 20 μm were recorded. The slopes of the load-deflection curves were estimated from an average value derived from three loading tests. Each specimen was also submitted to a temperature range between room temperature and 130 °C in order to determine the deposit CTE through bending curvature measurement. As indicated in Fig. 5, dwell times of half an hour were imposed to each steps of temperature increase, to insure a sufficient thermal stabilization of the testing set-up. The CTE of the graphite substrate was measured by a dilatometer and evaluated as $\alpha_s = 5 \ 10^{-6} \ °C^{-1}$.

Table 1 presents the measured values of Young's modulus and CTE of copper sheets. The results have been found in sufficient agreement with the values indicated elsewhere /8/ to consider the procedure and device as validated.

II. COATING CHARACTERIZATION

The procedures of characterization previously described and validated were applied to various types of coating processed by CVD on a thick graphite substrate heated by Joule effect and placed in a cold wall reactor. Various coatings reported in table 2 were obtained from a gazeous mixture of $TiCl_4$, C_4H_{10}, SiH_2Cl_2 in H_2 depending on the concentrations of the different species. Deposition and codeposition times of about half an hour were necessary to achieve coating thickness in the range of 150 - 200 μm. The related Young moduli and CTE values derived from testings are reported in table 2. Si and C elemental compositions were determined by electron probe microanalysis-wavelength dispersive spectrometry.

The measured Young's modulus of the silicon coating is rather low compared to the value of 164 GPa given in ref /9/. Higher inaccuracy is here induced by important geometrical irregularities produced by the coating polishing. Silicon carbide coating exhibits Young's modulus value close to 420 GPa as given in ref /9/. The deviation in the Young's modulus of these coatings as a function of the increasing value of the carbon content shows a sharp decrease. The same tendency was pointed out by ref /10/.

CONCLUSION

An experimental procedure was developped to evaluate Young's modulus and thermal expansion coefficients of coatings. The method is based on a three points bending test and on radius curvature measurements of thin films deposited on graphite substrates. The procedure was validated by testing thin copper sheets glued on the substrate. Various kinds of CVD coatings have been characterized. This experimental procedure will be used to investigate the thermo-elastic properties of intermetallic compounds.

ACKNOWLEDGEMENTS

The authors are indebted to SNECMA and MRE (Ministère de la Recherche et de l'Espace) for their financial supports and to JF.Sylvain from the Laboratoire de Chimie du Solide du C.N.R.S for his help in the specimen preparation.

REFERENCES

1   R. WERRON, 9th RISO Int. Symp. on Metal. and Mater. Sci., (1988) 233-235
2   P. MARTINEAU, R. PAILLER, M. LAHAYE, R. NASLAIN, J. Mater. Sci., 19 (1984) 2749.
3   K. BILBA, S.P.MANAUD, Y. LE PETITCORPS AND J.M. QUENISSET, Mater. Sci. and Eng. A (1991) 135-141.
4   Y. MIKATA, M. TAYA , J. Comp. Mat., 19 (1985) 554-578.
5   E. MARTIN, N. PIQUENOT, J.M. QUENISSET, K. DEBRAY AND A. DEGUEIL, to be published in Composites.
6   W.D. NIX, Metallurgical Transaction A, 20 A (1989) 2217.
7   T.F. RETAJCZYK, A.K. SINHA, Thin Solid Films, 70 (1980) 241-247.
8   T. SUGA, G. ELSSNER, S. SCHMAUDER, J. Comp. Mat. 22 (1988) 917-934.
9   E. SCAFE, G. GIUNTA, L. DI RESE, F. PETRUCCI, G. DE PORTU, Proceedings of 2nd European Cer. Soc. Conf., Augsburg, Germany, Sept. 91.
10   M. SASAKI Proceedings of the First Int. Symp., FGM, Sendai, (1990) 83-88.

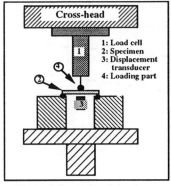 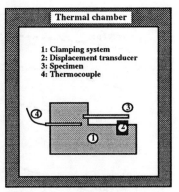

Fig 1: Schematic of the three points bending device.

Fig 2: Schematic of the curvature measurement device.

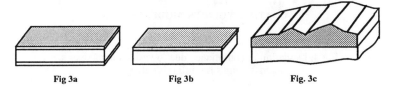

Fig 3a   Fig 3b   Fig. 3c

Fig 3: Schematic of the tested specimen:
(a) double coating, (b) simple coating, (c) actual coating surface.

Sample	Coating thickness (μm)	Young's modulus E (GPa)	CTE α $(10^{-6}/°C)$
n°1	25	107 ±10	20.5 ±2
n°2	100	127 ±12	-
n°3	100	114 ±11	17.5 ±1.7
Ref /8/	-	129	17

Table 1 : Characterization of copper sheets.

Sample	average thickness of the coating(s) (μm)	Young's modulus E (GPa)	CTE α $(10^{-6}/°C)$
Si	148	120 ±10	3.3 ±3
SiC	273 / 177	420 ±40	-
SiC + C (80%, 20%)	272 / 255	154 ±15	-
SiC + C (59%, 41%)	171	78±8	-

Table 2 : Characterization of CVD coatings.

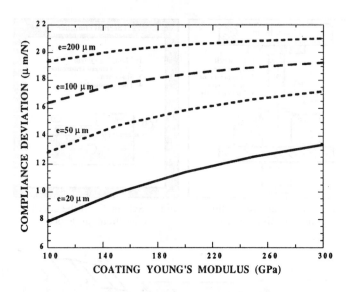

Fig 4: Deviation in compliance due to various thicknesses
and Young's moduli of coating.

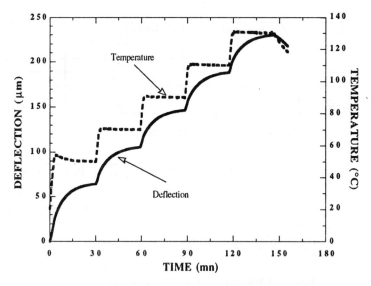

Fig 5: Procedure of thermal expansion coefficient measurement
for a graphite beam layered with a copper sheet.

DESIGN AND NUMERICAL MODELLING

FACTORS INFLUENCING THE SURFACE QUALITY OF DRILLED HOLES IN CARBON FIBRE REINFORCED POLYMER

A.R. CHAMBERS

University of Southampton - Dept of Engineering Materials
Highfield - Southampton SO9 5NH - UK

The aim of this research was to study the parameters which affect the surface quality and fatigue properties of drilled carbon/epoxy and carbon/BMI (bismaleimide). Increasing feed rate increased the cutting forces and tendency for delamination and fibre burst. Low feed rates generated high temperatures and thermally associated damage. Carbon/epoxy was more prone to these effects than carbon/BMI. Limited fatigue tests indicate that surface quality plays an important role in fatigue initiation. In this respect damage associated with high forces is more serious than that caused by high temperatures.

1. INTRODUCTION

Although in terms of bulk material removal, machining is less important with composites than with metals, it still causes significant problems for the composite manufacturing industry. In addition to the problems of tool wear and machining costs, it is very difficult to achieve the quality of surface needed for the accurate assembly of panels and structures. It is also important that the machined surface is not damaged as this may contribute to the initiation of a fatigue failure.

The major problems when machining composites are delamination, fibre fracture, fibre tearing, pull out and thermal damage. Clearly the incidences of any of these defects will detract from the surface quality and should be avoided wherever possible. Research has shown that for any given composite, the material damage is a function of the cutting parameters, tool geometry and tool material (1). Abrasion resistant tool materials such as polycrystalline diamond and cemented carbide are recommended for wear resistance and special geometries have been developed for cutting ductile aramid fibre composites.

The objective of this research were to study the factors which influence the surface quality of drilled holes in carbon/epoxy and carbon/BMI composites and to assess the effect of drilling upon the fatigue performance.

2.   EXPERIMENTAL

The composites used in this investigation were:-

> carbon/epoxy
> carbon/BMI (bismaleimide)
> carbon/BMI post-cured

With each composite, the reinforcement was a woven fabric with a 0/90 lay-up.

Drilling was selected for this study because it is one of the most important machining operation encountered with these materials. Drilling was performed using high speed pneumatic drills running at 23000rpm and 8500 rpm (Desoutter AFD40) with feed rate variable between 1.5 and 26mm/sec. The drills used were standard 6mmϕ 118° twist drills manufactured in HSS and cemented carbide. No coolant was used during machining.

Thrust and torque were measured using a Kistler Instrument piezo-electric quartz dynamometer (model 9257A). Temperature in the carbon/epoxy composite as a result of machining was measured by embedding thermocouples in the composite lay-up. Surface quality was assessed using optical and advanced microscopic techniques.

Fatigue testing under bending was conducted on a limited number of test pieces which were drilled with what was considered to be high quality and poor quality holes. The samples were fatigued for a fixed number of cycles prior to tensile testing. The residual strength provided a measure of fatigue damage.

3.   RESULTS AND DISCUSSION

3.1.   The Effect of Feed Rate upon the Cutting Forces.

The effect of feed rate upon the cutting forces when machining carbon BMI, carbon BMI (post-cured) and carbon epoxy with HSS and cemented carbides is shown in Figure 1. As increasing feed rate increases the cut taken in a revolution, the increase in force and torque (not shown) with feed rate was expected.

With both types of tool material, the force values were dependent upon the composite. The order of composites in terms of increasing force was carbon/epoxy, carbon/BMI and carbon/BMI post-cured. This is the same as that of the composites ordered in terms of increasing strength, stiffness and maximum service temperature. The differences in forces were greatest at low feed rates where due to frictional heating the temperature rises in the machined composite were the highest. This suggests that the temperatures generated in machining and the response of the resin system to those temperatures play an important part in defining the machinability of the composite.

The major difference between HSS and cemented carbide is in the magnitude of the forces with those of the cemented carbide being significantly lower, (Figure 1). As the tool geometries were identical, the force differences must result from differences in either the wear, frictional or thermal properties of the tool materials.

The effect of wear upon the forces and temperatures generated 1mm from the edge of holes in carbon/epoxy is shown in Figure 2. It can be seen that after 5 holes (10mm drilled depth)the forces with HSS had increased by 50% and the temperature in the composite increased by 70°C. Five drilled holes with cemented carbide gave no significant increase in either the force or temperature. Although the lower friction and thermal conductivity of cemented carbides could result in lower forces it is believed that the effect of wear is more significant.

3.2. Surface Quality

For all the composites, the quality of the hole on entry and exit and the surface within the hole is dependent upon the feed rate, drill material and drill wear. High feed rates increased the tendency for delamination and burst fibres on exit. This is consistent with the higher forces associated with the high feed rates. Low feed rates resulted in visible heat damage and resin burn off. Typical micrographs showing the different types of damage are given in Figures 3 and 4.

In terms of entry and exit damage, carbon/epoxy suffered more severe damage than carbon/BMI. For drills in good condition there was no significant difference between carbon/BMI and carbon/BMI post-cured.

The differences between the composites can be explained in terms of their properties. The loss of support as the matrix softens results in increased fibre bursting as the fibres are pushed out of the composite by the drill. This occurs more readily with the more temperature sensitive carbon/epoxy.

3.3. The Effect of Hole Quality on Fatigue

The effect of fatigue testing on the residual tensile strength of carbon/epoxy drilled at cutting speeds of 47 and 160 m/min at a feed rate of 1.5 mm.rev is given in Figure 5. It can be seen that at both speeds the residual strength reduces with increasing fatigue cycles. It can also be seen that the fatigue damage is worse with the low cutting speed.

The major effect of increasing the cutting speed at this feed rate was to reduce the forces and increase the temperature in the composite. It would appear from a fatigue standpoint that the damage dome to by high forces, outweighs that done by high temperatures.

4. Conclusions

1.   Increasing the feed rate increased the cutting forces with carbon/epoxy, carbon/BMI, carbon/BMI post-cured. The ranking order of composites in terms of force was the same as that of strength and stiffness.

2.   High cutting forces resulting from either high feed rates or drilling may result in delamination and fibre burst. Low feed rates generate high temperature and thermal damages to the hole surface.

3.   Carbon/BMI generally gives a better quality hole than carbon/epoxy for drills in good condition.

4.   Preliminary results indicate that surface quality influences fatigue initiation. Damage associated with high feed rates is more serious than that caused by low feed rates.

References

1.   S.J. Jamil and A.R. Chambers, "*Evaluation of Surface Quality of Drilled Holes in Composite Materials After High Speed Drilling*", Proc. Advanced Machining for Quality and Productivity, (1991), York, U.K., Institute of Metals.

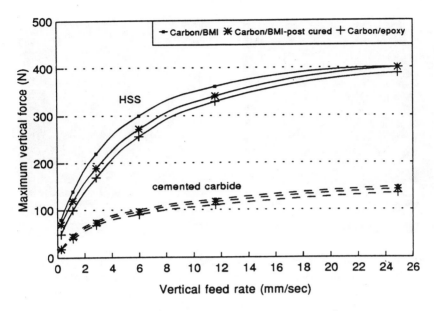

Figure 1   The effect of feed rate upon the cutting force.

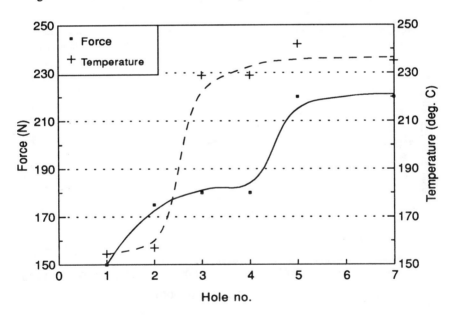

Figure 2   Increase in force and temperature when machining carbon/epoxy with HSS.

Figure 3   Resin burn-off with carbon/epoxy at 1.5mm/sec feed rate

Figure 4   F i b r e   b u r s t with carbon/epoxy at 20mm/sec feed rate

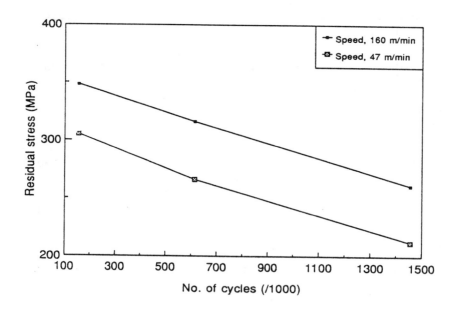

Figure 5   The effect of fatigue on residual strength of drilled carbon/epoxy.

DESIGN METHODOLOGY FOR A DAMAGE TOLERANT HAT-STIFFENED COMPOSITE PANEL

J.F.M. WIGGENRAAD

National Aerospace Laboratory - PO Box 90502
1006 BM Amsterdam - The Netherlands

ABSTRACT

A design methodology for a damage tolerant hat-stiffened panel was formulated, based on a design criterion and a damage tolerant design concept, and a panel was designed accordingly. Three stages in the failure process of a damage tolerant panel are recognized: initial failure, damage growth, and final failure, and this failure process was determined experimentally. The panel failure process was modelled based on a reduced stiffness approach.

1. INTRODUCTION

The design process of a stiffened wing panel made of composite material consists of finding a minimum weight configuration that satisfies certain constraints. In order to design a damage tolerant panel, the effect of the presence of a specified amount of damage on the (static) strength of the panel must be taken into account. A methodology to design a damage tolerant hat-stiffened panel is formulated and evaluated below.

2. DESIGN CRITERION

The amount of impact energy to be considered for the evaluation of the damage tolerance of a particular panel design is often related to Barely Visible Impact Damage, with a maximum of 136 J /1/, or specified by impact threat scenario's, as proposed by Whitehead /2/. In the study presented here, the design criterion for damage tolerance is formulated as follows. If the damage remains invisible up to Design Limit Load (DLL), it may not be discovered in service, hence, the damaged panel should not fail below Design Ultimate Load (DUL). If the damage in the panel grows and becomes visible below DLL, it can be discovered in time, and be repaired. In this case the damaged panel should not fail below DLL.

3. DAMAGE TOLERANT DESIGN

Two important characteristics of a damage tolerant panel are the ability to redistribute load, and the tendency to retard damage propagation. A damage tolerant panel design is often a multi-load path structure, with discontinuities incorporated to isolate damage (reducing "stress concentrations"). It was observed by Whitehead, that during static compression tests of impact damaged built-up structures, failure, in many cases, occurs in stages. At a certain applied load, local failure initiates at the damage site, followed by rapid damage propagation to a nearby load path, where it is positively arrested. After increased loading, final structural failure occurs. A similar observation was made in /3/.

4. DESIGN ANALYSIS

Design calculations are often made for damaged panels, with the damage represented by a region with reduced stiffness. The stiffness reduction factor that must be applied is difficult to obtain, and damage propagation can not be simulated, unless a reliable experimental database is established. Such a database is configuration dependent. Whitehead predicted *initial failure* using a semi-empirical elastic stiffness reduction formula combining energy level, laminate lay-up, laminate thickness, material toughness, support conditions, and impactor size. He recognized the influence of structural configuration on *damage propagation* and *final failure*. In his study, post-impact strength after damage propagation was determined by assuming that the impact damaged area between stiffeners was totally ineffective, causing a strain concentration in the adjacent structure, leading to final failure. In the study presented here, the reduced stiffness concept is evaluated for a hat-stiffened composite panel.

5. DESIGN CONCEPT

The design concept evaluated in this study is known as the soft-skin concept. It is a multi-load path structure as shown in figure 1. The panel is divided in high stiffness zones and zones with low stiffness. Damage tolerance of this concept is achieved when the damage does not propagate from one load path to the next, which is the case when the load redistribution due to damage does not lead to the occurrence of high stress concentrations. Ultimate damage tolerance corresponds to a stress concentration factor equal to 1.0, so the failure *load* drops in proportion to the loss of axial stiffness caused by the elimination of the damaged load path, while the failure *strain* remains the same as the failure strain of an undamaged panel.

6. PANEL DESIGN

The panel configuration used for this study was designed and analysed using optimization codes developed at NLR. With PANOPT /4/, a panel can be optimized to carry loads into the postbuckling region, based on the calculation of the initial postbuckling stiffness. The panel designed for this study was constrained to buckle at a gross strain of 0.003 or more, and to reach a gross strain at DUL of no more than 0.006. The constraints imposed on the design of the panel were limited in scope, which resulted in a rather extreme design. The distribution of axial stiffness is shown in figure 2.

7. EXPERIMENTS

Small, component size, "concept evaluation" specimens were tested first: three specimens representing skin zones, and six specimens representing short stiffener columns. Subsequently, three 4-stiffener panels were tested in compression, one without damage, the other two with damage either in a skin zone or in a stiffener zone.

7.1. Skin specimen tests

The specimens representing the skin zones between two stiffeners included the integral stiffener flanges on both sides, and were clamped at the sides, to represent the rotational restraint normally provided by the stiffeners. In figure 3 the results are shown of the compression strength tests on the skin zones. The impact energy level resulting in Barely Visible Impact Damage (BVID) for the skin (laminate thickness of 1 mm) corresponded to 7 J. The gross strain at failure did not drop below 0.006. Hence, impact damage in a skin area was not considered to be critical for the design.

7.2. Stiffener column tests

The stiffener column specimens included the flanges on both sides, and were clamped at the loaded ends. Two undamaged specimens reached failure gross strains of more than 0.006. Four specimens were impacted, at 7 or 15 J, and either at the skin-web intersection or at the stiffener centerline, where the laminate thickness is 3.12 mm. As shown in figure 4, impacts with 15 J resulted in reductions of the failure gross strain to values as low as 0.0035.

7.3. 4-Stiffener panel tests

The 4-stiffener panels (572 x 990 mm) had clamped loaded ends, while the sides remained free. One panel was loaded to failure in compression without damage. Local skin buckling occurred at a gross strain level of 0.0032, and failure was induced by a secondary global "Euler" mode. The panel failed at a load of 680 kN, and a gross strain level of 0.0063. Failure occurred near one of the clamped ends, induced by the high bending strains at the stiffener tops. The second panel was impacted at the center of a skin bay with 10J, resulting in a C-scan damage area of 100 mm². During the post-impact compression test the damage area increased to 1600 mm², but did not affect the adjacent stiffener zones. The panel behaved identical to the undamaged panel tested earlier, and buckled and failed at the same loads and strains.

The third panel was impacted with 15 J at the skin underneath a stiffener web. The resulting C-scan area was equal to 2400 mm², much larger than the 1100 mm^2 resulting from the corresponding impact applied to one of the short column specimens. The panel was loaded in compression to failure, which took place in three stages. First, the damaged stiffener base fractured at a gross strain of 0.0030, much earlier than the 0.0041 observed for the corresponding short column specimen. The C-scan damage area had increased to 6000 mm^2, and the damage had propagated into the adjacent skin bays. Increasing the compressive load, the webs and top of the damaged stiffener failed at a gross strain of 0.0041. Finally, the entire panel failed at a load of 523 kN, or 77 % of the undamaged strength (hence above DLL, which is equal to 67 % of DUL), and at a gross strain of 0.0063, equal to that of the undamaged panel. A loss of axial panel stiffness of 9 % was observed when the stiffener base failed, which increased to 17 % until the entire stiffener broke. After the stiffener broke, the panel stiffness reduction was 24 %, and increased even more until final failure, but this increase was partly due to overall bending of the panel.

8. REDUCED STIFFNESS APPROACH

From figure 2 it appears that one stiffener base (bottom and flanges) accounts for 15 % of the total 4-stiffener panel stiffness. The test on the panel with stiffener base damage revealed a stiffness loss of up to 17 % until the stiffener top broke. After fracture of the stiffener top the reduction of the panel stiffness observed in the test was 24 %, equal to the theoretical stiffness loss caused by the elimination of one stiffener zone. Hence, modelling the panel stiffness loss by eliminating the appropriate elements is a practical approach. It is concluded that the performance of the panel can be computed fairly well using the reduced stiffness approach to model the panel in a state of damage, as shown in figure 5.

9. EVALUATION

The damage tolerance of the panel design obtained in this investigation was demonstrated experimentally to be adequate according to the criterion formulated in Chapter 1. If Barely Visible Impact Damage (BVID) is applied to the skin zone of a panel, it remains "invisible" up to at least DLL, and the panel does not fail below DUL. If BVID is applied to the stiffener zone of a panel, it becomes visible below DLL (as the stiffener base fractures obviously), but the panel does not fail below DLL.

To determine the damage tolerance of a panel design by analysis, it is essential to predict the initial failure load (or strain), the damage propagation mode, hence the stiffness reduction and the resulting stress concentration in the adjacent structure causing final failure.

9.1. Initial failure

An effort was made to determine the initial local failure load with small "concept evaluation" specimens. Tests on the skin specimens indicated that damage in the skin zones does not propagate below DUL, which is correct. Tests on stiffener columns indicated that damage in these high load carrying regions would start to grow at 0.0041, which is not correct, as it started to grow at 0.0030 in the full panel. This discrepancy is caused by the larger damage area in the panel compared to that in the stiffener column, impacted at the same energy level. Apparently, the support conditions during impact of the concept evaluation specimens deviated too much from the corresponding conditions in the panel.

9.2. Damage propagation

The failure mode consisted of the collapse of the damaged stiffener base (first load path), followed by the collapse of the entire stiffener (second load path in the stiffener top), but apart from some growth into the adjacent low stiffness skin zones, the remaining part of the panel was not affected by damage growth until total failure occurred. The stiffness distribution as shown in figure 2 makes this scenario likely and predictable, hence the stiffness reduction approach can be applied by the designer before the outcome of verification tests reveals the actual behavior of a panel.

9.3. Final failure

The "stress concentration factor" in the structure adjacent to the damaged zone was not yet determined analytically. From the test results it is apparent that there is indeed a "stress concentration factor" larger than 1.0 within a stiffener column, as the second load path, the stiffener top, fails below 0.006, i.e., the gross strain at DUL. However, the stress concentration factor corresponding to the load redistribution between stiffener zones is 1.0, as the remaining part of the panel failed at the failure strain of the undamaged panel.

10. CONCLUSIONS

The soft-skin panel concept showed excellent damage tolerance properties. The "saw-tooth" failure behavior of a damaged panel can be predicted using the reduced stiffness approach, if the strains at which initial failures occur are determined first. Impact damage applied to concept evaluation specimens should closely represent the damage in actual structures. Damage propagation modes should be predictable, so they can be modelled for design purposes. This subject will be studied subsequently, using simplified Structure Relevant specimens.

REFERENCES

1. Demuts, E. "Barely visible damage threshold in graphite epoxy", 8th International Conference on Composite Matarials, ICCM VIII, Honolulu, 1991
9. Whitehead, R.S. & Kan, H.P., "Fatigue of aircraft materials", proceedings of the specialists' conference dedicated to the 65th birthday of J. Schijve, Delft, 1992
3. Labonté, S. & Wiggenraad, J.F.M., "Development of a Structure Relevant specimen for damage tolerance studies", 9th International Conference on Composite Materials (ICCM-9), Madrid, Spain, 12-16 July, 1993
4. Arendsen, P., & Wiggenraad, J.F.M., "PANOPT Users Manual", NLR CR 91255 L, 1991

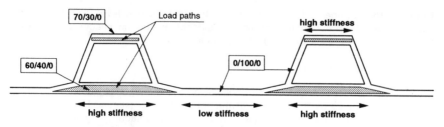

Fig. 1 Panel design concept

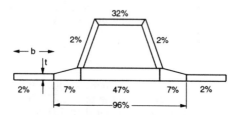

Fig. 2 Axial stiffness distribution Etb

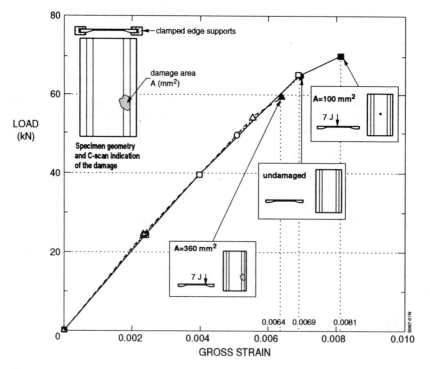

Fig. 3 Compression after impact test results for skin specimens

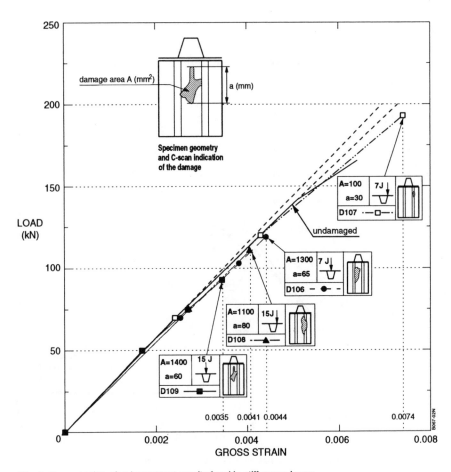

Fig. 4   Compression after impact test results for skin-stiffener columns

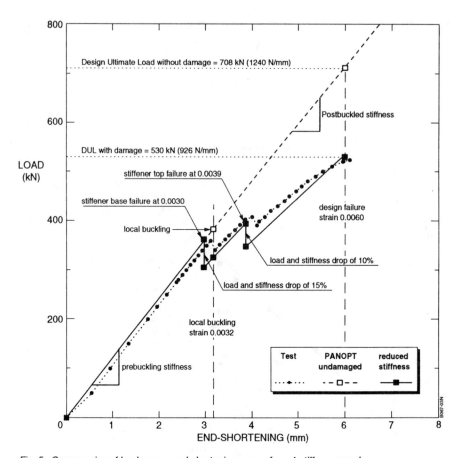

Fig. 5   Compression of load versus end-shortening curves for a 4-stiffener panel

THE ASYMPTOTICAL MODELLING OF THE DYNAMIC PROCESS IN THE COMPOSITE PLATE STRUCTURES

D. ZAKHAROV, I. SIMONOV

Institute for Problems in Mechanics - 101 Vernadskogo Avenue
117526 Moscow - Russia

ABSTRACT

For two groups of thin laminates the internal stress-strain state is investigated here in the long-wave approximation:
1. Arbitrary posed N plies with general anisotropy.
2. Transversely anisotropic plies with large difference in the properties.
The new asymptotic theories are deduced and classified, both separate and coupled bending and in-plane problems are considered. Then the analysis of engineering hypotheses is based on. The boundary conditions and the method of solving the static problems are suggested.

INTRODUCTION

In spite of technological tendency to deal with symmetric constructive elements the non-symmetric structures themselves are very interesting. The non-symmetry appears in delaminated sandwiches and optimal design of structures with a wide variety of isotropic layers.

Statics of anisotropic laminates of arbitrary structure was partially treated both by engineers (for example, N.A. Alfoutov and al., 1984) and mathematics (D. Caillerie, 1982). Below the regular dynamic generalization of Kirchhoff theory of plates is given.

Some results for plates of contrast isotropic plies (alternatively high and low modulus layers) were given by M.I. Gussein-Zade (1968) and V.V. Bolotin and al. (1980). Here the dynamic bending is investigated in general statement.

Methods based on the engineering hypotheses (S.G. Lekhnitski, 1963; R.M. Christensen, 1979) and on the asymptotical integration of 3-D equations of elasticity (K.F.Friedrichs and al., 1961; A.L. Goldenweiser and al., 1962) are typical to deduce the 2-D mathematical models for thin-walled structures. The asymptotic approach involves more complex considerations, but permits us to determine the limits when hypotheses are available and to calculate all the stresses. General problem to approximate the main stress-strain state is divided into internal and boundary layer effects.

1. ASYMMETRIC LAMINATE OF N ANISOTROPIC PLIES
1.1. General dynamic equations

Let a plate occupy a region $(\mathbf{x}, z \equiv x_3) \in \Omega \times [z_1, z_{N+1}] \subset R^3$ and consist of N anisotropic layers with full contact. Each ply is supposed to have a plane of elastic symmetry, parallel the plane $\mathbf{x} \equiv (x_1, x_2)$, and its thickness, mass density and matrix of stiffnesses are h_i, ρ_i, $G_i = \|g_{pq}\|_i$ respectively. The relation $\varepsilon = h/l$ is introduced now as a unique small parameter for the whole sandwich, where $2h$ is a thickness and l is a scale of process in the longitudinal plane $\mathbf{x}$ with the characteristic time $t_0 = O(\varepsilon^{-1})$. The plate is charged by the exterior surface loading, independent ε, regular with sufficiently weak variation

$$\sigma_{33}^{\pm}(\mathbf{x}, t) = \sigma^{\pm}, \qquad \sigma_{\beta 3}^{\pm}(\mathbf{x}, t) = \tau_{\beta}^{\pm} \qquad (z^{\pm} = z_{N+1}, z_1) \tag{1}$$

By asymptotic integration of 3-D equations of dynamic elasticity we deduce the consistent expressions for the displacements $\mathbf{U}, U_3$ which coincide with classical theory of plates. The displacements are independent on ply number i. Below we restrict the relative accuracy with members $O(\varepsilon^2)$, the indexes 0, 1 correspond to the normal and tangential loads. Then, on introducing the membrane $(k = 1)$, membrane-bending $(k = 2)$ and bending $(k = 3)$ rigidities d_{pq}^k and main operators the final equations result in the form (to sum with repeating greek indices $\alpha, \beta = 1, 2$)

$$U_3 = l\varepsilon^{-3}(W_0 + \varepsilon W_1 + \cdots), \quad \mathbf{U} = l\varepsilon^{-2}(\mathbf{V}_0 + \varepsilon \mathbf{V}_1 + \cdots), \quad \mathbf{V} \equiv (V_1, V_2)$$
$$l\varepsilon^{-3}(W_0 + \varepsilon W_1) = w(\mathbf{x}, t), \quad l\varepsilon^{-2}(\mathbf{V}_0 + \varepsilon \mathbf{V}_1) = \mathbf{u}(\mathbf{x}, t) - z\,\mathrm{grad}\,w, \tag{2}$$

$$\mathbf{a}_\beta(d_{pq}^1)\mathbf{u} - b_\beta(d_{pq}^2)w = (\tau_\beta^- - \tau_\beta^+), \quad \rho = \sum_i h_i \rho_i / 2h \tag{3}$$

$$\{2h\rho\partial_t^2 + b(d_{pq}^3)\}w - \mathbf{a}(d_{pq}^2)\mathbf{u} = (\sigma^+ - \sigma^-) + \mathrm{div}(z^+ \boldsymbol{\tau}^+ - z^- \boldsymbol{\tau}^-) \tag{4}$$

$$\mathbf{a}_1(\gamma_{pq}) \equiv \mathbf{i}_1[\gamma_{11}\partial_1^2 + 2\gamma_{16}\partial_{12}^2 + \gamma_{66}\partial_2^2] + \mathbf{i}_2[\gamma_{16}\partial_1^2 + (\gamma_{12} + \gamma_{66})\partial_{12}^2 + \gamma_{26}\partial_2^2] \tag{5}$$

$$b_\beta \equiv \mathbf{a}_\beta\,\mathrm{grad}, \quad \mathbf{a} \equiv \partial_\beta \mathbf{a}_\beta = \mathbf{i}_\beta b_\beta, \quad b \equiv \partial_\beta b_\beta = \mathbf{a}\,\mathrm{grad} \quad (1 \leftrightarrow 2)$$

$$d_{pq}^k = \frac{1}{k} \sum_i (z_{i+1}^k - z_i^k)\gamma_{pq}^i, \quad \gamma_{pq}^i = (g_{pq} - g_{p3}g_{3q}g_{33}^{-1})_i, \quad (pq = 11, 12, 22, 16, 66, 26)$$

$$e_{33} = e_{\beta 3} \equiv 0, \quad e_{\alpha\beta} = \varepsilon_{\alpha\beta} + z\kappa_{\alpha\beta}, \quad \varepsilon_{\alpha\beta} = \frac{1}{2}(\partial_\alpha u_\beta + \partial_\beta u_\alpha), \quad \kappa_{\alpha\beta} = -\partial_{\alpha\beta}^2 w$$

The quasistatic Equations 3 resemble a system for the generalized plane stress state (P), the dynamic Equation 4 looks like the bending (B) one in Kirchhoff theory. The influence of an arbitrary plate structure appears in the membrane-bending operators b_β, $\mathbf{a}(d_{pq}^2)$, i.e., the System 3, 4 is mixed.

1.2. Classification of coupled bending and in-plane problems

Generally the in-plane deformations $\varepsilon_{\alpha\beta}$ are no zeros at every longitudinal plane. So, the first hypothesis in Kirchhoff theory is not available because of the absence of neutral plane. As a criterion of connection between B- and P-problems the integral norm of operators b_β, $\mathbf{a}(d_{pq}^2)$ has to be introduced. It is determined by a modulus of the sum of two equivalent four-component membrane-bending vectors. Its minimum ξ gives a fixed position of plane $\mathbf{x}$ ($z = 0$) which coincides with neutral plane for special cases.

$$b_1(d_{pq}^2) \sim \mathbf{B}_1 \equiv \left(d_{11}^2,\, 3d_{16}^2,\, d_{12}^2 + 2d_{66}^2,\, d_{26}^2\right), \quad B(z^-) \equiv \left(|\mathbf{B}_1|^2 + |\mathbf{B}_2|^2\right)^{1/2} \geq 0$$

$$z^- \equiv \zeta: \ \frac{d}{d\zeta}B(\zeta) = 0, \quad \xi \equiv B(\zeta) \neq 0 \tag{6}$$

Consequence 1. Another consistent representation $U_3 = O(\varepsilon U)$ for stretching mode is possible only at $\xi = 0$ and "in-plane" loading /9/■

Consequence 2. A laminate of arbitrary structure is characterized by total thickness, mass density and 18 rigidities. Three layered anisotropic sandwich constitutes the more simple equivalent structure. A plate of three differently oriented orthotropic plies is generally not sufficient for this■

Consequence 3. Stresses, their integrands and energy depend on both kinds of deformations. Equations 3, 4 rewritten in terms of integrands coincide with classical theory■

$$\sigma_{\alpha\beta} = f_{\alpha\beta}(\gamma)e, \quad Q_{\alpha\beta} = f_{\alpha\beta}(d^1)\varepsilon + f_{\alpha\beta}(d^2)\kappa, \quad M_{\alpha\beta} = f_{\alpha\beta}(d^2)\varepsilon + f_{\alpha\beta}(d^3)\kappa$$

$$N_\alpha = \partial_\beta M_{\alpha\beta}, \quad 2(\Pi + T) = \int_\Omega \left(Q_{\alpha\beta}\varepsilon_{\alpha\beta} + M_{\alpha\beta}\kappa_{\alpha\beta} + \rho(w^\cdot)^2\right) d\Omega$$

$$f_{11}(\gamma)e \equiv \gamma_{11}e_{11} + 2\gamma_{16}e_{12} + \gamma_{12}e_{22}, \quad f_{12}(\gamma)e \equiv \gamma_{16}e_{11} + 2\gamma_{66}e_{12} + \gamma_{26}e_{22}$$

The expressions for transversal stresses are more cumbersome and presented in /9, 10/.
Then we can divide the main operator $\mathcal{D}$ in Equations 3, 4 into two parts $\mathcal{D}_{13}$, $\mathcal{D}_2$ ($\mathcal{F}$ is an operator of loading)

$$\mathcal{D}(\mathbf{u}, w) \equiv \mathcal{D}_{13}\mathbf{v} - \xi\mathcal{D}_2\mathbf{y} = \mathcal{F}$$

$$\mathcal{D}_{13} \equiv (\mathbf{a}_1,\, \mathbf{a}_2,\, \rho\partial_t^2 + b)^T, \quad \mathcal{D}_2 \equiv \xi^{-1}(b_1, b_2, \mathbf{a})^T, \quad \mathbf{v} \equiv (\mathbf{u}, \mathbf{u}, w)^T, \quad \mathbf{y} \equiv (w, w, \mathbf{u})^T$$

Theorem 1. The symbols of operators $\mathcal{D}_{13}$, $\mathcal{D}$ are of elliptic type■

When $\xi = 0$ the complete separation of B- and P problems is achieved and b_β, $\mathbf{a}(d_{pq}^2) \equiv 0$. For such laminate the separate equations are the same as for two homogeneous plates with equivalent bending and membrane rigidities (and Poisson's ratios and Young's moduli in the Cases 1-2). This improved classical statement takes place for the following structures:
1) *Orthotropic layered beam with common main axes.*
2) *Plate of plies with transversal isotropy.*
The transversal anisotropy has no influence on the equations. The only difference of the Cases 1 and 2 from homogeneous structure consists in the presence of membrane forces (B) and bending moments (P) at the plane $z = 0$. Two isotropic plies

can simulate these structures for both B- and P problems.

3) *Symmetrically posed anisotropic plies* $(d_{pq}^2 \equiv 0)$.

4) *The case of small ξ* . We suggest to search the solution in the recurrent form

$$(\mathbf{u}, w) = \sum_n \xi^n (\mathbf{u}, w)^{(n)}, \quad \mathcal{D}_{13} \mathbf{v}^{(0)} = \mathcal{F}, \quad \mathcal{D}_{13} \mathbf{v}^{(n+1)} = -\mathcal{D}_2 \mathbf{y}^{(n)}$$

where the operator $\mathcal{D}_{13}$ is familiar to usual separated B- and P problems to be solved at each step. Such iterative procedure can be investigated in detail by usual methods for perturbed operators, but for small enough ξ the sequence is convergent surely.

1.3. Potentials in statics

For statics the potentials for displacements can describe well the coupled fields and allow to use methods of theory of functions of complex variables. They are defined as follow

$$\mathbf{u} = \mathrm{Re}[\mathbf{u}^* \Psi(\mathbf{sx})], \quad w = \mathrm{Re}[u_3 \Phi(\mathbf{sx})], \quad u_\alpha^*, s_\alpha = \mathrm{Const} \in C, \quad \mathbf{s} = (s_1, s_2)$$

$$\mathbf{a}_\alpha \mathbf{u} = p_{\alpha\beta}^1 u_\beta^* \Psi'', \quad b_\alpha w = p_{\alpha 3}^2 u_3^* \Phi''', \quad \mathbf{au} = p_{\alpha 3}^2 u_\alpha^* \Psi''', \quad bw = p_{33}^3 u_3^* \Phi''''$$

$$p_{11}^k = d_{11}^k s_1^2 + 2 d_{16}^k s_1 s_2 + d_{66}^k s_2^2, \quad p_{12}^k = d_{16}^k s_1^2 + (d_{12}^k + d_{66}^k) s_1 s_2 + d_{26}^k s_2^2$$

$$p_{\alpha 3}^k = p_{\alpha\beta}^k s_\beta, \quad p_{33}^k = p_{\alpha 3}^k s_\alpha, \quad (s_1 = 1, \; s_2 = s)$$

$$\mathbf{Pu}^* = 0, \quad p \equiv p_0^1 p_{33}^3 - p_{11}^1 (p_{13}^2)^2 - p_{22}^2 (p_{13}^2)^2 + 2 p_{12}^1 p_{23}^2 p_{31}^2, \quad p_0 \equiv p_{11} p_{22} - p_{12} p_{21}$$

Equations 3, 4 are satisfied when $\Psi = \Phi'$, $p(s) = 0$. The function $p(s)$ is polynomial of 8-th order and has four pairs of roots $s, \bar{s}$ (including the multiple roots). Degenerated case is reduced to variants $p_0^1(s) = 0$, $\mathbf{au} = 0$ and $p_{33}^3(s) = 0$, $b_\alpha w = 0$. At $\xi = 0$ we obtain $p \equiv p_0^1 p_{33}^3$ and independent potentials Ψ, Φ coincide with classical results /1/.

1.4. Natural boundary conditions

The strict investigation of the interaction of boundary layer with internal stress-strain state is of great interest. A coupled model demands 4 boundary conditions at $\partial\Omega$. The most natural ones are combined of classical B- and P statements. From the engineering point of view they can be deduced easy by variational principles for energy $\Pi + T$ in the form ($\tau, \mathbf{n}$ are tangential and normal vectors at $\partial\Omega$):

$$u_\tau, u_n, w, \partial_n w \quad or \quad Q_\tau, Q_n, M_n, N_n + \partial_\tau M_\tau$$
$$or \; their \; mixed \; combinations \; at \; \partial\Omega \tag{7}$$

Theorem 2. The uniqueness of the solution in dynamics and statics is achieved at any of boundary conditions (7)■

2. BENDING OF PLATE OF INHOMOGENEOUS ISOTROPIC LAYERS

2.1. Physical assumptions and general 2-D equations

Now the plate consists of N isotropic elastic layers with arbitrary hierarchy across thickness. The high contrast in the properties of plies is defined by asymptotic equalities for Young's moduli, sound velocities and thicknesses:

$$e_i \equiv \varepsilon^{p_i} E_{max}/E_i = O(1), \quad C_i \equiv \varepsilon^{q_i} c_i^2/c_{min}^2 = O(1), \quad H_i \equiv h_i/h \sim \varepsilon^{r_i}, \quad t_0 \sim \varepsilon^\gamma$$

$$E_{max} = \max_{i \in I} E_i \equiv \max E_i, \quad c_{min} = \min c_i, \quad c_i^2 = E_i/\rho_i, \quad i \in I = \{i: 1 \le i \le N\}$$

where p_i, q_i, r_i are non-negative indexes of laminae contrasts and γ will be evaluated below. For Poisson's ratios the values $\nu_i, 1 - 2\nu_i$ are not assumed to be very small. By asymptotic method from the initial dinamic 3-D problem we establish the reccurent system of equations. Then we construct 2-D approximate equations with a relative error $O(\varepsilon^2)$ restricted by the first iteration, according to the accuracy of Kirchhoff's type theories. Naturally, the accuracy of different edge conditions for high inhomogeneous plate is too complicated question to answer and we do not touch it here. Before the reduction of the equations we have to limit the index values: indexes p_i, q_i can be restricted by practical considerations and the elimination of short waves yields the restriction for γ. The values r_i have no restrictions. As a result, we take into account the following estimations for indexes and unknown functions:

$$0 \le p_i \le 3, \quad 0 \le q_i \le 2, \quad \gamma \le 1/2 \tag{8}$$

$$U_3^i = w(\mathbf{x}, t) + O(\varepsilon^n w), \quad \mathbf{U}^i = O(\varepsilon^k w), \quad (k, n \ge 1) \tag{9}$$

Physically, the Relations 9 indicate a nondeflection of the normal across the whole section. The tangential displacement is asymptotically small in comparison with the normal one. In fact, they are necessary for the beginning of recurrent process and can be verified *a posteriori*. A complete system of approximate equations for the $2N + 1$ functions W and U_β^i and the relations for all the stresses are presented in /12/. In comparison with Timoshenko type theories only the terms corresponding to the influence of longitudinal shear appear in the new equations. The mutual influences of stretching, shear and bending connect the equations. The Limits 8 are very wide and therefore our 2-D theory possesses some excessive complication as the price of generality. At the particular sets p, r, q the system are variously simplified. As a rule, the essential simplifications are connected with possibility of separate description for the flexion and in-plane deformation.

2.2. Classification of degenerate-inhomogeneous structure models

On evaluating the local and mini-max asymptotic orders of functions we establish the restrictions for indexes p_i, r_i when Kirchhoff's assumptions either remain actual or break in turn. The absolute maximums of p_i, q_i, r_i, their sums and mutual position of various layers is important for structure behavior as a whole. We identify three classes of plates and their models of behavior.

First class. As known the very soft external laminae have no influence on the bending of a whole plate. It is possible to give a general description for hierarchies of soft plies surrounding a kernel of rigid laminae when the main equations are the same as in the improved classical theory /11,12/.

The second class is physically characterized by the existence of several kernels of rigid plies connected by not very soft laminae. Then some terms can be neglected in the main equations and they are divided into a six order equation for W and usual ones with perturbed right-hand side for **U**. Although the equations are separated the problem remains connected because of lateral boundary conditions.

The third class is formed by the plates with some very soft intermediate laminae. The equations remain connected and have the highest order but they can be simplified for the particular cases.

CONCLUSION

The 2-D mathematical models for the dynamics of non-symmetric high inhomogeneous or anisotropic elastic plates have been presented. In general, bending and stretching are strongly connected. Different modifications have been separated and classified and the improved classical and non-classical theories of laminates have been followed. That is a substantiation of the engineering approaches, but all the stresses have been determined in the present way. The suggested classification depends on the type of anisotropy, mutual lamina locations and the contrasts of layers. The classes of plates when the theory turns out no more complicated than the classical one are described.

REFERENCES

1. N.A. Alfoutov & P.A. Zinoviev & B.G Popov, "The Calculations of Multilayered Composite PLates and Shells". Mashinostroenie. Moscow (1984) (In Russian)

2. D. Caillerie, C. R. Acad. Sci. Paris, 294, Série 2 (1982) 159-162

3. M.I. Gussein-Zade, Appl. Math. Mech., 32 (1968) 232-243 (In Russian)

4. V.V Bolotin & U.N Novichkov, "Mechanics of Multilayered Structures". Mashinostroenie. Moscow (1980) (In Russian)

5. S.G. Lekhnitski, "Theory of Elasticity of Anisotropic Elastic Body". Holden Day. San-Francisco (1963)

6. R.M. Christensen, "Mechanics of Composite Materials". Wiley. New York (1979)

7. K.F. Friedrichs & R.F. Dressler, Comm. Pure and App. Math., 14 (1961) 49-65

8. A.L. Goldenweiser & A.V. Kolos, Appl. Math. Mech., 29 (1962) 141-155 (In Russian)

9. D.D. Zakharov, C. R. Acad. Sci. Paris, 315, Série 2 (1992) 915-920

10. D.D. Zakharov, Appl. Math. Mech., 56 (1992) 742-749 (In Russian)

11. I.V. Simonov, Int. J. Solids Structures, 29 (1992) 2597-2611

12. I.V. Simonov, C. R. Acad. Sci. Paris, 314, Série 2 (1992) 643-648

DESIGN STUDY FOR A HELICOPTER SUSPENSION ARM IN CFRP

A.F. JOHNSON, B. HINZ, M. REIPRICH

German Aerospace Establishment (DLR) - Institute for Structures and Design - Pfaffenwaldring 38-40 - 7000 Stuttgart 80 - Germany

ABSTRACT

The paper describes the development a prototype carbon fibre reinforced plastic (CFRP) component for a helicopter suspension system. The aim of this study was to investigate the potential of CFRP materials for the development of a suspension arm with the required stiffness and strength properties and with a lower weight than an existing forged aluminium component. The design study shows the feasibility of using carbon fabric and unidirectional (UD) carbon fibre reinforced epoxy resin for the arm and demonstrates relevant design procedures for determining the required laminate construction. A prototype CFRP arm was fabricated at the DLR for evaluation in loading tests.

1. INTRODUCTION

The paper describes the design and fabrication of a prototype CFRP suspension arm for the tail wheel of a helicopter. The development study was carried out at the Demonstration Centre for Composite Materials at the DLR Stuttgart, in close collaboration with Liebherr-Aero-Technik GmbH, Lindenberg, Germany.

Fig. 1 shows the original forged aluminium (Al) component from Liebherr-Aero-Technik together with the prototype CFRP component developed in this project. The suspension arm is a highly loaded component during landing, when it is subjected to combined bending and torsion loads applied through the landing wheel and the shock absorber. In order to achieve the required stiffness in the CFRP component it is necessary to change the component geometry from the Al beam structure to a double box-beam shell construction as shown in Fig. 1. The main shell components were hand laminated in the first prototype and the arm was assembled by adhesive bonding onto machined Al end fittings. These fittings match those of the Al component so that the CFRP arm can be fitted to an existing helicopter suspension for testing. The chosen construction is suitable for a series production using autoclave technology.

Design analyses were carried out on the CFRP suspension arm to first determine a suitable laminate construction for the CFRP shells, and then to assess the stiffness and strength of the component under a set of static design loads. For ease of manufacture UD and balanced CFRP fabric reinforcements were chosen with epoxy resin. A preliminary analysis based on a 2- and 3-celled box beam construction at various sections along the arm length was used to determine the maximum loads on the cell walls and hence suitable fabric lay-ups and thicknesses. Simplified analyses of load introduction through bonding and riveting, and an assessment of shell buckling were also made. This was followed by a detailed finite element analysis (FEA) of the CFRP structure using a new composites design software PERMAS-LA /1/, which is an integrated laminate analysis/FEA software. The analysis determined maximum deflections in the structure together with strength reserve factors for the composite laminates. The results show that under a hard landing flexural load case the suspension arm has adequate stiffness and strength properties and demonstrates the feasibility of using CFRP materials for the structure.

2. DESIGN CONCEPT AND PROTOTYPE FABRICATION

The suspension arm component has a length of 950 mm and pivots about a horizontal axis at the two bushes on the left end, Fig. 2, with a cylindrical mounting at the right end for a single wheel. At the position of 350 mm from the right end is the mounting bracket for a telescopic shock absorber. During a hard landing at 6 m/s the main impact energy is taken by the shock absorber and the design requirements for the component are based on stiffness and strength requirements at a maximum equivalent short term static loading. Long term static or dynamic fatigue loading are not considered to be significant. Additional factors which influence the choice of matrix resin and the protective surface coating for the CFRP arm are the temperature range under load - 30°C to 45°C, resistance to water and aviation fuels, stone impact and lightning strike protection.

Depending on the wheel position on landing, the arm may be loaded by axial, transverse and torsion loads. The design concept for the CFRP arm is a closed shell structure which, corresponding to the loads, has three main components: a torsion shell with fibres at $\pm 45°$ to the arm axis; unidirectional (UD) fibre reinforced flanges to take bending moments and longitudinal loads; central shear webs with $\pm 45°$ fibres and some quasi-isotropic fibre reinforcement for the transverse loads. The CFRP shells were fabricated from epoxy resin reinforced by two types of carbon fibre fabric, a balanced twill fabric and a fabric with 90% UD fibres, as described in more detail in /2/. Laminate thickness and ply lay-up were determined from the preliminary design calculations described in Section 3.1.

In order to fabricate a prototype arm by hand lamination it is necessary to split the shell into a number of subcomponents as shown schematically in Fig. 3. The torsion shells (4) with UD fibre flanges, the webs (2) and the end caps (6), (7) are fabricated in CFRP by contact moulding in one-sided tools. The remaining sub-components are machined in Al, these are the bushes (3), the wheel mounting (5) and the brackets for the shock absorber mounting (1). These Al fittings serve two purposes: they are identical to the fixing points on the Al arm so that the CFRP prototype can be tested in service conditions; and they serve as load introduction elements for the CFRP structure. Assembly took place on a steel fixing jig. The Al brackets (1) were both riveted and bonded to the CFRP webs (2), whilst the remaining subcomponents were bonded together with epoxy adhesive.

3. DESIGN ANALYSIS

3.1 Preliminary Design Studies
Design with composites is an iterative process and before a detailed FEA of the structure, it is necessary to first select ply materials, fibre orientations and laminate thicknesses in the structure. This was achieved from a simplified analysis based on box-beam elements in the arm cross-section. From the design loads on the arm maximum values of axial load, transverse load, bending moment and torsion moment at 100 mm intervals along the arm were determined. At these positions the arm cross-section has a either a two-cell form (between wheel mount and damper bracket), or a three-cell form as in Fig. 4 (between damper bracket and pivot). Simplified design formulae (see /3/,/4/) were then used within a spreadsheet program to compute properties such as bending strengths, torsion strengths and shear strengths for the two-cell and three-cell sections as functions of cell geometry, wall thickness and materials properties. The spreadsheet was used in conjunction with LAMICALC /5/, a PC laminate analysis software, for the calculation of the shell wall laminate properties.

In order to meet the section loading, each part of the section was designed with a specific function. Thus the outer shell is a torsion shell, with fabric reinforcement at 45° to the axial direction. The flanges at the corners of the two-cell and three-cell sections (Fig. 4) consist of UD fabric aligned axially to take the axial and bending loads. The shear webs consist also of 45° fabric for the transverse shear loads. The spreadsheet program was then used iteratively with LAMICALC to determine laminate thicknesses and the proportion of balanced and UD fabric for each part of the section required to avoid section failure under the main load conditions. Additional simplified calculations were carried out to determine the construction of the U-shell at the wheel mounting; to check that the webs and upper shell surface can withstand buckling, and to analyse load transfers from the metal inserts to the CFRP shell through rivets and adhesive joints.

From these analyses followed a set of laminate constructions for the various parts of the structure, which have been tailored to provide adequate strength under the full range of design loads. Table 1 summarises the main laminates in the CFRP arm structure. We see that there is a range of thicknesses from 1.4 to 9.1 mm, with between 5 and 35 fabric plies. The thicker laminates are required at the wheel mounting U-shell and in the two-cell section where the bending moments are a maximum. The structure has additionally taper regions between the two main shell constructions with a smooth drop-off in plies, and load introduction regions at the damper bracket and at the bushes, with extra 0/90° fabric plies to take the direct transverse loads, see /2/ for details.

3.2 FEA of the CFRP arm structure
A detailed FEA of the structure is required to determine structural stiffness and to confirm strength safety factors for the CFRP laminates especially under combined loads, which could not easily be estimated from the simplified analyses. Fig. 5 shows the FE model of the half suspension arm, which is symmetric about the middle line and contains about 5000 elements. The CFRP laminates are modelled by the SHELL4 elements required by PERMAS-LA /1/, which are orthotropic thick shell

elements with membrane, bending and transverse shear forces. The Al inserts are modelled by isotropic QUAD4 shell elements, with BAR2 beam elements used for the very stiff damper strut and for the loading rings in the wheel mounting. Coupling between the Al inserts and the CFRP shell is either by direct node coupling as at the wheel mount, or through an adhesive layer modelled by HEX8 volume elements.

Appropriate materials data for the model based on the various laminate constructions described above and materials property data for the two chosen CFRP ply materials are required. A feature of PERMAS-LA is the creation of a laminate databank in which the ply materials and laminate lay-up in the structure are defined. A pre-analysis computes the laminate stiffnesses which are then assigned to the SHELL4 elements in the FE model. This de-coupling of the laminate properties from the FE analysis allows laminate constructions to be easily changed during refinement of the structure. After completion of the analysis, nodal displacements and residual element stresses are computed which are then converted during post-processing to provide individual ply stresses. From these ply stresses reserve strength factors R for each ply, based on the Tsai-Wu failure criterion with first ply failure (FPF), are then computed for display. Material strengths used in the analysis were design allowable values for CFRP, which should be higher than measured values thus giving a conservative R-value. With this definition R<1 implies strength in reserve, whilst R>1 implies a failed ply in the laminate.

Results are presented here for one of the most severe load cases, the hard landing (HL) flexural load case in which the arm is symmetrically loaded through application of a transverse load F = 39 700 N and a bending moment M = 11 150 Nm about the y-axis at the wheel mounting, as shown in Fig. 5. The arm is free to rotate about the y-axis at the pivot bushes, but is constrained in the x and z directions at the pivots. The arm then bends about the y-axis under constraint from the shock absorber, which is modelled here as a very stiff lever arm. The HL is an ultimate load case which means the arm may be damaged by the loads but should not fracture.

Fig. 5 shows the computed structural deformations under this loading (not to scale), and Table 2 lists maximum displacements, maximum reserve strength factors R in the CFRP laminates and maximum computed von Mises stresses in the Al components. As the figure shows the arm bends about the damper bracket with a maximum vertical displacement w = 12.5 mm at the loaded end. The damper bracket displacement is only 4.2 mm, showing that most of the deformation takes place between the wheel mount and the damper bracket. In addition to this overall beam bending there are local shell deformations in the front shell upper surface because the outer box beams in the three-cell structure twist towards the centre line.

Table 2 shows typical average values for R in the CFRP laminate of 0.65 in the upper and lower faces of the arm, with a value 0.25 in the webs and side walls. Maximum R-values are 0.5 in the side walls and R = 1.2 in the front shell upper surface, where the local shell deformations take place. The results show that for most of the structure the CFRP laminate has more than adequate reserve strength, with some local damage but not complete failure predicted in a small region of the 1.4 mm thick front upper torsion shell. This damage would probably be prevented here by the local addition of a further fabric ply. The average von Mises equivalent stresses in the Al inserts listed in Table 2 are all below 100 MPa, with peak values in the bushes and wheel mount up to 250 MPa. These values are well below the design allowable stress values of 435-495 MPa for the alloy used. The maximum stress in the damper bracket of 550 MPa, which exceeds the allowable value, was a local value in the damper loading ring which thus needs further reinforcing.

4. CONCLUSIONS

The paper describes a design and fabrication study for a prototype CFRP suspension arm. Structural tests have not yet been carried out on the CFRP component. The detailed FEA has shown that, with some very minor modifications, the arm should have adequate stiffness and reserve strength in the HL flexural load case. The FEA results also show that in many areas the CFRP component is overdesigned, thus there is scope to selectively reduce the number of fabric plies and the size of the metal inserts.

In the first prototype CFRP arm the construction and materials were chosen for their suitability for hand-lamination. The fabricated component was found to weigh about 20% more than the Al suspension arm. Thus the primary objective of weight saving was not realised in the prototype. However, 40% of the prototype weight consisted of the Al inserts. Furthermore the hand lamination process causes excess resin which cannot easily be squeezed out. Tests showed that the fabricated CFRP shells had a fibre volume content of 35%, in contrast to typical values of 50% achieved in an autoclave. On this basis the CFRP arm would weigh about the same as the Al component, which has already been designed and optimised for low weight. Thus we conclude that the CFRP prototype has the potential to be optimised for lower weight, through better designed attachment points or within a total CFRP design concept based on better integration to neighbouring components.

ACKNOWLEDGEMENTS: The authors wish to acknowledge Liebherr-Aero-Technik GmbH for supporting the project, in particular Mr Abler and Mr Fass for their help, and their DLR colleagues U Reifegerste and G Rieger for fabricating the CFRP prototype.

REFERENCES
1. PERMAS-LA. Intes GmbH, Industriestr. 2, D-7000 Stuttgart-80, (1992)
2. Reiprich M. DLR Report, IB 435-93/05, (1993)
3. Megson T H G. Aircraft Structures for Engineering Students, Arnold, London (1972)
4. Johnson A F & Sims G D. Composite Polymers, 2/2,89-112, (1989)
5. LAMICALC. K Stellbrink, Drosselweg 7, D-7046 Gäufelden-2 (1992)

Table 1 Laminate constructions in the CFRP arm

	3-cell section			2-cell section			U-shell		
	Thickness mm	No. of plies Fabric 45°	UD	Thickness mm	No. of plies Fabric 45°	UD	Thickness mm	No. of plies Fabric 45°	UD
Shell	1,4	5	0	3,1	11	0	9,1	13	22
Flange	3,4	5	8	5,6	11	10	-	-	-
Web	1,4	5	0	2	7	0	-	-	-

Table 2 FEA results for CFRP arm under flexural load

	Max. vertical displ. w mm	Strength reserve factor R ave.	max.	von Mises equiv. stress MPa ave.	max.
CFRP shell top	4,2	0,65	1,2	-	-
CFRP shell sides	-	0,25	0,5	-	-
Al bushes	0	-	-	70	170
Al damper bracket	4,2	-	-	100	550
Al wheel mount	12,5	-	-	75	250

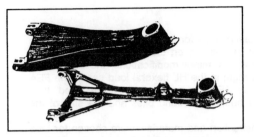

Fig. 1 Aluminium and CFRP suspension arms

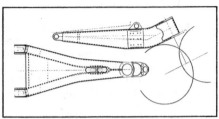

Fig. 2 CFRP arm geometry

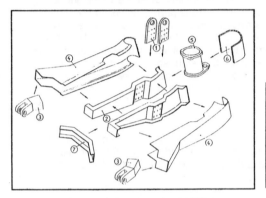

Fig. 3 Component parts of the CFRP arm

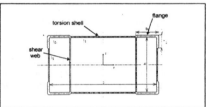

Fig. 4 Three-cell cross section

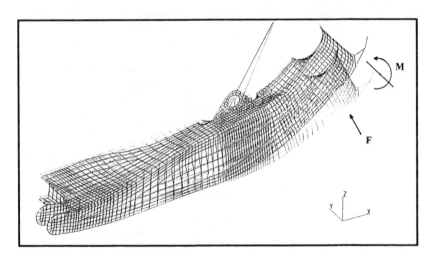

Fig. 5 FE model and deformations (not to scale) in the flexural load case

DETERMINATION OF HOMOGENIZED CHARACTERISTICS FOR COMPOSITE BEAMS

E. ESTIVALEZES, J.J. BARRAU, E. RAMAHEFARISON

Institut National des Sciences Appliquées de Toulouse - Lab. de Génie Mécanique (LGMT INSA) - Complexe scientifique de Rangueil 31077 Toulouse Cedex - France

ABSTRACT

The purpose of this paper is to determine the characteristics of homogenized composite beams formed of orthotropic materials whose orthotropic axis is not necessarily orthogonal to the section. To realize this objective, a simplified and precise theory based on the principle of virtual work is developed to calculate the strain energy and the work done by external forces.The resulting equations are solved by finite element analysis and satisfactory results are obtained.

INTRODUCTION

It is well known that the characteristics of homogenized composite beams, having orthotropic axis orthogonal to their cross section, can be determined by many adequate calculation methods [1]. However this orthogonalty condition is frequently too restrictive. It might be interesting for aeroelasticity problems of helicopter rotor blades, for example, to realize a composite skin with orthotropic materials whose orthotropic axis is oriented by an angle α with respect to the blade axis [2]. In the present paper, for simplicity purposes, the analysis is carried out for beams having simple cross sections .

I. DEFINITIONS

In this study, a simplified cylindrical composite beam having a rectangular cross section is considered as shown in Fig.1. The beam is formed of orthotropic materials with orthotropic axis oriented by an angle α with respect to the beam axis.

1.1 Assumptions

The stresses σ_{yy}, σ_{zz} and τ_{yz} are supposed to be negligible compared to the stresses σ_{xx}, τ_{xy},and τ_{xz}. To realize this assumption, the Poisson's ratios of the constituating materials should be near to each other. It is supposed also that the different phases are completly stuck.

1.2 Loadings - Boundary conditions

The analyses of the beam are carried out for different loadings such as, Tensile stress, Bending moment and Shear stress. For the boundary conditions, again reffering to Fig.1, the rigid body displacement is constrained at the root section (x = 0), but the section is free to warp.

II. THEORETICAL STUDY

In our case, the characteristics of the considered beam can be determined by using anisotropic elasticity theory for tensile and bending stresses [3] .The effect of these types of loadings would successively be studied in the following sections using the general stress-strain relation which can be expressed by:

$$[\Sigma] = [Q][\varepsilon] \quad \text{or} \quad [\varepsilon] = [S][\Sigma] \tag{1}$$

where $[Q]$ and $[S]$ are respectively the rigidity and the compliance matrices.

2.1 Tensile stress: N

In this case, the stress tensor can be represented by:

$$[\Sigma] = \begin{bmatrix} \dfrac{N}{A} & 0 & 0 \\ 0 & 0 & 0 \\ 0 & 0 & 0 \end{bmatrix} \tag{2}$$

which verifies the boundary, equilibrium and compatibility conditions. A is the cross sectional area of the beam.

By using equations (1) and (2), the strains can directly be obtained then, after integration, the displacement field can be determined.Therefore the components of the displacement vector $\bar{P}(u,v,w)$ due to tensile stress can be given by equation (3). From this equation we can see that, under this type of loading, the section undergoes not only a body displacement in the x-direction but also a body displacement in the y-direction and a rotation.

212

$$\begin{cases} u = \dfrac{N}{S}(S_{11}x + S_{16}y + S_{15}z) \\[2mm] v = \dfrac{N}{S}(S_{12}y + S_{14}z) \\[2mm] w = \dfrac{N}{S}(S_{13}z) \end{cases} \qquad (3)$$

2.2 Bending stress: M_z

In the present case, the stress tensor becomes:

$$[\Sigma] = \begin{bmatrix} -\dfrac{M_z}{I_z}y & 0 & 0 \\[2mm] 0 & 0 & 0 \\[2mm] 0 & 0 & 0 \end{bmatrix} \qquad (4)$$

As explained before, the component of the displacement vector would be:

$$\begin{cases} u = -\dfrac{M_z}{I_z}(S_{11}xy + S_{16}\dfrac{y^2}{2} + S_{15}yz) \\[3mm] v = \dfrac{M_z}{I_z}(S_{11}\dfrac{x^2}{2} + S_{15}\dfrac{xz}{2} - S_{12}\dfrac{y^2}{2} + S_{13}\dfrac{z^2}{2}) \\[3mm] w = -\dfrac{M_z}{I_z}(S_{15}\dfrac{xy}{2} + S_{14}\dfrac{y^2}{2} + S_{13}\dfrac{yz}{2}) \end{cases} \qquad (5)$$

Here the section undergoes a translation, a rotation and furthermore a warping.

III. UNIDIMENSIONNAL APPROACH
3.1 Problem formulation

For the different loading conditions, it is assumed that the section undergoes a translation, a rotation and a warping independent of x. However, the displacement component w is zero because of the symmetry of the section.

$$\begin{cases} u = u_0(x) - y\theta_z(x) + g(y) \\ v = v_0(x) \end{cases} \tag{6}$$

Since Tensile stress N and Shear stress T_y are constants, and Bending moment M_z is linear, then the following conditions are realized:

$$\begin{cases} a = \dfrac{du_0(x)}{dx} = cte \\ b = \dfrac{dv_0(x)}{dx} - \theta_z(x) = cte \quad \text{where a, b, c and d are constants (7)} \\ cx + d = \dfrac{d\theta_z(x)}{dx} \end{cases}$$

Using the assumptions of the paragraph 1.1, the stress-strain relation can now be written:

$$\begin{bmatrix} \sigma xx \\ \tau xy \end{bmatrix} = [Q] \begin{bmatrix} \varepsilon xx \\ \gamma xy \end{bmatrix}. \tag{8}$$

From equations (6) ,(7) ,and (8) and by using the theorem of virtual work, the problem can be solved through the calculation of the strain energy J and the work done by the

external foces $V_{F_{ext}}$

IV. RESOLUTION

The strain energy is determined with respect to the constants a, b, c, d and the warping function g. To determine these constants, finite element analysis is used where two node elements with linear interpolation functions are employed.

If g_1, g_2, ...g_n are the values of the warping function at the nodes, the vector of unknowns can be represented by:

$$\{q\}^T = \{a, b, c, d, g_1, g_2, ..., g_n\} \tag{9}$$

Therefore, the strain energy can be written as: $J = \dfrac{1}{2} \{q\}^T [K]\{q\}$ (10)

214

where $[K]$ is the global stiffness matrix and the work done by external forces can be given by:
$$V_{F_{ext}} = \{q\}^T [F_{ext}] \qquad (11)$$
Now it is necessary to note that the root section does not undergo neither body translation nor body rotation, but it is only free to warp. Also the work done by the warping is effectively zero and consequently:
$$\int_A g(y)dA = 0 \text{ and } \int_A yg(y)dA = 0 \qquad (12)$$
The final solution is obtained by using a Lagrange Multiplication technique where the following expression: U can be calculated:
$$U = \frac{1}{2} \int_V \sigma_{xx}\varepsilon_x + \tau_{xy}\gamma_{xy}dV + \lambda \int_V yg(y)dV + \beta \int_V g(y)dV + V_{F_{ext}} \qquad (13)$$
The minimization of U with respect to $\{q\}$, λ and β helps to obtain the solution of the problem as:
$$\{q\} = [B]\{F_{ext}\} \qquad (14)$$

V. RESULTS

Calculations have been carried out for a beam made of only one material Graphite/Epoxy with fiber orientation angle of 45° with respect to the beam axis.

5.1 Tensile stress case

The constants a, b, c and d are found to be completly agreed with the theory, and the warping function is found to be zero.

5.2 Bending stress case

In this case, the constants a, b, c and d are found similar to those predicted by the theory. With regard to warping, the average error between theoretical and calculated values is about 3% [Fig.2]. The previous analyses may be used to determine the characteritics of orthotropic homogeneous beams subjected to shear stress loadings. Consequently the shear coefficient can be obtained. As an example, we have found that, for a rectangular cross sectional beam made of an isotropic material, the coefficient k_y=1.1961 while the theoretical value is 1.2.

VI. CONCLUSIONS

Unidimensionnal study is carried out on homogenized composite beams having simple cross sections under different loading conditions. A calculation method using the principle of virtual work is developed and the obtained results show very good agreement with the theoretical ones. This technique could be extended for the calculation due to twisting moment and for the study of more complicated sections using bidimensionnal finite element method.

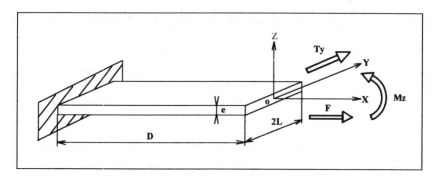

Figure 1. Loadings of the beam

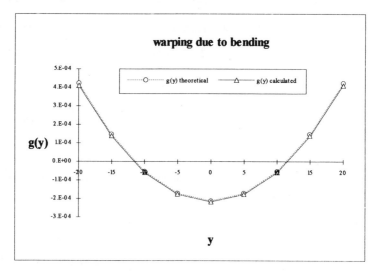

Figure 2 . Plotting of theoretical and calculated warping

REFERENCES

1. Wörndle R.:"Calculation of the cross section properties and the shear stresses of composite rotor blades", Vertica, Vol 6 p 111-129 (1982).
2. Valisetty R. R & Rehfield W.:"Simple theoretical models for composite rotor blade", Georgia Inst. of Tech. Final report (NASA-CR-175620) (1984).
3. Barrau J.J & Laroze S.:"Calcul des structures en matériaux composites", Ed.Eyrolles-Masson, Paris (1987).

FIBRE STRUCTURES

KNITTED CARBON FIBERS, A SOPHISTICATED TEXTILE REINFORCEMENT THAT OFFERS NEW PERSPECTIVES IN THERMOPLASTIC COMPOSITE PROCESSING

K. RUFFIEUX, J. MAYER, R. TOGNINI, E. WINTERMANTEL

Chair of Biocompatible Materials Science and Engineering
Swiss Federal Institute of Technology - ETH Zürich - Wagistrasse 13
8952 Schlieren - Switzerland

ABSTRACT

Two advantageous manufacturing techniques for advanced composite processing were studied for knitted fiber reinforced thermoplastics: net- shape pressing and organo sheet deep drawing. In a single processing step, the net shape forming process allowed the forming of 6-hole osteosynthesis plates with moulded-in holes, which results in a self reinforcement of the force induction zones, with improved plate stiffness, failure behavior and with the complete coating of the plate surface.
In diaphragm deep drawing of organo sheets, it could be shown that the hot drawing behavior is isotropic and deformation degrees up to 80% can be obtained by using weft knits. Based on the analysis of the fiber orientation distribution, the anisotropic distribution of the mechanical properties has been investigated.

1. INTRODUCTION

Weft knitted fiber reinforced thermoplastics can be processed by different manufacturing technologies depending on the textile pre-form or on the pre-consolidated semi-finished part /1/. Compared to the state of the art in composite processing, knitted structure reinforced thermoplastics offer an improvement of shaping techniques. The geometry needed can be knitted or draped into the final shape. Therefore holes and force induction zones are defined and reinforced in situ /2/. A waste reduced processing can be achieved.
Knitted carbon fibers as reinforcement are interesting for loaded parts with a complex geometry due to their drapability /2,3,4/. Additionally, a considerable advantage is seen in the coherence of the knit structure, which prevents uncontrolled fiber flow during thermoforming processes and allows strain hardening and strain stiffening during

controlled drawing /5/. Compared to common composite manufacturing processes this new process is more cost-efficient. The application as load bearing structure in medical implants is of special interest due to the possibility to define their mechanical properties close to cortical bone properties and to the adaptability of the composite implant to individual bone shapes /6/.

2. MATERIALS AND METHODS

2.1 Materials

Two different fiber-matrix systems were used in this study: 1) HT-fibers combined with polyethylenmetacrylate (T300/PEMA, powder bath impregnated). The advantage of PEMA for surgical applications is its low glass transition point of app. 65°C which allows to adapt the manu-factured part to a special individual geometry by heating it up in hot water /6/. 2) HT-fibers / polyetheretherketone (AS4/PEEK, co-mingled yarn) was studied as a high performance composite.

2.2 Net shape forming of osteosynthesis plates /2/

Net shape pressing is defined as thermoinduced forming of a raw material in one production step with no need of further processing. The matrix-impregnated, rolled knitting was pushed over the thorns of the form (Fig.1 right). After inserting all four sidewalls, a stamp was lowered onto the knitting. Due to this process, no fibers had to be cut and the complete surface is polymer coated.
The pressing cycle for the plates containing T300/PEMA was as follows: heat-up to 190 °C at a rate of 18 °C/min., holding period at 190 °C for 30 min., pressure 175 bar, cooling rate 10 °C/min. The same cycle was applied to the AS4/PEEK composite, except that the holding period was at 390°C.
Bending strength and modulus were determined by a 4-point bending test (DIN 29971, 2mm/min.) at 25°C . The fiber volume content was measured by gravimetrical and optical methods /3/. The failure mechanisms were observed by scanning electron microscopy (SEM).

2.3 Deep drawing of knitted carbon fiber reinforced organo sheets

Knitted carbon fiber reinforced organo sheets (T300/PEMA) were deep drawn in a diaphragm process into a cone-like form (Fig. 2). The forming parameters were as follows: temperature: 150°C (polymer melting temp.), pressure: 5 and 10 bar, vacuum, duration of the forming: 20 min.
The global deformation field was measured from the dislocation of beforehand applied silver dots (Fig. 3). To determine local stitch deformation a copper filament (diameter: 100 μm) was co-knitted. The deformation was observed by x-ray analysis.

3. RESULTS AND DISCUSSION

3.1 Mechanical properties and structure of the osteosynthesis plates

Plates made with the net shape pressing method show superior mechanical properties compared to machined plates (Fig. 4).
One of the main advantages of using knits in the net shape process is the possibility to reinforce load induction zones: by pressing the stitches over the thorns of the net shape pressing form, the fiber content around the holes is increased and fiber orientation is improved. An overall fiber volume content of 50% has been reached, whereas the

zones around the holes have a fiber volume content up to 63%. This results in an additional reinforcement of the otherwise weakest part of the plate, leading to plate failure outside its smallest cross section (Fig. 5 left).
By rolling the knits, the 2D structure is dissolved which results in a 3D fiber reinforcement of the part. The plates have a sealed surface which is necessary to prevent release of carbon particles and serves as a barrier of water uptake and hinders crack initiation during fatigue loads (Fig. 5 right).

3. 2 Organo sheet forming of pre-consolidated parts

The unreinforced polymer sheets showed an isotropic deformation behavior as expected. At a pressure of 5 bar, an unisotropic deformation behavior was observed by the reinforced organo sheets, where as at a pressure of 10 bar, it was found to be isotropic again (Fig. 6 left).
Opposed to woven fiber structures, knitted structures have no influence on the deformation behavior up to a deformation degree of 80%.The isotropic deformation behavior leads to an anisotropy in the mechanical properties of the cone (Fig. 7). In the direction of 0° (see Fig. 3) the stitches are drawn in the wale direction where as in the 90° direction the stitches are drawn in the course direction. This finding can be explained by the coherence of the knitted fiber reinforcement /5/.

4. CONCLSIONS

The possibility to produce load bearing implants featuring homoelastic mechanical properties, in situ reinforcements of the load inducement zones and the complete coating in one step (net shape) by using knitted carbon fibers offers a new potential to cut production cost as well as to create implants with unique features.
The reinforcement of organo sheets with knitted fibers offers new possibilities in the use of the deep drawing technique: the coherence of the knit structure guarantees the fiber orientation and the fiber content even during large plastic deformation. The unique drapability of knitted textile performs and the plastic deformation behavior of consolidated semi-finished parts allow an unhampered shaping.

5. REFERENCES

/1/ J. Mayer, P. Lüscher, E. Wintermantel, in VTT Congress, Textiles and composites '92, Tampere, Finland, 315-321.
/2/ K. Ruffieux, M. Hintermann, J. Mayer, E. Wintermantel in VTT Congress, Textiles and composites '92, Tampere, Finland, 326-332.
/3/ J. Mayer, E. Wintermantel, F. et al, in Euromat Cambridge UK, 1991, in Advanced Structural Materials, Ed. T.W.Cyne, 18-26.
/4/ C.D. Rudd, M. J. Owen , V. Middleton in Comp. Sci. Tech. 39, 1990, 261-277.
/5/ S.W. Ha, J. Mayer, et al, in ECCM 6, Bordeaux, Sept. 1993.
/6/ J. Mayer, K. Ruffieux, B. Koch, E. Wintermantel et al in 1992 Int. Symp. on Biomedical Eng. in the 21st Century, Taipei, Sept. 1992, 23-26.

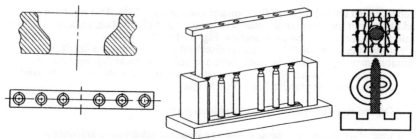

Fig. 1 Geometry of the osteosynthesis plate (left) and of the net shape pressing form (middle). The rolled knit is pushed over the holeforming thorns (right). The stitches are distorted to a circular fiber alignment around the thorns which has a self reinforcing effect.

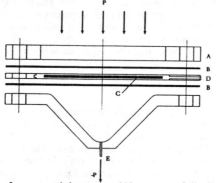

Fig. 2 Diaphragm form containing a ring (A), vacuum foils (B), knitted organo sheet, vacuum ring (D) and cone form (E)

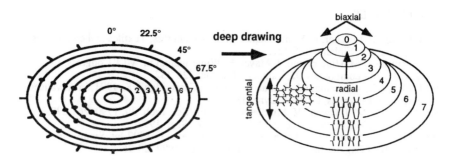

Fig. 3 Organo sheet forming of a pre consolidated sheet using a diaphragm process.
(left): Diaphragm mold and its setup and
(right): orientation and drawing of the knit during hot deformation.

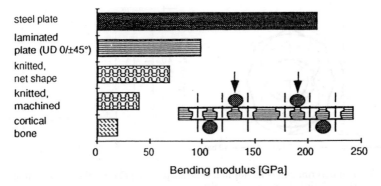

Fig. 4   Bending moduli of osteosynthesis plates, made from stainless steel, UD-laminated and machined, knitted net shaped and knitted machined (AS4/PEEK) were tested in a 4-point bending test. Young's modulus of cortical bone is added to the figure in order to demonstrate a possible mechanical approach with knits.

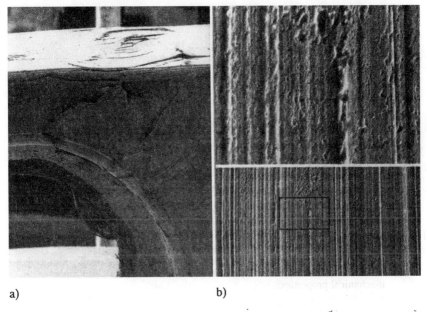

a)   b)

Fig. 5   (left): net shape pressed plate with rolled knit, SEM 20X, compression side in the critical section. The crack initiates beside the critical crossection
(right): fully polymer coated surface by net shape pressing, SEM XX X, no carbon fiber particles were observed.

223

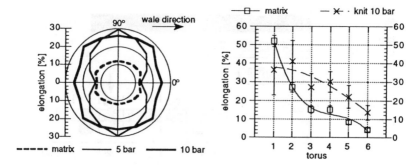

Fig. 6: Hot deformation behavior in a cone shape of a pre consolidated knit reinforced sheet during diaphragm deep drawing,
(left): circular distribution of elongation in torus 4, comparison with pure matrix and the influence of drawing pressure,
(right): comparison of the elongation between matrix and knit along the generating line of the cone. The error bar expresses the circular homogeneity[1] corresponding to figure on the left side.

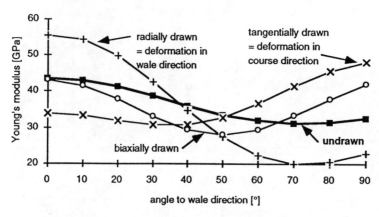

Fig. 7: Anisotropy of the Young's modulus (calculated with the short fiber approach /1/) of the knit reinforced organo sheet before and after deep drawing. The spots analyzed on the cone are shown in figure 3. In comparison with figure 6 it obvious that isotropic hot deformation behavior induces anisotropic mechanical properties.

3-D BRAIDED COMPOSITES - DESIGN AND APPLICATIONS

D. BROOKSTEIN

Albany International Research Co. - 777 West Street
02048-9114 Mansfield, MA - USA

ABSTRACT

3D braided composites are finding many applications for a variety of aerospace, civil engineering and biomedical applications. This paper focusses on the development of a unique 3D braiding process. Design parameters are reviewed along with example applications. These applications include rocket motor exit cones and inflatable airbeams.

INTRODUCTION

Braided composites joined the ranks of conventional composite materials almost 20 years ago. The braided fabric structure offered the composites designer a wide range of fiber orientation possibilities, thus enabling efficient structural reinforcement for a variety of high performance structures. Yet, while the braids offer essentially unlimited orientation possibilities in the plane, they suffer from the lack of the possibility to provide reinforcement out of the plane. Orthogonally-reinforced woven structures can easily be produced with trans-laminar reinforcement using conventional textile weaving looms; conventional braiding machines can not produce 3-D interconnected fabrics.

Recognizing the enhanced design potential of interconnected 3-D fabrics, Albany International Research Co. has developed a unique and proprietary system for producing 3-D braided composites. The inter-connectivity of the three-dimensional braid is accomplished by using a system of intermeshing and counter-rotating horn gears similar to those found on a conventional braider. In contrast to a conventional braider, each cylinder of horn gears is intermeshed not only with its circumferential neighbors, but also with those producing an adjacent layer of braid. The tracks that overlay the horn gears direct the braider yarn packages to move and transfer from one layer to the adjacent layer and back again. A schematic illustration that compares the path of the braider yarns for the 3-D braiding system developed by Albany International Research Co. and the path of the braider yarns for a conventional laminated braid, is shown in *Figure 1.*

At Albany International Research Co., we have built a five-layer interlock braiding machine which can produce tubular braids with up to 48 yarns per layer and 48 axial (or longitudinal) yarns in each layer. Carbon, aramid, fiberglass, silicon carbide and conventional textile (such as polyester and nylon) yarns have all been successfully braided on this equipment.

1.0 DESIGN

Braided fabric reinforcements are generally indicated when tubular axisymmetric structures are specified. The yarn orientations may vary from 15° to 89° with respect to the longitudinal axis. Further, the yarn orienta-tions can vary along the length of the structure. In addition to the bias braider yarns, it is relatively easy to add axial yarns which have an orientation of 0°. The tubular diameter may also vary along the longitudinal axis. The longitudinal axis can either be linear or curvilinear. Finally, the tubular shape can be either circular, curved, rectilinear, or non-specific as long as there are no re-entrant regions. A set of schematic illustrations which presents these design possibilities is provided in *Figure 2.*

When more than one layer of braid is required, as usually is, for reinforcing a composite structure, it is common practice to braid additional layers on top of existing layers. Accordingly, the stresses that are transferred from layer-to-layer, are transmitted through the relatively weak resin systems. Mechanical behaviors such as compression after impact strength and energy absorption are significantly diminished when the transfer of stress between layers is minimized, and thus can be increased whenever a mechanism is provided which interconnects adjacent layers of fibers. Data which show the property enhancement are shown in *Figure 3.*

In the introduction of this paper we described the multilayer interlock braiding process which produces braided structures with interconnected adjacent layers of fabrics. Essentially, all of the design parameters that are possible using conventional triaxial braiding technology are possible using multilayer interlock braiding.

2.0 APPLICATIONS

There are a wide range of reinforced structures that can be developed using the multilayer interlock braiding system. Both rigid and flexible composites have been investigated. In all cases, the design attributes of braided structures have been utilized. Some of the most recent applications are described below.

2.1 Rocket Motor Exit Cones

Rocket motor exit cones are seamless structures that transmit substantial thermokinetic energy from the combustion chamber of a rocket motor to the environment. Accordingly, they are subjected to intense thermal energy along with relatively high propulsion forces. For many solid rocket motor systems the intense thermal energy can only be withstood by carbon fiber based composites. In many cases, the impregnating resin is a temperature resistant carbon-based material. A schematic illustration of a typical rocket motor exit cone is shown in *Figure 4*. Note the substantial variation in cone diameter from the throat section to the aft section of the cone. It is the nature of this diameter variation that indicates braided textile preforms are suitable for the exit cone. In the past, traditional 2-D braided fabrics have been used for exit cones; however, there are two important contraindications that prohibit the optimal performance of a carbon-fiber based rocket motor exit cone produced from 2-D fabrics. The carbon-based impregnation system does not exhibit good interfacial bonding to the carbon fibers. In addition, many carbon fibers have a negative coefficient of thermal expansion. Accordingly, when the rocket motor is fired, the fibers in the throat section have a tendency to shrink and cause the 2-D structure to delaminate. In some instances there have been cases of "throat slippage" causing failure of the motor. The nature of the inter-connectivity of the multilayer interlock braided textile preforms obviates this problem. Further, since the interconnecting yarns do not pass from the exterior layer of the exit cone to the interior layer, any internal damage does not cause the entire exit cone to become dysfunctional.

2.2 Inflated Airbeams

Temporary fabric-based shelters for military and civilian applications can be designed in one of several ways. A common approach is to internally pressurize an impermeable fabric shell. While relatively low pressures are required to maintain inflation of the shell, it is impossible to have large openings in the shelter. This is a critical shortcoming, since aircraft maintenance shelters require easy equipment access. In addition, the shelters require a constant source of air pressure generation. Frame supported structures can be used to maintain the shelter functionality. These frames are generally fabricated from relatively heavy metal components and require substantial labor and equipment to erect. A new alternative to air-inflated and frame supported shelters is based on using an internally pressurized airbeam. For this case the airbeam acts in a fashion similar to the frame, yet it is significantly lighter in weight and can be used in "quick erect" situations.

Straight woven fabrics have been previously used for airbeams. While these are relatively lightweight, they require substantial support loads to force them into a curved shape. A new approach is to use the braiding process to produce an airbeam with a curvilinear axis. By producing the curved airbeam in a curved fashion we have found that the airbeam loses its tendency to straighten when internally pressurized. Accordingly, airbeams can be produced in any shape that is required as long as the braiding mandrel is configured in the required shape. The multilayer interlock braid is used to produce more than one layer of braid at a time. Since the layers are interconnected, it is only necessary to bond the layer adjacent to the mandrel.

Figure 5 shows the curved airbeam as it is being produced in the braider.

SUMMARY

Multilayer interlock braids can be used to produce essentially all of the items that have been previously produced using conventional 2-D braids. Yet, with increased damage tolerance and energy absorption, along with the inherent interconnectivity of adjacent layers, it is possible to produce both rigid and soft composites with enhanced attributes.

Track Paths and Horn Gears

Figure 1.

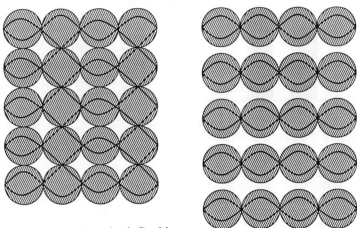

3-D Multilayer Interlock Braider

2-D Conventional Braider

Design Possibilities with Braids

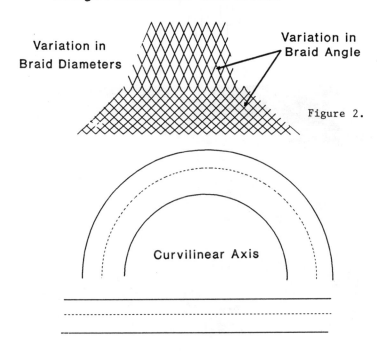

Variation in
Braid Diameters

Variation in
Braid Angle

Figure 2.

Curvilinear Axis

Straight Axis

Figure 2.

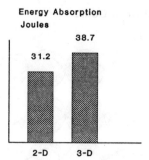

Energy Absorption
Joules

38.7

31.2

2-D   3-D

Compression After Impact
MPa

214

190

2-D   3-D

Braided Composites
+60/0/-60
50% volume fraction

Figure 3.

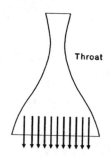

Throat

Rocket Motor Exit Cone

Figure 4.

Figure 5.

230

FIBRE PACKING IN THREE DIMENSIONS

A. KELLY, G. PARKHOUSE

University of Surrey - Dept of Materials Science and Engineering
Guildford - Surrey GU2 5XH - UK

ABSTRACT

Consideration is given to the geometry of packing of long straight fibres in a number of directions in space so as to minimise the usual extreme anisotropy of fibre composites. A structure showing strong resistance to a general deformation in 3 dimensions is described and its modulus calculated.

1.INTRODUCTION

The attainable greatest closeness of packing of long straight fibres or bundles of fibres, and the resultant elastic modulus and density of the array (with or without a matrix) are becoming increasingly important. This is because of the need to arrange fibres so as to prevent delamination of planar arrays; a second reason is to avoid the extreme anistropy of fibre arrangements.

Although many descriptions of 3-D arrangements of fibres appear in the literature, eg /1/, we are unaware of a systematic treatment of the most close packed arrangements. Further the elastic moduli of the various arrangements which have been described do not appear to have been measured and compared with simple theoretical estimates. It follows that what is attainable or not attainable is not clearly known. In what follows we shall make simple estimates of the elastic moduli of fibre arrays assuming that the fibres have no flexural stiffness so that each can transmit load only in tension. This approximation is a good one, giving a conservative estimate of the moduli and one little different from more complicated treatments. We assume that the fibres or fibre bundles are of uniform cross section of length l, modulus E_f and density ρ (which may also stand for specific gravity). The proofs of the new results appearing in this paper will be published elsewhere /2/.

2. 2-D ARRANGEMENTS

If square sectioned fibres are packed together these aligned arrays can attain a volume fraction (V_f) of 1. Circular sectioned fibres in the plane cannot attain a V_f greater than 79% ($\pi/4$). To attain isotropy in the plane is strictly speaking not possible since aligned arrays must be stacked one upon another. In order for any applied stress to be resisted by extension of at least some fibres, then fibres must be arranged in at least three directions parallel to the plane. Practically, so-called, quasi isotropic arrangements are produced eg /3/ which attain V_f = 0.60 or a little greater. The moduli in the plane are about ⅓ (in practical cases 0.3) of the fibre modulus and the density (or strictly the specific gravity) assuming no matrix is clearly 0.6 of the density of the fibre so that (E_f/ρ) is about ½ that of the fibre; it would be exactly ½ if V_f were ⅔.

The factor of ⅓ relating the Young's modulus of the fibres with that of the arrangement, provided the specimen thickness is much great than the diameter of the fibres, arises also for a so-called random arrangement of fibres parallel to a given plane /4/. In this case we have

$$E = \frac{E_f V_f}{3}; \quad G = \frac{E_f V_f}{8}$$

and Poissons ratio v = ⅓.

The question then arises what is the maximum value of V_f in this case? For very long fibres it is, strictly speaking, zero unless the fibre is in the form of a thin sheet! "Quasi isotropic" arrangements can be made but only for (l/d) not too large and then the spacing of fibres must not be too small, if they are sensibly to lie in closely spaced planes parallel to that in question. If (l/d) is small then the fibre modulus is not attained and stretching of the fibres depends very much on the properties of the matrix. In practice with flexible fibres of very large (l/d) a V_f of about 20% is attainable /5/ but the "in plane" modulus falls far short of what should be attained - partly no doubt due to the curved arrangements of the fibres. The attainable modulus is then no greater than $E_f/15$ in the plane and a matrix must be present in order to resist even a vanishingly small stress normal to the plane.

3.3-D

A bundle of aligned fibres can "in principle" attain a packing fraction of 100% for most fibre cross sections and attains 91% for circular sectioned fibres. Cox /4/ implies that if fibres are oriented at random in three dimensions then the array is elastically isotropic with

$$E = \frac{E_f V_f}{6}; \quad G = \frac{E_f V_f}{15}$$

and $v = 0.25$.

Again the question arises as to the maximum value of V_f. If solid fibres are arranged at random the attainable V_f again depends on the slenderness or aspect ratio (l/d). For very long fibres of large aspect ratio the attainable volume fraction tends to zero as (l/d) becomes very large. The proof of this will be published elsewhere /2/.

Figure 1 shows theory and experiment. In practice even with flexible fibres volume fractions of greater than a few percent (10% at the very most) are impossible to obtain. The Young's modulus is then reduced to $E_f/60$, which is not attractive; although, in principle, if no matrix were necessary E_f/ρ for the array would be as high as $E_f/6$. To obtain a large value of V_f either the fibres must be made short, which can greatly reduce the modulus or special 3-D arrangements must be sought.

Most 3-D arrangements described in the literature consist of fibres running parallel to the edges of a cube (call this cube arrangement). Such an arrangement attains a V_f of ¾(0.75) with one third of the fibres running in each of three directions. The maximum modulus is $E_f/4$ and in the cube direction $E_f V_f/\rho = E_f/3$ but the array "collapses" under almost any shear and must rely upon a matrix or additional fibres in other directions to resist this.

In order to resist any general stress in 3 dimensions by stretching of the fibres, these must be arranged in at least 6 directions and we return to the simplest arrangement which we have found to achieve this below. However, one which comes close to achieving this is a tetrahedral arrangement. If the fibres or fibre bundles are arranged with a triangular cross section and to lie along the normals to the faces of a regular

233

tetrahedron with one quarter of the fibres running in each direction a volume fraction of ½ may be obtained - illustrations of this will be given in the oral presentation - with a maximum modulus of $E_f/8$. Since the density is reduced by 2 a value of E_f/ρ is obtained of $E_f/4$. Such a tetrahedral array is very close packed containing closed voids forming a body centred cubic arrangement and it fails to resist an imposed deformation only in a small number of highly specific directions.

An arrangement which can resist a general deformation by extension of the fibres is obtained with a volume fraction of as high as ⅓ for infinitely long rigid fibres (or bundles) provided the fibres are of rhombic cross section and arranged to lie along the normals to the faces of a regular rhombohedron. We have made such a structure. The voids again form a regular 3-D array for equal numbers of fibres in the 6 directions The maximum modulus of the array is $E_f/15$ and the minimum $E_f/27$. No matrix is necessary to hold the array together. Since the density is only one third that of the fibre E/ρ lies between $E_f/5$ and $E_f/9$. The material has cubic symmetry and conventional moduli as follows.

Young's modulus parallel to the edges of the unit cube defining the rhombohedron is $E_f/27$ and this is the smallest value of Young's modulus. Poisson's ratio in a plane normal to this direction is (⅓). Young's modulus along the body diagonals of the unit cube is ($E_f/15$) and there are four such directions. Poisson's ratio in a plane normal to this is (1/5). The two principal shear moduli are ($E_f/36$) and ($E_f/72$) and the bulk modulus (in tension) is ($E_f/27$).

The import of these simple considerations is shown in the Table which compares the calculated values of E/ρ for a modern carbon fibre (of very modest properties) formed into the rhombic array described, with that for some conventional materials. It is striking to see that in principle a quasi isotropic arrangement of fibres can equal the properties of the isotropic metals so far as modulus per unit weight is concerned. The materials would show correspondingly high strengths in tension.

REFERENCES

1.L.E.McAllister and W.J.Lachman in "Fabrication of Composites" (Kelly A. and Mileiko S.T. Eds) (1983) 109-175 North Holland, Amsterdam
2.J.G.Parkhouse and A.Kelly (1993) to be published
3.S.W.Tsai and H.T.Hahn "Introduction to Composite Materials" (1980) 145 and 380, Technomic Westport CT
4.H.L.Cox, Brit.J.Appl.Phys.3 (1952) 72-79
5.S.A.Hitchen, S.L.Ogin, P.A.Smith and C.Souter, Composites (1993) in press

TABLE

Fibre Arrangement Material	V_f	Density (s.g.)	E_{max}	E_{min}	$\underline{E}(GPa)$ (s.g.)
Cube	¾	0.75$^+$	$E_f/4$	0	-
Tetrahedron	½	0.5$^+$	$E_f/8$	0	-
Rhombohedron	⅓	0.33$^+$	$E_f/15$	$E_f/27$	26.2, 14.5
Quasi-iso 2D	0.6	1.62		70	116
Al	1	2.7		70	26
Steel	1	7.9		220	28
Ti-6Al-4V	1	4.6		119	26

+   fraction of fibre

$$E_f = 230GPa \ (T - 300) \ \rho_f = 1.75 \ g.cm^{-3} \left(\frac{E_f}{s.g.} \right) = 131 \ GPa$$

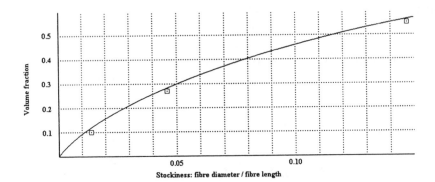

Variation of attainable volume fraction of a uniform "random" array of long straight fibres with "stockiness" of the fibres. Stockiness is equal to (d/l) where d is the diameter and l the length. The points are experimental values. The full line shows our theory which contains no adjustable parameter.

SPLIT FILM BASED WEFT INSERT WARP CO-KNITTING AND LOST YARN PREFORMING, TWO NEW ROUTES FOR COMPOSITE MATERIALS PREFORMING

C.H. ANDERSSON, K. ENG*

Dept of Production and Materials Engineering - Lund Inst. of Technology - PO Box 118 - 22100 Lund - Sweden
*Engtex - 56500 Mullsjö - Sweden

ABSTRACT

The high productivity and state of development of modern weft insert warp knitting machinery and the non crimped insert structures produced make warp knitting potentially very attractive for composite materials preforming. Two new route for composite materials preforming, Split-Film Co-Knitting and Lost Yarn Preforming are discussed. Keywords: Knitting, textile, drape, preforming, thermoplastic, metal matrix

I PRINCIPLES OF FABRIC FORMATION

A fabric is defined as an integrated fibrous structure produced by fiber entanglement or yarn interlacing, interlooping, intertwinning or multiaxial placement. The point of these structural features of fabrics is the arrangement of some kind of fibre to fibre contact for the subsequent build up of frictional stresses and plastic deformation. In weaving, braiding and knitting, the integrity of the structure is given by the crimp.

Textile fabrics are produced by weaving, braiding or knitting processes. Fundamentals of these processes for composite materials preforming use are reviewed in Refs /1-2,4/.

II TEXTILE HANDLING OF BRITTLE LOAD BEARING FIBRES

The textile techniques and thinking were developed for the use of rather soft viscoelastic fibres exhibiting stress relaxation and extremely ductile behaviour. Most load bearing fibres for composites however exhibit very little plastic deformation or relaxation of internal stresses.

237

The absense of stress relaxation processes and sensitivity to surface damage of the high modulus fibres severely limits the choise of textile structures, machine design and preform structures. Criteria for possible textile handling and conservative estimates of strength losses of reinforcement fibres can be estimated from the constitutive relations and experimental materials data for the different modes of local stress build up /2,3/:
- Tensile preload stresses
- Stress build up due to bending, due to the difference between the outer and inner radius of the fibre.
- Frictional stress build up due to sliding over bent surfaces, exhibiting exponential increase with the contact angle of the stresses obtained.
- Herzian contact stresses due to transverse compression giving radial tensile stresses at the edge of contact.

III PRINCIPLES OF KNITTED FABRICS

Knitted fabrics are interlooped structures in which the knitting loops are produced by introducing the knitting yarn either in the cross machine direction (weft knit) or along the machine direction, (warp knit).

Knitting can be used for production of tufted structures, double layered connected structures etc or be combined with inserts of non crimped inlays of weft and warp yarn in the knitted structures.

The principle of warp knitting with weft yarn application is illustrated in Fig.2. It involves the insertion of a straight yarn stretched across the whole width of the machine behind the needles. In this way the yarn is connected between the fabric face and the underlaps. The needles in this kind of machines are mounted on rocking bars, and all needles are thus in work simultaneously.

These insert structures and the technique of production exhibit some obvious merits for preforming:

1) Weft insert warp knitting techniques exhibit knitted structures giving the integrity to the fabrics and taking the crimp. By the interlooped nature of the stitch pattern, improved fracture toughness can be obtained by intermingling between the subsequent layers of fabrics in the final composite /10/. By controlling the stitch (loop) pattern and density, a wide range of pore geometries can be generated.

2) By weft insert, the load bearing fibres are introduced into the knitted structure without the small radius bending, crimp, of woven fabrics in the load bearing system. For the knitted structure, fibres exhibiting ductile behaviour can be used. Local off axis loading and internal stress build up are thus minimized. The potential yield of mechanical properties of composites made from unidirectional weft insert fabric preforms is thus in principle the same as that of unidirectional tape lay up structures, i.e. properties close to the Rule of Mixture predictions, as reviewed in Refs /5-9/.

3) By the use of systems with multiple simultaneously feed insert yarns into chains for feed into the knitting bars, very high productivity, 200-300 m/min at full 6m machine width can be obtained without too high speeds of the individual yarns.

4) The weft inserts can be introduced into the knitted structure in different ways giving in principle two different kinds of structures:
i) Impaled structures with stitches piercing the insert yarns - LIBA and Malimo machines.
ii) Non impaled structures have the stitchs around the insert yarns - Karl Mayer machines.

Drape of non impaled unidirectional weft insert warp knitted fabrics is dominated by three separate mechanisms, to be compared with the drape of woven structures as rewieved in Ref /12/:
i)   Shear of the knitted loops.
ii) Stretching of the knitted loops perpendicular to the reinforcement.
iii) Sliding of the reinforcement.

The shape of the final composite can thus be given to the fabric with very little stress build up in the load bearing fibre system and well controlled positioning.

5) Thermoplastic materials can be introduced as the fibres for the knitted structure. The knitted loop structure can in principle give any volume fraction.

6) Additional matrix material can be added in the weft insert system or by film or fleece stacking to be bonded to the fabric during the knitting process. The binding of the additional material to the load bearing fibre system fabric by the knitted structure is also done in the machine during the fabric formation process without additional processing steps /5-7/.

7) For the knitted structure of the preform fabrics, split film of thermoplastic matrix materials can be used instead of fibres, Split Film Based Co-Knitting.

8) Due to the possible absence of bending of the insert in the machines and the knitted structure, ultra high modulus carbon or ceramic filaments not possible to handle by other fabric forming textile methods has been handled.

9) The use of fabrics with knitted structures of non fusable consumable textile fibres i.e. aramides, acrylics or viscose is the base of the Lost Yarn Preforming of ceramic filaments with short fibres to green bodies for subsequent MMC and CMC use.

IV   EXPERIMENTAL RESULTS AND DISCUSSION

Co-Knitted Thermoplastic Preforms - Samples of aramide Twaron 1055 HM were knitted with PP yarn to unidirectional fabrics with 49 % aramide n:o 3 and 60 % aramide n:o 4-5 respectively, % by volume. Pressing was done at 270° C, 5 bar and 20 min with cooling down to RT under pressure.

	σ_f [MPa]			E [GPa]		
	min		max	min		max
N:o 3	678	810	923	29.4	44.0	53.7
N:o 4-5	840	930	980	24.4	40.6	67.2

The potential yield of mechanical properties of composites made from unidirectional weft insert fabric preforms is in principle the same as that of unidirectional tape lay up structures, i.e. properties close to the Rule of Mixture predictions. This was obtained for the measured elastic moduli values, with pronounced strain hardening of the fibres during the loading process. Adhesion problems, i.e. debonding however limited the tensile strength of the samples.

Properties obtained with unidirectional weft insert fabrics and thermosetting resin matrices are reviewed by /9/. Similiar tensile behaviour was reported for composites made by filament winding and unidirectional knitting.

Lost Yarn Preforming - Unidirectional knitted viscose acrylic or aramide fabrics with inserts of Nicalon SiC, Nextel Al_2O_3 or C-fibres were produced on machines of Raschel type with modified insert system. The fabrics were combined with random or ortotropic short fibre reinforcement, Saffil-type and silica binder to green bodies by ceramic wet forming processes. The binder yarn structure is removed during the subsequent heat treatment procedure, typically 900° C with silica sol binder.

The technical problems of the lost yarn preforming method belong mostly to the ceramic chemistry, i.e. phenomena like bad wetting and penetration of the ceramic slurry into the reinforcement and absorption of silica binder in the knitted system giving residues after the final heat treatment.

Split Film Based Co-Knitting. In the work going on, Split Film based Co-Knitting has shown some obvious merits:
- Improved distribution of the matrix material compared with conventional co-weaving.
- It is possible to use this kind of co-knitted structure as coated fabrics after calendering.

- Split film technique is highly developed and extremely productive and possible to combine with existing knitting machines used for industrial textile production. Productivity typically up to 200 m/hr is thus possible with E-glass rowing inserts.

A fundamental drawback of the techniques of handling thermoplastic matrix materials as fibres, i.e. in comingled yarns, co-weaving, co-knitting etc., lies in the release of internal crimping stresses in the polymeric fibre system when heated. The handling of heated preform blankets for stamping is thus not trivial and some kind of clamping has to be used.
- Film is usually produced less oriented than filaments, i.e. exhibit less shrinkage force release when heated.

-Unidirectionally oriented non crimped reinforcement preform structures for hot pressing have been produced on machines of Raschel type. By the use of split film improved and more even distributions of the material can be obtained

V   REFERENCES:

1. Ko F. : "Preform Fiber Architecture for Ceramic-Matrix Composites" Ceramic Bulletin 68 (1989) 401-414
2. Andersson C-H : "Mechanical Properties of Fibres and Mechanical Models for Preforming" Proc. Verbundwerk'90, Wiesbaden 1990, pp 28.1-14

3. Månsson O., Karlsson M. and Andersson C-H : "Stress distribution due to transverse compression of crossing fibres" Dept of Production and Materials Engineering, LTH 1993.
4. Backer S. : "Fibrous materials" in McClintock F., Argon A. (eds): "Mechanical Behaviour of Materials" Addison Wesley, Reading Mass. 1966
5. Raz S. : "Warp Knitting Production" Verlag Melliand Textilberichte GmbH, Heidelberg 1987
6. Raz S. : "The Karl Mayer guide to technical textiles" Karl Mayer Textilmachinenfabrik Obertshausen 1988
7. Dexter H.B., Hasko G.H. and Cano R.J. : "Characterisierung multiaxial kettengewirkter Verbundwerkstoffe" Karl Mayer Textilmachienenfabrik Obertshausen 1992.
8. Ackermann N.: "Multiaxiale Kettengewirke als Verstärkungsstruturen in Faserverbundwerkstoffen" Chemiefasern/Textilindustrie 41/93 (1991) T128-T130
9. DeMint Th., Van Schooneveld G. : "Fibre preforms and resin injection" Engineered Materials Handbook Vol I Composites, ASM INTERNATIONAL, Metals Park Ohio 1988, pp 528-533
10. Verpoest I. and Dendauw J. : "Mechanical properties of knitted glass fibre/epoxy resin laminates" Proc. ECCM 5, Bordeaux april 7-10, 1992, pp 927-932
11. Hensen F.: "Herstellung von Polyolefin-Folienbändchen" Chemiefasern/Textilindustrie 41/93 (1991) 1185-1190
12. Wulfhorst B., Hörsting K.: "Rechnergestützte Simulation der Drapierbarkeit von Geweben aus HL-Fasern für Faserverbundwerkstoffe" Chemiefasern/Textilindustrie 40/92 (1990) T118-T123

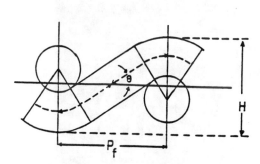

Fig 1. The principle of conventional textile crimped structure build up /1-4/.

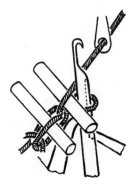

Fig 2. The principle of warp knitting with weft inserts /6/.

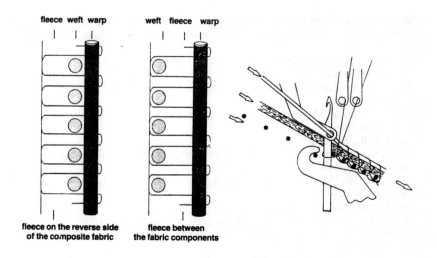

fleece weft warp   weft fleece warp

fleece on the reverse side   fleece between
of the composite fabric   the fabric components

Fig.3. Schematic illustrations of fleece bonded to composite knitted fabrics /6/.

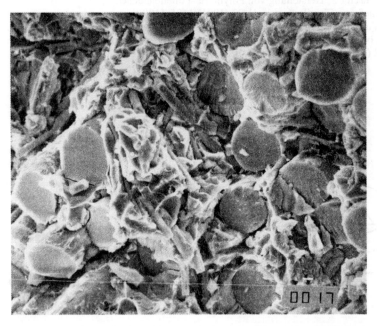

Fig.4. Illustrates the 3:D hybrid reinforcement structure with unidirectional Nicalon SiC and random Saffil Al_2O_3 in Al6Si matrix, impact fracture, SEM 1000X.

MECHANICAL PROPERTIES

C-FIBRE DEGRADATION, DEBONDING AND LAMELLA BUCKLING MEASUREMENT IN CMC BY NEW X-RAY TOPOGRAPHY

M.P. HENTSCHEL, K. EKENHORST, K.-W. HARBICH, A. LANGE

*BAM - 6.22 (Fed. Inst. for Materials Research a. Testing)
Unter den Eichen 87 - 12205 Berlin - Germany*

ABSTRACT

Unconventional and conventional x-ray scattering techniques have been developed and applied to meet the actual demand for improved nondestructive characterization of CMC (and other high performance composites). X-ray wide angle scattering reveals C-fibre degradation by high processing temperatures analysing the carbon reflection profiles. X-ray scanning microscopy of three-dimensional resolution images lamella architecture and buckling. X-ray refraction at very small scattering angles analyses the ratio of debonded fibres. Scanning refractometry visualises the projection of fibre debonding concentration at 20 μm spatial resolution.

INTRODUCTION

In research, processing and application of composite materials a strong demand for nondestructive characterization has developed. Since several years it has continuously been stated in composite material conferences that the performance of NDT techniques is dissatisfying . In attempting to bridge this technological gap, new methods of nondestructive characterization of composites by x-ray scanning topography have been created. They are capable of imaging micro structural properties by crystallographic contrast, interface sensitivity and orientation selectivity [1].

1. X-RAY SCATTERING REVEALS C-FIBRE DEGRADATION

X-ray wide angle scattering is traditionally an analytic method, employed in crystallography for finding atomic and molecular structures in solids [2]. Laue patterns and Scherrer diagrams are created by white and monochromatic x-ray diffraction respectively. In both cases the scattering is "elastic" (without energy shift). Typical x-ray energies applied are between 6 and 20 keV. Carbon fibres give oriented diffraction patterns when exposed to monochromatic x-rays. The 002- reflection of 0.35 nm spacing appears at a diffraction angle of 26 , when monochromatic x-rays of 8 keV (res. 0.154 nm) of Cu-target tube is used. The direct ("primary") beam hits a beam stop to prevent overexposure at the center of the x-ray film as it is three orders of magnitude stronger than the diffracted intensities.

It is well known that C-fibre SiC CMC can be manufactured by liquid Si infiltration into CC CMC at elevated temperatures. The C fibre degradation by this process is of essential interest. It is detectable by x-ray wide angle scattering. The quite unknown effect of the inverse relation between x-ray C-fibre 002reflection width and Young's modulus of the fibres [3] has been employed to quantify the C-fibre degradation in terms of a modulus shift, which can be drastic during processing of C-SiC fibre ceramics, CFRP pyrolysis and graphitisation of C-fibres in C-C CMC.

Fig. 1 shows the wide angle scattering pattern of a 0/90 degree multilayer of C-C CMC with two perpendicular pairs of 002-reflections of high tensile C-fibres (left). The densitometric plot of the film along the hatched arrow results in the intensity distribution over scattering angle (middle). The full width half maximum of the peak near 26 deg. as indicated by two arrows is 3.5 deg. which corresponds to Young's modulus of 230 GPa (as given by manufacturer). After silicon infiltration the full width half maximum reduces significantly to 2.3 deg (fig.1, right). HM fibres of 400 GPa perform a full width half maximum of only 1.5 deg.

Applying the inverse linear relation of peak width and modulus it is easy to calculate that the infiltration process has degraded the HT fibre severely to a higher modulus of 330 GPa. High modulus C-fibres are much more stable. Their modulus is raised by only 2 %. This kind of x-ray scattering analysis of C-fibres in any composite is easy to handle, effective and nondestructive. Furthermore it can be carried out by any crystallography laboratory.

2. X-RAY DIFFRACTION SCANNING MICROSCOPY IMAGES CMC

This kind of microscopy is a X-ray diffraction scanning topography of three-dimensional resolution [4]. It reveals section computer images at $50 \mu m$

lateral and 0.3 mm (actual state) transversal resolution with selective crystallographic contrast to fibre concentration, fibre orientation in C-fibre CMC or SiC distribution in C-SiC CMC. The physical effect is x-ray wide angle. The arrangement of two focusing (bent) crystal monochromators as shown in fig. 2 reflects 20 keV monochromatic x-rays of a rotation anode x-ray generator. The scattering plane with reference to the fibre direction determines the selection of layers of a laminate which are contrasted. The scattering angle is chosen to detect the 002-reflection of C-fibres (see fig.1). Setting another scattering angle would visualise another substance exclusively, for example SiC in C-SiC CMC.

Two- or three-dimensional arrays of sample data can be collected when the sample is scanned in the common focus of the two monochromator crystals. Figure 2 gives the surface cut of the 4.5 mm laminate of carbon- carbon CMC (left) and the transversal x-ray scanning image. This computer image represents a transversal section of the sample plate, contrasting the carbon fibre concentration and the smeared layer thickness and distances (of layers with equal fibre direction) (bright areas). Transversal channels can be observed. They might reflect gas escape or other materials flow during processing. The channels are suspected to be of importance for the diffusion resistivity of the CMC. Large pores are indicated by dark areas.

Image refinement is performed by a procedure known as deconvolution or unfolding: After a transversal reference scan with a single 0.1 mm thin C-fibre layer, which provides the "smearing" function, the deconvolution of the original image is calculated and plotted (fig. 2, right). Now the layer architecture is clearly visible at a resolution of 0.1 mm. The buckling of the layers is similar to the surface micrograph.

The distances of adjacent layers can be analysed by computer. A plot of the distance probability over distance results in a histogram of which the full width half maximum is the"buckling"" factor "[1]. Typical scan times are 0.1 to 1 sec per position, depending on the desired precision of intensities. The scanning time is between 1 and several hours. Computing time on a PC is several minutes.

3. X-RAY SCANNING REFRACTOMETRY REVEALS FIBRE DEBONDING

X-ray refractometry has turned out to be an effective tool for determining interfaces in composites. Xrays of a fine structure tube transmit the sample and are detected at small scattering angles. The signal of a scintillation detector performs high selectivity for the orientation and debonding of single fibres. Due to the short wavelength of X-rays refraction is effective even to nanometer separation between fibre and matrix. The oriented scattering of debonded fibres provides good separation from isotropic porosity.

Mean values of the fraction of debonded fibres are determined to +1% precision within a second. This is a much higher precision and speed than any mechanical testing could reach, a desirable situation when processing is to be optimised. The method requires a (commercial) small angle x-ray camera (like Kratky type) and a standard fine structure x-ray generator. X-ray refraction can be understood by analogy to the well known refraction of visible light which is governed by Snell`s law. It has not been exploited so far in material science. On the contrary continuous x-ray small angle scattering for particle size determination can be mistaken when refraction effects become predominant [5].

Scanning a sample in the x-ray beam and detecting the refracted intensities for each position under a characteristic scattering angle of typically one minute of arc provides the data for a two-dimensional computer image of the inner surface distribution. An example of such a refractometric scan of the carboncarbon CMC laminate of fig. 2 is given in figure 3, which resolves the concentration of single fibre debonding at 10 μm spatial resolution and 2% precision. One single debonded fibre can be detected. The homogeneity of fibre (de-)bonding is clearly an essential property of CMC and CFRP. The high scattering intensities provide the possibility of high scanning speed and imaging within several minutes.

Further experience in the characterization of various heterogeneous materials including HT superconductors, monolithic ceramics, catalysts, injection moulding composites, ferrofluids, video tapes, paper, x-ray films and pigments support positive expectations in x-ray refractometry.

4. DISCUSSION

Beyond basic research X-ray scattering, especially when scanning techniques are perfored, can be a useful tool in nondestructive evaluation of material properties and their design as the given example demonstrates. The development of the two methods of scanning topography mark a new field of imaging x-ray techniques with high selectivity to solid state properties, microstructure and flaws.

They have been developed specifically in order to find solutions to the complex problem of nondestructive evaluation of composites. Interesting applications to other light weight materials can be expected. Nevertheless their application for large structure inspection is not intended. At present the methods are available in our laboratory, where they are continuously adapted to the needs of collaborators in materials science and current problems of industrial customers.

REFERENCES

1. Hentschel,M.P.; Harbich,W.; Lange, A.: New X-ray Topographic
   Approaches to Nondestructive Evaluation of Composites, Proc. Int.
   Sympos.' Advanced materials for lightweight structures', Noordwijk, The
   Netherlands, 25-27 March 1992 (ESA SP-336 1992) pp 229-232

2. v.Laue,M.: Röntgenstrahlinterferenzen, chapter 3, Frankfurt am Main 1960
   p 90

3. Richter,H.: Röntgendiffraktometrisches Prüfverfahren zum Bestimmen der
   Faserart in kohlenstoffaserverstärktem Kunststoff, Materialprüfung 31
   (1989) 6, p 196-199

4. Hentschel,M.P.; Lange,A.: Microscopie RX à balayage: une nouvelle
   méthode d'analyse non-destructive des materiaux composites, Composites
   28 (1988) 3, pp 241-243

5. Hentschel,M.P.; Hosemann,R.; Lange,A.; Uther,B.; Brückner,R.:
   Röntgenkleinwinkelbrechung an Metalldrähten, Glasfäden und
   hartelastischem Polypropylen, Acta Cryst. A43, (1987) pp 506-513

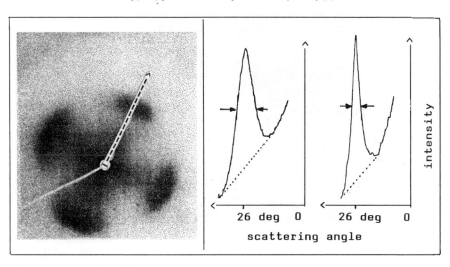

Fig.1: X-ray wide angle scattering of HT C-fibre prepreg (left). Photometric
plot along hatched arrow with indicated full width half maximum
(middle). Photometric plot of degraded fibres in SiC-C CMC reveals
reduced width (right).

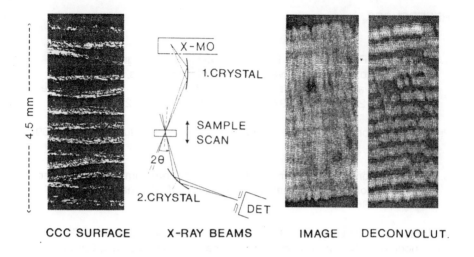

CCC SURFACE   X-RAY BEAMS   IMAGE   DECONVOLUT.

Fig. 2: X-ray diffraction scanning microscopy of C-C CMC

Fig. 3: X-ray refractometric scan of 1.6 by 4.4 mm C-C CMC laminate (as in fig. 2) revealing fibre depending variations at 10 μm spatial resolution

EFFECT OF FABRIC GEOMETRY ON DAMAGE RESISTANCE OF WOVEN FABRIC COMPOSITES

K.N. RAMAKRISHNAIAH, H. ARYA, N.K. NAIK

Aerospace Engineering Dept - Indian Institute of Technology Powai - 400 076 Bombay - India

ABSTRACT

The experimental investigations on the effect of fabric geometry on damage resistance of orthogonal plain weave fabric laminates have been presented. The special case of low energy impact barely visible impact damage has been considered. Percentage of impact energy absorbed, peak impact load, maximum deflection, damage area and impact duration are analysed for different fabric geometries. It is observed that damage resistance of E-Glass/epoxy composites increases with the decrease of strand tex at the same laminate thickness.

INTRODUCTION

Impact damage is generally recognised as the most severe threat to Fibre Reinforced Plastic (FRP) composites and in particular, Barely Visible Impact Damage (BVID) is a hidden menace in composites under low energy impacts. Concentrated efforts have been initiated in recent years to improve the impact performance of FRP composites. Woven Fabric (WF) composites which are an important class of textile composites are emerging as structural materials due to their key role in commercialisation of large, cost effective and damage resistant structures. Fibres in the textile form are used in composites to derive the advantages of textile structures to enhance the impact performance of FRP composites. The literature survey indicates that there is considerable work on impact behaviour of unidirectional (UD) composites [1,2], but similar work on WF composites is still limited.

EXPERIMENTAL INVESTIGATIONS

In the present work, experimental investigations have been carried out to study the effects of fabric geometry and impactor mass at different impact velocities and energies on impact behaviour of WF composites under low energy impacts. Accordingly, specimens fabricated from E-Glass orthogonal plain weave fabrics of three different fabric geometries and UD tapes in epoxy resin LY 556 and hardener HY 951 were tested in instrumented drop weight impact test apparatus. The load (P) / incident impact energy (E) vs time (t) and velocity (V) / deflection (δ) vs time (t) plots were obtained for each test. Further, damage area (A), percentage of impact energy absorbed (Ea/E), peak impact load (Pm), maximum deflection (δm) and impact duration (ta) were the characteristic impact parameters obtained from the above plots. These results were used in qualitative and quantitative damge resistance analysis of different material systems tested.

Three types of E-Glass orthogonal plain weave fabrics were used for laminate / specimen preparation, and they are designated as GLE2, GLE4 and GLE6 (Tables 1 and 2). The fabrics / laminates GLE2, GLE4 and GLE6 had strand widths of 0.39 mm, 0.62 mm and 0.84 mm respectively.

EFFECTS OF FABRIC GEOMETRY ON DAMAGE RESISTANCE

Fabrics used for material systems GLE2, GLE4 and GLE6 were the candidates for studying the effects of fabric geometry.

PEAK IMPACT LOAD (Pm)

The variation of Pm for different fabric geometries with impact energy is shown in Figure 1. The results indicate that the Pm curve for GLE2 is higher than for GLE4 and GLE6. The higher peak load indicates the ability of the material to resist impact. The peak impact load curve for GLE4 is below the GLE6 curve. Also, energy at Pm (Em) follows the same trend as Pm which is higher for GLE2 compared to GLE4 and GLE6. Em indicates more of elastic energy absorbed which is later transferred to the impactor as rebound energy (Er). So higher the Pm and Em, more is the damage resistance under given impact condition.

MAXIMUM DEFLECTION (δm)

The δm plotted for the three fabric geometries w.r.t. impact energy are shown in Figure 2. The results show that δm was the least for specimens made from GLE2. Higher δm was observed for GLE6 and the δm curve for GLE4 remained in-between GLE2 and GLE6. This agrees with the fact that material with more damage resistance shows less deflection because it offers more resistance to deformation. The ta follows the same trend of variation as δm. The transfer of energy to the impactor takes place quicker as observed from the sudden rise in rebound velocity. So less impact duration is another indication of a more damage resistant material.

PERCENTAGE OF ENERGY ABSORBED (Ea/E)

The variation of Ea/E for three fabric geometries indicates that laminate made from GLE2 absorbed least energy. The Ea/E curve for GLE6 seems to be higher than GLE4. The trend shows that more damage resistant material is able to resist the impact event by absorbing less energy. The energy absorbed directly appears in terms of quality and quantity of damage. So, under identical conditions of impact, the material which absorbs more of elastic energy and dissipates in the form of rebound energy without damage or sustaining least damage (less Ea/E) is more damage resistant.

DAMAGE AREA (A)

It was observed that in the range of impact energy used, the specimens made from GLE2 sustained least damage while GLE4 sustained maximum damage and GLE6 sustained damage lesser than GLE4 material system. Simple logic should show that damage sustained by GLE4 should be less than GLE6. But more damage sustained by GLE4 is probably due to its more thickness, which is 5.8 mm while the specimen thickness is 5 mm each for GLE2 and GLE6. Energy was normalised over the thickness assuming a linear relationship, as a result GLE4 was impacted with a higher energy. The linear relationship between specimen thickness and various impact parameters are reported in literature [3]. However, it is also observed that thinner specimens absorb more elastic-plastic energy leaving less energy for initiation and propagation of damage. The effect of thickness reported [4] on graphite/epoxy laminate showed a decrease in damage resistance with increase of thickness from 2 mm onwards and an increase in damage resistance with increase in thickness up to 2 mm. In this context, more damage area seen on GLE4 material system may be partly due to this thickness effect and also due to more impact energy with which it was impacted compared to GLE2 and GLE6. Intensity of damage is also an important parameter in quantifying the impact damage.

RESULTS, DISCUSSION AND CONCLUSIONS

From the above discussions it is clear that A, Ea, δm and ta were minimum for specimens fabricated from GLE2 and maximum for GLE6 and the values remained in-between for specimens fabricated from GLE4 material system. However, Pm and Em were higher at the same impact energy for specimens made from GLE2. This behaviour of increase in damage resistance with decrease of strand tex, strand width, fabric thickness and associated parameters is attributed to the very mechanism involved in distribution and dissipation of impact energy within the specimen. For consistent material behaviour, the minute structural cells must be properly integrated and unified within the assemblage. If the minute structural cells located through out the assemblage are not stressed uniformly, preferential strain areas become the nucleus of the damage. Accordingly, less the strand tex and strand width of the fabric, greater the number of unit cells per unit area on the surface and better is the quality to encounter and distribute the dynamic stress proportionately through out the composite specimen.

During impact, transfer of energy takes place in directions both perpendicular to the impactor and in the plane of specimen [5]. Dissipation of energy along the fibres of WF perpendicular to the impactor is a function of both fibre property and fabric geometry. This mode of transfer of energy is more predominant under low velocity impacts. Hence, fabric geometry plays a major role in distribution and dissipation of impact energy. More the yarns directly encountered during impact, better is the transfer of energy to the surrounding area, because directly impacted yarns share high impact load with the neighbouring yarns at the cross-over points (yarns intersections) in woven fabrics geometry. Therefore, decrease in strand tex increases the cross-over points. Also, lesser the fabric thickness under constant strand thickness to strand width ratio (h/a), better is the interlaminar shear strength. This results in more transverse loading ability. The material which absorbs low total energy but confines the impact damage to relatively small area may be desirable in primary structures provided that initiation energy (Ei), Pm and Em are relatively high. GLE2 materials are highy suitable for such applications. However, fabrication of laminates need more number of plies for given thickness compared to GLE4 and GLE6. Miner [5] also observed similar effects of increase of fragmentation resistance of plain weave Kevlar/epoxy armour protection composites with decrease of strand denier (strand fineness).

Within limitations, comparison of percentage of rebound energy at an impact energy of 6.54 J shows 15 % for carbon/epoxy plain weave composite as reported by Peijs et al. [6] and 3.5 % for carbon/PPE five-shaft satin weave composite as reported by Ghaseminejhad et al. [7]. Also, percentage of impact energy absorbed by E-Glass/epoxy plain weave fabric (strand tex 140 g/km) composite in the present study shows only 50 %, whereas carbon/PPE satin weave fabric composite [7] of the same thickness absorbed 88 % at 20 J of impact energy. This indicates the superior damage resistance of glass compared to carbon and possibly plain weave compared to satin weave. Besides, it was also observed in this work that all E-Glass WF composites displayed better damage resistance compared to UD composites.

REFERENCES

1. Hull, D. and Shi, Y.B., 1993. Composite Structures, 23:99-120.
2. Cantwell, W.L. and Morton, J., 1991. Composites, 22:347-361.
3. Strait, L.H. et al., 1992. J Comp Mater, 26:2119-2133.
4. Abrate, S., 1991. Appl Mech Rev, 44:155-191.
5. Miner, L.H., 1984. in Structural Impact Crashworthiness, Vol II, ed. Davies, G.A.O., Elsevier Applied Science Publishers, London, pp 532-543.
6. Peijs, A.A.J.M. et al., 1990. Composites, 21:522-530.
7. Ghaseminejhad, M.N. and Paravizi-Maid, A., 1990. Composites, 21:154-168.

Table 1. Strand and fabric properties.

MATERIAL	FABRIC WEIGHT (gm/m^2)	CRIMP, C (%)		NO. OF COUNTS (per Cm)		STRAND TEX (gm/Km)	
		FILL	WARP	FILL	WARP	FILL	WARP
GLE2	103	0.70	0.70	23	23	22.5	22.5
GLE4	210	0.96	0.96	14	14	51.0	55.0
GLE6	300	0.99	0.99	12	12	140.0	150.0

Table 2. Properties of Plain Weave Fabric Laminates.

MATERIAL	FABRIC THICKNESS, h_t (mm)	LAMINATE NO. OF LAYERS	LAMINATE THICKNESS, t (mm)	V_f^o	LAMINA THICKNESS, H_L (mm)
GLE2	0.094	40	5.00	0.29	0.13
GLE4	0.180	23	5.80	0.27	0.25
GLE6	0.220	16	5.00	0.33	0.31

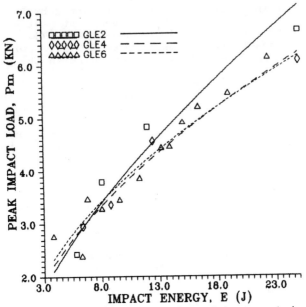

Fig. 1. Variation of peak impact load w.r.t. impact
energy for GLE2, GLE4 and GLE6

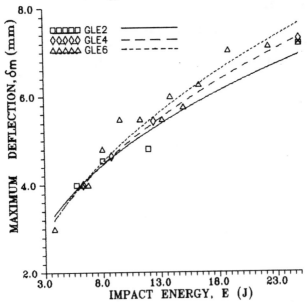

Fig. 2. Variation of maximum deflection w.r.t.
impact energy for GLE2, GLE4 and GLE6

IMPACT TOLERANCE OF SANDWICH PANELS WITH ADVANCED FIBRE FACE SKINS FOR MARINE APPLICATIONS

K. POTTER, P.E. NILSEN

SINTEF Production Engineering - Rick Berkelands Vei 2B
7034 Trondheim - Norway

ABSTRACT.

Foam cored sandwich panels have become a very common structural material for the construction of vessels and other marine applications. The face skins for these panels are generally based on glass fibres and polyester resin. Additional weightsaving would result from the use of higher stiffness and higher strength materials such as carbon fibres or other advanced fibres. One of the major design requirements for vessel constructions is that the sandwich panels shall have acceptable impact resistance. The work reported here was intended to provide baseline information for a comparison between the impact resistance of glass fibre and advanced fibre reinforced sandwich panels.

INTRODUCTION.

The performance of sandwich panels under impact conditions is a complex matter, involving both the materials responses of skin and core, the couplings between these and the overall structural response of the panels. The approach that has been taken here is to test panels of different face skin reinforcements and matrices at a size appropriate to the application, using materials forms and construction methods common in the boatbuilding industry. The equipment used to carry out the impact test work was a fully instrumented, high power, pressure augmented, drop weight rig capable of testing target structures up to 2.1*6*1.4m. The equipment, instrumentation and data processing approaches are fully described in ref1, and will not be repeated here. Tests were carried out at a range of available energies in order to determine the energy required to fully penetrate the panels.

In order to avoid differences in impact performance due to different structural stiffnesses between the panels the thickness of each reinforcement was chosen such that the panel's global bending stiffness was held approximately constant. Alternative approaches are possible such as a constant skin thickness and variable core thickness, but we wished to investigate the properties when using the minimum of costly fibres. The long term allowable design strains for glass fibres are approximately equal to those for carbon fibre, it was therefore considered that panels of equal stiffness would have equal long-term load carrying capability, although the ultimate strength of the glass skins in static testing would exceed that of the advanced fibres.

MATERIALS OF CONSTRUCTION.

The baseline was set by a panel construction used in earlier work, ref 1. This consisted of bidirectionally knitted glass fibre of $800g/m^2$, supplied by Devold Tekstil as DB800. Two plies were used for each face skin, each ply being oriented at +/-45° to the edges of the square test specimens. Contact moulding was used to produce the panels, giving a fibre volume fraction of about 33%. The matrix used was Jotun Polymer polyester resin, Norpol 20-80/catalyst 1. The core used was Barracuda Technologies' H200 PVC foam at 25mm thick.

For the work reported here the decision was taken to retain the use of two plies of reinforcement, to retain the use of 25mm H200 core for all panels, and to retain the use of contact moulding for panel production. To ensure compatibility with normal shipbuilding practices the panels were manufacured in a shipyard, rather than in the laboratory. Three matrices were used, the baseline polyester, Jotun Polymer vinyl ester resin, Norpol 92-40/catalyst 11, and Ciba Geigy epoxy, LY 5052/HY5052. Three reinforcements were chosen, the baseline glass reinforcement and specially knitted carbon fibre and carbon fibre/ultrahigh modulus polyethylene fibre hybrid cloths. The carbon fibre was HTA 5331, 200Tex, from Akzo, and the polyethylene fibre was Dyneema SK60, 133 Tex, from DSM.

The required weight/unit area was calculated for the two special reinforcements, with the aim of achieving equivalent tensile stiffness/ply in each material. The equivalent weight/unit area for the carbon cloth was calculated to be $185g/m^2$, to give an equivalent in plane stiffness to $800g/m^2$ glass cloth. At a tow weight of 200Tex, this required a cloth to be woven with 47.5 tows/10cm width, and this cloth was procured from Devold Tekstil. The use of the Dyneema fibre was intended to discern whether the apparent toughness of the fibre would be translated into good panel impact properties. The mechanical properties of Dyneema fibre are more complex than those of carbon fibre, as the material suffers compressive collapse above about 0.23% compressive strain. For this reason it was felt that the Dyneema content had to be limited to 50% of the total fibres, to avoid excessive reductions in compressive properties. It was estimated that the required total weight/unit area would be $218g/m^2$, breaking down as $140g/m^2$ carbon fibre and $78g/m^2$ Dyneema. These weights would lead to 36 tows of carbon/10cm and 29.5 tows of Dyneema/10cm. To simplify knitting, a cloth structure of 36 tows carbon and 30 tows Dyneema/10cm was chosen. This was knitted on an 11 tow repeat, eg -(C.D.C.D.C.D.C.D.C.D.C)-.

The two special reinforcements were of lower quality than the standard glass fibre cloth, being rather open and prone to misalignment. The hybrid cloth was rather better in this respect than the all-carbon cloth. An improved quality would have arisen if the special reinforcements had been woven at twice the weight and one ply had been used, but this may have created differences due to a different number of interlaminar interfaces between the baseline and special reinforcements. Flat laminates and sandwich panels were produced by Ulstein-Eikefjord. The sample quality was somewhat variable, being best for the well known glass/polyester system and worst for the carbon/epoxy system. Despite this variability all samples were considered to be testable. The use of a shipyard to make the samples has obvious limitations in respect of novel and unfamiliar materials, but is considered to be the best way of acquiring realistic specimens.

The flat laminates were statically tested in tension at 0.90 to the test axis with the following results. Results are expressed in terms of N/mm of sample width to eliminate variability due to volume fraction differences. Average values are quoted below, with coefficient of variation % in brackets.

Test type	Fibre	Polyester	Vinylester	Epoxy
Stiffness	Glass	28590 (3)	28790 (4)	31710 (6)
Stiffness	Carbon	25760 (1)	26840 (4)	28110 (5)
Stiffness	Hybrid	21850 (4)	24770 (2)	24900 (6)
Strength	Glass	608 (3)	606 (5)	570 (4)
Strength	Carbon	283 (9)	312 (7)	327 (11)
Strength	Hybrid	298 (10)	307 (7)	321 (8)
Fail strain %	Glass	2.7 (2)	2.6 (4)	2.4 (2)
Fail strain %	Carbon	1.1 (9)	1.2 (6)	1.2 (11)
Fail strain %	Hybrid	1.4 (10)	1.3 (5)	1.3 (7)

It is apparent from these results that the intended equality of in plane stiffness and thus global panel structural response has not been entirely achieved. The carbon fibre materials are about 10% down in stiffness on the glass baseline and the hybrid materials are about 20% down in stiffness on the baseline. Whilst this is disappointing, the panels will at least be close to each other in terms of their overall stiffness against panel deflection.

IMPACT TESTING.

Both static penetration and instrumented impact tests were carried out using the same 25mm radius hemispherical indenter. The static penetration tests allow the order of events to be considered, give curves free of the noise associated with instrumented impact and allow strain rate sensitivity to be established. A sample static force displacement curve is shown as fig 1. This exhibits an initial rise in load that is largely elastic, reflecting plate and contact stiffnesses. Approximately half way in displacement to maximum load we experience a drop in load. This drop is substantially higher for the glass laminates and is hardly visible for the carbon panels. This load drop is associated with front skin failure, and is followed by penetration of the core with the load still rising. When the core final failure occurs, the load peaks and back-face delamination begins. Results from these tests are shown below.

QUASI-STATIC PENETRATION RESULTS.

Material	Initial slope N/mm	Maximum load kN	Deflection at max load mm	Energy at max load Nm	Deflection maximum mm	Energy maximum Nm
Glass/poly	840	31.5	56	920	65	1060
Glass/vinyl	840	24.5	45	600	62	800
Glass/epoxy	850	24.5	45	580	52	750
Carbon/pol	610	19.5	51	570	89	920
Carbon/vin	650	16.5	39	380	55	590
Carbon/epo	640	17.5	43	430	43	430
Hybrid/pol	710	20.0	42	500	81	930
Hybrid/vin	670	18.0	43	450	65	700
Hybrid/epo	700	18.5	40	430	71	750

Results worthy of note are that in all cases the polyester matrices give the highest energy absorption, the totally brittle behaviour of the carbon/epoxy skinned panel after maximum load, the fact that energy at maximum load is very similar for the carbon and hybrid cases, and the fact that for carbon/polyester and all the hybrid cases substantial energy absorption occurs after maximum load.

SELECTED IMPACT PENETRATION RESULTS. (Using 52.7kg impactor and varying velocity to control available energy, velocities between 3.9 and 10.2m/s were used) Values quoted are in all cases for the lowest energy measured at full penetration.

Material	Available energy Nm	Maximum load kN	Deflection at max load mm	Energy at max load Nm	Deflection maximum mm	Energy maximum Nm
Glass/poly	2199	52	45	1090	78	1690
Glass/vinyl	2760	56	53	1340	78	1920
Glass/epoxy	2040	32	42	700	87	1080
Carbon/pol	773	20.5	41	500	74	670
Carbon/vin	932	20.5	38	460	65	760
Carbon/epo	489	20.5	38	450	47	490
Hybrid/pol	1793	23.5	40	590	58	750
Hybrid/vin	1068	23.0	52	710	90	1010
Hybrid/epo	1100	23.5	40	580	62	710

If we compare the results for quasi-static and dynamic penetration some interesting points emerge. The first is that for all the quasi-static cases the use of polyester resin gave the highest levels of absorbed energy, wheras for the dynamic cases the vinyl ester matrices gave the highest energy. Secondly the degree of "improvement" in energy absorption between the quasi-static and dynamic cases differed greatly in the various reinforcements, see below.

Material	Increase in energy absorbed in dynamic case %	
Glass/poly	59%	
Glass/vinyl	240%	
Glass/epoxy	44%	Average for glass = 80%
Carbon/poly	-27%	
Carbon/vinyl	29%	
Carbon/epoxy	14%	Average for carbon = -1%
Hybrid/poly	-19%	
Hybrid/vinyl	44%	
Hybrid/epoxy	-5%	Average for hybrid = 4%

Average figures are the total energy absorptions for the three matrices in dynamic penetration divided by the total energy absorption in quasi-static penetration. This is not a good figure of merit but does allow for some crude ranking, and one must of course, be very careful of excessive extrapolation from such very small data sets.

It can clearly be seen that the relatively slight advantage of the best glass fibre result in quasi-static penetration over the best carbon and hybrid results, is converted to a much greater advantage in dynamic conditions.

Why dynamic testing should lead to such great improvements over quasi-static testing for all glass reinforcements, and for vinyl ester resins with the other reinforcements, is far from clear. Impact velocities are higher for the glass reinforced plates which may lead to some additional energy absorption from non-linear behaviour of the foam core, and the generally higher toughness and strain of vinyl ester resins compared to polyesters may delay the onset of damage in the dynamic case.

Turning to the results for carbon fibre and hybrid panels, one might be tempted to ask why the carbon fibre panels are not worse with such thin skins and why a greater degree of improvement was not obtained from the use of the Dyneema fibre which is claimed to have high toughness. The carbon fibre face skins were only one third as thick as the glass skins, (0.63mm at 33% Vf), at this sort of thickness one would expect penetration of a single skin at no more than a few Nm, rather than the hundreds of Nm required to penetrate the front skin of the sandwich panel. This clearly demonstrates the importance of structural deflection of sandwich panels as an energy "absorption" process.

The hybrid results show some improvements over those of carbon fibre skins but are in the same region as those for carbon fibre rather than glass fibre. The appearance of failed plates was very different between fibre types. The carbon plates showed little front face damage other than a hole, and back face damage was usually limited to the locality of the exit hole. For glass plates, substantial back face delamination was seen, being greater for the polyester and vinyl ester than for the epoxy, front face damage was also apparent in the form of a cruciform pattern of compressive failure lines to the corners of the support structure. For the hybrids the exit face appearance was closer to that of the carbon case, but the front face showed many compression creases often more than ten individual damage areas without a very obvious preferred orientation, rather than the three or four creases seen in the glass case.

For real structures the impact performance is not entirely related to the maximum impact absorption, residual properties after impact and ease of repairing any damage are also of great importance. In these respects the performance of the carbon panels is rather better than that of the other types as the localised damages and lack of compression failure lines would be expected to lead to higher residual properties.

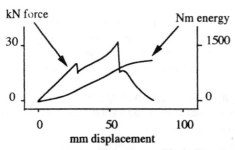

Fig 1. Quasi-static indentation, graph of load and energy against displacement.
Material of skins, glass/polyester. Note load drop half way to peak load. Front skin damage at about 300J applied energy.

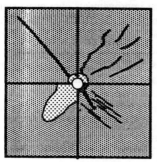

Fig 2. Sketch of the appearance of panels penetrated by impact.
1. Top left. Glass fibre skins, impact face. Delamination around entry point and single crease.
2. Bottom left. Glass fibre skins, exit face. Extensive delamination around exit hole.
3. Top right. Hybrid skins, impact face. Delamination around entry point and multiple creases.
4. Bottom right. Hybrid skins, exit face. laminate failure, core visible, some tows pulled off surface.

ACKNOWLEDGEMENTS. The assistance and support of the following organisations in this work is gratefully acknowledged:- Akzo Faser AG, Barracuda Technologies A/S, Devold Tekstil A/S, DSM High Performance Fibre, Jotun Polymer A/S, NTNF Fast Craft Programme, Ulstein-Eikefjord A/S.

REFERENCE. P.E. Nilsen, T. Moan, C-G Gustafson. Second Int Conf on Sandwich construction. Gainesville, FL, USA 1992.

ON THE LOCAL STABILITY LOSS IN
LAMINATED COMPOSITE STRUCTURES
(THREE-DIMENSIONAL PROBLEM)

I.A. GUZ

Institute of Mechanics - Nesterov Str. 3 - 252057 Kiev - Ukraine

ABSTRACT

A laminated composite of arbitrary structure was studied.
The phenomena of the local instability under beaxial compression in plane of layers was investigated. The research was performed in the frame of the three-dimensional
linearized theory of deformed bodies stability (TLTDBS).
In this research layers was simulated by elastic or elastoplastic, compressible or incompressible, orthotropic
bodies.

INTRODUCTION

The most exact and strong results in the theory of stability of laminated structures may be obtained within the
framework of the model of the piecewise-homogeneous medium with application of TLTDBS [1]. This strict statement
was used for investigation of the internal instability of
laminated composites with periodic structure, which consisted of elastic (compressible and incompressible) and
elastoplastic layers, for the cases of three-dimensional
and plane problems [2-4]; surface instability of laminated half-plane, half-space [5]. It is approved in all papers above, that the loss of stability occurs by the stability loss forms (SLF), which are periodic along boundaries of layers. But it is known from experimental resear-
ches that also local SLF may take place in laminated
structures, therefore, it is necessary to study this phe-

nomena accuratly. The present paper is devoted to strict investigation of local instability of composites with arbitrary laminated structure for the SLF, which are representable in the Fourier's integral form.

1. STATEMENT OF PROBLEM

The composite material of arbitrary laminated structure with parallel to the some plane boundaries of layers will be studied. It may consist from arbitrary combination of layers and half-spaces. Every package layer may have different characteristics (thickness and governing equations). For example, one of these structures is represented on Fig.1. local instability of one of the such structures under beaxial compression by "dead" loads in plane $x_1 0 x_2$ (see Fig.2) is discussed. Compressing stresses along axis $0 x_1$ and $0 x_2$ may be different. The research is performed in the unified general form for the finite strains theory and two versions of the small initial strains theory of TLTDBS [1] using Lagrangian coordinates for every layer, which coincide within precritical state with Cartesian coordinates. Layers and half-space are simulated by elastic or elastoplastic compressible or incompressible homogeneous along axis $0 x_1$ and $0 x_2$ bodies. Under "dead" loads sufficient conditions for the applicability of the static method of investigation of static TLTDBS's problems are fulfilled. That is the substantiation for its using in the present paper. In the case of elastoplastic bodies the continuing loading conception is applied and, therefore, changing of "loading – unloading" zones during the stability loss isn't taken into account. Let $Q(\beta)$ be number of layers in β -th package, index (ij) means "i-th layer of j -th package". Values of precritical state are marked by upper index "0". Main TLTDBS's equations for every compressible layer are

$$\frac{\partial}{\partial x_i} \left(\omega_{ij\alpha\beta}^{(\kappa\ell)} \frac{\partial u_\alpha^{(\kappa\ell)}}{\partial x_\beta} \right) = 0 \tag{1}$$

for every incompressible layer (with incompressibility conditions)

$$\frac{\partial}{\partial x_i} \left[\mathbb{a}_{ij\alpha\beta}^{(\kappa\ell)} \frac{\partial u_\alpha^{(\kappa\ell)}}{\partial x_\beta} + \left(\delta_{ij} + \frac{\partial u_j^{0(\kappa\ell)}}{\partial x_i} \right) p^{(\kappa\ell)} \right] = 0 \tag{2}$$

$$\left(\delta_{ij} + \frac{\partial u_j^{0(\kappa\ell)}}{\partial x_\beta} \right) \frac{\partial u_j^{(\kappa\ell)}}{\partial x_i} = 0$$

Relation between displacements and components of asymmetric tensor of stresses of Kirchhoff is (for compressible and incompressible layers, correspondingly)

$$t_{ij}^{(\kappa\ell)} = \omega_{ij\alpha\beta}^{(\kappa\ell)} \frac{\partial u_\alpha^{(\kappa\ell)}}{\partial x_\beta} \tag{4}$$

$$t_{ij}^{(\kappa\ell)} = \mathcal{ae}_{ij\alpha\beta}^{(\kappa\ell)} \frac{\partial u_\alpha^{(\kappa\ell)}}{\partial x_\beta} + \left(\delta_{ij} + \frac{\partial u^{o(\kappa\ell)}}{\partial x_i}\right)p^{(\kappa\ell)} \qquad (5)$$

Homogeneous boundary conditions are
- on the boundary between layers S_1 :

$$t_{3\beta}^{(\kappa\ell)}\Big|_{S_1} = t_{3\beta}^{(\kappa+1,\ell)}\Big|_{S_1} \; ; \; u_\beta^{(\kappa\ell)}\Big|_{S_1} = u_\beta^{(\kappa+1,\ell)}\Big|_{S_1} , \quad \beta = 1,2,3 \qquad (6)$$

- on the free surface S_2 (if it presents investigated structure)

$$t_{3\beta}^{(a(\ell),\ell)}\Big|_{S_2} = 0 \; ; \; u_\beta\Big|_{x_3 \to \infty} = 0, \quad \beta = 1,2,3 \qquad (7)$$

Main TLTDBS's equations and corresponding to structure homogeneous boundary conditions are the eigenvalue problem with respect to the loading parameters, which are included in components of tensors $w^{(\kappa\ell)}$ (for compressible layers) and $\mathcal{ae}^{(\kappa\ell)}$ (for incompressible layers). These tensors also depend on governing equations for layer and are determined by formulas [1] .

2. ANALYSIS OF THE EIGENVALUE PROBLEM

Let components of stresses and displacements and (for incompressible layers) also scalar $p^{(\kappa\ell)}$ for the local along axis Ox_1 and Ox_2 SLF be representable in the Fourier's integrals form

$$u_j^{(\kappa\ell)}(x_1,x_2,x_3) = \frac{1}{2\pi} \int\limits_{-\infty}^{+\infty}\!\!\int \frac{\mathcal{U}_j^{(\kappa\ell)}(x_3,\alpha_1,\alpha_2)}{exp[i(x_1\alpha_1 + x_2\alpha_2)]} d\alpha_1 d\alpha_2$$

$$t_{ij}^{(\kappa\ell)}(x_1,x_2,x_3) = \frac{1}{2\pi} \int\limits_{-\infty}^{+\infty}\!\!\int \frac{\mathcal{T}_{ij}^{(\kappa\ell)}(x_3,\alpha_1,\alpha_2)}{exp[i(x_1\alpha_1 + x_2\alpha_2)]} d\alpha_1 d\alpha_2 \qquad (8)$$

$$p^{(\kappa\ell)}(x_1,x_2,x_3) = \frac{1}{2\pi} \int\limits_{-\infty}^{+\infty}\!\!\int \frac{\beta^{(\kappa\ell)}(x_3,\alpha_1,\alpha_2)}{exp[i(x_1\alpha_1 + x_2\alpha_2)]} d\alpha_1 d\alpha_2$$

Then the eigenvalue problem (1)-(7) is reducable to the eigenvalue problem for usual differential equation with respect to x_3 (after substitution (8) to (1)-(7)):
- equations for compressible layers (from (1))

$$exp[i(x_1\alpha_1 + x_2\alpha_2)]\frac{\partial}{\partial x_i}\left[w_{ij\alpha\beta}^{(\kappa\ell)}\frac{\partial}{\partial x_\beta}\left(\frac{\mathcal{U}_\alpha^{(\kappa\ell)}(x_3,\alpha_1,\alpha_2)}{exp[i(x_1\alpha_1 + x_2\alpha_2)]}\right)\right] = 0 \qquad (9)$$

- equations and incompressibility conditions for incompressible layers (from (2))

$$exp[i(x_1\alpha_1 + x_2\alpha_2)]\frac{\partial}{\partial x_i}\left[\mathcal{ae}_{ij\alpha\beta}^{(\kappa\ell)}\frac{\partial}{\partial x_\beta}\left(\frac{\mathcal{U}_\alpha^{(\kappa\ell)}(x_3,\alpha_1,\alpha_2)}{exp[i(x_1\alpha_1 + x_2\alpha_2)]}\right)\right] +$$

$$+\left(\delta_{ij} + \frac{\partial u_j^{o\,(\kappa\ell)}}{\partial x_i}\right)\beta^{(\kappa\ell)}(x_3, \alpha_1, \alpha_2)\,exp[-i(x_1\alpha_1 + x_2\alpha_2)]\Big] = 0 \qquad (10)$$

$$exp[i(x_1\alpha_1 + x_2\alpha_2)]\left(\delta_{ij} + \frac{\partial u_j^{o\,(\kappa\ell)}}{\partial x_i}\right)\frac{\partial}{\partial x_i}\left(\frac{v_j^{(\kappa\ell)}(x_3, \alpha_1, \alpha_2)}{exp[i(x_1\alpha_1 + x_2\alpha_2)]}\right) = 0$$

- relations between displacements and stresses (for compressible and incompressible layers from (4) and (5), correspondingly)

$$\tau_{ij}^{(\kappa\ell)} = exp[i(x_1\alpha_1 + x_2\alpha_2)]\,\omega_{ij\alpha\beta}^{(\kappa\ell)}\frac{\partial}{\partial x_\beta}\left(\frac{v_\alpha^{(\kappa\ell)}(x_3, \alpha_1, \alpha_2)}{exp[i(x_1\alpha_1 + x_2\alpha_2)]}\right) \qquad (11)$$

$$\tau_{ij}^{(\kappa\ell)} = exp[i(x_1\alpha_1 + x_2\alpha_2)]\,\mathscr{X}_{ij\alpha\beta}^{(\kappa\ell)}\frac{\partial}{\partial x_\beta}\left(\frac{v_\alpha^{(\kappa\ell)}(x_3, \alpha_1, \alpha_2)}{exp[i(x_1\alpha_1 + x_2\alpha_2)]}\right) + \qquad (12)$$

$$+\left(\delta_{ij} + \frac{\partial u_j^{o\,(\kappa\ell)}}{\partial x_i}\right)\beta^{(\kappa\ell)}(x_3, \alpha_1, \alpha_2)$$

- boundary conditions (from (6), (7))

$$\tau_{3\beta}^{(\kappa\ell)}\Big|_{S_1} = \tau_{3\beta}^{(\kappa+1,\ell)}\Big|_{S_1}\,; \quad v_\beta^{(\kappa\ell)}\Big|_{S_1} = v_\beta^{(\kappa+1,\ell)}\Big|_{S_1}, \qquad (13)$$

$$\beta = 1, 2, 3$$

$$\tau_{3\beta}^{(Q(\ell),\ell)}\Big|_{S_2} = 0\,; \quad v_\beta\Big|_{x_3 \to \infty} = 0\,, \qquad (14)$$

In the case of internal instability of laminates by periodic along boundaries of layers SLF, representation of components of $\vec{u}^{(\kappa\ell)}$, $t^{(\kappa\ell)}$ and (for incompressible layers) $p^{(\kappa\ell)}$ may be written as

$$u_j^{(\kappa\ell)}(x_1, x_2, x_3) = V_j^{(\kappa\ell)}(x_3, \pi/\ell_1, \pi/\ell_2)\,exp[-i(\pi/\ell_1 x_1 + \pi/\ell_2 x_2)]$$

$$t_{ij}^{(\kappa\ell)}(x_1, x_2, x_3) = R_{ij}^{(\kappa\ell)}(x_3, \pi/\ell_1, \pi/\ell_2)\,exp[-i(\pi/\ell_1 x_1 + \pi/\ell_2 x_2)] \qquad (15)$$

$$p^{(\kappa\ell)}(x_1, x_2, x_3) = B^{(\kappa\ell)}(x_3, \pi/\ell_1, \pi/\ell_2)\,exp[-i(\pi/\ell_1 x_1 + \pi/\ell_2 x_2)]$$

where ℓ_1, ℓ_2 are lengthes of half-waves of SLF along axis $0x_1$ and $0x_2$. It must be underlined, that problems of internal instability by periodic along boundaries of layers SLF were solved for the particular cases of laminated structures in [2-5]. After substitution (15) to (1)-(7) the eigenvalue problem will be obtained for periodic SLF. This eigenvalue problem will coincide with eigenvalue problem (9)-(14), if only parameters α_1 and α_2 will be replaced by parameters π/ℓ_1 and π/ℓ_2. Therefore, values of critical loading parameters, which are solutions of corresponding eigenvalue problem, coincide for both problems above, namely, for the local instability and internal instability by periodic SLF of composites with arbitrary laminated structure.

3. NUMERICAL RESULTS

The above theoretical discussion may be illustrated by following numerical results for composite materials composed of alternating layers of aluminium and boron or aluminium and steel. Aluminium is modelled by an incompressible elastic body with power-mode dependence between intensities of stresses and strains ($\sigma_i^o = A(\varepsilon_i^o)^\kappa$); boron and steel - by a linear-elastic compressible body. Thickness of aluminium is h_m ; thickness of boron (or steel) is h_a. The solution of the internal stability problem by periodic SLF was represented in [3] for such structure. According to theoretical conclusions of the present article, we may use these results solving the problem of local instability of the same composite structure. Dependences of critical strain (ε_c) – on thickness ratio of layers (h_a/h_m) were obtained after corresponding analysis (see Fig.3). On Fig.3 continuous lines 1-3 correspond to aluminium-steel composites, 4,5 - to aluminium-boron composites. Values of parameters A, k (for aluminium) and E, ν (for boron and steel) are presented in Table 1.

CONCLUSIONS

It is regorously proved in the article that values of critical loading parameters for the arbitrary laminated composite structure obtained in the case of any local SLF and in the case of periodic along layers SLF coincide. This is proof for elastic or elastic-plastic compressible or incompressible isotropic or orthotropic layers. Numerical results are also presented in this paper.

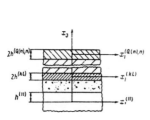

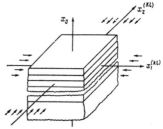

Fig.1. Laminated structure with   Fig.2. Scheme of
       approved coordinates   loading

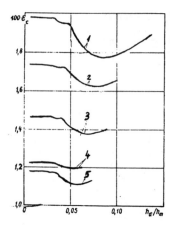

Fig.3. Dependences of
critical strain
on thickness
ratio of layers

Table 1. Values of parameters
for Fig.3

E ,GPa	γ	A,MPa	k	N
200	0,3	100	0,1	1
200	0,3	100	0,25	2
200	0,3	68	0,25	3
400	0,21	70	0,25	4
400	0,21	130	0,43	5

REFERENCES

1. Guz A.N. "Foundations of Three-Dimensional Theory of
   Deformable Bodies Stability". Vyshcha Shkola. Kiev.
   (1986) (In Russian).
2. Guz I.A. Prikladnaya Mekhanika. Vol.25. N 11. (1989)
   pp.26-31. (In Russian).
3. Guz I.A. Prikladnaya Mekhanila. Vol.25. N 12. (1989)
   pp.35-41. (In Russian).
4. Guz I.A. Mekhanika Kompozitnykh Materialov. N 6.
   (1990) pp.1051-1056. (In Russian).
5. Chekhov V.N. Prikladnaya Mekhanika. Vol.20. N 11.
   (1984) pp.35-42. (In Russian).

THE APPLICATION OF PROGRESSIVE DAMAGE MODELLING TECHNIQUES TO THE STUDY OF DELAMINATION EFFECTS IN COMPOSITE MATERIALS

M. PILLING, S. FISHWICK

AEA Technology Consultancy Services - Wigshan Lane
Culcheth - WA3 4NE CHESHIRE -UK

ABSTRACT

The analysis and prediction of the development of damage in composite materials represents an important area of study. The application of the techniques of progressive damage modelling, using finite element analysis, has proved to be a valuable methodology for investigating the propagation of various damage mechanisms.

The present work focuses on applying these techniques to the study of delamination effects in carbon fibre reinforced composite materials. In particular, modelling work has been carried out to investigate the aspect of free edge delamination in specimens loaded in tension, using the ABAQUS finite element analysis code.

INTRODUCTION

The analysis and prediction of the development of damage in composite materials represents a most important area of study at the present time. The application of the techniques of progressive damage modelling, using finite element analysis, has demonstrable value as a future design tool for routine use by materials engineers [1 - 4].

1. SPECIMEN GEOMETRY

The present work has focused on applying progressive damage modelling techniques to the study of free edge delamination effects in carbon fibre reinforced epoxy based composites. The modelling work has been carried out using the ABAQUS finite element analysis code [5] using techniques developed at SRD [6]. The specimen lay-up considered during the investigation, was based on that used by Schellekens and de Borst [7], having a [+25 / -25 / 90]$_s$ lay-up. One particular area of interest was the resin-rich layer, found between the laminates in a multi-ply lay-up, these being modelled as a separate layers, assuming mechanical properties appropriate for the resin system. The importance of utilising suitable material failure criteria and material property degradation rules have also been considered.

The analyses considered a three-dimensional model of a specimen having the dimensions shown in Fig. 1. Considerations of symmetry dictated that one-eighth of the specimen would adequately represent the behaviour, and the finite element model adopted is shown in Fig. 2. The model consisted of 1662, 8 node plane strain elements, to model the three plies and three resin-rich interface layers. An increased area of mesh refinement was used in the centre section of the specimen.

A cross sectional representation of the specimen showing the three interfacial, resin-rich layers is presented in Fig. 3.

2. PERFORMANCE OF THE MODEL

Incrementally increasing strains were applied to the model and failure criteria based on the following formulation were applied to individual elements in the model:

$$\sqrt{(S_{11}/S_{UTS})^2 + (S_{22}/S_{UTS})^2 + (S_{33}/S_{UTS})^2 + (S_{12}/S_{SS})^2 + (S_{13}/S_{SS})^2 + (S_{23}/S_{SS})^2}$$

$$> = 1 \quad , \text{ leads to failure of the material of the element} \qquad (1)$$

where: S_{nn} = stress in the directions indicated in Figs. 1 and 2, n = 1,2 or 3
             S_{UTS} = ultimate tensile failure stress
             S_{SS} = shear failure stress

Suitable material property degradation rules were applied and the effect on the predicted evolution of damage was investigated.

3. RESULTS

The damage propagation predicted by the model, as a function of applied strain for the central resin-rich interfacial layer, is shown in Fig. 4. The failure is essentially Mode I, being driven by the S_{33} component of stress. The area of delamination is seen to grow inwards from the edge of the specimen, and some internal delamination is seen prior to total failure of the layer.

The corresponding development of damage in the 90° ply is shown in Fig. 5. This is

270

essentially transverse ply cracking, driven by the S_{11} component of stress.

The results obtained may be compared with those presented in [7], in particular with reference to the growth of free edge delamination, and the comparison is shown in Fig. 6, where reasonable agreement is seen.

4. CONCLUSIONS

The present work has shown that the mechanical behaviour of and damage propagation in multi-ply composite materials may be adequately represented using finite element analysis methods and the techniques of progressive damage modelling. The importance of modelling the resin-rich interfacial layer has been demonstrated.

At the present time the methodology discussed here is being applied to various specimen, defect geometries, such as the Double Edge Notch (DEN) geometry, in order to predict the accumulation of damage under load.

In future work, it is proposed to examine the aspects of the probabilistic approach to damage development in composite materials, in order to account for the effects of voids and localised variability in materials properties.

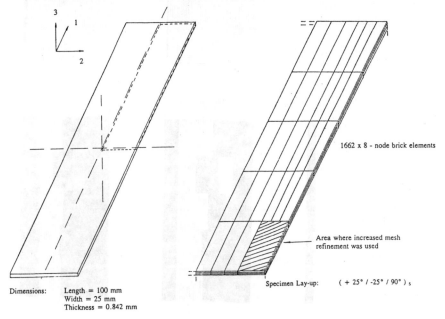

Dimensions: Length = 100 mm
               Width = 25 mm
               Thickness = 0.842 mm

1662 x 8 - node brick elements

Area where increased mesh refinement was used

Specimen Lay-up: (+ 25° / -25° / 90°)ₛ

Fig. 1   Specimen Geometry   Fig. 2   Finite Element Model Used for Delamination Study

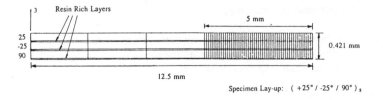

Specimen Lay-up: (+25° / -25° / 90°)ₛ

Fig. 3   Cross Section of Model Showing Resin-Rich Layers

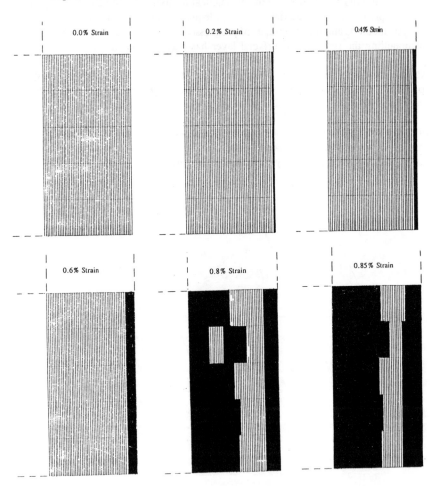

Fig. 4   Development of Damage in the Central Resin-Rich Layer

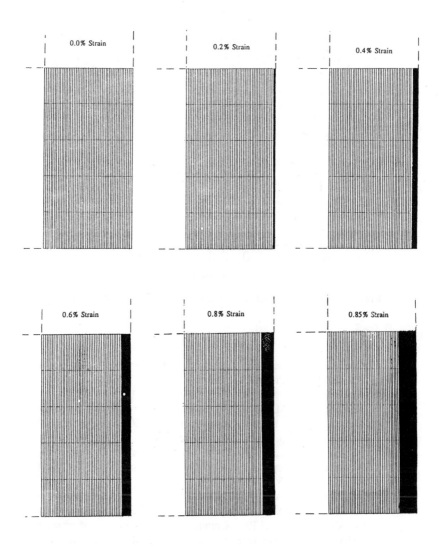

Fig. 5 Development of Damage in the Transverse Ply

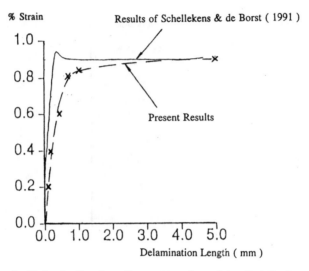

Fig. 6   Delamination Length as a Function of Applied Strain

REFERENCES

1. Lessard, L., Poon, C. J., Fahr, A., "Composite Pinned Joint Failure Modes Under Progressive Damage", ECCM-5, Proc. Fifth Int. Conf. on Comp Mats., 49-54, April 1992.

2. Billoet, J.-L., Matheron, G., "Generalised Progressive Damage Model for Composite Multilayered Structures", ECCM-5, Proc. Fifth Int. Conf. on Comp. Mats., 71-76, April 1992.

3. Chang, F. K., Chang, K. Y., "A Progressive Damage Model for Laminated Composites Containing Stress Concentrations", J. Comp. Mat., 21, 834-855, 1987.

4. de Rouvary, A., Dowlatyari, P., Haug, E., "Validation of the PAM/FISS/bi-phase Numerical Model for the Damage and Strength Prediction of LFRP Composite Laminates", Mechanics and Mechanisms of Damage in Composites and Multi-materials, ESISII (Ed. D. Baptiste), Mechanical Engineering Publications, London, 1991.

5. ABAQUS Version 5.2, Hibbit, Karlsson and Sorenson Inc.

6. Gamble, K. A., Wilson, A. M., MacInnes, D. A., "The Modelling of Fracture Processes in Fibre-Reinforced Composites", The Second European Conf. on Advanced Materials and Processes, EUROMAT 91, Univ. of Cambridge, UK, 22-24 July 1991.

7. Schellekens, J. C. J., de Borst, R., "Numerical Simulation of Free Edge Delamination in Graphite-Epoxy Laminates under Uniaxial Tension", Composite Structures 6, Proc of the 6th Int Conf on Comp Structures, Paisley College, Scotland, Sept 1991.

EFFECTS OF FIBRE SURFACE TREATMENT AND TEST TEMPERATURE ON MONOTONIC AND FATIGUE PROPERTIES OF CARBON FIBRE EPOXY CROSS PLY LAMINATES

Y. MATSUHISA, P.J. WITHERS*, J.E. KING*

Toray Industries Inc. - 1515 Tsutsui - Masaki-cho Iyo-Gun Ehime 791 31 - Japan
*University of Cambridge - Dept of Materials Science and Metallurgy Pembroke Street - Cambridge CB2 3QZ - UK

ABSTRACT

The effect of fibre surface treatment on the damage accumulation behaviour of carbon fibre epoxy cross ply laminates has been studied in monotonic and fatigue loading at room temperature. The surface treatments involved were untreated, half, standard and double treatment. For standard treatment fibres the effect of test temperature between -40°C and 100°C has also been investigated. Preliminary results on temperature effects are also presented as a function of surface treatment levels. The damage accumulation behaviour was found to vary significantly as a function of both surface treatment and test temperature.

INTRODUCTION

One of the most important internal boundaries in a composite material is the interface between fibre and matrix. This interface plays a major role in controlling mechanical and physical properties. In composite systems designed for aircraft applications the environmental temperature is an important variable to be considered. Aircraft skin temperatures can fluctuate typically between -40°C and 100°C. In order to optimise material processing for composites for use in aircraft, it is therefore important to understand the effect of interface strength as a function of temperature on mechanical properties. This paper presents an investigation of the effects of surface treatment and test temperature on the monotonic and fatigue properties of carbon fibre reinforced epoxy cross ply laminates.

1. EXPERIMENTAL
1.1. Materials
"Torayca" T800H, a high tensile strength, 5.49 GPa, and intermediate modulus, 294 GPa, type carbon fibre, and Toray resin system #3631, a semi-toughened 180°C cure epoxy resin, were used as the basic material for this study. The fibre volume fraction

was 0.6. The laminae of T800H/#3631 were stacked in a $[0_2/90_4]_s$ configuration. The test specimens were 1.6 mm thick, 25 mm wide and had a 120 mm long gauge length. Experimental carbon fibres based on the T800H with three different surface treatment levels, untreated, half and double treatment, were used to study the effect of surface treatment. Increasing the level of surface treatment increases the density of functional groups on the fibre surface, producing stronger fibre/resin chemical bonding.

1.2. Test Methods
Monotonic and fatigue tests were carried out at room temperature to study the effect of surface treatment using carbon fibres with four different surface treatment levels including standard. Monotonic and fatigue tests were also conducted at low (-40°C) and elevated temperature (100°C) to study the effect of test temperature using carbon fibres with standard surface treatment. Preliminary tests to examine the interaction between surface treatment and test temperature were run at 100°C using carbon fibres with three different surface treatments. Monotonic tests were carried out at a crosshead speed of 1.0 mm/min. Fatigue tests were conducted in tension-tension loading at a constant stress ratio, R=0.1, with a sinusoidal waveform at 5Hz. The maximum stress was 500MPa. Optical microscopy was employed to detect the damage accumulation on the polished edges. X-ray radiography was also conducted using zinc iodide dye.

2. RESULTS AND DISCUSSION
In both monotonic and fatigue tests, the first damage to appear was transverse cracks in 90° plies. With further increase in load or number of fatigue cycles, these cracks grow and increase in number. In some test conditions this is followed by longitudinal cracking in 0° plies, delamination between 0° and 90° plies and fibre fractures.

2.1. Effect of surface treatment
2.1.1. Monotonic tests
Cracking density increased significantly with decreasing surface treatment, as shown in Figure 1(a). Untreated carbon fibre laminates generated transverse cracks during the curing/cooling process before mechanical testing due to the effects of thermal residual stress and the weak fibre/resin adhesion. The initiation stress increased and saturation density of transverse cracks decreased with higher surface treatment level. The number of longitudinal cracks was highest for untreated fibres. Very little delamination was observed in any of the monotonic tests.

2.1.2. Fatigue tests
Cracking behaviour in fatigue also varied significantly as a function of the surface treatment, as Figure 1(b) shows. The number of cycles to transverse crack initiation and saturation for the standard treatment was almost the same as for the double treatment, and much higher than for untreated and half treated carbon fibre laminates. As for the monotonic loading, the number of longitudinal cracks was largest in untreated fibres and almost no delamination was observed in all tests.

2.1.3. Comparison of monotonic and fatigue behaviour
For the untreated, half and standard treated fibres, approximately the same saturation crack density was attained for both types of loading, (around 1.1 cracks/mm for untreated and half treated, 0.5 cracks/mm for standard), despite the maximum fatigue stress being below the saturation stress level for monotonic loading in all cases. The final saturated crack density, the so called characteristic damage state (CDS), predicted by Reifsnider /1/ was calculated to be 0.7 cracks/mm. This CDS value should be constant for these laminates which have the same stiffness, ply thickness and stacking sequence. Nevertheless in the case of these experiments the actual saturation densities were quite different. This indicates that the CDS value also depends on other parameters, such as fibre/resin adhesion and test temperature. For the double treated

fibres, a larger number of transverse cracks develop in fatigue because of the local resin plastic flow during repeated loading.

2.2. Effect of test temperature
2.2.1. Monotonic tests
The damage accumulation behaviour varied significantly as a function of the test temperature, as shown in Figure 2(a). Because of the reduction in thermal residual stress, the number of transverse cracks at 100°C was expected to be the lowest, the trend observed by Boniface et al. /2/. In the present experiments the actual number of transverse cracks was largest at 100°C at all stress levels. Predictions of the transverse crack densities at -40 and 100°C based on Bailey's equation /3/, which considers the mismatch in coefficient of thermal expansion between 0° and 90° plies, are shown as dotted lines in Figure 2(a). Both predictions lie some way from the actual data. Applying Withers' model /4/, based on Eshelby's equivalent elastic inclusion approach, which takes into account the thermal residual stress between each filament and matrix as well as the thermal expansion difference between the plies, the fit between the predictions and the data improve markedly as shown by the solid lines in the same figure, especially at -40°C. This suggests that transverse ply cracking is controlled by the local tensile stress at the fibre/matrix interface, and that no significant change in interfacial strength occurs between room temperature and -40°C. However, the data at 100°C still show the opposite trend to the prediction. SEM observation of the transverse crack surfaces showed that the amount of resin adhesion on fibre surfaces was less at 100°C than at the lower temperatures and crack propagation path was different from those at the lower temperatures. These changes appear to explain the large number of transverse cracks at 100°C. These change could arise both from changes in chemical bonding and matrix properties.

2.2.2. Fatigue tests
The damage accumulation behaviour was a strong function of the test temperature in fatigue as in monotonic tests, but exhibited a different trend, as shown in Figure 2(b) /5/. The number of cycles to transverse crack initiation and saturation was largest at room temperature. Early initiation and saturation at 100°C are attributed to the changes in fibre/resin adhesion and crack propagation path as in the monotonic tests. The number of transverse cracks was largest at -40°C in fatigue, although the data lie between that for room temperature and 100°C under monotonic loading.

2.2.3. Comparison of monotonic and fatigue behaviour
The main differences between monotonic and fatigue tests were in the number of transverse cracks at -40°C and the degree of delamination at 100°C.
In both monotonic and fatigue loading the damage accumulation behaviour at room temperature and -40°C shows the effect of temperature observed by Boniface et al. /2/, and expected from the increase in thermal residual stress at -40°C. However at 100°C transverse ply cracking develops earlier and to a higher density than would be expected, apparently because of changes in interfacial strength and crack propagation path at this temperature. These changes would be expected to increase the longitudinal crack density in a similar way to the transverse ply crack density. This effect is shown clearly for the untreated fibre at room temperature, where both types of cracking increase together. This suggests that the absence of longitudinal cracking at 100°C is associated with a change in transverse ply crack tip geometry. Resin flow at 100°C is thought to blunt the transverse ply crack tips, making them less effective stress concentrators from the point of view of initiating longitudinal cracking in 0° plies.
Almost complete delamination along the entire specimen occurred in fatigue at 100°C, at which temperature no delamination was observed in monotonic tests. Resin plastic flow will occur to a larger extent at high temperature in the resin rich region between 0° and 90° plies because of the low yield stress of neat resin /6/. This appears to be an

277

important factor for delamination under fatigue loading. Creep due to the long period of loading at elevated temperature could also be a factor encouraging delamination in fatigue.

2.3. Interaction between the effects of surface treatment and test temperature
Preliminary tests to examine the interaction between surface treatment and test temperature showed that the effect of elevated temperature is dependent on the surface treatment level. The effect of elevated temperature was significant with standard and double treated fibres and almost negligible with untreated and half treated fibres. The results suggest that the effect of elevated temperature on fibre/resin adhesion is not a simple mechanical clamping effect of the resin, but that there appears to be an interaction with chemical adhesion. Further tests to examine the interactions between surface treatment, test temperature and resin properties will be the theme of future research.

3. CONCLUSIONS
1) The damage accumulation behaviour was found to vary significantly as a function of both the surface treatment and test temperature.
2) The effects of the surface treatment and temperature on the damage accumulation behaviour were found to be different for monotonic and fatigue loading, particularly in the case of test temperature.
3) Surface treatment affected fibre/resin adhesion markedly, but did not affect delamination behaviour at room temperature.
4) Elevated temperature accelerated resin plastic flow in the resin rich region between 0° and 90° plies, causing delamination in fatigue loading.
5) Elevated temperature appeared to change interfacial strength and crack propagation path, such that damage accumulation was most severe at 100°C.
6) Preliminary tests to examine interaction between surface treatment and temperature suggested that the effect of elevated temperature on fibre/resin adhesion is not a simple mechanical clamping effect of the resin.
7) By taking account of micro thermal residual stress between each filament and the matrix, the fit between the prediction of transverse cracking behaviour and the experimental data was markedly improved.

REFERENCES

1. Reifsnider K.L., "Some Fundamental Aspects of the Fatigue and Fracture Response of Composite Materials", Proc. 14th Annual Meeting of Society of Engineering Science, (1977) p.373
2. Boniface L., Ogin S.L., Smith P., "The effect of temperature on matrix crack development in cross ply polymer composite laminates", Proc. ECCM5, (1992) p.141
3. Bailey J.E., Curtis P.T., Parvizi A., "On the transverse cracking and longitudinal splitting behaviour of glass and carbon fibre reinforced epoxy cross ply laminates and the effect of Poisson and thermally generated strain", Proc. R. Soc. Lond. A366, (1979) p.599
4. Withers P.J., Stobbs W.M., Pedersen O.B., "The application of the Eshelby method of internal stress determination to short fibre metal matrix composites", Acta Metall., Vol.37, No.11, (1989) p.3061
5. Matsuhisa Y., King J.E., "Effects of test temperature on monotonic and fatigue behaviour of carbon fibre cross ply laminates", Proc. Int. Conf. on Fatigue, (1993) p.1373
6. Ishibashi S., Masters thesis, Kyoto Institute Technology, (1990)

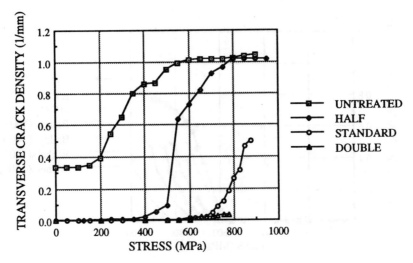

(a)

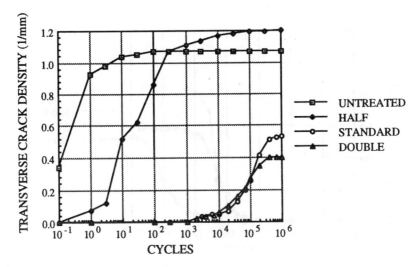

(b)

Figure 1. Effect of surface treatment on transverse cracking behaviour
at room temperature in (a) monotonic and (b) fatigue tests.

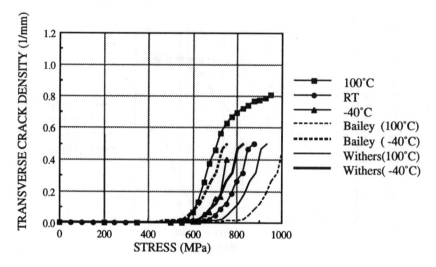

(a)

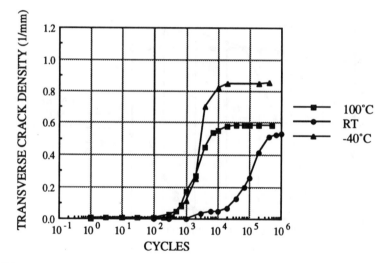

(b)

Figure 2. Effect of test temperature on transverse cracking behaviour for material with standard surface treatment in (a) monotonic and (b) fatigue tests.

PROBLEMS OF THE DETERMINATION OF THE INTERFACE STRENGTH WITH THE AID OF THE SINGLE FIBER PULL-OUT TEST

C. MAROTZKE

Federal Institute for Materials Research and Testing (BAM)
Unter den Eichen 87 - 1000 Berlin 45 - Germany

ABSTRACT

An analysis of the stress field arising in single fiber pull-out test specimen is performed using the finite element method in order to estimate the reliability of the recorded data under various test conditions. To this end, the interfacial stresses are calculated for a glass fiber embedded in a polycarbonate matrix. Furthermore, the influence of the shape of the matrix at the fiber entry (meniscus), of the fiber type (glass or carbon fiber) and of the fiber aspect ratio on the stresses governing the failure initiation is analysed.

INTRODUCTION

The micromechanical test methods used for the determination of the interface strength, like the single fiber pull-out test, the microdroplet test, the indentation test and the fragmentation test suffer from the problem that the stress transfer behaviour is very complex and does not allow the determination of material properties in terms of a continuum mechanical material theory. Some of the main problems are outlined in the following :

- polymer matrices only behave linear elastically within a small range of strains. In general, they exhibit a nonlinear elastic, viscoelastic or even viscoplastic behaviour

- the material properties in the vicinity of the fiber are varying with the distance from the fiber (interphase problem)

- the adhesion between fiber and matrix may be inhomogeneous

- the stress field arising in the vicinity of the fiber possesses severe concentrations, which locally give rise to a material behaviour outside the frame of the continuum mechanics, e.g. crazing

- the specimen may be prestressed by self-stresses due to the mismatch of the thermal extension coefficients or due to the restrained curing in the neighbourhood of the fiber

However, even if all these effects of complex material behaviour are disregarded and an ideal linear elastic material behaviour as well as a perfect bond between the two phases is presumed, it is not possible to evaluate a material property, for example an 'interfacial shear strength' from the measured test data. This will be illuminated in the following on the example of the single fiber pull-out test.

The model used for the stress analysis consists of a cylindrical fiber embedded in a halfspace of matrix (fig. 1). There does not exist a complete analytical solution of this boundary value problem. The only nearly exact solution within the linear theory of elasticity was given by Muki and Sternberg /1/, who obtained a Fredholm integral equation, which has to be solved numerically. There have been made many attempts to obtain a simplified solution for the axial fiber stresses and the interfacial shear stresses in form of a simple formula using the so called shear lag theory /2,3/. However, this approach is based on very restrictive assumptions about the stress field around the fiber, neglecting the stress concentrations at the fiber entry as well as at the fiber end. This formula only gives a crude estimation of the interfacial shear stresses and is limited to very special fiber length to diameter ratios as well as stiffness ratios, as shown by comparative finite element studies /4/. In this paper, the finite element method is utilized for the solution of the respective boundary value problem, reducing the halfspace to a cylinder.

1. INTERFACIAL STRESS DISTRIBUTION

In order to gain some insight into the elastic stress transfer arising in the single fiber pull-out specimen before the onset of failure, the interfacial radial and shear stresses of a glass fiber-polycarbonate matrix system with a stiffness ratio of 32 are shown in figure 2 for a fiber length to diameter ratio of 15. The interfacial radial stresses as well as the shear stresses are distributed very inhomogeneously with a distinct maximum at the fiber entry. Here, the radial stresses appear as tensile stresses, promoting a mode I fracture. They nearly vanish within a distance of one fiber diameter apart from the matrix surface. At the fiber end, a small zone of compressive stresses arises. The shear stresses decrease towards the central part of the fiber and increase again up to the fiber end, where a second, less intensive maximum arises. The distribution of the interfacial stresses reveals some noticable features:

- the stresses are distributed very inhomogeneously along the interface with concentrations at the fiber entry and at the fiber end, this is, at the singularities

- the main loading of the interface occurs at the fiber entry, where high radial stresses, superimposed by shear stresses, are active

- a mixed mode failure, dominated by mode I, is likely to occur, accordingly

2. INFLUENCE OF A MENISCUS ON THE STRESS FIELD

In the model utilised in the preceeding analysis, a sharp transition from fiber to matrix surface, i.e. a 90° corner, is assumed. However, in actual specimen, different kinds of meniscusses can be observed. In order to investigate the influence of this geometrical parameter, a 90° corner is compared with a meniscus of circular shape, which has the radius of the fiber. This study is performed on the example of a glass fiber embedded in a polycarbonate matrix. The shear stresses in the region around the fiber entry exhibit strong gradients in the axial and in the radial direction in case of the sharp corner (fig. 3a). The maximum, which is located in the corner, influences the interface as well as the fiber. In case of a meniscus, the stress maximum is decreased and shifted from the interface into the meniscus, i.e., the interface is unloaded (fig. 3b). A further analysis shows that, in addition to the diminution of the shear stresses, the peak of the radial stresses is also diminished by about 75% and, accordingly, reaches the same level as the radial stresses. These results show that:

- the existence of a meniscus has an important influence on the stress distribution in the interface and, therefore, on the pull-out force

- the neglect of this geometrical parameter in the evaluation of the test data will result in an additional scatter

3. COMPARISON OF A GLASS FIBER AND A CARBON FIBER

In this section, the influence of the fiber type on the interface loading is studied comparing a glass and a carbon fiber, both embedded into the same polycarbonate matrix up to a length to diameter ratio of 15. The stiffness ratio of the glass fiber system is 32. The axial stiffness ratio of the high modulus carbon fiber system, which is the essential ratio in case of the anisotropic carbon fiber, is about 170. The interfacial shear stresses show a quite different behaviour for the two fiber types (fig. 4). In case of the glass fiber, two almost equal maxima are reached at the fiber entry and at the fiber end. In case of the carbon fiber, a sharp, although much lower maximum arises at the fiber entry, followed by a minimum within a distance of less than 10% of the fiber diameter apart from the fiber surface. Near the fiber end, the interfacial shear stresses increase again and reach a second maximum of nearly the same height for both fiber types. This means that, in case of the more compliant glass fiber, a large amount of fiber force is transferred into the matrix in the vicinity of the fiber entry, while in case of the stiffer carbon fiber, the stress transfer increases up to the fiber end. In case of the carbon fiber system, the zone around the fiber entry is unloaded, not only concerning the shear stresses, but also concerning the radial stresses, which only reach half of the level of the glass fiber system. The analysis reveals:

- the stress maxima at the fiber entry are diminished in case of the carbon fiber, compared with the glass fiber

- this results in a higher pull-out force, if all other parameters would remain unchanged

4. INFLUENCE OF THE FIBER LENGTH TO DIAMETER RATIO

Finally, the influence of the length to diameter ratio can be estimated, comparing the shear stresses for a ratio of 15 (fig. 2) and a ratio of 5 (fig. 4) for the glass fiber system. The comparison reveals the essential influence of the length to diameter ratio on the interfacial shear stresses. It has to be noted that the same external fiber force is prescribed in both cases. This is, the average of the shear stresses is three times as high in case of the short fiber length. While the maximum at the fiber entry remains nearly unchanged, the stress concentration at the fiber end is significantly increased when the fiber length is decreased. From this result we can conclude the following:

- the interfacial shear stress distribution becomes less inhomogeneous, when the fiber length is decreased

- the ratio of the pull-out force divided by the fiber surface, i.e. the apparent shear strength, depends on the fiber length to diameter ratio

REFERENCES

1. R. Muki, E. Sternberg, Int. J. Solids Struct., 6 (1970) 69-90
2. M. R. Piggott, Load bearing fibre composites, Pergamon Press, Oxford, 1980
3. C. H. Hsueh, J. Mater. Sci. Lett., 7 (1988) 497-500
4. C. Marotzke, Composite Interfaces, 1 (1993) 153-166

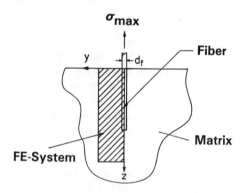

Figure 1   Sketch of the pull-out model

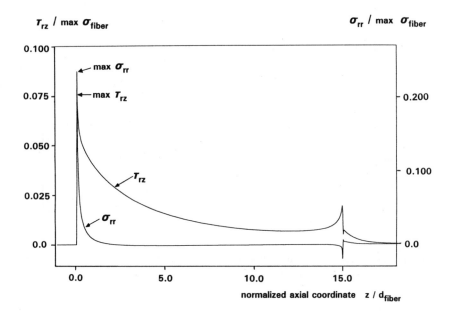

Figure 2   Normalised interfacial shear and radial stresses (l/d = 15)

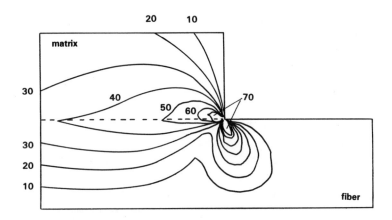

Figure 3a   Shear stress distribution τ_{rz} [N/mm^2] at the fiber entry,
sharp transition (90° corner)

285

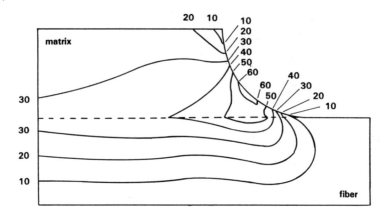

Figure 3b   Shear stress distribution τ_{rz} [N/mm^2] at the fiber entry,
smooth transition (meniscus)

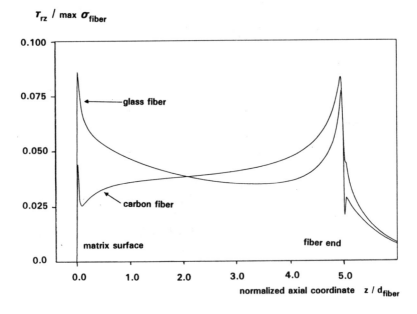

Figure 4   Normalised interfacial shear stresses for a glass fiber and a
carbon fiber system (l/d = 5)

286

FROM LABORATORY TESTS TO NUMERICAL DESIGN TOOLS : A COHERENT APPROACH OF DEAL WITH DYNAMICS

B. GARNIER, R. de MONTIGNY

*Metravib R.D.S. - 64 Chemin des Mouilles - BP 182
69132 Ecully Cedex - France*

ABSTRACT

We present hereafter a complete and coherent methodology to design composite structural elements with some vibro-acoustic functional requirements (transparency, damping, absorption, etc.).

Three examples are shown to demonstrate practical applications in a variety of projects:
- optimisation of vibration damping by car's internal trim panels,
- optimisation of flow noise reduction and acoustic transparency of sonar windows,
- development of anechoic coatings for submarines from rubber matrix composites.

INTRODUCTION

The design of industrial structural applications of composites and advanced materials requires a precise understanding of their dynamic behaviour: as far as they may be subjected to dynamic loads (vibrations from attached machines, aerodynamic excitations, high ambient noise levels, shocks and transient loads), they generally behave much more differently from static than classic metallic materials: Young modulus and shear modulus appear strongly temperature and frequency dependent, internal damping may be high, creeping and cyclic fatigue may bring important disorders, etc.

1. ABOUT THE DIFFICULTY TO DESIGN AN ADEQUATE VIBRO-ACOUSTIC SOLUTION

To illustrate the difficulty of the adequate design of vibro-acoustic elements, we introduce directly now a recent case history of METRAVIB R.D.S.: discussing with this important composite structures designer, on vibrations damping by visco-elastic layers, he objected: "we have already proved that such visco-elastic damping layers are useless for our composite structures". In fact, they had a 10 mm thick carbon-epoxy composite plate to damp, for a maximum efficiency at 200 Hz and 25° C. They had requesteded from a well-known visco-elastic material supplier the adequate product to create a constrained damping layer (cf. fig. 1) and had manufactured a 3 mm carbon-epoxy counterplate. They were supplied with the visco-elastic material n° 1, with the damping characteristics shown in fig. 3.a: the maximum internal loss of this material is over 0.4 (hysteretic damping notation, η = tan δ = 2 C/Cc.

However, the result appeared definitely disappointing... and it is easy to show why!

The graphs shown in fig. 3 provide the complete spectrum of the material versus temperature at 200 Hz, and the evidence that, at this frequency, the maximum damping corresponds to a temperature well below 0° C... METRAVIB R.D.S. could immediately identify in this database a better candidate, and the spectrum of this visco-elastic material n° 2 is shown in fig. 3.a: the effective damping at 25°C/200 Hz is now .9 and to bring now to the composite plate (cf. fig; 3.b) a global composite damping of 0.17 instead of the previous 0.04 - that means that the resonance peaks are reduced from a magnification of 1/0.04 = 25 to a magnification 1/0.17 = 6 = a nice 12 dB vibration level reduction!

It is now evident that the goal is not to provide a good damping material but to design the adequate damping solution.

2. DYNAMIC DESIGN METHODOLOGY

As a leading company in vibro-acoustic engineering, METRAVIB R.D.S. has developed since 10 years a complete and coherent approach to deal with these important dynamic aspects (figure 2):
- the precise experimental identification of the dynamic mechanical parameters of composites and advanced materials is made from samples on our set of laboratory equipment for mechanical spectrometry (Viscoanalyser, Viscostrain). The spectra shown in fig. 3 are obtained from these standard equipment, which can be operated from 10^{-2} to 1000 Hz and from - 170° C to + 350° C. Customised developments bring us to deal with particular environments (e.g. high pressures for underwater applications, or cyclic fatigue for aeronautic applications);
- all this information on any available material is stored in a specific data base; this data base helps us to identify the most adequate solution in any case in full independence from any particular manufacturer;
- a set of exclusive numerical models (generally based on analytic models) offers the correspondence of the elementary properties and geometry of fibres, resins, multilayered systems, etc., and the global dynamics of the design, and helps to optimise it to reach a given performance; the following examples will illustrate some of these specific tools;
- these global dynamic properties are entered into more classical dynamic FEM codes (NASTRAN, SYSTUS, etc.) to design globally complex applications; this "homogenisation" type approach bring to accurate results at very low computation cost - no more than classic materials design.

3. EXAMPLES

3.1 Dampening internal decorative panels for cars

The internal decorative lay-out of the car body may either improve or degrade the acoustic comfort of a vehicle, depending on its precise dynamic behaviour in this peculiar environment (cf. stiffness and flexural wave speed very different from ceiling to base plate, door panels, etc.). We show hereafter a significant improvement (x 2 means + 6 dB) of the steel plate damping from the composite decorative panel.

The design principle was to maximise the flexural vibration damping of the car's ceiling to reduce the internal noise. We suggested the following design steps:
- global characterisation of the 3 layers decorative panel on our Viscoanalyser,
- damping calculations with our AMOCO software of the global system steel/ interface/material/decorative panel,
- selection of the optimum interface material for a maximal damping,
- laboratory verification (mechanical + acoustic),
- final acoustic evaluation on the vehicle prototype.

The test set-up and the results are presented in fig. 4 and 5. The final optimised system increased by <u>more than twice</u> the ceiling damping introduced by the initial system.

3.2 Built-in damping of a sonar dome

It was proven that the turbulent flow mechanical excitation of the sonar domes is a major mechanism of background noise on underwater sonars (cf. both fishing and naval vessels). As a consequence, the mechanical flexural damping of the "acoustic window" is directly related to this background noise at high speed, and may provide a decisive performance improvement.

From many years, GRP domes are used as a significant improvement of the previous ribbed thin steel domes: their natural damping is higher (from a few ‰ for steel to several % for GRP), and they provide a better because "smoother" design.

But additional performance may be now gained from additional complexity of the dome:
- built-in internal damping from visco-elastic layers embedded during construction in the composite core,
- external rubber coating (generally polyurethane) designed to conjugate a perfect acoustic transparency (longitudinal acoustic impedance identical to the sea-water one), and the damping of turbulent flow excitation (highly damped/low shear modulus).

That brings us now to design a composite structure of four layers:

external coating/GRP/visco-elastic/GRP.

The visco-elastic layer may be introduced either as polymer sheets (1 to 2 mm thick), or as a spray - the ultimate solution for a perfect integration in these complex 3-D shapes.

Figure 7 presents the effect of such built-in visco-elastic layer on a GRP composite dynamic stiffness: the result is spectacular, as flexural resonance are now completely smoothed. In this case, the final gain on background noise from the turbulent flow was over 20 dB - promising a decisive detection range extension.

3.3 Rubber matrix composites

Our aim is now to develop particularly efficient sound absorbents "acoustic stealthiness" of naval vessels.

The detection of adverse vessels is made by two ways:
- passive sonars, which listen to the adverse ship's noise,
- active sonars, which emit a powerful sound wave and listen to the echo of this wave on the adverse hull (like radars with E.M. waves).

As a consequence, we have to develop coatings:

- to avoid the creation of sound from the hull vibration,
- to minimise the return echo from the external sound wave impinging the hull.

The application on submarines brings to an additional severe requirement of low compressibility and minimal acoustic performance variation versus depth.

One of the most useful physical principles of acoustic efficiency of such materials is to introduce a well controlled combination of dense fillers and hollow closed inclusions, with an optimum distribution versus material thickness.

Most of the time, such materials are manufactured as square tiles (typical 1 x 1 foot) to be glued to the hull. But a very promising technique (developed by METRAVIB R.D.S. through a European joint venture with the U.K. composites and urethanes manufacturers W. & J. TOD Ltd. at Weymouth) is again a spray technique supposing a perfect control of materials parameters all along the operation in a relatively uncontrolled environment (humid and cold dockyards).

The particularities of each Navy and each vessel class bring to design for each one a specific coating from a global modular technological concept, with the following steps:
- precise material design, by using dedicated METRAVIB software:
  • PACMAN, to calculate the global acoustic performance of non-homogeneous micro-inclusion structures,
  • MULTIC, to deal with multi-layered media;
- corresponding materials selection (matrix, fillers, etc.) and laboratory testing if necessary to extend our materials data base (cf. supra);
- prototyping of the global complex and acoustic laboratory characterisation (pressurised acoustic cells);
- naval acceptance testing in pressurised tanks (performance versus operational configurations, cyclic pressure fatigue, long term ageing...) and global process qualification (including the gluing process, etc.).

Of course, the best solutions are classified; we can show anyway some results on a relatively classic concept:
- figure 6.a shows the PACMAN result and figure 6.b the MULTIC result of a material composed of visco-elastic polyurethane matrix with 1 % dense filler and 5 % hollow inclusions;
- this material offers a final 15 dB echo reduction in the torpedoes homing systems, frequency range up to 50 bars.

4. CONCLUSIONS

Although vibro-acoustic performance prediction still appears a matter of particular expertise, we hope to have demonstrated here that both the "tool-box" (numerical + laboratory test apparatus) and dedicated technologies exist:

- to include in the design of structural composite specific vibro-acoustic control requirements,
- to introduce built-in specific layers to absorb/reflect/damp noise and vibrations at a minimum cost (money + time + weight) compared to further "corrective material" approaches.

More and more applications will probably arise in the close future in partnership with our customers - specially in the transportation area (cars/trains/planes).

- fig. 1 - CONSTRAINED LAYER

visco-elastic film   counter-plate

structure

shear —> damping

Young modulus E

E

optimum temperature range

η

Loss factor η

temperature

-fig.2- COMPOSITE SYSTEMES CONCEPTION

Vibro-acoustic fonctions researched
• Barriers
• Damping
• Anechoism / Absorption
• Decoupling

COMPUTATION MODELS
SAM   CAPS
AMOCO   PALETTE
MULTIC   PIE
DADY   CROMO

MATERIALS

DATA BASE

Dynamic measurements and characterization of materials

VISCOANALYSEUR
MICROMECANALYSEUR
VISCOSTRAIN
OEDOMETRE
FATIGUE
TENSION
CREEPING

Prototypes fabrication

for principle validation

metravib
RDS

3.b) dynamic spectrum at 200 Hz visco-elastic product (2)

1.4
1.2
1
.8
.6
.4
0

LOSS FACTOR η

η

E

temperature °C

0°   10°   20°   30°

3.c) global composite damping

counterplate
visco
structure

(2)

(1)

3.a) dynamic spectrum at 200 Hz visco-elastic product (1)

1
.8
.6
.4
.2
0

LOSS FACTOR η

E

η

temperature °C

- 20°   - 10°   0°   10°

-fig.3 - DAMPENING BY CONSTRAINED LAYER

291

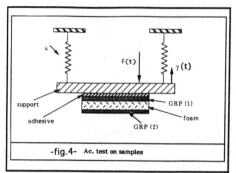

-fig.4- Ac. test on samples

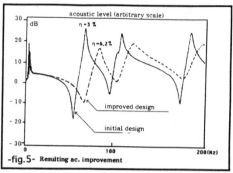

-fig.5- Resulting ac. improvement

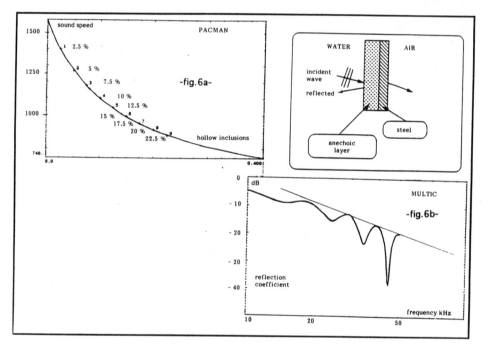

-fig.6a-

-fig.6b-

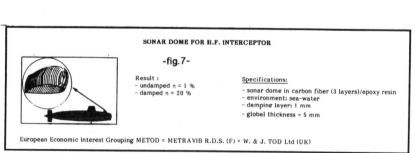

SONAR DOME FOR H.F. INTERCEPTOR

-fig.7-

Result :
- undamped η = 1 %
- damped η = 20 %

Specifications:
- sonar dome in carbon fiber (3 layers)/epoxy resin
- environment: sea-water
- damping layer: 1 mm
- global thickness ≈ 5 mm

European Economic Interest Grouping METOD = METRAVIB R.D.S. (F) + W. & J. TOD Ltd (UK)

292

DYNAMIC PROPERTIES OF A CFRP UNDER CYCLIC LOADING AT VARIOUS STRESS LEVELS

M. CHAPUIS, M. NOUILLANT*

L.A.M.G.E.P. Ensam - Esplanade des Arts et Métiers
33405 Talence - France
*CERMUB - 351 Cours de la Libération - 33405 Talence - France

ABSTRACT

Various models of the dynamic properties (mainly damping) of a carbon/epoxy composite (Hercules IM6 G 12 K / Ciba-Geigy LH-LY 5052) used for an experimental flexible robot arm are presented. In the linear viscoelastic domain, the independence of damping from frequency (at room temperature) leads to the consideration of stress as a fractional derivative of strain in a wide range of frequencies. Micromechanical models from various authors have been applied by our team and provide acceptable results, except for longitudinal damping. For higher strain levels, an elasto-viscoplastic model is proposed and predictions compared with tests results.

INTRODUCTION

During the last few years, extensive work on carbon/epoxy composite damping characterization has been performed at Bordeaux I University /1/2/3/, especially for flexible robot arms made of Hercules IM6 G 12 K / Ciba-Geigy LH-LY 5052 filament wound tubes. This paper is intended to present the various models of behaviour that were established for this composite. Particular attention has been paid to [± 45°]$_s$ laminates that provide an optimal passive damping for flexible robot arms.

I. LINEAR VISCOELASTIC DOMAIN

Measurements show that the material exhibits linear viscoelastic behaviour at strain levels below a limit of about 0.1 %. Within this domain, the complex moduli measurements agree with most micromechanical predictions proposed by various authors /3/4/5/6/7/. However, for the longitudinal damping capacity:

293

$\Psi_L = 2\pi \dfrac{E_L^{''}}{E_L^{'}}$, where $E_L^* = E_L^{'} + i.E_L^{''}$ denotes complex longitudinal modulus, the measured damping capacity is much higher than predictions, as pointed out by R.D.Adams /8/.

Because the behaviour of fibres can be considered as almost purely elastic, the frequency dependence of composite damping capacities (except for Ψ_L) is analogous to that of the matrix alone. For Ciba-Geigy LH-LY 5052 resin at room temperature, damping is almost independent of frequency within a large interval (10^{-3} Hz, 10^3 Hz), which is accurately represented by a law of the form:

$$\sigma(t) = \overline{G} . D^n \varepsilon(t) \tag{1}$$

where D^n is the fractional derivative operator, so that stress σ appears to be a fractional derivative of strain ε /3/.

More precisely, assuming a constant specific damping capacity (S.D.C.) Ψ means that the phase shift φ between harmonic stress and strain is a constant within the aforementioned interval of frequencies:

$$\varphi = \arctan \frac{\Psi}{2\pi} \tag{2}$$

Applying the D^n operator to an harmonic function introduces a constant phase shift:

$D^n \sin \omega t = \omega^n .\sin \left(\omega t + n \dfrac{\pi}{2} \right)$, so that, for stress and strain in the complex form:

$\sigma = \sigma^{\circ} . e^{i\omega t} \quad (\sigma^{\circ} \in \mathbb{C})$

$\varepsilon = \varepsilon^{\circ} . e^{i\omega t} \quad (\varepsilon^{\circ} \in \mathbb{C})$ we obtain: $\sigma^{\circ} = \overline{G} .(i\omega)^n. \varepsilon^{\circ} \tag{3}$

Then, separating the real and imaginary parts and using equation (2) yields:

$$n = \frac{2}{\pi} . \arctan \frac{\Psi}{2\pi} \tag{4}$$

$$\overline{G} = \frac{G^{'}(\omega_0)}{\omega_0^{\ n} . \cos \dfrac{n\pi}{2}} \tag{5}$$

where ω_0 is an arbitrary frequency taken as a reference.

An example approximation for the storage modulus is given in figure 1, where:

$$G^{'}(\omega) = \overline{G}.\omega^n . \cos \frac{n\pi}{2} \tag{6}$$

for a constant S.D.C. $\Psi = 10\%$.

The values of $\overline{G}$ and n for symmetric laminates with a fibre volume fraction of 0.6 are given in table 1. Let us note that fractional derivatives have been used by other authors to establish constitutive laws in viscoelasticity, e.g. /9/.

II. ELASTO-VISCOPLASTIC BEHAVIOUR

At higher strain levels (i.e. within a strain range of 0.1% to 0.2% for a [± 45°]$_s$ laminate), permanent residual strain appears, although dynamic measurements do not show any irreversible increase in damping (S.D.C. at low strain takes the same value before and after high strain cyclic loading). Moreover, at this strain level where the behaviour is essentially nonlinear (damping increases with strain /2/), the material experiences no significant reduction of its fatigue life /3/. This is particularly clear in longitudinal shear ([± 45°]$_s$ specimens in alternate tension-compression), where the material behaviour is described by an elasto-viscoplastic model (figure 2). At low strain level, the model behaves like a generalized Maxwell model with a relaxation time distribution function H(1/ω) being almost independent of frequency ω, according to equation 1. At higher strain, the nonlinear behaviour cannot be represented only by a plastic slip, but necessitates the introduction of a nonlinear (softening) spring. The nonlinear model parameters were identified from residual strain measurements following successive strain pulses, alternately under tension and compression (figure 3) /3/.

Finally, the results of cyclic loading simulation are compared with damping measurements (figure 4). Whereas the predicted S.D.C. follows approximately a square law: $\Psi = \Psi_0 + \alpha.\varepsilon_0^2$, the measured S.D.C. increases like the third power of strain. This discrepancy, if confirmed by other experiments, suggests that not all the nonlinearity can be attributed to time-independent (i.e. plastic) behaviour. Note that a similar result can also be obtained by using a series model instead of a parallel model. Therefore, a more sophisticated model involving nonlinear viscoelasticity would probably be more adequate, at least for stress levels over 25 MPa (0.15% strain).

III. DAMAGED MATERIAL

At high strain levels (over 0.2%), there is an irreversible increase in damping, which means that damping does not recover its initial value after strain is reduced. Various authors have shown that the material behaviour in this domain is plastic or viscoplastic, due to microslip along cracks parallel to the fibre axis /10/11/12/13/. For the purpose of robot arm damping control, it was not considered necessary to study the damping behaviour of the material over 0.2% strain.

CONCLUSIONS

As a conclusion, we can schematically distinguish three domains of behaviour depending on strain level:
- linear viscoelastic (< 0.1%)
- elasto-viscoplastic (0.1% to 0.2%)
- damaged (> 0.2%)

The first has been extensively studied, although no satisfactory explanation or prediction model could be found for longitudinal damping Ψ_L.

The third implies a short fatigue life of the material, which is of little use for practical purposes of structural vibration control.

The second, where nonlinear mechanisms provide a significant increase in damping, seems to be promising for high damping material development. An elasto-viscoplastic model has been proposed, and test results suggest that a part of the nonlinearity should be attributed to time-dependent (i.e. viscous) mechanisms.

	resin	$[\pm 15°]_s$	$[\pm 45°]_s$	$[\pm 75°]_s$
n	10^{-2}	$1.9\ 10^{-3}$	$8.1\ 10^{-3}$	$6.3\ 10^{-3}$
$\bar{G}$ (GPa.s^n)	3.122	131.37	16.135	7.682

Table 1   Parameters of the constitutive law (equation 1) for symmetric laminates.

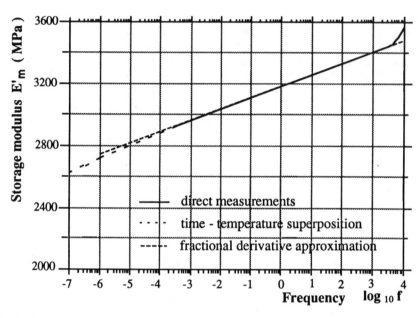

Fig 1   Storage modulus of the matrix (LH-LY 5052) versus frequency f (Hz) at 20°C.

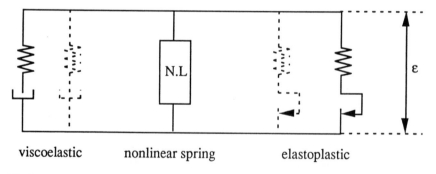

Fig 2   Elasto-viscoplastic model.

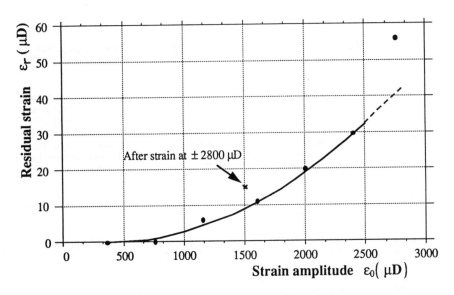

Fig 3   Residual strain ε_0 of a $[\pm 45°]_s$ tubular specimen, versus strain pulse amplitude ε_r.

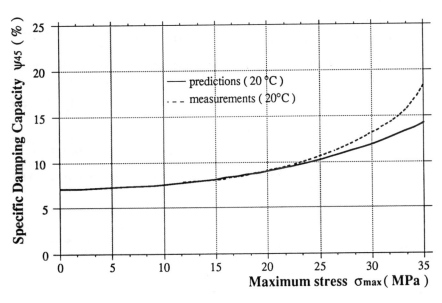

Fig 4   Comparison of measured and predicted S.D.C. for a $[\pm 45°]_s$ laminate.

REFERENCES

1. Nouillant M., Chapuis M., Delas J.M. & Miot P. "Experimental equipment and methods for measuring damping capacity". Experimental Techniques, pp.27-32, (nov-dec 1992).

2. Nouillant M., Chapuis M. "Specific damping of a carbon-epoxy laminate under cyclic loading". Euromech 269 conference, St Étienne, 6-9 dec 1990, in "Mechanical identification of composites", Ed. A.Vautrin, H.Sol, pp.406-414, Elsevier (1990).

3. Chapuis M. "Modélisation du comportement d'un composite carbone-époxyde dans le domaine linéaire et non-linéaire. Applications au contrôle de l'amortissement dans les structures sous sollicitations cycliques". Thèse de Doctorat de l'Université Bordeaux I, n° 720, spécialité "Génie Mécanique" (1992).

4. Hashin Z. & Rosen B.W. "The elastic moduli of fiber-reinforced materials". Journal of Applied Mechanics, ASME, pp.223-232 (1964).

5. Theocaris P.S., Spathis G. & Sideridis G. "Elastic and viscoelastic properties of fibre-reinforced composite materials". Fibre Science and Technology, vol.17, pp.169-181 (1982).

6. Hashin Z. "Complex moduli of viscoelastic composites. II : Fiber reinforced materials". International Journal of Solids & Structures, vol. 6, pp.797-807 (1970).

7. Adams R.D. & Bacon D.G.C. "The dynamic properties of unidirectional fibre reinforced composites in flexure and torsion". Journal of Composite Materials, vol.7, pp.53-67 (1973).

8. Adams R.D. "Mechanisms of damping in composite materials". Journal of Physics, Colloque C9, supplément au n°12, tome 44, pp.29-37 (1983).

9. Oustaloup A., Nouillant M. "Distribution parameters systems and non integer derivation". European Control Conference, ECC, Grenoble (1991).

10. Surrel Y. & Vautrin A. "On a modeling of the plastic response of FRP under monotonic loading". Journal of Composite Materials, vol.23, pp.232-250 (1989).

11. Sun C.T. & Chen J.L. "A simple flow rule for characterizing nonlinear behavior of fiber composites". Journal of Composite Materials, vol.23, pp.1009-1020 (1989).

12. Vaziri R. , Olson M.D. & Anderson D.L. "A plasticity-based constitutive model for fibre-reinforced composite laminates". Journal of Composite Materials, vol.25, pp.512-535 (1991).

13. Krempl E. & Tyan-Min Niu "Graphite/epoxy [± 45]$_s$ tubes. Their static axial and shear properties and their fatigue behavior under completely reversed load controlled loading". Journal of Composite Materials, vol.16, pp.172-187 (1982).

EXPERIMENTAL APPROACH OF UNIDIRECTIONAL COMPOSITE SHEAR BEHAVIOUR UNDER MULTIAXIAL STRESS

B. PALUCH

IMFL - Onera - 5 Boulevard Paul Painlevé - 59000 Lille - France

ABSTRACT

The response of a composite material is strongly non linear and dependant on the existing two-axis stress field. A test setup designed to subject the material to shear stress coupled to any plane stress was therefore developped. In this case, a test involving a tubular specimen was found to be the best fitted in view of the possibilities it offers. In addition to the setup itself, the guidelines proposed to make the specimens as well as the relevant dimensioning procedures were also analysed. Preliminary tests on T300/914 showed that the experimental concept was valid and that work on characterizing this material could begin.

INTRODUCTION

The response of a composite material to shear stress can be modified by the presence of a plane 2-axis stress filed, particularly in the case of compression. Testing tubular specimens was shown to be the best way to characterize this response, since it makes it possible to induce any 2-axis stress state by applying, independently or simultaneously, a traction or compression force $\mathbf{F}$ along the tube axis (stress σ_{zz}) and a pressure $\mathbf{p}$ inside or outside its wall (stress $\sigma_{\theta\theta}$), the torsional moment $\mathbf{M}$ generating the shear stress $\sigma_{z\theta}$ (Fig. 1). In this configuration, which has been used for many years on metallic tubes, applying stress correctly remains the most difficult problem to solve in the case of composite material tubes. For this reason, the tubes must be fitted with tabs designed to properly transfer the loads from the testing machine to the central zone of the specimen, called the test zone, where the stress field needs to be as uniform as possible and directly proportional to the load $(\mathbf{F,p,M})$. Accordingly, the specimens must be dimensionned in such a way that the tabs do not induce prohibitive stress concentrations, and that no buckling phenomena occur, if possible, before the rupture of the material in the test zone. In addition, the central wall of the specimen needs to be thin enough, compare to its radius, in order to minimize the radial stress gradient, and this may conflict with the no-buckling condition. Also, the cost of the specimen manufacturing must be reasonable. So although testing tubular composite material specimens may appear simple at first, it does raise a number of problems that need to be solved.

I. SPECIMEN DESIGN

The specimen and the system used to fasten it to the setup includes (Fig.2):
- a T300/914 unidirectional carbon tube 120 mm in diameter and 450 mm total length,
- E/914 fiberglass fabric tabs, uniformly thick inside and tapered outside, with about 5 % conicity,
- and 2 sleeves clamped using a low elastic modulus fusible alloy.

Loads are gradually applied to the test zone via the tapered tabs in order to minimize the stiffness variation in the tube/tab transition area, and as a result, the stress concentration phenomena /1/. Also, as shown by Sullivan *et al.* /2/, fastening the specimen by means of material with a small elastic modulus offers the advantage of transferring **F** and **M** correctly from the grips to the tube. The tabs are directly integrated at the time the tube is manufactured and after curing, boring the inside tabs produces high-precision cylindrical and concentric surfaces thanks to which the metallic sleeves are properly aligned with respect to the testing machine grips. The space between the outside tabs and the metallic sleeves is filled with a low fusion temperature alloy /4/, which makes the sleeves easy to reuse after each test. The use of an epoxy resin, as recommanded by some authors /3/, was found to be too difficult to handle, the shrinkage of the resin after its polymerisation is not negligible, unlike the fusible alloy which has a dilatation coefficient of about 6 ‰ after a few hours cooling period, ensuring that the specimen is securely clamped inside the sleeves. Confinement rings are placed on the sleeves to prevent the fusible metal from extruding during the tests, particularly in traction.

The stress field was calculated using an axisymmetric finite element code in order to check the field's uniformity and the absence of stress concentrations. The mesh of the figure 3 shows the boundary conditions and the distribution of stresses σ_{zz}, $\sigma_{\theta\theta}$, and $\sigma_{z\theta}$ along the unidirectional carbon tube, generated respectively by **F**, **p** and **M**. A study of the various load cases shows that the stress concentration phenomena are negligible, and that the field ($\sigma_{zz}, \sigma_{\theta\theta}, \sigma_{z\theta}$) is uniform over at least two thirds of the gauge length. However the finite element approach is nothing but a means to check the solution used, leaving it up to the experimenter to determine the optimal size of the fasteners /5/. To avoid this problem, analytical dimensioning procedures were also used /6/, which produce similar design guidelines.

Instability phenomena must be avoided during the tests as much as possible. These phenomena, which can occur when the tube is subjected to compression, torsion, internal or external pressure, or to a linear combination of the theese three types of loads, are mainly evidenced by buckling /7/. The procedure we developped to calculate the critical buckling loads aims not so much at calculating the exact loads but rather at getting approximate load values without which it would be impossible to quickly dimension the specimens. The critical membrane forces ($N_{zz}, N_{\theta\theta}, N_{z\theta}$) are found by cancelling the determinant in the following general relation :

(1) $[T] \{X\} = 0$ with $|T| = 0$

[T] being an operator obtained after integrating Timoshenko's equilibrium differential equation system /8/ using a Galerkin resolution method, and taking into account the displacement field $\{X\}$ relevant to the three load cases (F, p, M). The code yields accurate results, since the deviations observed between the calculated loads and those measured for example in /9/ and /10/ are below 5 % in both cases. Predicting critical loads with a small margin error therefore helps to quickly and accurately dimension the specimens. However, using a single geometry will not suffice to characterize shear behaviour over the entire stress domain ($\sigma_{zz}, \sigma_{\theta\theta}$) because buckling may occur in some parts of this domain. In view of the initial results yielded by the code, we found it necessary to use sets of specimens having the same external size but different wall thicknesses and gauge lengths depending on the stress domain being explored.

II. TEST SETUP

In order to characterize the shear behaviour of composites under a 2-axis stress state, we had to use a setup where force F (F_{max}=1000 kN) and pressure p (p_{max}=60 bars) are generated by a hydraulic unit (Fig. 4). The torsional moment (M_{max}=3000 mN) is applied using a low clearance speed reducer (with an overall speed reducing ratio of 1:500,000) piloted by a step by step motor in order to accurately control the specimen torsional angle which is always very small, on the order of a few degrees. The torsional moment mechanism is located on the upper fixed crossbeam, while the force F is transferred by the lower moving crossbeam. Two half-rings clamp the metallic sleeves on each grip of the machine through loading rings (Fig. 2). Water is uzed to pressurize the tube via a distributor located inside the lower grip, coming in contact with the side of one of the aluminum plug. The two plugs inside the specimen can slide with respect to each other, so that F and M are only transmitted by the test zone.

III. PRELIMINARY TESTS

A set of five specimens was tested for different load cases corresponding to the application of constant-level axial forces in traction and compression during each test, and without pressure. After applying the axial force F, each specimen was subjected to load-unload cycles of inceasingly high shear strain until the material failure. Two of the specimens were tested until buckling in compression, with deviations of about 10 % between the calculated and measured critical loads which confirmed the validity of the critical buckling load calculation code. The non linearity of the shear response, shown in figure 5 for σ_{zz}=300 Mpa, is clearly visible for values of stress $\sigma_{z\theta}$ ranging from 15 to 25 MPa (the various curves were shifted for easier reading).

301

At each cycle the shear elastic modulus decreases very slightly. The next step will therefore consist in characterizing the relation $\sigma_{z\theta} = f\,(\varepsilon_{z\theta}, \sigma_{zz}, \sigma_{\theta\theta})$ in more detail.

CONCLUSION

Initial tests helped to start characterizing T300/914 and to fully demonstrate the usefulness of the procedure used in this study for dimensioning the specimens and the validity of the experimental concept. All the opportunities offered by the test setup were not followed up systematically here, but will be in the next phase of this work.

ACKNOWLEDGEMENTS

This work was supported by the french agency Services Techniques des Programmes Aéronautiques, section Etudes Générales.

REFERENCES

/1/ I.M. Daniel, Handbook of Composites, Vol. 3,Elsevier, (1985), pp. 277-373
/2/ T.L. Sullivan, C.C. Chamis, ASTM STP 521, (1973), pp. 277-292
/3/ S.R. Swanson et al., J. of Comp. Mat., Vol.20, (1986), pp. 457-471
/4/ R.H. Marloff et al., SAMPE Quarterly, Vol. 17, (1985), pp. 40-45
/5/ R.R. Rizzo, A.A. Vicario, ASTM STP 497, (1972), pp. 68-88
/6/ J. Highton et al., J. of Strain Analysis, Vol. 17,N°1, (1982), pp. 31-43
/7/ J.M. Whitney, C.T. Sun, J. of Comp. Mat., Vol. 9, (1975), pp. 138-148
/8/ S.P. Timoshenko, J.M. Gere, Théorie de la stabilité élastique, Dunod, Paris, (1966)
/9/ M. Uemura, H. Kasuya, ICCM 4, Tokyo, (1982), pp. 583-590
/10/D.E. Marlowe et al., ASTM STP 546, (1974), pp. 84-108

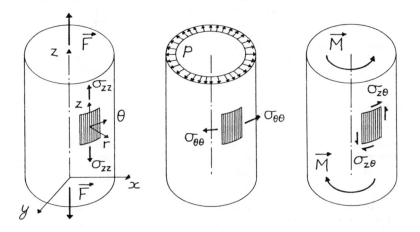

Fig. 1 : The three types of loads (**F,p,M**) applied on the specimen.

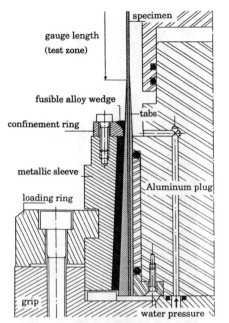

gauge length
(test zone)

specimen

fusible alloy wedge

confinement ring

tabs

metallic sleeve

Aluminum plug

loading ring

grip

water pressure

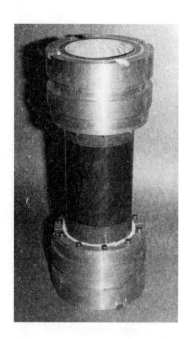

Fig. 2 : Specimen view and cross-section.

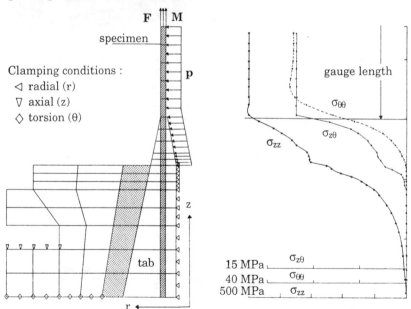

F M

specimen

Clamping conditions :
◁ radial (r)
▽ axial (z)
◇ torsion (θ)

p

gauge length

$\sigma_{\theta\theta}$

$\sigma_{z\theta}$

σ_{zz}

z

tab

r

15 MPa $\sigma_{z\theta}$
40 MPa $\sigma_{\theta\theta}$
500 MPa σ_{zz}

Fig. 3 : Finite element mesh and evolution of the stress field ($\sigma_{zz}, \sigma_{\theta\theta}, \sigma_{z\theta}$) along the unidirectional carbon tube.

303

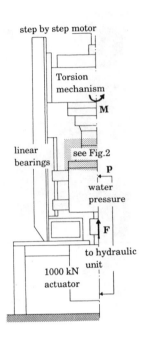

Fig. 4 : Test setup view.

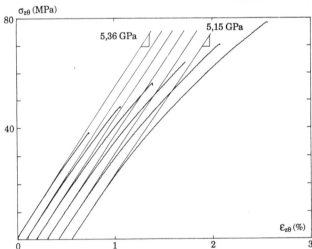

Fig. 5 : T300/914 stress-strain curve.

ANALYSIS OF GEOMETRIC IMPERFECTIONS IN UNIDIRECTIONALLY REINFORCED COMPOSITES

B. PALUCH

IMFL - Onera - 5 Boulevard Paul Painlevé - 59000 Lille - France

ABSTRACT

The compressive strength of unidirectionally reinforced composites is limited partly by the presence of geometric imperfections, essentially characterized by an undulation and a global misalignment of fibers. Optical microscopic photographs of a certain number of cross-sections of material samples were processed by an image processing program to extract the positions of the centers of the fiber sections. Their paths were reconstructed from this, and their rate of undulation analyzed. This method thus opens the way to computer analysis of the compressive properties of this type of material.

INTRODUCTION

Although the tensile strength and Young's modulus of carbon fibers currently available on the market is ever on the rise, little progress has been made in the compressive strength of unidirectionally reinforced composites manufactured using these fibers. Aside from the high cost of the fibers themselves, the great difference between the tensile and compressive properties of the final material is a handicap in use and restricts the development of high-performance carbon fibers in structural elements, where compressive stresses play a far from negligible role in service. Since the compressive strength domain is limited partly by the presence of geometric imperfections within the material, essentially due to an undulation of the fibers, these imperfections must first be quantified before we can perform any mechanical analysis of this phenomenon. One of the first predictive models of the compressive behavior of unidirectionally reinforced composites was developed by Rosen [1], showing that the failure is initiated by in-phase buckling of the fibers on a microscopic scale. To explain the great difference that exists between the failure stress, σ_c, as suggested by Rosen's theory, and experimental values, many authors, including /2//3//4/ have introduced the initial curvatures of the fibers as well as the nonlinear shear properties of the matrix. We assume that the fibers are affected by an undulation of periodic form such that :

(1) $V = f_o \cos \pi x / L$

f_o and L being the amplitude and semi-wavelength of the undulation, respectively. According to these models, it is found that the failure stress decreases extremely fast as a function of the ratio f_o/L, and then tends toward an asymptote, reflecting the fact that the presence of waviness, even of small amplitude, has a very strong effect on the compressive strength. So it is imperative to know the fiber curvature. An indirect approach to identification of this parameter, according to analog computer models, was contributed by Frost /5/, who developed a semi-empirical procedure for estimating the fiber waviness. The basic idea was to compare the stress values computed in a given curvature range with those measured experimentally, for different types of composites. Although this is interesting, this approach yields no more than a mean estimated value of the curvature, closely dependent on the computation model. In order to get an idea of the different phenomena affecting the value of σ_c, it is indispensible to quantify the geometric imperfections intrinsic to the material, independently of any computer model.

I. DETERMINATION OF FIBER WAVINESS

Yurgartis /6/ used a method for working back to the local misalignment angles of fibers by inclining the section a few degrees with respect to the mean direction of the fibers, so that the fiber sections are elliptical. Assuming that the fibers are circular to begin with, this author was able to work back to the local misalignment angle by measuring the length of the major axis of the ellipses. But for materials other than the one he analyzed, the results may be biased by two types of phenomena. Firstly, the fibers are assumed to be circular (which was verified in the material he was studying), while in most cases they are not. Secondly, the fibers are assumed to be rectilinear, whereas they are actually slightly curved. The apparent local misalignment angle therefore does not correspond to the real local angle. However, the major problem with this approach does not reside so much in the bias phenomena as in the impossibility of calculating the value of f_o/L appearing in (1).

Although more complex than the above method, the one we have developed does make it possible to calculate the fiber curvature. On a single sample, we perform N cross-sections perpendicular to the mean direction of the fibers, spaced as equidistant intervals. The precision of a few microns in the position of the optical microscope sampleholder is' good enough to insure that the section plane superimpose on each other, with just slight systematic shifts that we have no way of knowing beforehand. For each cross-section, the microscope pictures are processed by image processing to extract the coordinates of the center of each fiber.

The paths followed by the fibers are then calculated in two phases. Taking two contiguous sectional planes P_j and P_{j+1} (Fig.1) with their centers of gravity G_j and G_{j+1}, equal to the arithmetic mean of all of the coordinates of the centers of each plane, determined after eliminating those fiber sections that are truncated at the edges of the images. The fiber is followed from one plane to the next by looking for the shortest distance d_j separating the point O' from the center of the fiber of index k the closest to O'. The centers of indices j and k are considered to be part of the same fiber when d_j is less than a certain exploration radius limit r_{lim}. Because of the precision of the positioning mentioned above, there exists a systematic shift symbolized by the vector $\vec{b}$ between two contiguous planes, affecting all of the coordinates of the centers contained in P_{i+1} with respect to P_i. The vector $\vec{a}$, on the other hand, expresses the shift due to the waviness of the fibers. This is the vector we want to evaluate. In fact, the norm of $\vec{a}$ is altered slightly by the imprecision in the position of the center of each fiber section as determined by the image processing program. So we can say equivalently :

$$(2) \qquad \vec{d_j} = \vec{b} + \vec{a_j}$$

as the undulations are randomly distributed, we may consider that, within the same sectional plane, the resultant of the shifts due to the waviness is null or even zero, which yields :

$$(3) \qquad \lim_{N_c \to \infty} \sum_{j=1}^{j=N_c} \vec{a_j} = \vec{0} \; => \; \vec{b} = \frac{1}{N_c} \sum_{j=1}^{j=N_c} \vec{d_j}$$

in which N_c is the number of centers in each plane. The vector $\vec{b}$, which according to (3) should result from the vectorial combination, is in reality taken to be equal to the vector $\vec{d_j}$ having the greatest occurrence. Once the systematic shifts between planes are eliminated, the second phase consists in starting from the median plane and plotting out the correlations of the fibers upstream and downstream of this plane with a value of r_{lim} that is less than before. Once the correlations are made, knowing the coordinates of the centers of a fiber, expressed in an axis system (x, y, z) referenced to the base plane, we can then calculate its waviness and quantify it. We first determine the main plane of inertia P_p (Fig. 2) associated with the N_p pairs of points M_i belonging to a given fiber, and such that the sum of the squares of the distances between the real points and the points projected on P_p is minimum, and that the sum of the squares of the distances separating the projected points on P_p is maximum. The main directions are given by $\vec{U}$, whose components correspond to the eigenvectors associated with the three eigenvalues, arranged in decreasing order, of the following system :

$$(4) \qquad \mathbf{V U} = \lambda \mathbf{U} \quad \text{with} \quad \mathbf{U} = \{ \vec{U_1}, \vec{U_2}, \vec{U_3} \} \quad \text{and} \quad \mathbf{V} = \mathbf{X}^t \mathbf{I} \mathbf{X} - \mathbf{g} \, \mathbf{g}^t$$

$\mathbf{V}$ being the variance-covariance matrix /7/, $\mathbf{I}$ the identity matrix and $\mathbf{g}$ the matrix associated with the center of gravity of the coordinates $\mathbf{X}$ of the centers of a fiber, such that :

$$(5) \qquad \mathbf{X} = \begin{bmatrix} x_1 & y_1 & z_1 \\ \cdots & \cdots & \cdots \\ x_{N_p} & y_{N_p} & z_{N_p} \end{bmatrix} \qquad \mathbf{g} = \frac{1}{N_p} \mathbf{I} \mathbf{X}$$

The coordinates $\mathbf{X_p}$ of the points M_i projected into the (X, Y, Z) axis system attached to the plane P_p is finally found from the projection matrix $\mathbf{P}$ by

$$(6) \qquad \mathbf{X_p} = \mathbf{V}^t \mathbf{P} \mathbf{X}^t \quad \text{with} \quad \mathbf{P} = \mathbf{U} (\mathbf{U}^t \mathbf{U})^{-1} \mathbf{U}^t$$

The new (X, Y, Z) axis system is defined from the direction vector $\overrightarrow{U_1}$ giving the general direction of the fiber. Then, in a second phase, we use the coordinates of the points projected on P_p to find the curvature of the neutral line of the fiber in the form (1) using a least squares method. The orientation of the X axis with respect to the (x, y, z) axis system constituting the observation frame of reference can then be represented by two angles δ and β such that:

$$(7) \qquad \delta = \text{arctg}\left(\frac{u_{11}}{u_{12}}\right) \quad , \quad \beta = \text{arctg}\,\frac{\sqrt{u_{11}^2 + u_{12}^2}}{u_{13}} \quad \text{with} \quad \overrightarrow{U_1} = \{u_{11}, u_{12}, u_{13}\}$$

β is the global misalignment angle of the fibers.

II. EXAMINATION OF A REAL MATERIAL

A sample of AS4/PEEK composite was examined by polishing 40 cross-sections spaced from 20 μm. The mean diameter of the fibers is 6.9 μm. The paths (Fig. 3) of 60 fibers were tracked starting at the position of the 91 centers contained in the median section plane. Figure 4 shows the undulations, smoothed according to (1) with respect to the coordinates projected into the main plane of inertia P_p. The angle δ follows a uniform distribution through the interval from 0° to 360°, which confirms relation (3), while β follows a three-parameter (a = 0°, b = 0.8°, c = 1.2) Weibull distribution such that:

$$(8) \qquad P(f_n = F) = 1 - \exp\left(-\left(\frac{F-a}{b}\right)^c\right) \quad , \quad P(F<a) = 0$$

The normalized curvature distribution $f_n = f_o/L$ also follows a three-parameter Weibull law (Fig. 5). Kolmogoroff-type comparisons /7/ indicate that, for a threshold of 5 %, we can accept the hypothesis that the β and f_n distributions follow a Weibull distribution. As for the wavelengths, they fall more or less uniformly into the interval of 500 μm to 840 μm. A χ^2 test also shows that, for 50 of the fibers analyzed, out of 60 found, we can liken their undulations to the form of equation (1).

CONCLUSION

The method we have explained for finding the curvature and global misalignment of fibers can be extended with no problem to other fiber-matrix associations. The fact of being able to reconstruct the fiber paths in three dimensions experimentally, and quantifying their geometric imperfections, opens the way to computations of both the analytical and finite-element types, for the purpose of studying the compressive properties of unidirectionally reinforced composites.

ACKNOWLEDGEMENTS

This work was supported by the french agency Services Techniques des Programmes Aéronautiques, section Etudes Générales.

REFERENCES

1. B.W. Rosen, Fiber Composite Materials, American Society for Metals (1965), pp. 37-75.

2. H.T. Hahn, J.G. Williams, NASA Technical Memorandum No. 85834. (1984)
3. P.S. Steif, Int. Jour. Solids Structures, Vol. 23 (1987), pp. 1235-1246.
4. P.S. Steif, J. of Composite Materials, Vol. 22 (1988), pp. 818-828.
5. S.R. Frost, J. of Composite Materials, Vol. 26 (1992), pp. 1151-1172.
6. S.W. Yurgatis, Comp. Sc. and Tech., Vol. 30 (1987), pp. 279-293.
7. B. Saporta, Probabilités, analyse des données et statistiques, Editions Technip, Paris (1990).

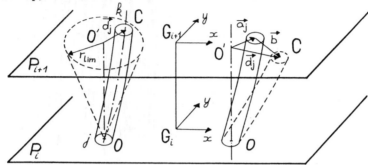

Fig. 1: Fiber path research method by proximity analysis and shifting vectors composition.

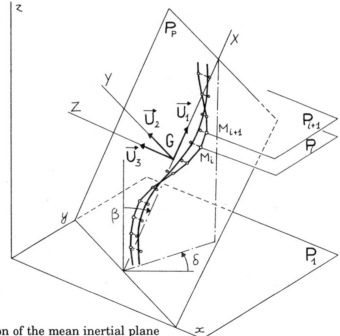

Fig. 2: Position of the mean inertial plane associated to the fiber coordinates.

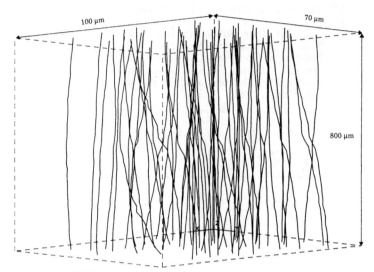

Fig. 3: Visualisation of fiber paths.

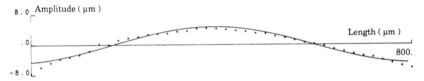

Fig. 4: An example of fiber waviness smoothing.

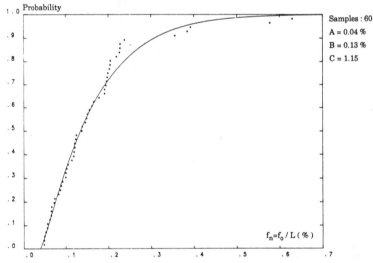

Fig. 5: Weibul distribution of normalizd curvature f_o.

A RATIONAL APPROACH TO MODE I INTERLAMINAR FRACTURE TOUGHNESS TESTING

C.J. VERBERNE, C.G. GUSTAFSON

*Norwegian Institute of Technology - c/o Raufoss A/S - PO Box 2
2831 Raufoss - Norway*

ABSTRACT

An improved way to perform mode I Interlaminar Fracture Toughness testing with the DCB-specimen is presented. By using a one step loading/unloading procedure and an energy based calculation of Fracture Toughness data, the test is performed easier and more consistently. The method is much less sensitive to fiber bridging than compliance based methods. It results in unambiguous Fracture Toughness data on a conservative basis.The results are supported both by Finite Element simulations and experimental results. The method can be automated easily.

1. INTRODUCTION

Testing of the Mode I Interlaminar Fracture Toughness is generally performed on a Double Cantilever Beam (DCB) specimen. It will test the composite material in its weakest direction, delamination under opening. The test itself is time consuming and yields inconsistent results.

Differences between specimens of one batch indicate that the test method itself is too difficult to perform uniformly. Differences between laboratories indicate that the test standard leaves room for different interpretations.

To reduce the variability in DCB test data, a more rational approach to mode I interlaminar fracture toughness testing is presented here. The method is closely related to the Area Method but does not attempt to create R-curve data for every few millimeters of crack growth. A one step loading/unloading concept is used.

2. BACKGROUND

Generally an initial and a propagating value for the mode I interlaminar fracture toughness are identified. The toughness shows an increase with the crack length. This is often explained by the effect of fiber bridging. Initial values depend on the starting conditions of the crack. The propagating values are of more interest for design purposes since a safe design has to assume the presence of already initiated cracks. The design will therefore have to focus on the assurance that these cracks will not grow under a normal service condition.

As with every other phenomenon concerning fracture, the resulting fracture toughness data is rather irregular. Also the determination of the crack length is inaccurate, since it can be extremely difficult to identify the exact position of the crack tip throughout the process of cracking. Add to this the effects of fiber bridging, specimen geometry, surface waviness, etc. and it becomes virtually impossible to get identical load/deflection curves and crack length data for two identical material samples. Even if they are produced from the same plate of composite material, significant variability is inevitable.

To compare the proposed method with other existing methods, a consistent set of data is needed. These are extremely difficult to obtain for the above mentioned reasons. Therefore, a finite element simulation /1/ of the DCB-testing was developed with the main purpose to simulate the effects of fiber bridging. This tool is extremely helpful in understanding the influence of fiber bridging on the apparent fracture toughness data. It produces consistent data enabling the variation and control of one variable at the time. It also produces a significant amount of valuable data which is not available from a laboratory testing, e.g. strain energy and virtual crack extension.

3. IMPROVED METHOD OF MODE I INTERLAMINAR FRACTURE TOUGHNESS TESTING

The following procedure is proposed :

Loading sequence

1. Monitor load and deflection accurately and preferably with a PC that can perform integration of the external work on-line.
2. Follow existing protocols like the one from ESIS /2/ to determine the initial values for G_{Ic}.
3. Continue the test until a fully developed crack of agreed length is formed. Crack propagation is not monitored during this process.
4. Stop the cross-head but do not unload.

Fully developed crack

5. Register the total external work, U_{ext}, performed up to this point of maximum deflection.
6. Measure carefully the length of the crack at both sides of the specimen. Calculate the total crack growth, Δa, from the known position of the starter foil.

Unloading sequence

7. Unload the specimen while again monitoring the load/deflection curve. The integral from the point of maximum deflection down to zero load is the elastic energy of the specimen, U_{elas}, at the point of maximum deflection.
8. Calculate $G_{Ic,\,init}$ as usual /2/.
   Calculate $G_{Ic,\,fe}$ according to :

$$G_{Ic,\,fe} = \frac{U_{ext} - U_{elas}}{\Delta a \cdot b}$$

Where b is the width of the specimen. At this stage the term, $U_{frac} = U_{ext} - U_{elas}$, the Fracture Energy is defined. The Fracture Energy plays a central role in this paper.

312

If one wants to indicate an R-curve, it can be constructed as a linear interpolation between the initial value $G_{Ic, init}$ and $G_{Ic,fe}$ for a fully developed crack. This is a conservative approach. For pure energy balance, the R-curve should have been a straight line at the level of $G_{Ic,fe}$. $G_{Ic,fe}$ is based on the integral of the load/deflection curve and any energetic effects coupled to initiation phenomena are already included in the level of $G_{Ic,fe}$. Therefore, the here proposed linear interpolation between the initiation point and the final $G_{Ic, fe}$-value will always lie underneath (read on the conservative side) of the proper material R-curve.

As an example of the degree of simplification, figure 1 shows the results for a glass/epoxy specimen for both the Corrected Beam (CB) method /2/ and the proposed Fracture Energy (FE) method. Data is taken from ESIS directed round-robin testing at SINTEF, Norway /3/.

4. THE FINITE ELEMENT SIMULATION

Earlier, a finite element simulation /1/ of the DCB-testing was developed with the main purpose to simulate the effects of fiber bridging. As presented in /1/, the simulation confirmed well with expected and observed behavior of a DCB-specimen under mode I loading. An important result from /1/ was the conclusion that only a small amount of fiber bridges lead to a large increase in the apparent propagation value of G_{Ic} when it was calculated by the CB-method /2/. The CB-method is compliance based. The present work shows that this effect is much less pronounced if an energy based method is used.

4.1. The principle of the simulation

If fracture toughness is to be an intrinsic material property, it should be single valued. It might show some irregularness at initiation, but steadily increasing propagating values should not be found. For an intrinsic material property, it should be sufficient to concentrate only on the area close to the crack tip and the local strain state as to monitor crack growth.
    The simulation is based on the general theory of Linear Elastic Fracture Mechanics (LEFM). In agreement with the LEFM, the value of the J-integral at fracture is considered equivalent to G_{Ic}, the critical strain energy release rate. The value of the J-integral at the crack tip is used as the local fracture criterium. It is ideal as a crack growth criterium since it can be chosen not to include the influence of the fiber bridges and only describe the local situation at the crack tip.

4.2. The generation of load data.

To be able to explain the effect of fiber bridges on the apparent G_{Ic}, the load data for different lengths of fiber bridges is given in figure 2. The effective area of the fiber bridges per mm² of crack surface is kept constant. A whole specter of fiber lengths would have been more realistic, but is more difficult to understand since interference occurs.
    As expected, the load increases when the fiber bridges are active. The effect of short fibers is already seen early in the process of cracking. The longer fibers do not show their effect until the crack length is sufficiently long. The simulation confirms the creation of a stable zone when the outer most fibers are strained to failure. This zone will then move with the crack front and no additional effect on the load level is found.

4.3. The Resulting Fracture Toughness Data

If the simulated load data are transferred into fracture toughness data for both the CB-method /2/ and the here proposed FE-method, a marked difference is seen between the two R-curves. In figure 4 it is clearly seen that the effect of the fiber bridges on the CB-method is much larger than on the FE-method.

313

It is worth noting that G_{lc} does not show a direct increase after initiation. The crack obvious must advance a few mm´s to activate the fiber bridges. In reality energy is also required to create the fiber bridges, but this is difficult to model.

The reason for the difference between the two methods is found in the influence of the fiber bridges on the external work and the measured load. Most theories, including the CB-method, are compliance based. Compliance is directly coupled to the measured load. The FE-method is energy based. In figure 3 a simulation based on a specter of multiple fiber lengths and a simulation without fiber bridges are compared. The influence of the fiber bridges on the external work is smaller than their influence on the load. Inherent to the formula involved, the influence increases when fracture toughness data are calculated :

Influence on :	Load	U_{ext}	G_{lc}
Corrected Beam method	12.8 %	-	22.0 %
Fracture Energy method	-	4.6 %	6.0 %

The FE-method is strictly coupled to energy balance. Only the amount of energy absorbed by the broken fiber bridges appear in the fracture toughness data. Any elastic energy within the remaining fiber bridges is retrieved and does not influence fracture toughness data. The CB-method is directly coupled to the measured load and includes always the complete elastic contribution of the fiber bridges on the load. The influence of the fiber bridges on the load is substantially larger than their influence on the external work. This results in an exaggeration of the effect of fiber bridges on fracture toughness data.

5. THEORETICAL AND EXPERIMENTAL OBSERVATIONS

In the latter chapter it was shown by the simulation that higher values for the apparent fracture toughness data are found with the CB-method than with the FE-method. This is confirmed by the experimental results from figure 1. Since the FE-method is strictly coupled to energy balance, it is investigated more thoroughly if the CB-method really violates energy balance.

5.1. Theory of Energy Balance

According to the theory of Energy Balance, the change in the external work is used to increase the elastic energy of the specimen and to propagate the crack, $U_{ext} = U_{elas} + U_{frac}$. It is assumed, as usual for the LEFM, that dissipation is a part of the fracture energy and the kinetic energy is ignored. The external work is calculated as the integral of the load : $U_{ext} = \int F \cdot d\delta$. The elastic energy is calculated similarly for the unloading curve. The Fracture Energy can then be defined in two ways :

$$U_{frac} = U_{ext} - U_{elas} \quad \text{and} \quad U_{frac} = \int G_{lc} \cdot da$$

From these equations it can be seen that an integration of the R-curve will result in the fracture energy. Together with the external work and the elastic energy, balance should occur.

5.2. Experimental Verification

The experimental data from 3 glass/epoxy specimen, including those from figure 1, are used to verify energy balance. The external work is calculated from the load/displacement data observed at those points where crack propagation was registered. The elastic energy left in the specimen at the end of cracking is approximated by assuming linear unloading.

314

The following results are obtained :

Specimen	$\int G_{Ic,cb} \cdot da$	$G_{Ic,fe} \cdot \Delta a$	Ratio
2-2	445.3 Nmm	441.4 Nmm	1.01
2-4	371.9	380.5	0.98
2-6	443.9	391.7	1.13

The experimental verification actually shows that energy balance can be violated if the CB-method is followed. However, the difference between the two methods is not as large as the simulation indicated. Only specimen 2-6 shows evidence of fiber bridging. Effects other than fiber bridging, like surface waviness and multiple crack paths play a role. Their contribution on the apparent G_{Ic} is identical for both methods and energy balance will be preserved.

To check the improvement in consistency, the statistics for each method are compared :

Specimen	$G_{Ic,init}$	$G_{Ic,cb}$	$G_{Ic,fe}$	$G_{Ic,fe}$ (full)
2-2	0.215	0.53	0.424	0.451
2-4	0.153	0.43	0.365	0.396
2-6	0.188	0.50	0.374	0.449
mean	0.185	0.487	0.388	0.432
std.dev.	16.8 %	10.5 %	8.2 %	7.2 %

All units : kJ/m^2

As expected, the initial values show the largest variability. Compared to the CB-method, the FE-method reaches more consistent results based on exactly the same measurement points. Since the FE-method includes initiation effects in the propagating value of G_{Ic} ,the mean value is more conservative. The full potential of the FE-method is demonstrated by $G_{Ic,fe}$ (full) which is based on the exact loading/unloading curve from the tensile testing machine. In this case virtually no information is lost and a further improvement of the variability is reached.

CONCLUSIONS

A more rational approach to the testing of mode I Interlaminar Fracture Toughness testing is proposed. The method is based on the total Fracture Energy and is performed as a one step loading/unloading procedure. It requires only a minimum of operator attention. The test produces unambiguous and consistent results. The method can be automated easily.

The Fracture Energy method is less sensitive to the influence of fiber bridging. Finite element simulations indicate that compliance based methods exaggerate the effects of fiber bridging resulting in too high propagation values. In the case of significant fiber bridging, the here proposed method yields more correct results than compliance based methods.

REFERENCES

1.   C.J. Verberne and C.-G. Gustafson. The effect of fiber bridging on the mode I interlaminar fracture toughness analyzed by finite element technique. ECCM-CTS proceedings September 1992, Amsterdam.

2.   European Structural Integrity Society, Polymers & Composites Task Group. Protocols for Interlaminar Fracture Testing of Composites, Rev. May 1992.

3.   T. Schjelderup, Fracture Mechanics testing of Glassfiber Reinforced Epoxy Specimen (in Norwegian), STF20 F92168, SINTEF, Trondheim, Oct. 1992.

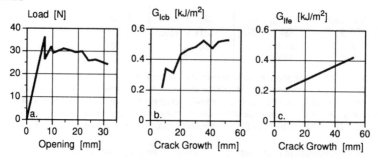

Fig. 1 A typical load/deflection curve (a) results in ambivalent fracture toughness data (b) but becomes notably less ambiguous if the Fracture Energy method (c) is used.

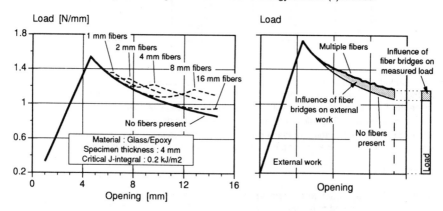

Fig. 2 The simulated load/deflection curve on a per unit width basis.

Fig. 3 An illustration of the influence of the fiber bridges.

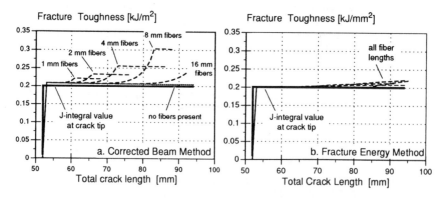

Fig. 4 The simulated fracture toughness data for both methods.

A COMPARISON OF TEST METHODS FOR EVALUATING THE BONDING OF A COMPOSITE REPAIR PATCH TO A COMPOSITE SUBSTRATE

M. KEMP

*DRA - Materials & Structures Dept - R50 Building
GU14 6TD Farnborough Hants - UK*

ABSTRACT

The hot-bonding of carbon fibre/epoxy repair patches under various conditions has been evaluated by mechanical tests. The following specimens were used; integrally-machined single lap shear (SLS); double cantilever beam (DCB); wedge test; and interlaminar shear test (ILSS). Acceptable repairs were achieved on a 180°C cure parent material using 120°C cure pre-preg under atmospheric pressure. The DCB test gave a reliable quantitative measure of bond line resin toughness and the wedge test was satisfactory for testing the environmental sensitivity of the bond. The SLS and ILSS specimens were found to be acceptable as screening tests.

INTRODUCTION

Composite aircraft structures are routinely repaired *in-situ* according to the manufacturers service repair manual (SRM) and general guidelines [1]. This paper compares different mechanical test specimens for use in patch system development or quality assessment of an *in-situ* repair. The effects of cure pressure and post-cure environmental conditioning were assessed for two systems curing at 120°C and 180°C respectively.

1. EXPERIMENTAL DETAILS

1.1. Materials

The substrate (parent) material was 16-ply T800H-924C carbon fibre reinforced epoxy and the lay-up of both the substrate and patch was: $[(+45,-45,0_2)_2]_s$. Baseline data was obtained from autoclaved material: (a) a patch of T300-913 on T800H-924C; (b) 32-ply T800H-924C; (c) 32-ply XAS-913; in each case with 1 layer of 913 film resin between plies 16 and 17. A comparison of curing pressure (autoclave (690kPa) or hot-bond controller (96kPa)) was performed using patches of T800H-924C (180°C cure) and

XAS-913 (120°C cure). A limited number of tests on a unidirectional patch of T800H-924C were included.

1.2. Mechanical Tests

The materials were tested parallel to the 0° ply direction in tension and compression using specimens 250mm × 20mm according to the CRAG standard [2]. Single lap shear (SLS) specimens were integrally-machined 187.5mm long and 20mm wide with a 12.5mm overlap. Initial tests did not have a 0.8mm deep under-cut notch as shown in Fig 1a which was incorporated for subsequent tests.

Interlaminar shear strength (ILSS) tests used a standard specimen geometry [1]. Additional tests were carried out on a holed specimen (HILSS) as shown in Fig 1b which allowed rapid access of environment to the bond line. Specimens were either conditioned (75%RH; 60°C for 1000 hours) or immersed in deionised water at 21°C for 24 hours. The 3-point bend test roller spacing was 20mm roller (substrate resting on rollers) and the load at which the first interlaminar failure (load drop) occurred was recorded. The interlaminar shear strength was calculated from:

$$ILSS = (0.75.P)/(W.t) \qquad (1)$$

where P is the failure load (N); W is specimen width (mm); t is specimen thickness (mm). Wedge test specimens 110mm long and 25mm wide were manufactured with a 10mm PTFE insert. A 3mm thick steel wedge was driven into the bond line at the insert and the initial crack length was measured on both edges. The crack length was monitored after immersion (at 21°C). The double cantilever beam (DCB) specimen geometry was 150mm long and 25mm wide with a 25mm long PTFE insert. Load was introduced via bonded hinges, and edge crack length was measured optically and recorded against load. Values of mode I strain energy release rate (GIC) were calculated from the following formula (no correction factor);

$$GIC = (3.P.d)/(2.B.a) \qquad (2)$$

where P = load to extend crack (N); a = crack length (mm); B = specimen width; d = crosshead displacement (mm).

2. RESULTS

2.1. Patch Properties

The mechanical strengths of the T800H-924C and XAS-913 patch materials cured at two pressures are given in Table 1. The tensile properties of T800H-924C were reduced by 10% at the lower pressure and the compressive strength was reduced by 20%. The properties of XAS -913 were similar at both pressures.

2.2. Bond Strength

The single lap shear results plotted in Fig 2 show much higher values for specimens without the undercut notch. The six baseline tests performed on an autoclaved T300-913 repair (variant D) showed low scatter, and failure in each case was in 0° plies at the base of the notch. The majority of tests with a notch failed in this way.

ILSS results for baseline tests on autoclaved XAS-913 and T300-913 patches.on T800H-924C are given in Table 2. Conditioning of XAS-913 produced no change in ILSS results and a small effect on HILSS results which was similar for water immersion tests. For a unidirectional (U/D) T800H-924C patch the low pressure cure reduced ILSS from 62MPa to 45MPa; and from 62MPa to 25MPa for multi-directional (M/D) material. For XAS-913 the effect was less severe, the low pressure cure reducing ILSS from 63MPa to 53MPa. Wedge test results for the patch variants (Fig 3a) show a large variation in crack extension values. Data from six replicates of the T300-913 patch material also showed large scatter and no particular trend (Fig 3b). Mode I fracture toughness (GIC) results for the patch variants (Fig 4a) showed slightly higher toughness values for the T800H-924C patch. Baseline data (Fig 4b) for the parent (32-ply) material showed higher values for T800H-924C than for XAS-913. The data for five replicate tests on a T300-913 patch showed good reproducibility and low scatter (data points enclosed in shaded region), with fracture occurring uniformly along the bond line. The T800H-924C material showed the highest toughness values, and T300-913 the lowest values.

3. DISCUSSION

The effect of curing pressure on laminate mechanical properties was found to be material-dependent, being less significant for a 120°C curing system (913) than the 180°C cure system (924). Visual comparison of the 924 laminates cured at low pressure showed large interlaminar voids which were not evident in the 913 material. Voidage is related to enhanced thermal expansion effects at the higher cure temperature, being exacerbated between dissimilar ply orientations (the unidirectional patch showed less voids). The presence of large voids reduced the tensile, compressive and ILSS properties of the T800H-924C patch laminate, but the bond strengths as measured by SLS, wedge and DCB tests were not significantly affected.

3.1. Comparison of Test Methods

The tests were assessed on their sensitivity to small differences in bond performance and their reliability (ie high sensitivity/low scatter required). The DCB test has received some attention for bond assessment [3], and in the present work shows that when the fracture mode is consistent (ie in T300-913) the data shows low scatter (Fig 4b). Crack branching led to anomalously high toughness due to crack bridging effects with an apparently high scatter. The 'true' toughness of the adhesive, however, was represented by the lowest values measured, since any branching implied that the adhesive was at least as tough as the matrix. A minimum toughness value could be defined as a minimum patch performance quality criterion. The test was sufficiently sensitive to show that the adhesion between the T300-913 patch and the T800H-924 substrate was poorer than in either parent material (Fig 4b). The wedge test is widely used [4] to assess the environmental susceptibility of adhesives. The considerable scatter in the results (Fig 3) made precise comparison difficult, but the lower susceptibility of the 32-ply T800H-924C material was readily identified by zero crack extension. This test may yield quantitative data if a sufficient number of replicates is used.

The single-lap-shear test is widely used [5] but appeared to be relatively insensitive. This may result from the dominant effect of the notch geometry which was deeply undercut. A shallower notch is recommended. The results are also dependent on the degree of bending which is related to the length of the recess, in this case 12.5mm (Fig 1a). ILSS tests showed failure between various ply lay-ups and hence tested the shear strength of the overall patch+bond system. Presumably the test would show up a significantly poor bond on a pass/fail criterion. The HILSS test has potential as a screening test to assess the environmental sensitivity of the bond although is dependent on the positional accuracy of the drilled hole.

For patch system development tests, the DCB test has potential since it shows good reproducibility and tests a large area of patch bond. A mode II (shear) test would be desirable, and the end-notch-flexure test may be more suitable than the ILSS for this purpose. The wedge test is a useful test for environmental susceptibility. The ILSS, HILSS, and SLS tests are suitable as screening tests.

5. CONCLUSIONS

The G_{IC} (DCB) specimen has been found to be a sensitive and reliable bond integrity evaluation test for composite patch repairs. The wedge test is suitable for determining environmental susceptibility of the adhesive bond. The ILSS, holed ILSS and single lap shear specimens are suitable as screening (pass/fail) tests.

ACKNOWLEDGEMENTS

This work was supported by the Ministry of Defence.

REFERENCES

1. R.E. Trabocco, T.M. Donnellan, J.G. Williams: in "Bonded Repair of Aircraft Structures" (A.A. Baker, R. Jones, eds) (1987) 175-211, EAFM Martinus Nijhoff Publishers
2. P.T. Curtis (ed): "CRAG Test Methods for the Measurement of the Engineering Properties of Fibre Reinforced Plastics", DRA Technical Report 88012, February 1988
3. S. Mostovoy, P.B. Crosley, E.J. Ripling: Journal of Materials, 2 (3) 1967, 661-681
4. B.M. Parker: in "Bonding and Repair of Composites", Proc. Seminar Birmingham July 1989, Butterworth & Co. 1989, 51-56
5. K.B. Armstrong: Int J. Adhesion & Adhesives, January 1983, 37-52

Table 1 - Patch properties for different cure pressures

Patch	Pressure, kPa (psi)	Tensile strength (MPa)	Av TS	Compressive strength (MPa)	Av CS
T800-924	96 (14)	1087,1188,1178	1151	-603,-495,-461	520
T800-924	690 (100)	1326,1228	1277	-664,-666	665
XAS-913	96 (14)	1052,998,1074	1041	-647,-568,-753	656
XAS-913	690 (100)	917,1119	1018	-596,-754	675

Table 2 - ILSS baseline data tests (690kPa cure pressure)

Material	ILSS, dry	ILSS, H/W	HILSS, dry	HILSS, H/W	HILSS, water
XAS-913 (32-ply)	82,84,60,53, 77,82,84 (75)	77,78,77, 80,78 (75)	81,74,81 (77)	79,59,76, 77,70 (72)	81,74,56,75 73 (72)
T300-913 patch	65,83,75,74, 67,68 (72)		73,71,67,68 (70)		29,43,57,35, 56 (44)

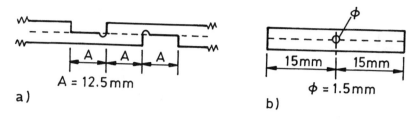

a)

b)

Fig1 a) Single lap shear (SLS); b) Holed ILSS (HILSS)

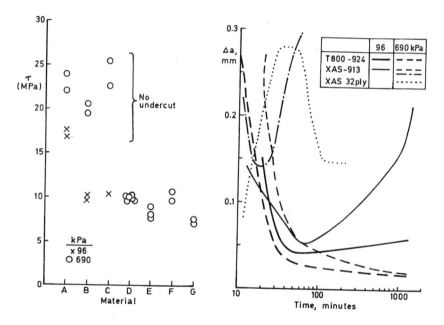

Fig2   Single - lap - shear data for patches of T800 - 924 : U/D (A); M/D (B);32 - ply (F) & : XAS - (913) (C); (E); 32 - ply (G) : T300 - 913 (D)

Fig3   Wedge test results a) Patch variants

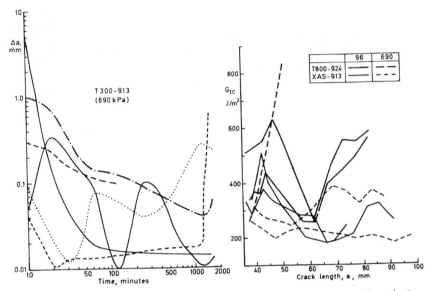

Fig3   Wedge test results b) T300 - 913/T800H   Fig 4 G_{IC} results a) Patch variants
        - 924 patch

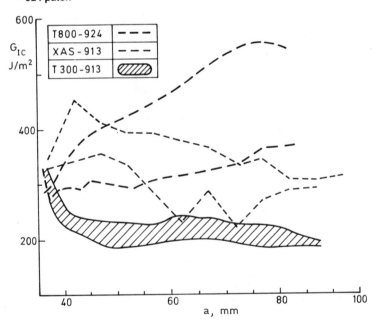

Fig 4 G_{IC} results b) Baseline data

MECHANICAL BEHAVIOR OF EPOXY RESINS IN UNIAXIAL AND TRIAXIAL LOADING

L. ASP, L.A. BERGLUND, P. GUDMUNDSON*

Lulea University of Technology - Dept of Materials and Production Engineering - 95187 Lulea - Sweden
*Royal Institute of Technology - 10044 Stockholm - Sweden

ABSTRACT

The transverse strain at failure of composites is much lower than what is expected from uniaxial tensile tests of neat resins. It has been shown that in glass/epoxy, a triaxial stress-field exists in the matrix. A similar stress-field as in the composite is applied on neat resins by the *poker-chip* test method. The strains at failure measured on four epoxy resins were found to be in the same range as for transversely loaded composites.

INTRODUCTION

Although final failure of composite structures usually involves fiber fracture, transverse crack development parallel to the fiber direction is also of great importance. Transverse cracks lead to reduced stiffness and, for pressure vessels and pipes, leakage of liquids or gases may occur. In addition, the presence of transverse cracks may induce other types of damage such as local delamination and fiber fractures. Transverse failure is troublesome because of its early occurrence in the deformation process. It is not unusual that transverse cracks develop at strains of 0.2%, although final structural failure occur at much higher strains /1/.

An important property in the context of this problem is the transverse tensile strength. This property has been found to be relatively insensitive to the uniaxial stress-strain response of the pure matrix loaded in tension /2,3,4/. It is somewhat surprising that the transverse strain-at-failure is in the order of 0.2% for composites based on resins that have demonstrated strain at failure of 5% in uniaxial tension. Explanations such as effects from flaws, fiber/matrix debonding, stress and strain magnification in the resin due to the presence of fibers and triaxial stress state in the resin have been put forward /3/. The relative importance of these effects is unclear and will depend on the material system.

Theoretical treatments of transverse failure have been summarized by Hull /3/. Kies estimated the non-uniform strain distribution in the matrix using a square array of fibers /5/. From his equation, the strain magnification can be obtained as a function of fiber volume fraction. Garrett and Bailey have employed the theory by Kies to explain the insensitivity of uniaxial failure strain on that of the matrix /4/. Adams has used finite difference analysis to describe the stress and strain fields within the composite /6/.

The objective of this paper is to estimate the importance of the triaxial stress state in the matrix on the low transverse strain at failure of composites. This is possible by comparison of data from uniaxial tensile tests of neat resins with those from triaxial ones. As a triaxial test, the pokwr-chip method was chosen, previously used for Rubbers /7,8/.

1. EXPERIMENTAL

1.1. Materials

For epoxy systems were used. In three systems the epoxy component is based on diglycidyl ether of bisphenol A (DGEBA), commercially available as D.E.R.332 (DOW Chem. Co.). In addition three different curing agents were used; (i) diethylene triamine (DETA) sold as D.E.H.20 (DOW), (ii) a methyltetra hydrophtalic anhydride (MHPA), HY 917, and a methyl imidazole (MI) accelerator, DY 070 (both by Ciba-Geigy) and (iii) polyoxy propyleneamine (APTA) also referred to as Jeffamine T-403 (Texaco Chem. Co.). System (iv) consists of TGDDM, tetraglycidyl 4, 4' diaminodiphenyl methane epoxy, sold as MY 720, cured with DDS, 4, 4' diaminodiphenyl sulfone, market name HY 976, both manufactured by Ciba-Geigy.

1.2. Specimen fabrication and testing methods

The resin systems were cured according to the suppliers recommendations. The specimens for uniaxial testing were designed according to ASTM D638M-81. The strains in these tests were measured by an extensometer with 50 mm gauge length.

A the multiaxial test method, the so called *Poker-chip* method /7, 8/, was used for triaxial testing. Here a thin circular specimen, with diameter 30 and a thickness 4 mm, is adhesively bonded between two aluminium holders and pulled until failure. The specimens were bonded to aluminium fixtures using 73M OST epoxy film from American Cyanamid Co. The strains in the multiaxial tests were measured with a sensitive extensometer (max displacement ±0.5 mm), also with a gauge length of 50 mm. Strain-rates for uniaxial tests and poker chip tests were 1% min^{-1} and 0.2% min^{-1} respectively. All tests were performed at room temperature.

The mechanical properties in Table I are based on experimental data from 30, out of the over all 65 specimens. Specimens that failed at the interface were discarded. However, for DGEBA/APTA did 3 out of 7 specimen fail in the adhesive. These three specimen were all taken into account as they all were stronger than the specimen that failed in the epoxy.

2. THEORETICAL ANALYSIS

The finite element method (FEM) was used to determine the stress state in a transversely loaded glass fiber/epoxy composite. The FEM-analysis was performed using an ABAQUS system, with generalized plane strain elements. The calculations show that a triaxial stress state acts within the glass-fiber reinforced epoxy. At the position subjected to maximum stress is the magnitude ratio between stress components 1:1:2 (x:y:z), where the stress component in the loading direction is the largest. Our results were in agreement with those presented by Adams /6/.

The poker-chip method was used for multiaxial testing. To ensure that we actually tested the material in a triaxial stress field were the stress components within the poker-chip specimen calculated. These calculations were performed using previously derived equations by Schapery /9/. For a chosen aspect ratio (diameter/thickness) of 7.5 and the experimental Young's modulus of DGEBA/DETA is a stress component relationship of 1:1:2 achieved. Hence does the poker-chip method provide a good experimental set-up for simulating the effect of a transverse load on a composite matrix.

3. RESULTS AND DISCUSSION

3.1. Uniaxial test

Uniaxial testing was performed on four epoxies expected to show different stress-strain characteristics. The results are presented in Table I. The most brittle behavior is shown by TGDDM/DDS, the strain at failure is only 1.77 %. This epoxy has a densely cross linked network, low G_{IC} (60-90 J/m^2) and a high T_g. Young's modulus is the highest for this system (3.77 GPa) due to efficient packing of aromatic segments of the molecules /10/.

The other three systems show strains at failure of 6-7%. DGEBA/MHPA has the highest strength (85.9 MPa). Before failure, DGEBA/ APTA showed significant localized yielding in the form of necking. This system has low crosslink density and the lowest T_g of the investigated systems. DGEBA/DETA has the lowest modulus (2.07 GPa) of the investigated systems. This indicates inefficient molecular packing and maybe some creep effects. The strain at failure is 7.0 % which is lower than the 14% reported by Morgan et al. /11/. This may be due to the specimen geometry. For DGEBA/MHPA the strain at failure is larger than expected.

3.2. Triaxial testing

The poker-chip test method has previously been used for rubbers by Gent and Lindley /7/ and for polyurethanes by Lindsey /8/. It was of interest to find out if this test could be used to study glassy polymers. Four epoxies were tested. The results and are presented in Table I. The behavior of the epoxies were similar for DGEBA/DETA and DGEBA/APTA which failed at strains up to 0.8%. Also TGDDM/DDS and DGEBA/MHPA showed almost identical stress-strain behavior, with strains at failure of 0.5%. All four epoxy systems exhibited a linear stress-strain relation until failure.

A comparison between the stress-strain behavior of DGEBA/ MHPA from the uniaxial and triaxial tests is presented in Figure 1. Comparing the experimental results from the poker-chip tests with the uniaxial ones a decrease in strain at failure and strength is clearly shown for all four systems, see Table I.

3.3. Fractography

Similar fracture surfaces are found for DGEBA cured with the aliphatic hardeners, DETA and APTA. Both epoxies exhibit plastic deformation before total failure. In DGEBA/DETA many cracks have coalesced and in DGEBA/APTA only one crack has developed causing total failure. For both of these systems fractography confirms that crack initiation took place within the specimens, and not at the interface or edge. Resemblance is also found between DGEBA/MHPA and TGDDM/DDS when studying the fracture surfaces. They both show very rough fracture surfaces, highly branched cracks, with small or undetectable initiation points. Also the cracks grew with an angle to the plane perpendicular to the load direction and continued along the interface, causing total failure. For these reasons is it difficult to state whether the crack nucleation took place within the specimen or not for these two epoxies.

The fracture texture for DGEBA/DETA and TGDDM/DDS was discussed by Morgan and O'Neal [16]. Their observations are confirmed by our experiments; DGEBA/DETA fails after a high degree of plastic flow as the TGDDM/DDS fails catastrophically. This is what one expects as there is a large difference in critical strain energy release rates (G_{IC}) for the two epoxies. However, the fracture surface of the DGEBA/DETA specimens show a striking resemblance to that of polyurethanes tested with the poker-chip method by Lindsey /8/. Thus, different glassy polymers show different failure mechanisms when subjected to a triaxial stress-field.

4. CONCLUSIONS

A test method which induces a triaxial stress state in the material, the poker chip test, has been successfully applied to glassy polymers. The method has previously been used for rubbers. Similar stress states were found to exist in the poker chip test and in the mostly stressed point of the matrix of a transversely loaded glass fiber/epoxy composite with a fiber volume fraction of 0.5. Four epoxies commonly used as composite matrices showed uniaxial strains at failure from 1.77 to 7 percent. In the poker chip test, the strains at failure were reduced to in between 0.5 and 0.8 percent. This is in agreement with observations of the effect of uniaxial matrix strain at failure on transverse failure strain in composites. Irrespective of uniaxial matrix strain at failure, transverse strain at failure falls in a narrow strain region. In the light of the present data, it is possible that the triaxial stress state has a major influence on the low transverse strain at failure of unidirectional polymer composites.

ACKNOWLEDGEMENTS

Financial support from SICOMP, Swedish Institute of Composites, is greatfully acknowledged.

REFERENCES

1. Spencer B and Hull D, "Effect of winding angle on the failure of filament wound pipe," *Composites,* (1978), pp 263-271.

2. Joneja SK, "Influence of Matrix Ductility on Transverse Fatigue and Fracture Toughness of Glass Reinforced Composites," *SAMPE Quarterly,* (1984), pp 31-38.

3. Hull D, "*An introduction to composite materials,*" Cambridge University Press, 1981, pp 145-154.

4. Garrett KW and Bailey JE, "The effect of resin failure strain on the tensile properties of glass fibre-reinforced polyester cross-ply laminates," *J. Mater. Sci.,* 12, 1977, pp 2189-2194.

5. Kies JA, "Maximum strains in the resin of fiberglass composites," *US Naval Research Laboratory* , (1962), Report NRL 5752.

6. Adams DF and Doner DR, "Transverse normal loading of a unidirectional composite," J Comp Mat, 1 (1967), pp 152-164.

7. Gent AN and Lindley PB, " Internal rupture of bonded rubber cylinders in tension," *Proc. Roy. Soc.* (London) 249A (1959), pp 195-205.

8. Lindsey GH, "Triaxial Fracture Studies," *J. Appl. Phys.,* 38 (1967), pp 4843-4852.

9. Lindsey GH, Schapery RA, Williams ML and Zak AR, *Aerospace Research Laboratories Rept.* ARL 63-152 (September 1963).

10. Oleinik EF, "*Epoxy-Aromatic Amine Networks in the Glassy State Structure Properties,*" in K. Dusek, Epoxy Resins and Composites IV, Berlin (Springer Verlag), 1986, pp 49-99.

11. Morgan RJ and O'Neal JE, "The Durability of Epoxies," *Polym. -Plast. Technol. Eng.,* 10(1), 1978, pp 49-116.

Table I. Strength data and standard deviations from triaxial and uniaxial tests. The variations are the calculated standard deviations. For DGEBA cured with APTA is the yield stress given instead of strength from the uniaxial tests.

epoxy system	Uniaxial strength (MPa)	Multiaxial strength (MPa)	uniaxial strain at failure (%)	multiaxial strain at failure (%)	Young's modulus (GPa)
(i) DGEBA/DETA	69.0±5.4	29.1±4.5	7.00±1.5	0.85±0.1	2.07±0.15
(ii) DGEBA/MHPA	85.9±3.8	26.9±5.7	6.50±1.0	0.57±0.2	2.92±0.12
(iii)DGEBA/APTA	73.1±1.2	32.0±2.2	6.14±0.5	0.79±0.1	2.93±0.13
(iv)TGDDM/DDS	59.9±12	26.6±7.7	1.77±0.4	0.55±0.2	3.77±0.07

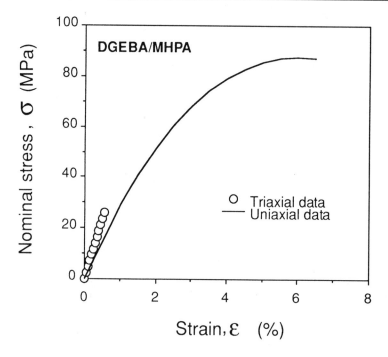

Figure 1. Comparison between stress-strain relations for DGEBA/MHPA from uniaxial and triaxial tests.

THE INFLUENCE OF STACKING SEQUENCE ON IMPACT DAMAGE IN A CARBON FIBRE/EPOXY COMPOSITE

M. KEMP, S. HITCHEN

DRA - Materials and Structures - Building R50
GU14 6TD Farnborough Hants - UK

ABSTRACT

The effect of stacking sequence on impact damage in a carbon fibre/epoxy composite was studied. During the impact event the energy absorbed was monitored which was subsequently related to damage development. The energy absorbed in the initiation of delamination extension was affected by the stacking sequence and was higher for panels containing a larger number of dissimilar interfaces and having 45° plies in the surface layers. The residual energy absorbed in delamination extension was found to increase linearly with increasing delamination area.

1.0 INTRODUCTION

Composites offer benefits over other structural materials in terms of their specific mechanical properties. These properties can be further optimised by tailoring the stacking sequence /1/. A limitation in their use is the susceptibility to localised impact damage imparted by a dropped tool or by runway debris. A few studies have been concerned with the effect of stacking sequence on the impact performance. Hong and Liu /2/ and Strait and al. /3/ found the stacking sequence affected the impact resistance although other studies /4,5/ concluded that the stacking sequence had little or no effect on the energy absorption or damage extent. Hence the effect of stacking sequence on the impact resistance of composite laminates is not yet fully understood.

The current work investigates the effect of stacking sequence on the impact damage in a carbon fibre/epoxy composite comprising of equal numbers of 0° and 45° plies.

2.0 EXPERIMENTAL

The carbon fibre/epoxy composite system T800H/924C was studied. Laminates having a range of stacking sequences were fabricated from pre-impregnated unidirectional sheets, 0.125 mm thick, according to the manufacturer's recommended cure cycle. The six stacking sequences are listed in Table 1 with each laminate comprising of equal numbers of 0° and 45° plies.

The composite panels were impacted using a Rosand Instrumented drop weight machine to a level of 7 joules. During the impact event the panel was clamped between two steel rings, 100 mm diameter and the projectile captured after the first impact to prevent secondary strikes. The energy absorbed during the impact event was monitored by the instrumented impactor with the primary output being force vs time data which can be analysed into energy values. The impacts were repeated at 100 mm intervals across the panel. After impact the panels were scanned by through transmission ultrasonics from which the maximum delamination area was measured with 4 C-scans analysed for each panel.

The distribution of damage through the panel thickness was assessed using a de-ply technique. The specimens were impregnated with a solution of gold chloride, placed in a furnace to partially 'burn-off' the matrix and separated into the individual plies. The delaminations were indicated by a gold residue from which the area of each delamination was measured.

3.0 RESULTS

The position and extent of the delaminations through the panel thickness were assessed using a de-ply technique and the results are shown in Figure 1a&b. Due to the destructive nature of the de-ply technique only one impact was evaluated for each panel. Delaminations, which initiated at almost every interface throughout the panel thickness, tended to increase in size towards the back-face. The total delamination area for each panel is listed in Table 1. The stacking sequence clearly influenced the extent of delamination with less damage sustained by panels containing 45° surface layers, ie panels P1, P2 and P3.

During the impact event the energy absorbed was measured by the instrumented impactor. In all six panels the force increased to a peak value and then gradually fell to zero. The energy absorbed, however, continued to increase to a maximum value after the peak force had been reached and then decreased. Hence both the peak energy (energy at peak force) and the maximum energy were determined and the mean value of the 4 curves analysed are listed in Table 1.

The stacking sequences used in this investigation comprised of different numbers of dissimilar interfaces. In Figure 2 the peak energy and maximum energy are plotted against the number of dissimilar interfaces. Both the orientation of the fibres in the surface layers and the number of dissimilar interfaces influenced the peak energy which was increased by placing 45° fibres in the surface layers and by increasing the number of dissimilar interfaces. The maximum energy was approximately constant for all six panels and appeared to be independent of the stacking sequence.

4.0 DISCUSSION

The present work has shown that altering the stacking sequence influenced both the peak absorbed energy and the delamination area. It is proposed to discuss these effects in terms of the energy required to initiate delamination and the energy absorbed during delamination extension and to relate these to the delamination area.

Previous work by Srinivason and al. /6/ noted a step in the load displacement curve, which remained constant as the impact energy increased, that appeared to correspond to a threshold load for delamination to occur. In the present work, this threshold load corresponded to the peak force value which will, therefore, be taken to represent the energy absorbed in delamination initiation (E_i). The stacking sequence had a marked effect on the delamination initiation energy, being increased significantly by the presence of 45° surface plies, and increased slightly by increasing the number of dissimilar interfaces (Figure 2).

Since the energy absorbed during the impact event continues to increase after the peak load has been reached, it is proposed that the energy absorbed between the peak (E_i) and maximum (E_{max}) energy values produces delamination extension. This 'residual' energy (E_r) is therefore defined as follows:

$$E_r = E_{max} - E_i . \tag{1}$$

The residual energy was influenced by the stacking sequence and was higher for stacking sequences comprising of 0° surface plies (Table 1). The maximum energy (E_{max}) was independent of the stacking sequence for the laminates used in the present work which comprised of similar numbers of 0° and 45° plies.

The stacking sequence had a marked influence on both the total delamination area (measured from de-ply) and the maximum delamination area (measured from C-scans). Since the residual energy was also influenced by the stacking sequence, a correlation between these two parameters was apparent, ie higher residual energies should correspond to larger delamination areas. This effect is observed in Figure 3 where the total delamination area is plotted against the residual energy (E_r). An increase in maximum delamination area was also observed with increasing residual energy as shown in Figure 4, supporting the residual energy concept. A least squares fit (solid line) indicated an approximately linear relationship between the delamination area and the residual energy as shown in Figures 3 and 4. Work is proceeding on other composite systems to determine whether the initiation and extension energies for delamination can be similarly correlated.

5.0 CONCLUSIONS

The force-time history of impact events in the T800H/924C system has been related to the delamination area. Altering the stacking sequence was found to change the peak absorbed energy (E_i), but not the maximum absorbed energy (E_{max}). The peak energy has been related to the energy required for delamination initiation and a 'residual' energy term (E_r) has been proposed (where $E_r = E_{max} - E_i$) which defined the energy absorbed during delamination extension. Resistance to delamination initiation was enhanced by the use of 45° surface plies and by increasing number of dissimilar interfaces.

ACKNOWLEDGEMENTS

This work was supported by the Ministry of Defence.

REFERENCES

1. W.S. Chen, C. Rogers, J.D. Cronkhite and J. Martin. J. American Helicopter Society 32 (1987) 60-69

2. S. Hong and D. Liu. Experimental Mechanics (1987) 115-120

3. L.H. Strait, M.L. Karasek and M.F. Amateau. J. Comp. Mat. 68 (1992) 1725-1740

4. J. Morton and D. Goodwin. Composite Structures 13 (1989) 1-19

5. C.K.L. Davies, S. Turner and K.H. Williamson. Composites 16 (1985) 279-285

6. K. Srinivason, W.C. Jackson and J.A. Hinkley in 36th Intern. SAMPE Symposium, April 1992, 850-862

Panel	Stacking Sequence	Max. Del. Area mm^2+	Total Del. Area mm^2*	Peak Energy J+	Max. Energy J+	(Max.-Peak) Energy J+
P1	$[(+/-45,0_2)_2]_s$	510	1350	6.31	6.65	0.34
P2	$[(+/-45)_2,0_4]_s$	870	1360	6.23	6.60	0.37
P3	$[(+45,0,-45,0)_2]_s$	650	870	6.41	6.55	0.14
P4	$[(0_2,+/-45)_2]_s$	765	1510	6.06	6.63	0.57
P5	$[(0_4,(+/-45)_2]_s$	1620	2070	5.98	6.70	0.72
P6	$[(0,+45,0,-45)_2]_s$	360	2170	6.11	6.66	0.56

+ average of four values

* value relating to specimen subsequently assessed using the de-ply technique

Table 1: The total and maximum delamination areas together with the energy values obtained during impact of T800/924 laminates.

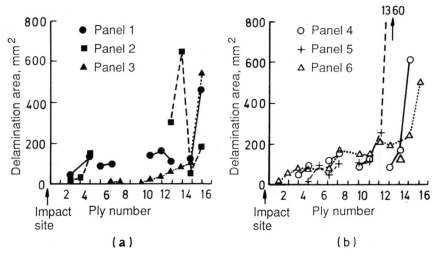

Fig 1 Distribution of delamination through the panel thickness
a) Panels 1, 2, 3   b) Panels 4, 5, 6

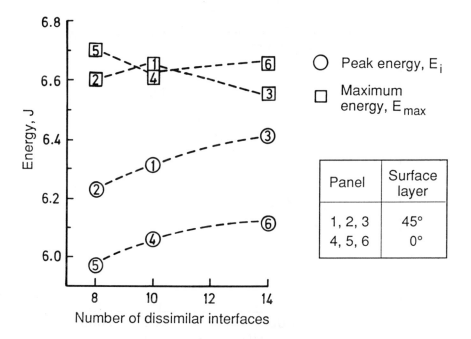

Panel	Surface layer
1, 2, 3	45°
4, 5, 6	0°

Fig 2 The effect of the number of dissimilar interfaces on the peak and maximum energy values

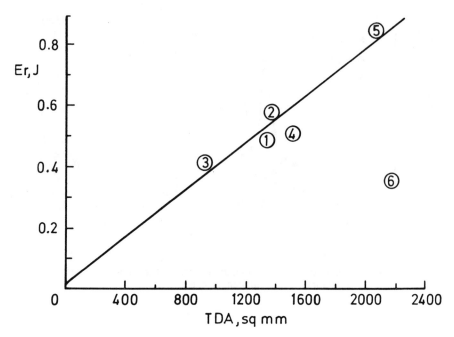

Fig 3 The relationship between the total delamination area (TDA) and the residual absorbed energy

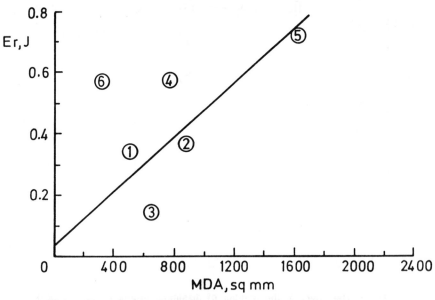

Fig 4 The relationship between the maximum delamination area (MDA) and the residual energy (E$_r$)

INCREASE OF WEAR RESISTANCE OF GFRP WITH HARD SUPERFICIAL LAYERS

A. LANGELLA, I. CRIVELLI VISCONTI

Dipartimento Di Ingegneria Dei Materiali E Della Produzione (DIMP)
Piazzale Tecchio - 80125 Naples - Italy

ABSTRACT

The wear behaviour of composite materials with resin filled with hard particles were investigated. Some different hard particles as: SiC, TiC, WC, SiO_2, Cr_3C_2 were used to fill the epoxy matrix of the composite materials.

Sliding wear tests against smooth steel surface, pin on disk method, was conducted under dry condition.

Weight loss percentage and wear rate were evaluated during the tests.

Scanning electron investigations (SEM) to individuate the wear mechanisms were carried out.

1. INTRODUCTION

It is certainly known that all cases when singularities are present in the design are critical for the choice and the use of composite materials.

In this paper possible means to increase composite wear and abrasion resistance are evaluated, obtaining that the external surface of the component are harder than normal.

Ideally, this can be achieved in two ways:
- using harder matrix (i.e. metals or ceramic matrix)

- depositing harder coatings.

These two criteria can be sinergically integrated, trying at the same time to avoid the drawbacks of each of the criteria, that is:

- To avoid the use of a metal and ceramic matrix, still having a hard final surface;

- To avoid a dangerous interface between coating and substrate.

In particular an attempt is made to create Functional Gradient Composite Materials (FGCM) by using a polymeric matrix filled with the largest possible amount of hard ceramic particles, with the feature that only the external layer of the composite is made in this way, while the amount of the particles filling the resin should decrease toward the inner part of the piece.

The advantages of doing so are twofolds:

- to preserve a single materials;

- to avoid the large loss in the properties, if large quantities of the particles were used for the entire piece.

2. SAMPLES MANUFACTURING

The first step was to choice an epoxy resin, and to fill it up with the maximum amount of hard particles. The following hard particles: SiC (meshes 320 and 600), TiC, WC, SiO_2, Cr_3C_2 were used.

The apparent good miscibility conditions to achieve for all particles types have been evaluated: for ponderal percentage more than 30 the appearance of air bubbles and of bottom deposits has been noted.

Eight layers laminates have been manufactured by manual technique starting by E glass unidirectional fibres and epoxy resin (Ciba-Geigy type Araldite XB 5082); the polymerisation phase was carried-out in a hot-platen press with a temperature of 120^0 C for 2 hours under a pressure of 4 bar.

The two external layers, top and bottom layers, have been impregnated by resin filled with ponderal percentage, on global matrix weight, of 25%.

The tensile tests conducted in according to ASTM standards have shown that while the elastic properties have not appreciable variation, the maximum values of strength slightly decrease.

3. WEAR TESTS

The wear tests were done on a pin-on-disk apparatus; a schematic sketch of the experimental arrangement is given in fig.

1. This wear machine has a steel disk C40 with Vickers hardness of 200 N mm², and can rotate at constant angular speed of 1415 rpm.

The sliding specimen angular speed can be changed by varying its radial distance from the disk center.

Two series of specimens, with square form and apparent contact area of 100 mm², have been tested.

In the first series (five specimens for each types of powders) the wear resistances have been evaluated by progressive loss weight of the specimens, with a precision balance, every three minutes for a total time of 15 minutes; in the second series (three specimens for each types of powders) the progressive loss weight of the specimens have been evalueted every ten minutes for a total time of 60 minutes for a samples with unfilled resin and with SiC (meshes 320 and 600) and Cr_3C_2 filled resin.

The specimens were glued on metallic cylinders of 1 c m diameter in order to be put up correctly in the specimen-holder of the test machine.

The other details of the tests are:

periferic sliding speed (v)= 10 m/s

sliding radius track = 68 mm

speed rotation of the disk = 1415 rpm

apparent contact pressure(p) = 0.35 N mm^{-2}

p*v product= 3.5 Nmm^{-2} ms^{-1}

All the tests were performed in a parallel sliding direction.

The medium value of the weight loss percentage as a function of test time for the first and second series of the specimens is reported, respectively, in figure 2 and figure 3.

In these figures it is possible to observe that the wear resistance, evaluated as weight loss percentage, of the specimens with resin filled is improved respect to the not filled resin.

The lower value of the weight loss percentage is obtained with resin filled with Cr_3C_2 powder; in this case the weight loss percentage is about 3 times lower than that of the not filled resin.

With loss weight (Δm) it is possible to evaluate an adimensional parameter w, "wear rate", defined as:

$$w = \frac{\Delta m}{\rho * v * t * A}$$

where: Δm= loss weight (g)

ρ = average density of laminate (gcm^{-3})

v = periferic sliding speed (m/s)

t = test time (s)

A = apparent contact area (mm^2)

In fig. 4 the wear rate evaluated for the second serie of specimens with time test of 60 minutes are reported. In this figure it can be seen that the lower wear rate is obtained for the specimens made with resin filled with Cr$_3$C$_2$ powder. It is important to observe that for the specimens made with resin filled with SiC 320 mesh the wear rate is higher than that for not filled resin.

Microscopic examination of the tested specimens were made by scanning electron microscope (SEM). This analysis, similarly to what has been found in bibliography, has enabled to individuate four wear mechanisms:

- matrix wear
- fiber sliding wear
- fiber fracture
- interfacial fiber-matrix debonding.

In figures 5 e 6 SEM micrographies show the above wear mechanisms. In particular figure 5 show the wear of the matrix after a time of 15 minutes in specimens with resin filled with Cr$_3$C$_2$ powder.

In the same figure it is possible to see the fractures of the fibers. This fractures could be caused by the lateral impacts on the fibers due to wear debris and to the asperities present on the disk.

Figure 6 shows the face of a fractured fiber and the debonding of the fiber from the matrix; it is possible to see, also, the particles of resin and of powder produced during the wear.

4. CONCLUSIONS

The wear tests have demonstrated that using some types of hard powders it is possible to improve the wear resistance, evaluated as weight loss percentage of the composite material laminates. In particular important results have been obtained using Cr$_3$C$_2$ powder.

A Functional Gradient Composite Material (FGCM) by using a polymeric matrix filled with the largest possible amount of hard ceramic particles seem to be worthwhile to have:

- hard surface;
- polymeric (soft) core;
- gradual variation of the mechanical properties.

It is possible to individuate several sectors of industrial application of this type of composite materials:

- tribological couples;

- space application.

In the first case the mechanical elements are subjected, frequently, at extreme lubrication conditions that should cause the removal of materials from tribological elements.

In the second case special effect due to the bombing of spatial powders (IPD) on the composite structures can cause the degradation of the composite material.

From the technological point of view it is possible to image a technology with large application capability as the pultrusion. With this technology it is possible to realise with sufficient easiness and precision the variation of the percentage of the powders in the thickness. This assures the gradual variation of the mechanical properties from the core towards the external surfaces of the laminate without abrupt interfaces.

Achnwledgements

The Authors are grateful to the C.N.R for the financial support of the research. Special thanks are also due to the ing. Palazzi of the University of Salerno for his help in performing the wear tests.

References

1. LANCASTER J.K.,in Friction and Wear of composite materials, ed. K. Freidrich, Elsevier, Amsterdam, 1986, Chapter 12
2. FREIDRICH K., in Friction and Wear of polymer composites, ed. K. Freidrich, Elsevier, Amsterdam, 1986, Chapter 8
3. TSUKIZOE T., OHMAE N., "Friction and wear of advanced composite materials", Fibre Sci. Technol. 18 [1983] pp 265-286
4. CIRINO M., FRIEDRICH K., PIPES R.B., "Evaluation of polymer composites for sliding and abrasive wear applications", in Composites, vol 19, n° 5, Butterworth & Co. LTD, 1988
5. CIRINO M., PIPES R.B., FRIEDRICH K., The abrasive wear behaviour of continuous fiber polymer composites", J. Mater. Sci. 22 [1987] pp 2481-2492
6. TURNER R.M., COGSWELL F.N., "The effect of fiber characteristics on the morphology and performance of thermoplastic composites", 18th International SAMPE Techinical Conf., Seattle, Washington October 7-9 [1986] pp 32-44
7. FRIEDRICH K., "Friction and wear of polymer composites", Elsevier Scientific Publishers, Amsterdam, 1986
8. VISHWANATH B., VERMA A.P., RAO C.V.S.K., "Wear study of glass woven roving composite", Wear 131 [1989] pp 197-205

9. CIRINO M., PIPES R.B., "The wear behaviour of continuos fiber polymer composites", in Wear 5 [1987] pp 302-310

10. LHIMN C., TEMPELMEYER K.E., DAVIS P.K., "The abrasive wear of short fiber composites", In Composites, vol 16, n° 2, April 1985

11. SUNG N.H., SUH N.P., "Effect of fiber orientation on friction and wear of fiber reinforced polymeric composites", in Wear 53 [1979] pp 129-141

12. VOSS H., FRIEDRICH K., "On the behaviour of short fibre reinforced PEEK composites", in Wear 116 [1987] pp 1-18

13. FRIEDRICH K., MALZAHN J.C., in Proc. Int. Conf. on Wear of materials, Washington, USA, 1983

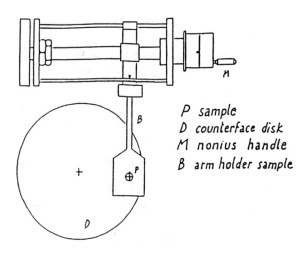

P sample
D counterface disk
M nonius handle
B arm holder sample

Fig. 1 - Pin on disk apparatus for wear test.

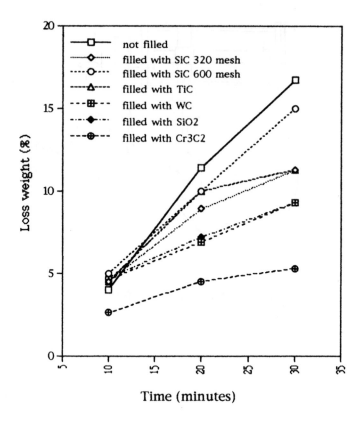

Fig. 2 - Loss weight (%) of specimens with not filled resin and filled resin as a function of the test time for the first series of specimens (total time test 30 minutes).

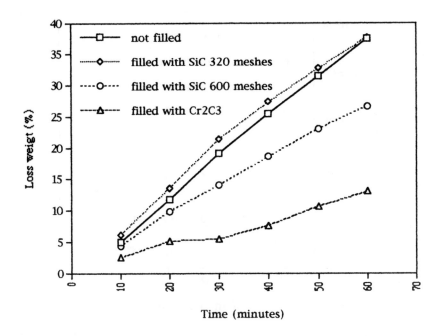

Fig. 3 - Loss weight (%) of specimens with non filled resin, filled with SiC (320 and 600 mesh) and filled with Cr3C2, as a function of the test time for the second series of specimens (total time test 60 minutes).

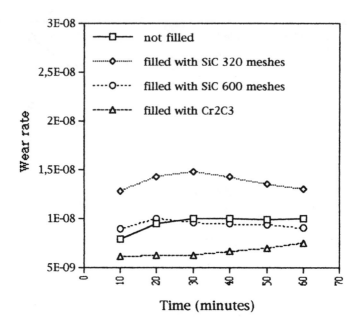

Fig. 4 - Wear rate of specimens with non filled resin, filled with SiC (320 and 600 mesh) and filled with Cr_3C_2, as a function of the test time for the second series of specimens (total time test 60 minutes).

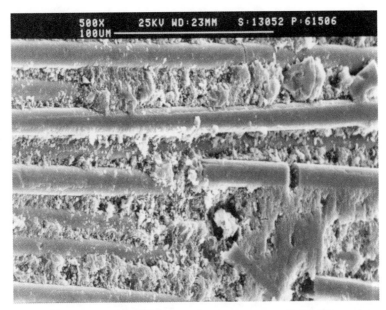

Fig. 5 - SEM micrograph 500x: Matrix wear and unaffected zone.

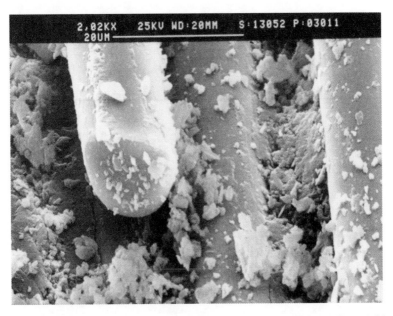

Fig. 6 - SEM micrograph 2000x: Fracture fiber and interfacial fiber-matrix debonding

345

AGEING AND DAMAGE TOLERANCE

A COMPARATIVE STUDY OF THE MECHANICAL BEHAVIOR OF COMPOSITES CURED USING MICROWAVES WITH ONES CURED THERMALLY

S.L. BAI, V. DJAFARI, M. ANDREANI, D. FRANCOIS

Ecole Centrale de Paris - Lab. MSS/MAT ECP - Grande Voie des Vignes 92295 CHATENAY-MALABRY - France

ABSTRACT

Unidirectional glass fiber/epoxy composite materials were fabricated separately using thermal curing and microwave curing, and mechanical tested. Microscopic observations showed that the microwave cured composites contained more voids than the thermal cured ones. The influence of void content on the elastic moduli was taken into account using an analysis based on the Mori-Tanaka model /1/. Another analysis gives the average interfacial tensile strength (AITS). By comparing the experimental results and the corresponding fracture surfaces, it appears that the fiber/matrix interface of the microwave cured composites is stronger than that of the thermally cured ones.

INTRODUCTION

Microwave curing is widely used for industrial applications /2/. However, the use of microwave curing for composite materials is not so wide spread as for other industrial applications. The microwave curing of composite materials was devised early in 1984 by Lee and Springer /3/. Other studies were done on optimizing the curing process /4//5/. However, little research has been conducted on the final mechanical properties, although studies on microwave cured resins have showed that they exhibited satisfactory mechanical properties /6/ /7/.

Microwave curing is a fast and efficient technique when applied to homogeneous materials such as epoxy resin. For composite materials with two or three phases, the situation becomes more complicated. If the curing is too fast and the pressure is too low to force the voids to get off they are blocked in the materials.

The presence of voids has a significant influence on the mechanical properties /8//9/. It reduces the elastic moduli, and creates stress concentrations which can lead to premature failure. In the present paper, both the influence of the voids on the mechanical properties and the mechanical characterization of fiber/matrix interface were studied. Longitudinal and transverse tensile tests were carried out. A comparison of mechanical properties was made between the microwave cured composite (MCC) and thermally cured composite (TCC).

CURING OF COMPOSITE MATERIALS

A unidirectional E-glass/epoxy composite was preimpregnated using a filament winding technique. The matrix DGEBA-3DCM is a three-dimensional, thermosetting epoxy resin. Its mechanical characteristics and those of E-glass are given below :

| DGEBA-3DCM | E=2.8Gpa | $\upsilon=0.35$ | σ_r=90Mpa |
| E-glass | E=74Gpa | $\upsilon=0.22$ | σ_r=2100Mpa |

E, υ, σ_r are Young's modulus, Poisson's ratio and tensile strength, respectively.

The curing cycle is given in Table 1. Mettler TA 3000 DSC was used to measure the glass transition temperature, Tg. The same Tg was found for the two composites according to /10/. It should be noted that different pressures were applied during the two curings and that the microwave curing is much more rapid than the thermal one. The experimental device for the microwave curing is shown in Fig.1. The microwaves cross the composite specimen in the fiber direction. The thermal curing was performed in an autoclave using a final cure temperature of 140°C for 1 hour.

The fiber volume fraction was 73% and 75% for the MCC and TCC, respectively. SEM observation revealed an 5% void content in the MCC, versus an almost zero percentage in the TCC. The two photos in Fig.2 show the void distribution.

EXPERIMENTAL PROCEDURE

Test specimens were sectioned from the composite plates using a diamond cut-off wheel and the end tabs were donded to the grip sections(Fig.3). The longitudinal tensile tests were carried out using a crosshead speed of 4mm/min on a hydraulic MTS machine and the transverse tensile tests on a motor driven testing machine which is installed in the interior of a SEM. A dynamometer gives the applied load.

RESULTS AND ANALYSIS

(a) Longitudinal tensile tests

Figure 4 shows the longitudinal tensile curves for the composites. The two curves represent the average results of three tests for each composite. The two composites exhibit elastic behavior. During loading, longitudinal cracks were created and caused the final failure of the specimens. The experimental data are the following :

	E_l(Gpa)	υ	σ_r(Mpa)
MCC	$52^{+6.6\%}_{-5.9\%}$	0.28	$1163^{+13.8\%}_{-7.7\%}$
TCC	$56^{+6.5\%}_{-4.7\%}$	0.29	$1376^{+0.4\%}_{-0.4\%}$

The mechanical properties of composites depend on their constituents and on the fiber/matrix interface. If the fiber fraction is large, the load in the fiber direction is carried mainly by the fibers. In this case, the comparison becomes meaningless since the fibers are not modified during the curing. The longitudinal elastic modulus of composite materials is a linear function of the fiber fraction according to the Rule of Mixtures. Here, the TCC has a 2% greater fiber volume fraction than the MCC, the modulus E_l should increase by 2.8% (1.53Gpa), and the modulus E_t by 7.1% (0.9Gpa).

We can analyze the influence of the voids on the elastic moduli by using the Mori-Tanaka model which in itself is based on Eshelby's theory of inclusions /11/. In our model, the voids are considered to be inclusions with zero moduli. All voids are assumed to be aligned in the fiber direction and to be distributed homogeneously within the matrix. The Mori-Tanaka equation is for the elastic tensors given below.

$$C_c = C_m[I+V_fQ(I+V_fQ(S-I))^{-1}]^1 \tag{1}$$

$$Q = [(C_m-C_f)(S-I)-C_f]^{-1}(C_f-C_m)$$

Where C_m, C_f, C_c are the elastic tensors of matrix, fiber and composite, respectively
S is Eshelby's tensor, V_f is the fiber volume fraction

The results of these calculations are shown in Fig.5. First, with a given void aspect ratio l/d, the moduli of the composites decrease as the void content increases. It should be noted that the modulus in the fiber direction is almost not affected by the void content. This is logical since the load is borne mainly by the fibers when their volume fraction is high. In the transverse direction, the matrix and the interface play the dominant role. Since voids in the matrix will reduce its modulus, they will also affect the transverse modulus. Furthermore, the void shape has a slight influence on the moduli of the composite. By correcting experimental results with the analytical results, we found that the difference of moduli between two composites is smaller than 3%.

The composite tensile strength is also influenced by the voids. The voids create stress concentrations and, leading to premature fiber failure. Thus, this explains why the MCC has a lower tensile strength and why the experimental data are scatter.

(b) Transverse tensile tests
The average transverse tensile curves are showed in Fig.6. Because of the small dimension of the specimens, the only stress-time response was recorded. The results are listed in Table 2. E* represents the slope of the tensile curve and is written as $E*=d\sigma/dt$ (Mpa/minute). Using the same strain rate allows us to obtain the ratio of the transverse moduli of the MCC and TCC composites :

$$E^*_M/E^*_T = E_M/E_T \tag{2}$$

This ratio is 0.82 for our composites. If this value is compared with that of Fig.5 by assuming that TCC contains no voids ($E_T=E_0$), it is found that the experimental result is greater. This difference comes from the fact that, on the one hand, the test dispersion is large, and on the other hand, that the TCC contains voids ($E_T<E_0$). With a 1% increase in void content, the transverse modulus decreases by 6.2%,.

The transverse tensile strength σ_t is the same for the two composites. In fact, the transverse strength of the composite is affected directly by that of the interface and also by that of the matrix. In general, σ_t decreases as the fiber fraction increases since the interface is weaker than the matrix. One can estimate theoretically the composite transverse strength by applying a Rule of Mixtures as follows /12/ :

$$\sigma_t = \sigma_m(1-\sqrt{\frac{4V_f}{\pi}})+\sigma_i\sqrt{\frac{4V_f}{\pi}} \tag{3}$$

Where σ_m is the matrix tensile strength and σ_i is the average interfacial tensile strength (AITS) needed to separate the fiber from the matrix under transverse loading. If σ_t is known from mechanical tensile tests, one can obtain σ_i with this equation (3). The results obtained are reported in Table 2.

It was found that σ_i of the TCC is 8% greater than that of the MCC. The equation (3) gives only an estimation based on the Rule of Mixtures. The influence of voids is not taken into account. We can modify this equation by two different methods. First, the matrix can be replaced by one containing voids. This change decreases σ_m. Secondly, the matrix volume fraction is reduced owing to the presence of the voids. The second method gives the modified equation :

$$\sigma_t = \sigma_m(1 - \sqrt{\frac{4V_f}{\pi}})\frac{V_m - V_b}{V_m} + \sigma_i\sqrt{\frac{4V_f}{\pi}} \qquad (4)$$

Using this last equation (4), the modified AITS (σ_i') of the MCC is also given in Table 2. The modified σ_i' is almost equal to that of the TCC. By taking into account the stress concentration, it is possible that MCC has a greater AITS than that of the TCC.

The observation of the fracture surface in the SEM shows that the fibers are partially covered with resin for the MCC. In contrast, the fibers are completely exposed for the TCC. Near the edges of the fracture cracks, the matrix is damaged by microcracks and the fiber/matrix interface remains intact for the MCC (Fig.7b). This is indicative of a strong fiber/matrix interface. For the TCC, matrix damage is not seen, however, the fiber/matrix interface has decohered (Fig.7a). It is thus concluded that the fiber/matrix interface of the MCC is stronger than that of the TCC.

CONCLUSIONS

- microwave curing is much faster than conventional thermal curing.
- microwave curing induces a higher void volume fraction due to a lower external applied pressure and shorter curing time. By modifying the microwave curing process (i.e. increasing pressure), it is hoped that the void content can be reduced.
- the presence of voids decreases the elastic moduli and causes premature failure of the composite specimens. The elastic moduli are identical for the two composites after taking into account the differences of void content and of the fiber fraction.
- by analysis of the experimental results and observations of the fracture surfaces, it appears that the average interfacial tensile strength is greater for the microwave cured composites than for the thermal cured ones.

ACKNOWLEDGEMENT

This study was completed under the programe "Microwaves-Polymer", sponsored by CNRS of French, Thiais.

Table 1   The curing process of composites studied

curing	mold	specimen (mm)	process	pressure	Tg(°C)
MCC	Teflon	250×36×1	1000W, 20 minutes	10 bar	144
TCC	metallic	250×180×1	140°C, 1hour	14bar	144

Table 2   Experimental and analytical results of transverse tensile tests

N°	MCC				TCC		
	E^*	σ_r Mpa	σ_i Mpa	σ_i' Mpa	E^*	σ_r Mpa	σ_i Mpa
1	1.28	14.19	11.37	11.99	1.84		
2	1.7	14.32	11.50	12.12	1.64		
3	1.14	15.57	12.80	13.42	1.29	17.30	15.60
4	1.18	18.03	15.35	15.97	1.30	14.13	12.36
5	1.12	14.50	11.69	12.31	1.78	14.50	12.74
average	1.28	5.32	12.54	13.16	1.57	15.31	13.57

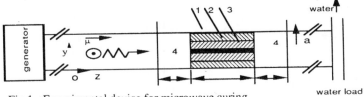

Fig.1   Experimental device for microwave curing
1. 2 : mold materials 3 : composite specimen   4 : matched transitions

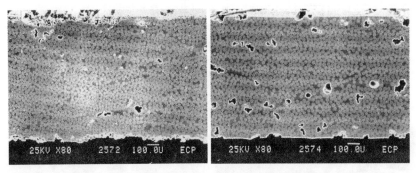

(a)   Thermal cured composite   (b)   Microwave cured composite

Fig.2   Micrographs of the two studies composite materials

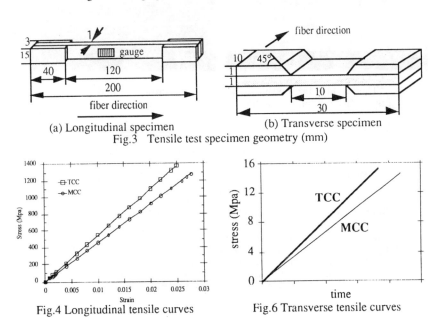

(a) Longitudinal specimen   (b) Transverse specimen
Fig.3   Tensile test specimen geometry (mm)

Fig.4 Longitudinal tensile curves   Fig.6 Transverse tensile curves

353

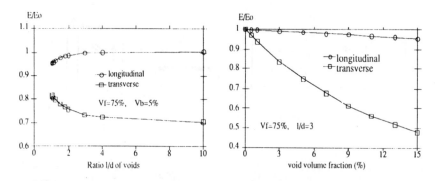

(a) Void shape   (b) Void content

Fig.5   Analytical results of the influence of void content on the composite moduli.
l and d are the voids length and diameter, respectively
V_b is the void volume fraction, E_0 is the composite modulus without voids

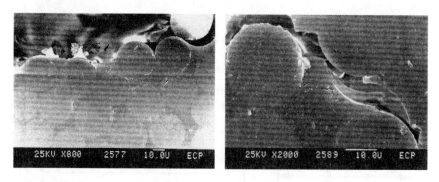

(a)   Thermal cured composite   (b)   Microwave cured composite

Fig.7   Fractography of the fracture surfaces after transverse tensile tests

REFERENCES

1. Mori T. & Tanaka K., *Acta Met*. Vol.21 (1973), p.571-574
2. <<Technologie & Stratégie >>, French, April 1992
3. Lee W. I. and *al.*, *J. of Comp. Mater.*, Vol.18, July 1984, p.387-409
4. Wei J. and *al.* SAMPE J., Vol.27, 1991, p.33-39
5. Boey F. Y. C. & Lye S. W., *Composites*, Vol.23, No.4, 1992, p.265-270
6. Singer S. M.and *al.*, SAMPE Quaterly, Vol.20, 1989, p.14-18
7. Bai S. L., ph.D, Ecole Centrale de Paris, French, September 1993
8. Hancox N. L., *J. of Materials Science* 12 (1977), p.884-892
9. Bowles K. J. and *al.*, *J. of Comp. Mater.*, Vol.26, No.10, 1992, p.1487-1509
10. Jordan C., ph.D, INSA of Lyon, French, December, 1992
11. Eshelby J. D., *Proc. Roy. Soc.*, London, Serie A. 241, 1957, p.376-396
12. Cooper G. A. & Kelly A., ASTM STP 452, 1969, p.90-106

RELIABILITY OF REINFORCED PLASTICS UNDER THERMAL LOADING

M.R. GURVICH

Riga Technical University - 1 Kalku Street - LV PDP 1658 Riga - Latvia

ABSTRACT

Methodology of prognostication of random thermal expansion of reinforced plastics, taking into account conditions of maintenance, their possible vagueness and anisotropy of materials properties, has been elaborated. Independent statistic characteristics, defining random thermal expansion of materials, have been proposed. Condition of thermostable reinforced plastics reliability has been formulated and expression for its quantitative estimation have also been proposed.

INTRODUCTION

Production of thermostable structures, being under the conditions of considerable changes of exploitation temperature, leads to actuality of creation of materials with low thermal expansion. From this point of view, modern reinforced plastics (RP) and, in particular, graphite-plastics are high effective materials. Features of RP thermal expansion, in comparison with traditional construction materials, are in nonlinear dependence on temperature /1,2,etc./ and as well as in heightened instability of thermal properties /2-5/. Conditions of maintenance have also the character of some uncertainty and in a quantitative analysis may be considered as random. Therefore estimation of thermostable structures reliability may be realized only on the basis of the methods of prognostication of materials reliability, taking into account random nature of materials properties and parameters of loading. Here, reliability is determined as a probability of carrying-out of necessary requests (size-stability, strength, etc.).

1. RANDOM THERMAL EXPANSION

In accordance with deterministic models /1,2,etc./ let's introduce random function of thermal expansion $\tilde{\alpha}(\tau)$, which may may be quantitatively defined by the function of mean values $\bar{\alpha}(\tau)$ and the correlation function $K_\alpha(\tau,\tau')$. (Here and farther sign " ~ " above variable or function shows its random character; sign "-" - mean value). Hence, random thermal strain under temperature change from T_0 up to T may be determined as

$$\tilde{\varepsilon}(T_0,T) = \int_{T_0}^{T} \tilde{\alpha}(\tau)d\tau \tag{1}$$

statistic functions of which are written as

$$\bar{\varepsilon}(T_0,T) = \int_{T_0}^{T} \bar{\alpha}(\tau)d\tau \; ; \tag{2}$$

$$K_\varepsilon(T_0,T,T') = \int_{T_0}^{T}\int_{T_0}^{T'} K_\alpha(\tau,\tau')d\tau d\tau' \tag{3}$$

Let approximation of $\tilde{\alpha}(\tau)$ in accordance with /3,5/ be expressed in the following form:

$$\tilde{\alpha}(\tau) = \bar{\alpha}(\tau) + \tilde{\mu} \tag{4}$$

where $\tilde{\mu}$ - random variable with zero mean value ($\bar{\mu}$ = 0) and standard deviation σ_α. The function $\bar{\alpha}(\tau)$ by analogy with /1,2/ may be written as

$$\bar{\alpha}(\tau) = \sum_{i=1}^{n} \bar{\alpha}_i \cdot \tau^i \tag{5}$$

where $\bar{\alpha}_i$ - coefficients, calculated on the basis of statistic analysis of experimental investigations. From here, in accordance with expressions (2), (3) we have:

$$\bar{\varepsilon}(T_0,T) = \sum_{i=1}^{n} \bar{\alpha}_i(T^{i+1} - T_0^{i+1})/(i + 1) \; ; \tag{6}$$

$$K_\varepsilon(T_0,T,T') = \sigma_\alpha^2(T - T_0) \, (T' - T_0) \tag{7}$$

and standard deviation of thermal strain may be found by the rule:

$$\sigma_\varepsilon = (K_\varepsilon(T_0,T,T' = T))^{1/2} = \sigma_\alpha \cdot |T - T_0| \tag{8}$$

It is necessary to note, that if distribution of $\tilde{\alpha}(\tau)$ is determined by a normal law, the distribution of $\tilde{\varepsilon}$ will also be a normal one. Therefore it is enought to use parameters $\bar{\alpha}_i$ (i = 1,...,n) and σ_α in this case for numerical estimation of probability density function $p_\varepsilon(\varepsilon)$.

Let's consider, as an example, the experimental results of thermal expansion of unidirectional graphite-plastics (fiber - ЭЛУР-4П; binder - polymer matrix ПАИС-104) in transversal direction /4/. The area of random deformation ($T_0 = 30^{\circ}C$; $T \leq 250^{\circ}C$) for sample of 19 uniformity specimens is shown in Figure 1. The results of approximation are

$$\bar{a}(\tau) = 0,2904 \cdot 10^{-4} - 0,1200 \cdot 10^{-7} \cdot \tau \quad [^{\circ}C]^{-1} ;$$

$$\sigma_a = 0,2493 \cdot 10^{-5} \quad [^{\circ}C]^{-1} . \tag{9}$$

Histigram of experimental values of $\tilde{\mu}$, calculated by the rule

$$\tilde{\mu} = \Delta\tilde{\alpha} = \frac{\varepsilon_k - \bar{\varepsilon}(T_0, T_k)}{T_k - T_0} \tag{10}$$

is shown in Figure 2. (Here, ε_k - experimental value of thermal expansion for " k " score; $\bar{\varepsilon}(T_0, T_k)$ - calculation by expression (6) with $T = T_k$). For analysed sample ($1 \leq k \leq 817$) there is a good agreement with normal law (statistic $\chi^2 = 3,504$; $P(\chi^2) = 0,32$). It may be also served as an argument of validity of normal law hypothesis.

Let's note that in general there are possible more complex distributions of $\tilde{a}(\tau)$ than (4): for example in /3/ model, taking into account the dependence of $\tilde{a}(\tau)$ instability on temperature, has been proposed.

2. RANDOM CONDITIONS OF MAINTENANCE

Conformably to thermostable structures, vagueness and uncertainty of conditions of maintenance consist of random character of boundary temperatures T_0 and T. Let distributions and necessary parameters of random value T_0 and T be determined, as a result of statistic analysis of conditions of maintenance: $\bar{T}_0, \bar{T}$ - mean values; σ_{T0}, σ_T - standard deviations; $k_{T0,T}$ - possible correlation between its /5/. Hence on the basis of (1), (4) we have:

$$\bar{\varepsilon}(T_0, T) = \bar{\varepsilon}(T) - \bar{\varepsilon}(T_0) + \tilde{\mu} \cdot (T - T_0) \tag{11}$$

Using the method of linearization and taking into account that

$$\frac{d\bar{\varepsilon}(T_0, T)}{dT} = \bar{a}(T) \tag{12}$$

we obtain:

$$\sigma_\varepsilon^2(T_0, T) = \bar{a}(T)^2 \cdot \sigma_T^2 + \bar{a}(T_0)^2 \cdot \sigma_{T0}^2 +$$
$$+ 2 \cdot \bar{a}(\bar{T}) \, \bar{a}(\bar{T}_0) \cdot K_{T0,T} + \sigma_a^2 \cdot (\sigma_{\Delta T}^2 + \Delta \bar{T}^2) , \tag{13}$$

where

$$\sigma^2_{\Delta T} = \sigma^2_{T0} + \sigma^2_T + 2K_{T0,T} \; ; \quad \Delta\overline{T} = |\overline{T} - \overline{T_0}|$$

Mean values of strain $\overline{\varepsilon}(T_0,T)$ are calculated by expression (6) with $T = \overline{T}$; $T_0 = \overline{T_0}$. These statistic characteristics ($\sigma_\varepsilon, \overline{\varepsilon}$) define thermal strain distribution (under rigorous analysis not normal distribution even for $\overline{\alpha}(\tau)$ normal law), but however for accuracy of engineering practice it is usually enough for probabilistic calculations.

3. RANDOM STRAINS STATE OF PLASTICS

For analysis of thermal deformation of real structures made of RP, it is usually necessary to take into account the whole of thermal strains field. Let the axes 1,2 be the axes of "thermal" symmetry and let's introduce the following matrixes by the rules:

$$[\overline{\alpha}(\tau)_*]_t = [\; \overline{\alpha}_1(\tau) \quad \overline{\alpha}_2(\tau) \quad o \;] \; ; \tag{14}$$

$$[K_\alpha(\tau,\tau')_*] = \begin{bmatrix} K_{\alpha1,1}(\tau,\tau') & K_{\alpha1,2}(\tau,\tau') & 0 \\ K_{\alpha2,1}(\tau,\tau') & K_{\alpha2,2}(\tau,\tau') & 0 \\ 0 & 0 & 0 \end{bmatrix} \tag{15}$$

Here, the indexes show the direction of deformation. Therefore for arbitrary axes x,y, turned on an angle of θ , we have the following values of statistic characteristics (indexed by sign "*") /4/:

$$[\overline{\alpha}(\tau)] = [m] \cdot [\overline{\alpha}(\tau)_*] \; ;$$
$$[K_\alpha(\tau,\tau')] = [m] \cdot [K_\alpha(\tau,\tau')_*] \cdot [m]_t \tag{16}$$

where [m] - matrix of transformation; t - sign of matrix transposition. Consequently the field of RP random thermal strains may be determined on the basis of the following characteristics:

$$[\overline{\varepsilon}(T_0,T)] = [m] \int\limits_{T_0}^{T} [\overline{\alpha}(\tau)_*]d\tau \; ;$$

$$[K_\varepsilon(T_0,T,T')] = [m] \int\limits_{T_0}^{T} \int\limits_{T_0}^{T'} [K_\alpha(\tau,\tau')_*]d\tau \cdot d\tau' \cdot [m]_t \tag{17}$$

In case of space strains state analysis the similar equations may be written by introduction of third dimension. It may be also noted that in case of random conditions of maintenance, calculation of random strains field may be easily carried-out by analogy with one-dimensional analysis.

4. CRITERION OF RELIABILITY

Statistic characteristics (17) give a possibility to approximate the probability density function $p_\varepsilon(\varepsilon_x,\varepsilon_y,\gamma_{xy})$. In case of normal law choice, a quantitative estimation of $p_\varepsilon(\varepsilon_x,\varepsilon_y,\gamma_{xy})$ is especially a simple one. Let Ω be an area of possible deformation. Hence a probability of not exceeding it may be considered as a reliability R (condition of size-stability):

$$R = \iiint\limits_{\Omega} p_\varepsilon(\varepsilon_x,\varepsilon_y,\gamma_{xy}) \; d\varepsilon_x d\varepsilon_y d\gamma_{xy} \tag{18}$$

358

Condition of design of thermostable structures or their elements by the criterion of reliability may be written as

$$R \geq R_* \qquad (19)$$

where R_* - some standard (limiting) value of reliability. Problem of RP optimal design may be also formulated on the basis of the reliability conception:

$$R \longrightarrow \max \qquad (20)$$

This formulation permits to take into account random nature of materials properties and parameters of loading.

CONCLUSION

Elaborated methodology allows to predict principal statistic characteristics of random field of RP thermal deformation depending on deterministic or random conditions of maintenance. Obtained results may be used in the following applications: - numerical estimation of independent statistic characteristics of materials thermal expansion, including elaboration of experimental investigations plan; - carrying-out of probabilistic analysis of thermal expansion of structures made of RP and estimation of their reliability (by size-stability criterion); - basis for creation of prognostication structural methods for composites random thermal properties investigation and, consequently, for optimal design their structures by maximum reliability criterion.

REFERENCES

1. Skudra A.M., Bulavs F.Ya., Gurvich M.R. and Kruklinsh A.A. "Structural Analusis of Composite Beam Systems". Technomic Publishing Co. Lancaster, USA. Chapter 3. (1991).
2. Gurvich M.R., Sbitnev O.V., Sukhanov A.V. and Lapotkin V.A. Mechanics of Composite Materials. 26 N1. (1990) 32-36.
3. Gurvich M.R., in "Mechanics of Composite Materials" (1990) 86-99, Riga, Latvia.
4. Gurvich M.R., Lokshin V.A. and Lepikash E.R., in "Mechanics of Reinforced Plastics" (1991) 58-68, Riga, Latvia.
5. Gurvich M.R., Sbitnev O.V., Sukhanov A.V. and Lapotkin V.A. Mechanics of Composite Materials. 28 N6 (1992) 778-786.

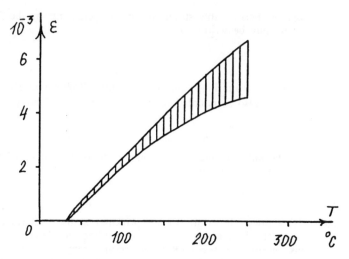

Fig. 1. Area of experimental distribution of graphite-plastic random thermal strains.

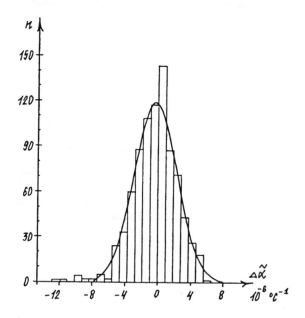

Fig. 2. Experimental histogram and theoretical approximation (by normal law) of $\tilde{\mu} = \Delta\tilde{\alpha}$.

DAMAGE TOLERANCE OF NON-CRIMPED FABRIC COMPOSITES

M.L. KAY, P.J. HOGG

Queen Mary & Westfield College (University of London)
Dept of Materials - Mile End road - EA 4NS London - UK

ABSTRACT

The recent introduction of non-crimped fabric (NCF) materials where layers of woven fibres are fixed in place by the presence of a thermoplastic polyester stitching yarn , so reducing the crimping effect of the fibres, has led to new potential for composites. In the following study both NCF prepregged and hand lay-up composites have been examined and non - crimped fabric prepregged composites were seen to have a poorer resistance to impact damage than their equivalent unidirectional F913G prepregs in regard to the amount of damage sustained. However, the NCF laminates prepared by hand lay-up exhibited superior resistance to damage formation in impact. Normalising the compression after impact results indicates a greater damage tolerance in the NCF materials. The differences in performance between unidirectional prepreg and NCF materials may be linked to the different nature of the damage formed in the two systems.

INTRODUCTION

The increased usage of composite material in structural applications at the expense of metallic alloys, has been achieved to a large extent by their high strength, stiffness and potential for weight savings. However, an inherent susceptibility to low energy impact damage can be a major problem, particularly when incorporating composites into aircraft design. For low energy impact damage, which may not be visible at the impacted surface, the residual compressive strength of test coupons can be reduced by up to 40% as compared to undamaged coupons (1). For this reason, the compression after impact test (CAI) has been developed (2,3 & 4) as a means of assessing qualitatively different fibre and resin composite systems. The CAI test, which has been used in the following work, is often seen as a damage tolerance test but in actuality can be separated into two distinct elements. These elements consist of the ability to measure the resistance of the material to impact damage and the resistance to the propagation of this damage in subsequent compression. It is the latter of the two parts which can be correctly described as measuring the damage tolerance of the system.

The importance of developing composite materials with both increased damage resistance and damage tolerance is evident. Considerable effort has been made by manufacturers to develop woven fabric composites which provide beneficial damage tolerance with the advantage of increased ease of manufacture and reduced production time. Although weaving has gained increased prominence it does suffer from the problem of fibre crimping in the weave and this can result in a reduction in its in plane properties in comparison to material with straight fibres. This has led to the introduction of non-crimped fabrics (NCF) where the fibre tows are arranged in specific orientations and the crimping effect is removed by incorporating a polyester stitching yarn into the material, figure 1.0. The stitching yarn travels around the fibre tows and ideally is integrated into the resin on curing.The stitching yarn influences the microstructure of the composite by retaining fibres in discrete bundle form. NCF's have been found to exhibit surprisingly high in-plane mechanical properties, often in excess of those predicted for equivalent materials based on unidirectional prepregs (5). The impact properties of NCFs are particularly noteworthy with high forces and high energies required to cause full penetration of a laminate during drop weight impact (6).

The objective of the work reported here was to explore the potential of NCF-based composites to provide a route to highly damage tolerant materials. The CAI of a number of NCF based composites have been examined. The materials include a simple glass fibre - polyester laminate produced by hand-lay up (representative of marine applications) and high volume fraction glass fibre-epoxy prepregged systems (appropriate for aerospace).

1. EXPERIMENTAL
1.1 Materials

For the purpose of the study three types of material were used, as listed in the table below. These consisted of two materials based on the same NCF , EQX 1168, a quadriaxial fabric produced by TechTextiles Ltd. One fabric was prepregged with an epoxy resin, Cycom 759 by Cyanamid Aerospace Products and the other was hand laminated at QMW with a vinyl ester resin, Derakane 411-45 produced by Dow Europe. A third material, a unidirectional glass fibre-epoxy prepreg Fibredux 913G-E-5-30% manufactured by Ciba-Geigy Plastics was also used for comparative purposes. Both prepregged materials were cured according to the manufacturers' specifications in an autoclave at 120°C and the hand lay-up material was cured at room temperature with a post cure at 80°C. Post-manufacture non-destructive inspection was carried out by Ultrasonic C-scanning. Glass fibre volume fractions and void contents were measured using ASTM standards D-3171 and D-2734 .

MATERIAL	FIBRE	Void Content (%)	FIBRE V_f (%)	MATRIX	Abbreviation
Cycom 759 / EQX1168 (NCF)	Woven E-glass & polyester stitching yarn	4.55	51.3	Epoxy	NCF 759
DERAKANE 411 / EQX 1168 (NCF)	Woven E-glass & polyester stitching yarn	n/a	55	Vinyl Ester	NCF EQX 1168
Fibredux 913G	E-glass	1.77	58.9	Epoxy	F913G

The fibre stacking sequence for the non-crimped fabrics EQX 1168 / Cycom 759 and EQX 1168 / Derakane 411 was $[-45,90,+45,0]_2$. This is not a symmetric lay-up. The F913G was laminated with both a symmetric quasi-isotropic lay-up of $[45,0,+45,90]_{2s}$ and an asymmetric lay-up of [-45,-45,90,90,+45,+45,0,0,90,90,+45,+45,0,0,-45,-45] which simulated the NCF material more closely. Two plies were positioned at each orientation to match the double thickness layers in the NCF prepreg. From each plate produced, specimens were cut of dimensions 89 x55 mm with a nominal thickness of 2mm.

1.2 Experimental Method

Testing of laminates using a process developed at QMW and described in detail elsewhere , (7) was carried out at room temperature. Impacting of the specimens was performed on an instrumented dropweight impact machine which had a hemispherical impactor of mass 3.96kg. The machine had been previously statically calibrated. All specimens were clamped between two plates which each had a 40mm diameter circular opening. The impact energy and velocity could be altered by varying the height from which the hemispherical tup was released.

The incident energies used for the impact testing ranged from 2 to 8 Joules. Here 8 Joules was the upper limit for low velocity impact testing as above this energy the spread of damage was restricted by the diametral support conditions of the impactor.

For the compression part of the test a specially designed miniaturised test jig was used and this was equipped with an anti-buckling guide used to support the specimen and prevent plate buckling. A loading rate of 0.3mm / min was used in compression and all specimens were filmed using a video camera system. Measurements of impact damage width were taken from scaled ultrasonic c-scan plots and this enabled further plotting against residual compressive strength results.

2.0 RESULTS

The compression after impact strengths for the four materials tested are shown in figure 3.0 plotted as a function of striker kinetic energy. Only specimens that failed at the impact damage sites during compression are included in this plot for clarity. A number of additional specimens failed during compression at the unsupported region of the plate. The two subsets of specimens produced from the unidirectional glass fibre prepregs, F913G, are plotted separately but the best fit line has been drawn considering the two subsets to form one overall common data set.

The compression strengths recorded from the NCF based materials show an inferior compression strength retention as a function of increasing impact energy compared to the UD prepreg based systems.

The extent of the damage region caused by impact for all materials was measured prior to the compression testing. The prepreg based laminates exhibited a well defined damage area consisting of delaminations at various positions through the thickness of the plate, as has been described widely for similar materials by other authors (8,9). In these cases the width of the delaminated zone could be measured optically using transmitted light. In the case of the NCF laminate, the damage formed was very diffuse in nature and the true extent could not easily be determined by eye. The damage area was accordingly measured from C-scan plots which showed a well defined region. The damage zone defined by the C-scan related to an area within which some fibre fracture could be identified visually on specimens. However some diffuse damage was apparent outside this zone and some uncertainty exists as to the exact meaning of the damage zones recorded for the NCF specimens.

The damage widths recorded for all specimens as a function of impact energy are plotted in figure 2.0. The two subsets of specimens for the F913G show similar results but the NCF base prepreg system exhibits a larger damage zone for a given energy while the NCF based laminate produced by hand-lay-up shows a smaller damage zone and therefore a greater resistance to impact damage.

DISCUSSION

The initial results from the post-impact compression tests show that non-crimped fabrics have inferior damage tolerance compared to the equivalent unidirectional glass fibre prepregged composites. However, the trends in the two groups of materials, NCF and UD prepreg based composites are different. The data suggests that if bigger specimens were used, where larger damage areas could be sustained, then the NCF materials would become superior to the UD based systems when higher energy impacts were involved. In an attempt to identify the cause of the difference in behaviour, i.e differences in either impact response or subsequent compression , the data has been replotted as compression strength versus damage width. Earlier studies of UD prepreg base composites with very different overall post impact compression performance (when judged on the basis of a given impact energy) have shown very similar behaviour when compression strength was plotted versus damage width. This is not the case for the NCF based materials. Figure 4.0 shows compression strength plotted against damage width along side similar data for the F913G specimens. The differences in trends observed in figure 3.0 are still apparent. As the resin in F913G and Cycom 759 are qualitatively similar, then this observation suggests that there is a fundamental difference in the nature of the impact damage in the two groups of materials.

The use of a damage parameter based on damage width is valid for UD prepreg base systems, as in all cases the predominant damage consists of well defined delaminations. The difficulty in identifying the extent of the damage in the NCF materials by eye and the lack of correspondence between C-scan damage area and visual observation as to the extent of damage must make the validity of such a damage parameter for NCFs open to question. The C-scan results suggest that a well defined area of delamination does occur. However, microscopic examination of polished sections taken from impacted NCF laminates, shows that cracking tends to occur at and around the discrete fibre bundles formed by the fabrics and the development of a pseudo delamination can take place by the coalescence of these cracks. A photograph obtained from a separate experiment, showing such apparent delaminations in a relatively low volume fraction (40%) glass fibre NCF- polyester laminate is shown in figure 6.0. Scanning electron micrographs , shown in figure 7.0 reveal cracking in the prepregged NCF growing around a fibre bundle. In the F913G specimens, the fibres are uniformly dispersed throughout the laminate and delamination cracks propagate across the specimen between plies for long distances and spread from ply to ply via cross ply shear cracks. The indications are that two distinct failure mechanisms are therefore being exhibited in the two types of material examined.

A direct comparison between the F913G laminates and the NCF prepreg laminates is slightly misleading when comparing the efficiency of the different reinforcing forms used, as the volume fraction of fibres is different is each case. If it is assumed that the compression properties are directly proportional to the fibre volume fraction, then the NCF data may be adjusted to give results for an equivalent volume fraction to the F913G specimens. This is shown in figure 5 where the results are plotted in all cases as a fraction of the compression strength of undamaged F913G, and the NCF performance now appears closer to that of F913G. Alternatively, if it is assumed that the comparatively poor properties of the NCF laminates are due to the inherently poor compression properties of the material (undamaged) then normalising the results will provide a better indication of the relative damage tolerance of the two systems. The results, normalised on the basis of undamaged strength, are shown in figure 5 where the NCF materials are shown to be clearly superior to that of the UD prepreg based composites .

It should be noted that the compression after impact results for the non crimped fabrics do not take into account the higher void contents present (average NCF V_v=4.55% as compared to average F913G V_v=1.77%) which may contribute to the low undamaged compressive strength of the laminates. Future work may allow the production of non-crimped fabric prepregs with higher compressive properties due to greater uniformity and consolidation leading to a lower percentage of voids. The degree of bonding of the fibres and resin are also thought to be likely to improve the damage tolerance of this material as it is not clear that the NCF fabrics have been optimised in this respect.

CONCLUSION

The object of this study was primarily to make an initial comparison between non-crimped fabric prepregs and unidirectional fibre prepregs in terms of the post- impact compression test. Evidently, in the case of the materials tested the non - crimped fabric prepregged composite was seen to have a poorer resistance to impact damage than its equivalent unidirectional prepreg F913G in terms of damage area sustained for a given blow. In contrast, the NCF laminates prepared by hand-lay-up exhibited superior resistance to damage formation in impact. In absolute terms the compression after impact properties of the NCF systems were in all cases inferior to those of the UD prepreg based laminates, but the normalised results indicated a greater damage tolerance in the NCF materials. The studies of microcracking indicate that this difference may be linked to the different nature of the damage formed in the NCF laminates. This consists primarily of cracking around fibre bundles which coalesce to form larger cracks which have many of the characteristics of conventional delaminations but which appear more resistant to subsequent propagation.

ACKNOWLEDGEMENTS
This work was financially supported by SERC in collaboration with the National Physical Laboratory (UK).

REFERENCES

1. Lee S.M. , " Compression after Impact of Composites with Toughened Matrices" SAMPE Journal (March / April 1986) pp 64 - 68

2. Anon. , "Standard Tests for Toughened Resin Composites (Revised Edition)" NASA Reference Publication 1092 (1983)

3. Anon. , "SACMA Recommended Test Method for Compression After Impact Properties of Oriented Fibre-Resin Composites" Recommended Method SRM 2-88 (1988)

4. Anon. , "Advanced Composite Compression Tests" Boeing Specification Support Standard 7260 (1982)

5. Guild F.J., "A Model For The Reduction in Compression Strength of Continuous Fibre Composites After Impact Damage" Composites (1993) To Be Published

6. Hogg P.J., Ahmadnia A. & Kay M.L., "The Mechanical Properties of Non-crimped Fabric Composites with Integral Through-Thickness Stitching" Fifth International Conference on Fibre Reinforced Composites (FRC) (1992)

7. Prichard J.C. "Post-impact Compression Behaviour of Continuous Fibre Composite Materials" Ph.d. Thesis , Q.M.W.,University of London (1991)

8. Bishop S.M. & Dorey G., "The Effect of Damage on the Tensile and Compressive Performance of Carbon Fibre Laminates" AGARD Conf. Proc. No. 355 (1983) pp 10-1 to 10-10

9. Lesser A.J. & Filippov A.G., "Kinetics of Damage Mechanisms in Laminated Composites" 36th International SAMPE Symposium (1991)

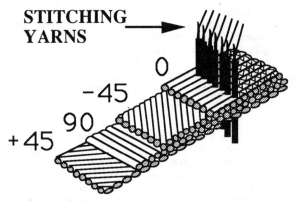

Figure 1.0 - Schematic illustrating the construction of multiaxial non-crimp fabric

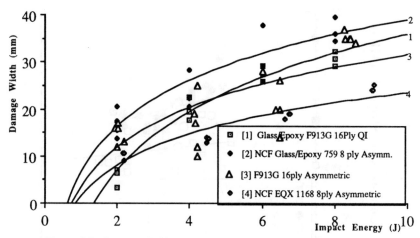

Figure 2.0 - Damage Width versus Impact Energy Graph

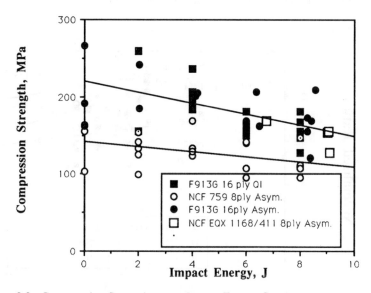

Figure 3.0 - Compression Strength versus Impact Energy Graph

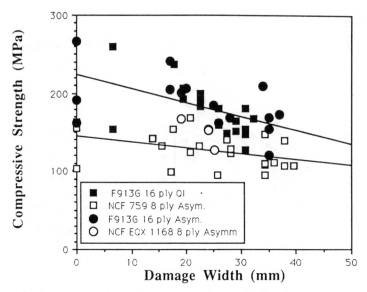

Figure 4.0 - Compression Strength versus Damage Width Graph

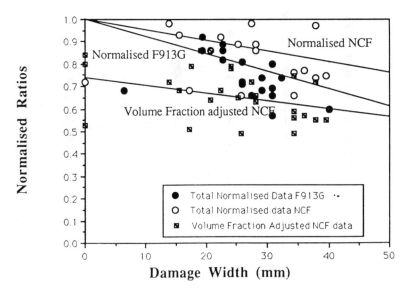

Figure 5.0 - Normalised & Adjusted Compressive Strength versus Damage Width

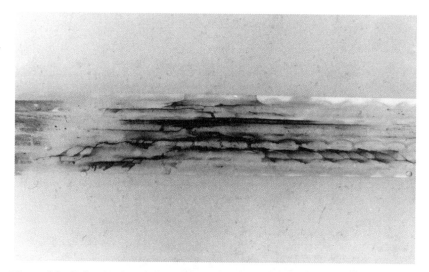

Figure 6.0 - Delaminations in low volume fraction (40%) glass fibre NCF- polyester

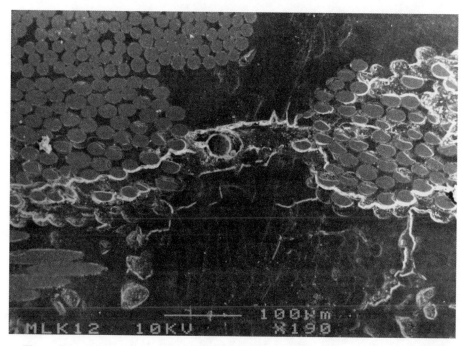

Figure 7.0 - S.E.M. Micrograph showing the crack path around fibre bundles in NCF
EQX 1168 / Cycom 759 (impacted at 4J followed by compression).

AGEING BEHAVIOUR OF CERAMIC COMPONENTS USED FOR THERMAL PROTECTION SYSTEMS

U. RIECK, L. KAMPMANN

Dasa/Erno - Huenefeldstr. 1-5 - 2800 Bremen - Germany

ABSTRACT

Amongst the Thermal Protection Systems for space vehicles the flexible blanket type insulation represents a novel solution combining attractive lightweight properties with the adaptability to various thermal and structural requirements.

The prolonged structural coherence of the blanket assembly - consisting of a microfiber insulation core, embedded between cover fabrics and sewn together using a glass/silica thread system - is depending on the proper performance of each single component. The investigation performed aimed at the mechanical performance of the different component materials upon thermal and structural ageing conditions.

INTRODUCTION

Space vehicles need to be protected against thermal loads arising during ascent and reentry into the earth atmosphere.
The specific kind an design of such a **Thermal Protection System (TPS)** is mainly driven by the heat evolution and distributuion at the spacecraft's surface and the maximum temperature tolerated by the primary structure.
The use of a flexible blanket TPS combines attractive lightweight properties with adaptability to complex substrate shapes and the extension to higher service temperatures by exchanging the component materials.

I. FLEXIBLE BLANKET DESIGN

Flexible blanket type insulations - being under development for the thermal protection of parts of the HERMES primary structure and certain locations at ARIANE 5 - in principle consist of the components, depicted in figure 1.

The insulation core which has to reduce the arising thermal loads to a level acceptable for the primary structure material, consists of a multilayer stack of silica microfiber fleeces embedded between a silica outer cover fabric and a glass inner fabric sewn together using a silica (hot surface) and glass (cold surface) thread system.
Additionally, the temperature application range of the blanket can be extended by substituting the outer silica fabric and thread components by corresponding aluminoborosilicate (ABS) components.

The outer surface of the blanket is protected against particle and plasma erosion by a silica based coating which in addition improves the surface stiffness of the assembly. A water proofing is added to the core to prevent excessive moisture absorption during stay at the launch site and ascent.

II. MATERIAL CHARACTERISATION
2.1 Fabrics and threads

The structural coherence of a flexible blanket type insulation is mainly governed by the property profile of the components used for the blanket assembly, i.e. the flexible insulation core, the embedding fabrics and the threads used for the sewing process.
The knowledge of the properties in the as manufactured as well as in the aged state is of vital importance for the successful performance of the blanket assembly. Therefore extensive characterisation has been performed determining the material behaviour after thermal and mechanical ageing.

Threads and fabrics have been subjected to tensile testing at elevated test temperatures and after annealing at temperatures as high as 800C in order to gain insight into the glass, silica and ceramic material performance.

The various strength behaviour is depicted in figures 2 and 3 -revealing the typical course of tensile loadability of the different threads and fabrics as a function of test temperature. The steep decrease at approximately 500C is attributed to the decomposition of the thread sizing which serves as a protective layer reducing frictional induced notch effects of the filaments when sliding against each other.

It is clearly visible that the strength behaviour of the aluminoborosilicate components at elevated temperature is far superior compared to the silica components.
However, in view of thermal insulation properties and manufacturability this material has to be further analysed.

In terms of frequent use of such TPS, e.g. for reentry vehicles, the material behaviour upon thermal cycling is of great importance. Figure 4 depicts the strength behaviour of different fabrics (silica and ABS) being exposed to thermal cycling up to 50 cycles.
As can be seen the load bearing capability of the silica and ABS material does not follow a monoton decrease but runs through a minimum.
This partial material recovery is attributed to a kind of self healing effect within the material upon cyclic annealing.

2.2 Insulation core

During mission the insulation blanket is assumed to experience various loads resulting from acoustic noise, differential pressure and precompression of the insulation core already introduced during the manufacturing process (in order to keep the bobbin threads always under tension), see figure 5.

. Therefore the core compression behaviour is of importance in view of damping characteristics and the resulting fatigue behaviour of the assembly.

Figure 6 presents the compression behaviour of core materials with different fiber diameter after thermal and mechanical ageing. The latter was carried out by deforming the core sample within certain load limits in compression mode for several thousand cyles in order to simulate the oscillating loads acting during mission.

The effect of the different aging mechanisms is quite significant, reducing the compression strength visibly. Further core development will aim at the improvement of the compression behaviour using suitable mixtures of fibers with different diameters.

2.3   Coating/blanket surface interaction

As already mentioned the coating applied to the blanket surface serves as an erosion protection and increases surface stiffness.

The coating material has to fulfill the following criteria:

- chemical and mechanical compatibility with the
  blanket surface components
- excellent high temperature resistance
- processability at room temperature
- suitable thermooptical properties

The mechanical compatibility between the coating and the blanket surface is of great importance, as during acoustic noise loading induced vibrations at the surface can lead to relative movement between coating particle agglomerates and the filaments of the thread system.

If unsuitable coating constituents are used, the abrasive effect between the agglomerates and the filaments can lead to notch effects, see figure 7, with subsequent early failure of the complete thread, leading to blanket malefunction.

In order to avoid such behaviour, coating constituents were optimised regarding particle size and ability of film formation.

An effective method to check the interaction between coating, fabric and thread is the fast compression test, which is schematically depicted in figure 8.

The cyclic loading of the coated blanket simulates an oscillating surface eventually leading to the damage mechanism described above.

Table 1 indicates the maximum number of cycles being tolerated by differently coated blankets before first thread failure occured.

The morphology of the coating which endured the most compression cycles is shown in figure 9. No particle agglomerates are observed but a rather coherent film is achieved on the blanket surface.

373

III. Summary and Conclusion

The development of the flexible blanket type insulation for space vehicles implies the characterisation of its different components w.r.t. their mechanical performance under certain ageing conditions.

The test program performed so far revealed the significant influence of thermal and structural impacts on the blanket assembly components. As can be deduced from the results obtained the aluminoborosilicate material exhibits better high temperature strength capabilities compared to the silica and might therefore be a suitable alternative in thermally and structurally highly loaded areas.

The reduction of possible damage potential caused by the surface coating was achieved using special constituents which help to avoid agglomerate formation between the filaments and lead to a more coherent surface film.

Based on the know how gathered during the present program and the material characteristics determined so far a further flexible blanket design refinement will be performed.

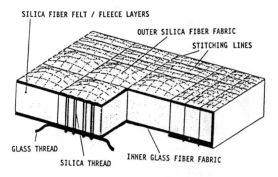

Fig.1 Flexible blanket assembly

Thread Break Loads
at elevated Temperatures

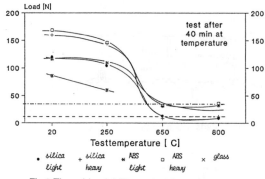

Fig.2 Thread loadability vs. test temperature

Fabric Strength
at elevated Temperatures

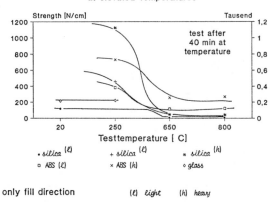

Fig.3 Fabric strength vs. test temperature

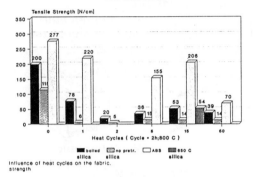

Influence of heat cycles on the fabric.
strength

Fig.4 Fabric strength development as a function of
thermal cycles

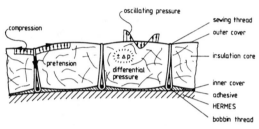

Fig.5 Loads acting on the blanket assembly

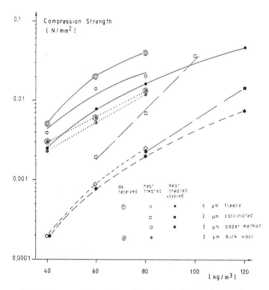

Fig.6 Core compression strength evolution depending
on the ageing conditions

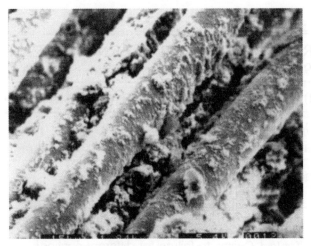

Fig.7 Coating agglomerates deposited on/between filaments

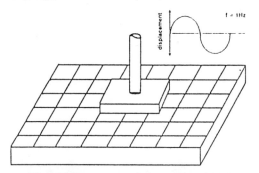

Fig.8 Fast compression test set-up

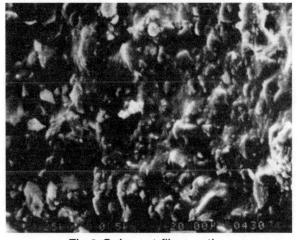

Fig.9 Coherent film coating

ERNCZ 59	ERN CA 4	ERN CA 4a	ERN CA 3
80.000	50.000	30.000	100.000
slight damage	medium damage	severe damage	no damage
back up	skipped	skipped	1st choice

Table 1 Cycles to failure for different coatings

RESIDUAL DELAMINATION PROPERTIES OF CARBON FIBRE REINFORCED COMPOSITES AFTER DROP-WEIGHT IMPACT

J.-K. KIM, P.Y.B. JAR, Y.-W. MAI*

Australian National University - Engineering Program
The Faculties - 0200 Canberra - Australia
*University of Sydney - Dept Mech. Eng. - Sydney NSW 2006 - Australia

ABSTRACT

The residual interlaminar fracture properties are evaluated for carbon fibre reinforced composites (CFRPs) containing several different polymer matrices after drop–weight impact loading. The residual stiffness of the damaged composites (normalised with the values for undamaged composites) measured in double–cantilever–beam (DCB) tests is higher in the order of the composites with poly(etheretherketone) (PEEK), rubber modified epoxy and pure epoxy matrices, which is consistent with the residual mode I interlaminar fracture toughness. This finding is a direct manifestation of the energy absorption capability of the matrix materials. The advantage of using the post–impact DCB test is also described.

1. INTRODUCTION

It is well known that low–energy, low–velocity impact is a potential danger to polymer matrix composites because it may produce extensive sub–surface delamination and transverse cracking which may not be visible on the surface. These barely visible impact damages (BVIDs) are known to be detrimental to mechanical/structural performance of the composites /1,2/. A number of different techniques have been applied to assess the damage tolerance of composite laminates, which are either non–destructive or destructive tests. The former technique is invariably intended to measure the extent of damage by visual and microscopic examination, or by using X–radiography, ultrasonic C–scan and holographic interferometry. Of the various destructive techniques which include tensile, compressive, flexural, Iosipescu, short beam shear, multi–span beam shear tests, measurement of compressive strength after impact (CAI) has been the most popular due to its high sensitivity to the degree of damage under predominant mode I failure /3/.

In the present study which is an extension of our earlier investigation /4/, preliminary results are reported on the residual delamination properties of carbon fibre reinforced composites (CFRPs) containing several different matrix materials. A particular emphasis is placed on the novel technique developed to study the mechanical response of damaged composites with respect to reduction in stiffness and interlaminar fracture toughness (IFT) using double–cantilever–beam (DCB) specimens after impact.

2. EXPERIMENTS

2.1. Materials

Four different matrix materials were employed for the CFRPs which include pure epoxy, rubber–modified epoxy (R–EP) and poly(etheretherketone) with two different processing temperature 380°C and 400°C (PEEK–380 and PEEK–400). Thermoset matrix composites were made by hand lay–up of eight plies of unidirectional carbon fibre rovings (Torayca T300, Toray) which were impregnated with the epoxy resins. Thermoplastic matrix composites were prepared from 24 unidirectional plies of IM6/PEEK prepregs (APC–2, ICI Fiberite) which were consolidated using a hot press. Further details of the composite constituents and the fabrication procedures used are described in the previous papers /5–7/. During lay–up, a pre–crack was introduced by placing an aluminium film (or a Teflon release film for epoxy matrix composites) in the mid–plane of the laminates.

2.2. Tests

The impact tests were performed on 120 mm x 120 mm square specimens which were fixed between circular rings with a test window 80 mm in diameter (Fig. 1). A hemispherical punch of 40 mm in nose diameter was guided to drop in the centre of the window, and the bouncing punch was caught by a catch. Incident energy levels 6.1 J and 8.0 J were applied to carbon–epoxy and carbon–PEEK composite systems respectively, in proportion to the thickness of the laminates (which were 2.3 mm and 3.0 mm, respectively). This allows the composites of different thickness to absorb a similar level of impact energy per unit volume. Total delamination area is shown to be approximately proportional to input impact energy for a given volume of composite /8/.

Damaged laminates were cut into rectangular strips of 120 mm long and 20 mm wide (Specimens A, B and C, Fig. 1) from the edge to the impact centre to be used as DCB specimens. In the mode I interlaminar fracture test, load, P, and displacement, δ, were recorded for a given crack length, a, and the potential energy release rate, G_R, was calculated from the equation:

$$G_R = (P^2/2B)(dC/da) \qquad (1)$$

where C (= δ/P) is the compliance which was measured experimentally from the reciprocal slope of the loading/unloading line of the P–δ curves. The differential, dC/da, was determined using an appropriate polynomial function:

$$C = k\,a^n \qquad (2)$$

Based on Eq. (2), the residual stiffness of the damaged composite relative to the undamaged composite, C_0/C_r, was calculated:

$$C_0/C_r = a^{\Delta n} (k_0/k_r)$$ (3)

where $\Delta n = n_0 - n_r$, and the subscripts o and r refer to undamaged and damaged states respectively.

3. RESULTS AND DISCUSSIONS

Compliance values are plotted as a function of crack length in Fig. 2 for composites with different matrix materials. It is clearly demonstrated that for the two epoxy matrix composites that the compliance of specimen C (which is taken from the impact centre) is significantly greater than the other specimens A and B at the initial stage of crack propagation near the pre–crack, and the difference in compliance between these specimens diminishes as the crack length is further increased. In sharp contrast, compliance values are almost identical, within data scatter, for all three specimens of PEEK matrix composites over the whole range of crack lengths studied, suggesting little damage has taken place (at least in the major delamination plane) due to the impact loading of a low input energy. These results are summarised in Fig. 3 in terms of residual stiffness (which is the reciprocal of compliance) of damaged specimen (C) relative to undamaged specimens (A). Considerable reductions in stiffness are noted for the epoxy matrix composite, the maximum reductions obtained near the pre–crack tip being approximately 32% and 22% for the unmodified (EP) and rubber–modified (R–EP) systems, whereas there is no appreciable reduction observed for the PEEK matrix composites. This means that after impact loading of a given input energy, a higher level of stiffness is retained (or, damage tolerance is higher) in the order of the composites with PEEK, R–EP and EP matrices.

The crack growth resistance in terms of mode I IFT, G_R, is shown for these composites in Fig. 4. When G_R values are compared between specimens A and C, particularly for the rising part of initial several measurements, it is clear that the fracture resistance has been degraded much more in the EP and R–EP systems than in the PEEK systems. For the EP system in particular, G_R values for specimen C are retained only about 20 % of those of specimen A near the pre–crack where damage is most severe. No remarkable reduction in IFT is observed for the PEEK matrix composites with two different processing temperatures. All the above results are a direct manifestation of the energy absorption capability of the matrix materials: the critical strain energy release rates $G_{IC} = 0.23$ kJ/m^2, 1.88 kJ/m^2 and >4 kJ/m^2 (depending on loading rate), respectively for EP, R–EP and PEEK /5/.

In general, a slight difference in the processing temperature for the PEEK matrix composites did not much alter both the damage tolerance and subsequent interlaminar fracture behaviour. Scanning electron microscopic (SEM) examination also shows little distinction between these specimens, except at the very pre–crack tip where damages might have occurred but were not big enough to be reflected by a reduction in stiffness and IFT values. Fig. 5 taken from the pre–crack tip strongly suggests that delamination actually has taken place in predominantly mode II shear due to impact loading. The main difference between the two processing temperatures is that PEEK–400 system has a small number of tiny resin particles adhering to the debonded fibre surface whereas the PEEK–380 system consists of very clean fibre surface, an indication of relatively strong interface bond for the PEEK–400 composite system as suggested previously /7/.

The mode I delamination test is of value, whether loaded in edgewise compression (i.e.

381

CAI test) or in transverse tension (i.e. DCB test), because it is one of the principal modes of failure in laminate composite components in actual service. However, the post–impact DCB test has an important advantage over the CAI test in that both the residual compliance (and thus stiffness) and IFT in mode I can be simultaneously measured during progressive crack propagation. This enables quantitative evaluation of the extent of damage (i.e. damage tolerance) as well as damage resistance in interlaminar fracture for a prescribed impact loading condition. In addition, degradation in mechanical properties can also be systematically quantified across the laminate plane using the specimens cut from different locations relative to the impact centre. This feature is particularly useful when the test is conducted on a large piece of laminate along with C–scan examination to measure the damage extent. The CAI test gives only one compressive strength/modulus value per impact specimen.

4. SUMMARY

A drop–weight impact procedure was employed to produce damage on carbon fibre reinforced composites (CFRPs) containing several different polymer matrices. The post–impact mechanical properties including the residual stiffness and the residual IFT were measured on DCB specimens using a sectioning technique. It is shown that the residual properties are better retained in the order of the composites with PEEK, rubber modified epoxy and pure epoxy matrices. This finding is essentially consistent with the energy absorption capability of the matrix materials. Within the restriction of limited experimental data presented in this paper, it is demonstrated that the post–impact DCB tests on the laminates with pre–crack is a useful tool to evaluate the damage tolerance and resistance in interlaminar fracture. Advantages of this technique are also described compared to the conventional CAI tests.

ACKNOWLEDGMENTS

The authors wish to thank the Australian Research Council (ARC) for the continuing support of this project, and JKK is particualrly grateful for the Australian Postdoctoral Research Fellowship awarded by the ARC when part of this work was performed at the University of Sydney.

REFERENCES

1. W.J. Cantwell, P.T. Curtis and J. Morton, Comp. Sci. Techn. Vol. 25 (1986) pp 133–148.
2. H.T. Wu and G.S. Springer, "J. Comp. Mater. Vol. 22 (1988) pp 518–532.
3. D.A. Wyrick and D.F. Adams, J. Comp. Mater. Vol. 22 (1988) pp 749–765.
4. J.K. Kim, D.B. Mackay and Y.W. Mai, Composites Vol. 24 (1993) in press.
5. J.K. Kim, C. Baillie, J. Poh and Y.W. Mai, Comp. Sci. Techn. Vol. 43 (1992) pp 283–297.
6. P.Y.B. Jar, W.J. Cantwell, P. Davies and H.H. Kausch, J. Mater. Sci. Lett. Vol. 10 (1991) pp 640–641,
7. P.Y.B. Jar, R. Mulone, P. Davies and H.H. Kausch, Comp. Sci. Techn. Vol. 46 (1993) pp 7–19.
8. S. Hong and D. Liu, Exper. Mech. Vol 13 (1989) pp 115–120

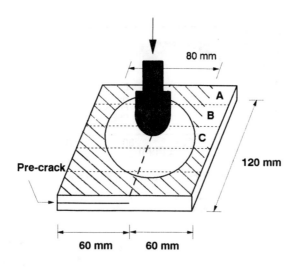

Fig. 1 Schematic diagram of the drop–weight impact test on a pre–delaminated composite.

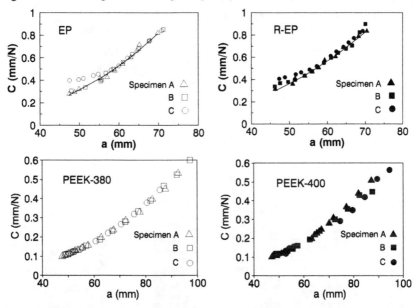

Fig. 2 Compliance, C, as a function of crack length, a., for CFRPs with four different matrix materials: pure epoxy (EP); rubber–modified epoxy (R–EP); poly(etheretherketone) processed at 380°C and 400°C (PEEK–380 and PEEK–400).

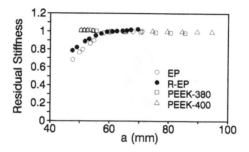

Fig. 3 Residual stiffness of specimen C relative to specimen A.

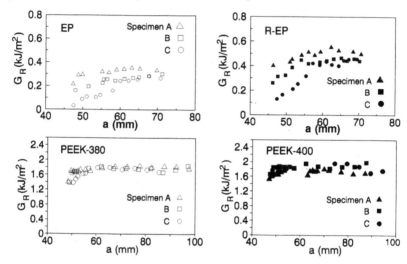

Fig. 4 Potential energy release rates, G_R, plotted as a function of crack length, a.

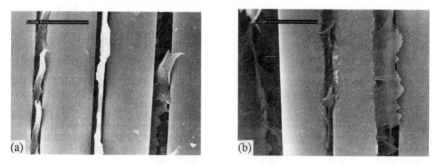

Fig. 5 Scanning electron micrographs of fracture surfaces representing the mode II shear damages for composites with (a) PEEK–380 and (b) PEEK–400.

THE EFFECT OF HYGROTHERMAL AGEING ON DICYANIAMIDE CURED GLASS FIBRE REINFORCED EPOXY COMPOSITES

R. HARVEY, N. MARKS, T. COLLINGS*

*Westland Helicopters Ltd - Materials Laboratory - Box 103
BA20 2YB Yeovil Somerset - UK
*Colcom Consultants - 25 Ferndrive - Chruch Crookham
fleet Hants GU13 0NW - UK*

ABSTRACT

The influence of unreacted dicyandiamide hardener on the hygrothermally aged 0 degree tensile strength of some glass reinforced epoxy composites has been investigated with respect to fibre type, sizing, hardener content and cure cycle. It was found that for all materials investigated, the percentage reduction in tensile strength with ageing, increased with increased residual unreacted hardener. There was no notable affect of fibre type, sizing or matrix on the percentage reduction, but the cure cycle did influence the amount of unreacted hardener, the glass transition temperature and moisture absorption. It is suggested, therefore, that an alkaline solution formed from the unreacted hardener and moisture, attacks all fibres similarly and/or the fibre/matrix interface, thereby reducing the tensile strength.

INTRODUCTION

The effects of the environment, particularly hot and wet, have been of concern to the aviation authorities for sometime with respect to fibre reinforced polymeric composites. It is not surprising therefore, that considerable research has been performed, and is still ongoing in this field. It has been found in general, that degradation of these materials usually has the greatest bearing upon the matrix, causing a reduction in those properties dependant upon it. However, previous research at Westland Helicopters Ltd [1], in conjunction with the D.R.A. (Farnborough), indicated that in some fibre/resin combinations, the tensile strength property can also suffer severe degradation under given hot/wet environmental conditions.

The 0 degree tensile strength of E-glass fibre reinforced epoxies was found, under both natural and accelerated hot/wet environments, to degrade to a greater extent than those of carbon fibre reinforced epoxies. It was postulated that this may have been due to attack upon the fibres and/or fibre/matrix interface. Research at both Kingston Polytechnic [2] and Surrey University [3] indicated that the fibres were being degraded by an alkaline solution formed when the absorbed moisture dissolves unreacted Dicyandiamide (Dicy) hardener.

The aim of the research reported here was to investigate more fully the effect of residual unreacted dicy in the matrix and its dominance on the fibre degradation and also to explore the influence of other mechanisms which may be responsible.

1. EXPERIMENTAL PROCEDURES
1.1. Materials

Eight different unidirectional continuous glass fibre/epoxy resin combinations were assessed during this programme. All were Fibredux pre-pregs supplied by Ciba Geigy Plastics Ltd U.K.

i.	913 GE5	Equerove E-glass fibre	(Fibreglass Ltd.)
ii.	913 GE5	Gevetex E-glass fibre	(Vetrotex Ltd.)
iii.	913 GR5	P109 standard sizing	(Vetrotex Ltd.)
iv.	913 GR5	K45 E-glass sizing	(Vetrotex Ltd.)
v.	913 G-S2-5	Type 463 sizing	(Owen Cornings.)
vi.	DLS1058 GE5	25% reduced dicy, Gevetex fibre.	
vii.	DLS1059 GE5	25% excess dicy, Gevetex fibre.	
viii.	920 GE5	Gevetex E-glass fibre	

The matrix for all materials, except 920 GE5, were based upon Araldite MY720 resin.The 920 GE5 material was based upon MY750.

1.2. Cure cycles

The two cure cycles used were:

      a) 1 hour at 125°C with no intermediate dwell (N.D).

      b) 1 hour dwell at 90°C followed by 1 hour at 125°C (D).

1.3. Hygrothermal conditioning

Specimens were hygrothermally conditioned to the accelerated technique of Collings and Copley /4/ and to the mean worst world humidity level described by Collings /5/. The ageing cabinets were set up for conditions of 45°C/84% relative humidity (R.H). The specimens were monitored for moisture uptake by the use of traveller specimens, and the 95% equilibrium moisture content was determined when the change in weekly weighings was less than 0.005%. The tensile specimens were then tested and the travellers were sliced to determine the mean through thickness moisture content, and the equilibrium moisture content ($M\alpha$).

1.4. Testing of materials

The 0 degree tensile strength was determined from a sample size of 6 specimens of 1mm nominal thickness, width 12.75mm and 150mm gauge length, tested at a constant strain rate of 2mm per minute. Investigations into the residual dicy content and the glass transition temperature (Tg) of the materials, were conducted by Ciba Geigy (B.S.D.) using High Perfomance Liquid Chromotography (H.P.L.C.) and a P.L. Dynamic Mechanical and Thermal Analyser (D.M.T.A.) respectively.

2.FIBREDUX 913 GE5 - EQUEROVE

It was found that the Fibredux 913 reinforced with the Equerove E-glass rovings which had been used for the former research /1/ gave significantly lower unaged tensile strengths to those previously attained. The material used was obsolete and the rovings were of old stock, it is therefore believed, that the sizing on the fibres had been degraded by hydrolysis. This would reduce the interfacial strength between the the fibres and matrix in the cured laminate and hence the ability to transfer loads effectively. It was decided, however, that the material was still useful for comparison of the effects of residual dicy content and ageing.

3. EFFECT OF RESIDUAL DICY ON 0 DEGREE TENSILE STRENGTH

The effect of residual/unreacted Dicy content on the 0° tensile strength is presented in Figure 1. The reduction in strength from unaged to aged at room temperature (R.T.) increased with Dicy content until an upper limit was achieved of 31% reduction which correlated to 1% residual Dicy in the matrix. Further increases in residual Dicy did not have any significant effect on the strength. The reduction in strength when tested at 70^oC in the aged condition indicated a similar trend with the reduction in strength increasing until it stabilised at 43%, which again corrolated to a 1% residual Dicy level. This nominal 12% increase in the reduction of 0° tensile strength when evaluated at 70^oC, compares well to the percentage reduction in strength of unaged specimens tested at 70^oC with reference to R.T, Table 1.

There is evidence that all the glass reinforcements, with their respective sizings, are affected to a similar extent by the combination of moisture in the presence of residual unreacted Dicy in the matrix. Hence the phenomenon that had been noted in the 913 GE5 Equerove material in previous research /1/ was not unique to to this fibre/sizing system when unreacted Dicy is present in hot/wet environments.

4. EFFECT OF CURE CYCLE ON MATRIX, Tg AND MOISTURE CONTENT

All the materials were cured to two different cure cycles; one with an intermediate dwell, and one without. It was found that the cure with the 90^oC dwell produced different matrix characteristics to that cured without a dwell. The dwell had the effect of reducing the glass transition temperature (Tan δ measured at 10Hz) of all materials, Table 2, and in the MY720 based matrices, a single phase structure was observed (Plate 1) as opposed to the two phases present when cured without a dwell (Plate 2). However, no two phase structure was observed with either cure cycle for the MY750 based 920 material. The residual unreacted dicy content was higher in the materials cured with a dwell when compared to those without, and this has been shown by other researchers /2/ to be a result of early encapsulation of the dicy hardener during the dwell period. The dwell also reduced the equilibrium moisture content and this was to be expected since increased unreacted dicy would result in reduced cross-linking, thereby, directly affecting the capability to retain hydrogen-bonded moisture.

The cure without a dwell, for a given material, absorbed more moisture than that cured with a 90^oC dwell. This supports work that had been conducted at Ciba Geigy /6/,which displayed the same trend although under very different ageing conditions of 14 days immersion in de-ionised water at 71^oC. In tables 1 and 2, data are presented giving: the moisture content in the matrix, the amount of Dicy reacted, the reduction in Tg, and the cure cycle. From this data it can be seen that the greater the quantity of dicy that has reacted during the cure, the higher the absorption of moisture into the matrix. This was predominantly greater with the materials cured with no intermediate dwell, the exception being with the reduced Dicy material, DLS1058, where the full extent of cross-linking may not have occured due to lack of readily available Dicy hardener. There is also a general trend of the percentage reduction in Tg (unaged to aged) increasing as the amount of reacted Dicy increases and therefore, the absorbed moisture. This was not as found by Shah *et al* /3/ who reported that the greater the concentration of unreacted Dicy, the higher the equilibrium moisture content and the lower the Tg. It is possible that this difference, in the effect on moisture content, may be a result of the different ageing conditions, this previous work having been immersed in water at 50°C.

5 . EFFECT OF MATRIX

The 920 resin system differed from the other seven materials as it was based upon the Diglycidyl Ether of Bisphenol A (DGEBPA) epoxide, MY750, which is bi-functional. This is in contrast to the Tetraglycidyl 44' Diaminodiphenyl Methane (TGDDM), MY720, which has four functional groups. Both resin systems when mixed with an accelerator and hardener are able to cure by one or more of the following reactions:
    i .Epoxide-dicyandiamide (hardener) reaction .
    ii .Epoxide-accelerator rection
    iii .Epoxide auto polymerisation (TGDDM only)

The MY750 based 920 system is thought to have a lower cross-link density than that of the 913 system due to it being only bi-functional. This produces a lower Tg in the 920 system and reduces the number of sites available for the hydrogen bonding of absorbed moisture.

6.CONCLUSIONS

i. The incorporation of a 90°C dwell in the cure increases the amount of residual dicy present and induces the formation of a second opalline phase in the MY720 based materials.

ii. The reduction in tensile strength after hygrothermal ageing increases as the residual dicy content increases, stabilising with a 31% reduction at 1% residual dicy.

iii. At 70°C after ageing the reduction in tensile strength stabilises at 43%, again at a 1% residual dicy level.

iv. There appears to be a relationship between the reduction in 0 degree tensile strength of G.F.R.P. and the residual dicy content if the latter is lower than 1%.

v. All types of glass fibre and sizing investigated appear to be affected to a similar extent by the effects of residual dicy and moisture content.

vi. The 12% increased reduction in strength when tested at 70°C after ageing compares to a similar 12% reduction at 70°C when tested unaged.

vii. As the reduction in strength was comparable for all materials with a dicy content of 1% or greater, then the alkaline solution formed by the residual dicy must attack all fibres similarly or act to degrade the fibre/matrix interface reducing the ability to transfer loads effectively.

ACKNOWLEDGEMENTS

This research was conducted at Westland Helicopters Ltd under Ministry of Defence funding and guidance (contract number H22A/43). We would like to thank Ciba Geigy (Bonded Structures Division) for their technical support and advice.

REFERENCES

1. "Static and fatigue testing of composite materials after exposure to accelerated ageing conditions" Westland Helicopters Materials Laboratory Reports LR 84,383 and LR 86,118 .

2. K.A. Kasturiarachchi and G. Pritchard . "Free dicyandiamide in cross linked epoxy resins" Journal of material science letters 3 (1984) pp 283-286 .

3. M.A. Shah ,F.R. Jones F.R. and M.G. Bader . "Residual dicyandiamide (DICY) in glass fibre composites" Journal of material science letters 4 (1985) pp 1181-1185 .

4. T.A. Collings and S.M. Copley . "Accelerated ageing of C.F.R.P's" Composites. Volume 14 , No. 3. (July 1983) pp 180-188 .

5. T.A. Collings . "The effect of observed climatic conditions on the moisture equilibrium level of fibre reinforced plastics" Composites. Volume 16 , No. 1. (January 1986) pp 33-41

6. "Fibredux 913 residual dicy phenomenon investigative programme" Ciba Geigy (B.S.D.) January / April 1986 .

TABLE 1

REDUCTION IN 0 DEGREE TENSILE STRENGTH

X REDUCTION IN STRENGTH	EQUEROVE		GEVETEX		920		DLS 1058		DLS 1059		GR5 (STAN)		GR5 (E-FIN)		S2	
	D	ND	D	ND	D	ND	D	ND	D	ND	D	ND	D	ND	D	ND
RT U/A - A	-1.5	3.1	18.1	18.5	19.7	28.7	26.0	33.7	30.8	33.3	27.6	24.8	32.1	32.6	30.8	22.5
70° U/A - A	12.8	16.2	27.0	20.3	37.2	32.7	32.7	34.7	36.0	40.9	30.9	34.4	34.7	36.3	26.3	34.4
U/A RT - 70	12.9	13.5	11.9	10.8	10.1	19.7	7.8	13.8	8.7	9.6	10.0	10.7	12.8	12.6	13.2	4.3
A RT - 70	25.2	25.2	21.5	12.9	29.7	24.2	16.2	15.2	15.6	19.9	14.2	22.2	16.0	17.4	7.5	18.9
RT - 70°C U/A - A	24.1	27.4	35.7	28.9	43.6	46.0	38.0	43.7	41.6	46.6	37.8	41.5	43.0	44.4	36.0	37.2
M α (%) i	.972	1.036	.997	1.125	.737	.875	1.009	1.061	1.002	1.159	1.023	1.10	1.055	1.302	1.12	1.102
ii	.932	1.030	1.006	1.044	0.717	0.783	0.906	0.989	0.964	1.156	0.978	1.078	0.957	1.023	1.051	1.184
iii	2.329	2.575	2.516	2.610	1.794	1.958	2.264	2.471	2.410	2.891	2.446	2.696	2.392	2.557	2.628	2.960
Mean Moisture (%) i	0.871	0.904	0.859	0.957	0.698	0.786	0.873	0.897	0.840	1.037	0.837	0.850	0.911	1.102	0.876	0.834
Absorption ii	0.835	0.899	0.867	0.888	0.680	0.703	0.784	0.835	0.808	1.035	0.800	0.841	0.830	1.000	0.822	0.896
iii	2.086	2.247	2.167	2.226	1.699	1.758	1.959	2.089	2.020	2.587	2.001	2.103	2.065	2.499	2.056	2.239
X wt of Reacted Dicy	4.97	5.20	4.94	5.35	4.60	5.10	3.63	3.79	5.33	5.79	4.60	4.78	4.84	4.97	4.99	5.23

i) Measured from traveller slices) For composite material
ii) Normalised for 40% resin volume)
iii) For matrix only

TABLE 2

GLASS TRANSITION TEMPERATURE (Tg) - D.M.T.A. (CIBA GEIGY ANALYSIS)

MATERIAL	CURE	Tg (LOG E') °C		Tg (TAN δ) °C	
		UNAGED	AGED	UNAGED	AGED
913 GE5-30% (EQUEROVE)	DWELL	151	115	180	143
	NO DWELL	153	120	184	150
913 GE5-30% (GEVETEX)	DWELL	147	116	177	146
	NO DWELL	151	123	179	153
920 GXE5-34% (EQUEROVE)	DWELL	95	82	108	100
	NO DWELL	95	85	108	106
DLS 1058 GE5-34% (GEVETEX)	DWELL	130	114	156	140
	NO DWELL	137	114	166	144
DLS 1059 GE5-34% (GEVETEX)	DWELL	125	112	150	138
	NO DWELL	131	111	160	139

FIGURE 1

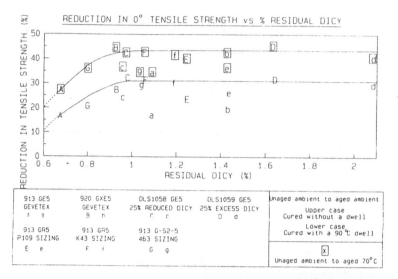

REDUCTION IN 0° TENSILE STRENGTH vs % RESIDUAL DICY

913 GE5 GEVETEX	920 GXE5 GEVETEX	DLS105B GE5 25% REDUCED DICY	DLS1059 GE5 25% EXCESS DICY	Unaged ambient to aged ambient
A a	B b	C c	D d	Upper case Cured without a dwell
913 GR5 P109 SIZING	913 GR5 K43 SIZING	913 G-S2-5 463 SIZING		Lower case Cured with a 90°C dwell
E e	F f	G g		x
				Unaged ambient to aged 70°C

PLATE 1

913 GE5 GEVETEX DWELLED FOR
1 HOUR AT 90ºC AND CURED
FOR 1 HOUR AT 125ºC.

- SINGLE PHASE STRUCTURE

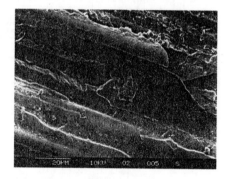

PLATE 2

913 GE5 GEVETEX CURED FOR
1 HOUR AT 125ºC WITHOUT
A DWELL.

- TWO PHASE STRUCTURE

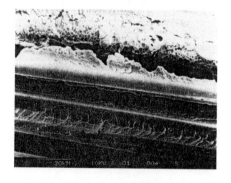

390

EFFECTS OF ENVIRONMENTAL CONDITIONS ON PROGRESSIVE CRUSHING BEHAVIOURS OF GLASS CLOTH/EPOXY COMPOSITE TUBES WITH DIFFERENT SURFACE TREATMENT

H. HAMADA, S. RAMAKRISHNA, M. NAKAMURA, A. FUJITA, T. MORII*
Z. MAEKAWA, D. HULL**

Kyoto Institute of Technology - Matsugasaki - Sakyo-Ku
606 Kyoto - Japan
*Shonan Institute of Technology - 1-1-25 Tsujido - Nishikaigan Fujisawa
Kanagawa 251 - Japan
**University of Liverpool - PO Box 147 - Liverpool L69 3BX - UK

ABSTRACT
Effects of hydrothermal ageing on two kinds of glass/epoxy tubes have been investigated. One type of tubes were reinforced with glass cloth treated with amino silane coupling agent and the other type of tubes were reinforced with glass cloth treated with acryl silane coupling agent. Tubes were immersed in hot water at 80^0C for different periods ranging from 100 to 3000 hours. The energy absorption capability of both kinds of tubes decreased with increasing immersion time. The crushing performance of amino tubes is higher than that of acryl tubes. This was attributed to the different modes of crushing. The amino tubes were crushed by splaying mode whereas the acryl tubes were crushed by fragmentation mode.

INTRODUCTION
One of the most important requirements placed on composite materials when they are being considered for crashworthy structural applications in cars is that they absorb energy in a controlled manner under crash conditions. Much data has been produced concerning the material, geometrical and testing parameters which influence the levels of energy absorption attainable by progressive crushing[1,2]. Structural composites are often exposed to hydrothermic environments in the course of service. Pafitis and Hull[3] have investigated the effect of long term exposure to moisture on the energy absorption of glass/epoxy tubes. Hamada et al[4] have studied the effect of fibre surface treatment on the energy absorption of glass/epoxy tubes. In this study attempts have been made to identify the

effects of both the fibre surface treatment and the moisture on the energy absorption behaviour of glass/epoxy tubes.

MATERIALS
The materials used were woven glass cloth/epoxy composite tubes with different glass fibre surface treatment. Two types of prepreg were used in this investigation. One contained glass cloth treated with amino silane coupling agent and the other used glass cloth treated with acryl silane coupling agent. The prepreg sheets were obtained from Arisawa Mfg Co Ltd, Japan. Tubular sections of 50 mm internal diameter and 55 mm external diameter were manufactured by Toyo Lite Co Ltd, Japan. Tubes were fabricated by mandrel wrapping of prepreg sheets, glass volume fraction 43%, with the warp and weft directions parallel to the hoop and axial directions of the tubes.

EXPERIMENTAL
Composite tubes of 55 mm long with a 45^0 chamfer at one end were dried in vaccumm at 100^0C for 72 hrs. Tubes were subjected to hydrothermal ageing by immersing in hot distilled water at 80^0C using a thermostatically controlled bath. At various time intervals 100, 300, 1000 and 3000 hrs, tubes were removed from the water bath, dried using absorbant tissue, and weighed. The apparent weight gain was computed from the weight of the tubes before and after the immersion test. Tubes were axially compressed over a total distance of 20 mm, at a rate of 1 mm/min using a 250 kN capacity, twin-screw driven testing machine. During crushing, load-displacement traces were recorded and the nature of the crush process observed. Crushed specimens were finally mounted in a cold curing resin and sections of the crush zone removed for optical microscopy.

RESULTS AND DISCUSSION
Composite Tubes without Hydrothermal Ageing
The specific energy absorption values for amino and acryl tubes without hydrothermal ageing are listed in Table 1. The crushing performance of amino tubes is approximately 25% higher compared to the acryl tubes. Both the tubes were progressively crushed at a constant load as illustrated Fig. 1. The serrations of the load-displacement (P-d) traces are mainly due to microfracture of tube wall during crushing. Typical appearance of crush zones of both the tubes are shown in Fig. 2 and detailed photographs of sections cut through the crush zone are shown in Fig. 3. Amino tubes were crushed progressively by splaying of tube wall into internal and external fronds, known as 'splaying' mode of crushing [1]. A completely different form of crush zone developed with acryl tubes in which the formation of thin rings of material can be seen. This is called as the

'fragmentation' mode of crushing. This change in crushing mode with fibre surface treatment is attributed to the different frictional forces acting between the crush zone and the platen [4].

Composite Tubes with Hydrothermal Ageing

The relationship between the apparent weight gain (W_a) and the immersion time for both amino and acryl tubes is shown in Fig. 4. In general, amino tubes gained less weight compared to the acryl tubes. This may be attributed to the good bonding between the epoxy matrix and the amino silane treated glass fibres, which results in lower amounts of moisture diffusion into the composite. For amino tubes the W_a increased rapidly upto 300 hrs and later increased marginally. For acryl tubes the W_a reaches a peak value at 1000 hrs and after that it decreases remarkably. During hydrothermal ageing, moisture diffuses into the material and at the same time degradation of composite occurs [5]. In acryl treated tubes, because of the poor bonding between the fibres and matrix resin, the ingressed moisture degrades the fibre/matrix interface more severely and resulted in decrease of apparent weight gain.

The relationship between the energy absorption and immersion time for both the tubes is shown in Fig. 5. For 3000 hours of hydrothermal ageing the specific energy of amino tubes decreased by 30% compared to the 45% decrease observed for acryl tubes. This indicates that amino silane treated tubes are less suceptable to hydrothermal ageing compared to the acryl silane treated tubes. In order to relate the energy absorption with mechanical properties, the compressive strengths of square ended tubes were measured and summarised in Fig. 6. The compressive strength of acryl tubes decreased rapidly with immersion time comapared to the amino tubes. This indicates that parallels can be made between the changes in the compressive properties and the changes in energy absorption.

Typical load-displacement traces for 100 and 3000 hours of immersion time are shown in Figs. 7 and 8 respectively. The mean crush load and the serration amplitude decreased with increasing immersion time. The cross-sections of typical crush zones of amino and acryl tubes are shown in Figs. 9 and 10 respectively. The crush zone of amino tubes (Fig. 9) contained central wall crack and debris wedge separating the internal and external fronds typical of splaying mode. The fronds were smaller compared to those observed in amino tubes without ageing. The crush zone of acryl tubes contained internal and external rings typical of fragmentation mode. It can be summarised that the crushing performance is dependent on the hydrothermal ageing condition of the tubes whereas the crushing mode is not.

CONCLUSIONS

The amino silane treated tubes were progressively crushed by splaying mode whereas the acryl silane treated tubes were crushed by fragmentation mode. The mode of crushing was independent of immersion time. The crushing performance decreased with increasing immersion time. The acryl tubes were more succeptable to hydrothermal ageing compared to the amino tubes. In general the crushing performance of amino tubes is superior to acryl tubes.

Table 1 Energy absorption performance of tubes made from glass cloth/ epoxy material

	Mean Crush Load (kN)	Mean Crush Stress (MPa)	Specific Energy Absorption (Es) (kJ/kg)	
Amino-Silane (M)	47.0	107.7	61.0	as
	51.5	118.4	66.7	averaged
	46.6	106.3	60.6	
	50.0	121.1	73.2	(65.4)
Acryl-Silane (C)	37.0	84.5	49.1	as
	37.8	87.2	50.2	averaged
	38.9	90.2	51.8	
	40.0	96.8	56.9	(52.0)

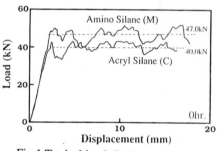

Fig.1 Typical load-displacement traces.

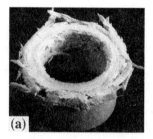
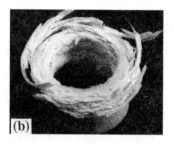

Fig.2 Crush zone of:(a) an amino-silane-treated tube and (b) an acryl-silane-treated tube.

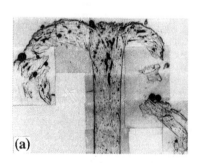

Fig.3 Photomicrographs of cross-section of the crush zone of:(a) an amino-silane-treated tube and (b) an acryl-silane-treated tube.

394

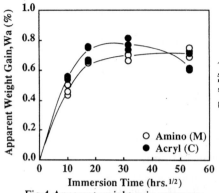

Fig.4 Apparent weight gain vs square
root of immersion time graphs.

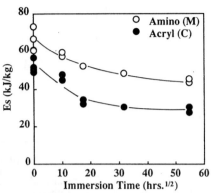

Fig.5 Relation between Es and square
root of immersion time.

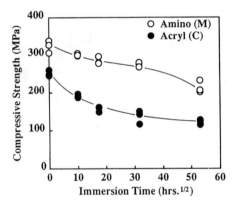

Fig.6 Variation of compressive strength
of square ended tubes.

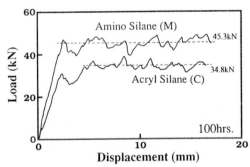

Fig.7 Typical load-displacement traces,
hydrothermally aged for 100hrs at 80°C.

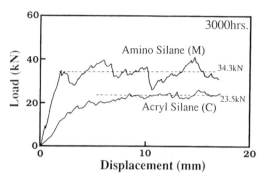

Fig.8 Typical load-displacement traces,
hydrothermally aged for 3000hrs at 80°C.

Fig.9 Crush zone of amino-silane-treated
tubes, hydrothermally aged for
3000hrs at 80°C.

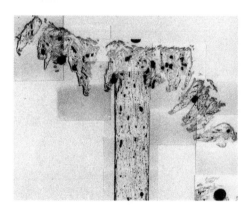

Fig.10 Crush zone of acryl-silane-treated
tubes, hydrothermally aged for
3000hrs at 80°C.

REFERENCES
1. D. Hull, Composites Science and Technology, 40 (1991) 377-421.
2. S. Ramakrishna, "Knitted Fabric Reinforced Polymer Composites" Ph.D Thesis (1992), University of Cambridge, UK.
3. D.G. Pafitis and D. Hull, in "Proceedings of ICCM-VIII" (1991) paper 16-L-1, Honululu.
4. H. Hamada, J.C. Coppola and D. Hull, Composites, 23, No.2 (1992)93-99.
5. H. Hamada, A. Yokoyama, Z. Maekawa and T. Morii, in "Proceedings of ICCS-V" (1989) 259-277, Scotland.

FATIGUE PERFORMANCE OF COMPOSITE STRUCTURAL I-BEAMS

M.D. GILCHRIST, A.J. KINLOCH, F.L. MATTHEWS, S.O. OSIYEMI

Imperial College of Science, Technology and Medicine
South Kensington - London SW7 2BY - UK

ABSTRACT

This paper discusses experimental aspects of a generic procedure which is being developed by the authors for predicting the strength and fatigue life of full scale notched and unnotched structural I-beams. Circular and diamond shaped cutouts in the web regions of the I-beams are considered. The approach adopted within this work is equally applicable to the analysis of other components. Various results which have been obtained using a commercially available carbon/epoxy composite are presented in this paper.

1. INTRODUCTION

The dimensions and manufacturing procedure of these structural I-beams, which are typically used as support members in the wings of aircraft and the tailcones of helicopters, are detailed in Fig. 1. Continuous fibre-reinforced carbon/epoxy prepreg is stacked using a particular combination of 0° and ±45° plies to provide two C-channels of 12 plies and two rectangular strips or caps, also of 12 plies. The beams are prepared by curing the two C-sections back to back together with the top and bottom flange caps in one single operation in accordance with the material manufacturer's specified curing cycle. Two wound prepreg tows of the same material are used to fill the void in the triangular channel which exists between the flange and web areas behind the fillet radii on the C-sections. Further details of the manufacturing process are given elsewhere /1/.

Both unnotched and web notched beams are being tested. A circular notch of 46mm diameter and a diamond notch of 46mm diagonal lengths and

11/64" corner radius (≈ 8.73mm diameter) are located along the neutral axis of the beam as shown in Fig. 2. The dimensions of these notches were selected such that the maximum notch dimension did not exceed 40% of the internal web depth. These web cutouts are representative of lightening holes or access ports within an I-section component. It is worth noting that if the primary objective of such holes is weight reduction, the circular notch is more effective since it's cutout area is larger by some 26%. However, in practical circumstances it is always important to consider whether such a lightening hole is likely to reduce the component strength and also to establish from which notch failure is more likely to propagate.

Different modes of damage influence the behaviour of different regions of the I-beams during the stages of damage initiation, development and ultimate failure. The complex interaction between the various modes which occurs within the beams can be elucidated using appropriate fracture mechanics test geometries. In particular, delamination, a predominant fatigue damage mechanism, can be characterised using either delamination growth rate or delamination onset data to obtain stable growth rate or damage tolerant information. The growth rate data associated with Modes I and II and various mixed ratios of I/II which exist throughout the I-beams can be examined separately using different specimen geometries: Double Cantilever Beam /2/, End Loaded Split /2/ and Mixed Mode Bending /3/. By relating the fatigue data obtained from these fracture mechanics coupons to appropriate regions of a beam it is possible to establish damage initiation data and rates of damage evolution within the beams.

2. TESTING PROCEDURE

Each I-beam is tested under four-point flexure as shown in Fig. 2. Load control is used for fatigue tests whilst static tests are performed under displacement control. The loading is transferred through the web and flange regions of the beams using eight loading pads (four on either side) which are adhesively bonded to the I-beam. The four central loading pads are connected to the machine actuator whilst the two pairs of outer loading pads are attached to the base of the machine. Under this loading, the flanges are subject to tension and compression whilst the web is principally subjected to shear. Unlike other test methods /4, 5/, where loading holes are drilled through the beam, this procedure does not damage the integrity of the beam prior to testing.

The evolution of damage within the beams is monitored by interrupting a fatigue test after a certain number of fatigue cycles or a certain percentage drop in beam stiffness (increase in compliance), and using ultrasonic C-scanning to identify the location and extent of damage. Ultrasonic A-scans can further identify the inter-ply location of the damage (i.e., through thickness) whilst edge replication can establish the presence of different damage mechanisms along the edge of a flange or notch. Repeated testing, interruption and nondestructive evaluation prior to final component failure is necessary if damage development is to be properly monitored.

3. FATIGUE DAMAGE CHARACTERISTICS

The authors have previously shown /1/ that initial damage within an unnotched beam, preceeding catastrophic failure, emanated outwards from the channel region between the compression flange and web between the central loading pads where a uniform state of maximum strain exists. This particular beam had been subjected to some 1.2×10^6 cycles at a frequency of 5Hz and a load ratio of R=0.1, with maximum recorded direct strains of approximately 0.3%.

The region of damage initiation within this unnotched beam, namely the channel along the middle of the compressive flange, was sectioned after final beam failure for fractographic analysis. A T-section of this compressive flange and web region was prepared and polished for optical microscopic investigation specifically to identify the principal delaminations and the presence of damage in the curved sections of the beam. Figure 3 presents a schematic of the various delaminations, transverse ply cracks and matrix cracks which were observed. The principal delaminations lie between the backs of the two C-sections and the flange caps and the C-sections. It appears as though the two pieces of tow separated from the adjacent plies with delaminations usually being present in the resin rich areas, which exist at the three corners of the fillet, and extending to the extremes of the polished sections. This information, together with an examination of polished sections further along the channel region, indicates that the delaminations initiate from the fillet region of the beam and propagate out towards the flange edges.

Tests have also been performed on two web-notched beams. Schematic representations of ultrasonic C-scans showing the extent of delamination around the circular notch prior to testing, after static failure, and after some 0.54×10^6 cycles of fatigue loading at a frequency of 5Hz and an applied actuator load range of 5-50kN are shown in Fig. 4. The primary and induced secondary shear loadings which surround the circular notch due to the arrangement of the loading pads are detailed in Fig. 4(a). These can be resolved into tensile and compressive loads acting around the notch circumference. Comparing these loadings with the static failure of Fig. 4(b), the major damage, extending diagonally across the notch diameter from compressive to tensile flange, appears to be due to tensile loads whilst the secondary damage, not extending from flange to flange across the notch, is due to compressive loads. Visual examination of the web surfaces around the notch identifies some web buckling surrounding the secondary damage and a general appearance which is consistent with compressive fracture. A complete description of the fractography of this failure is beyond the scope of the present article and is to be published elsewhere. The initial fatigue damage, detailed in Fig. 4(c), initiates around the circumference of the notch in a manner which is consistent with the static failure as shown in Fig. 4(b).

It is worth noting that there was no evidence of damage surrounding the diamond notch in the static or fatigue cases of Figs. 4(b) & (c). This is in direct agreement with the results of Ahlstrom & Bäcklund /6/ who used a 2D finite element technique to predict optimum notch shapes in different laminates subjected to both biaxial tension and shear loadings. Shape optimization aimed to minimise structural weight without increasing the stresses above an acceptable level. The optimum shape for notches under shear was found to be approximately rectangular, rotated by 45° to the directions of shear. In other words, it was predicted /6/ that a circular notch was likely to fail at lower shear loads than a diamond notch.

4. CONCLUSIONS

Delamination was seen to be the most significant damage mechanism within the I-beams although transverse ply cracking, matrix cracking and fibre breakages were also observed. In unnotched beams critical delamination occured along the interfaces between the separate components which comprise the I-beams: namely the flange caps and the C-sections and the backs of the two C-sections. Furthermore, it was noted that this damage tends to propagate towards the flange edges (in some cases interacting with edge delamination damage) and down the depth of the web. These results are important for design purposes because they indicate that a process which is used to manufacture load bearing components can directly influence the site of damage initiation and the subsequent failure within such components. With regard to web notched beams, it was found that failure due to shear was more likely to be associated with circular notches than with diamond notches.

ACKNOWLEDGEMENTS

The authors acknowledge the financial assistance of British Aerospace Ltd., the DTI (via the LINK "Structural Composites" programme), Du Pont, Rolls Royce plc, the SERC, Shell Research, Westland Helicopters Ltd., as well as the advice of both Dr. P. Curtis of DRA and Prof. B. Harris of Bath University.

REFERENCES

1.   M. D. Gilchrist et al.   Paper 38, Deformation and Fracture of Composites, UMIST, Manchester, UK, 29-31 March (1993).

2.   S. Hashemi et al. Proc. R. Soc. Lond., A427, (1990), 173-199.

3.   J. R. Reeder & J. H. Crews Jr. NASA TM 102777, (1991).

4.   K. Hollmann. Engng Fract. Mechs., 39, (1991), 159-175.

5.   D. Purslow. Composites, 15, (1984), 43-48.

6.   L. M. Ahlstrom & J. Bäcklund, Composite Structures, 20, (1992), 53-62.

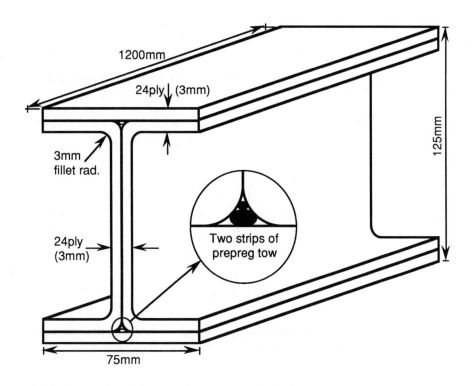

Figure 1.   I-beam construction from two C-sections and two flange caps.

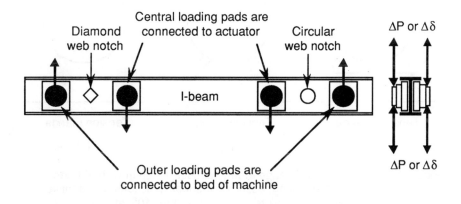

Figure 2.   Schematic of 4-point bending testing arrangement for I-beams.

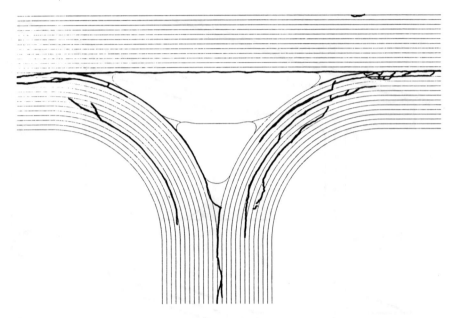

Figure 3.   Unnotched I-beam fatigue damage observed in channel region
at the intersection of the compression flange and web.

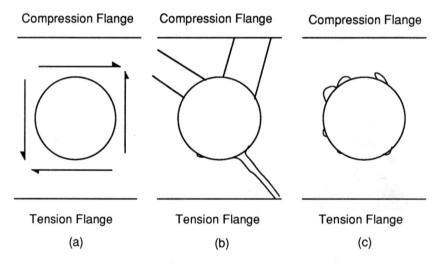

Figure 4.   C-scans of damage around circular notch in web of beam.
(a) before testing; associated shear directions are identified,
(b) after static failure and (c) after 0.54×10^6 fatigue cycles.

A UNIAXIAL TEST FACILITY FOR THE STUDY OF CMCs UNDER TENSILE AND CREEP LOADING

M. STEEN, J.-N. ADAMI*, J. BRESSERS

*Institute for Advanced Materials - Joint Research Centre - PO Box 2
1755 ZG Petten - The Netherlands
Ingold Messtechnik - Urdorf - The Netherlands

0. ABSTRACT

An experimental setup to perform tensile and creep tests on CMCs at high temperatures is described. The intrinsic material behaviour is studied by performing the tests under high vacuum, thus avoiding environmental ingress to the fibre-matrix interface. The uniaxial testing mode is selected because of its advantages in terms of interpretation and evaluation of the results. Flat untabbed specimens are clamped in watercooled grips which are incorporated in a rigid load train. Local heating of the gauge part of the specimen is achieved by induction through a susceptor. The performance of the test setup with respect to accuracy of alignment as well as time and spatial temperature gradient is documented.

1. INTRODUCTION

The common types of test used nowadays to characterise the mechanical behaviour of continuous fibre reinforced ceramic matrix composites (CFRCMCs) rely on flexural loading: strength testing in four point bending with a sufficiently large span to depth ratio, short beam three point bend testing for the determination of interlaminar shear strength, and notched beam testing to quantify damage tolerance, are but a few examples. As thoroughly discussed in a number of recent publications /1, 2/, flexural loading of monolithic and fibre reinforced ceramics presents a number of serious drawbacks, specifically at high temperatures. In order to overcome these limitations, particularly with respect to the interpretation of the test results, a uniaxial test facility has been set up in which a uniform and stationary stress distribution is imposed over the specimen's cross section. Suitable transducers are mounted at appropriate locations on the specimen to measure its mechanical and thermal response. This setup provides highly reliable test results which can directly be used for design and modelling purposes. In the present paper the considerations leading to the selected solutions for constructing the experimental setup are discussed, and its performance is illustrated.

2. CONSIDERATIONS FOR THE REALISATION OF THE TEST FACILITY

The concept of a uniaxial high temperature test facility for CFRCMCs is based on a number of considerations which are outlined below. In order to study the intrinsic long term mechanical behaviour at high temperature, the tests must be performed under a protective environment. An ultra-high vacuum best meets the criteria of purity and ease of monitoring its quality during testing. The final concept of the facility is the result of a number of compromises that have to be made when deciding on the specimen geometry, the load train configuration, the heating method, the number and position of sensors and transducers, the gripping system, etc. The solutions adopted for these items mutually influence each other, making a brake-down of the selection of the final concept in single items very difficult. The requirements for the separate individual items are discussed first, followed by the criteria to be considered with respect to their interaction.

2.1. Specimens
Unidirectionally and bidirectionally reinforced materials are virtually exclusively available in low thickness product forms, and flat rectangular cross section specimens are therefore used in the majority of cases. The gauge length cross section has to include a representative number of reinforcement plies. In order to make full use of the uniformity of the stress distribution under uniaxial loading, one of the reinforcement directions should be aligned with the loading axis.

2.2. Loading train configuration
The failure mode of CFRCMC specimens is influenced by the machine stiffness. When a specimen starts to fail from one edge, the crack normally propagates horizontally across the specimen. In the case of a laterally compliant machine frame, the crack can deflect, imposing a tearing action rather than a uniaxial stress on the remaining ligament, and causing an extraneous failure. In view of the importance of alignment, a stiff loading frame is also required to limit induced bending and to decrease its dependence on the axial position of the actuator or piston.

2.3. Gripping
The gripping system should enable adaptation of the clamping force exerted on the specimen's ends relative to the axial load, in order to prevent slipping or failure in the clamping ends. Bonded tab load introduction may provide a partial solution to the latter problem at low temperatures where organic adhesives can be used. However, adequate adhesive bonding materials are not available for high temperature use. In view of the importance of alignment, another requirement for the grip system is that it should allow adequate reproducibility in positioning and clamping of specimens of nominal identical geometry.

2.4. Heating equipment
In uniaxial high temperature testing a small spatial temperature gradient along the gauge length of the specimen is required because the deformation and damage mechanisms occurring within the material under load are thermally activated. Temperature differences cause differential mechanical responses which complicate the evaluation of the test result. The uniformity of the temperature distribution depends on the heating method, the testing environment, the specimen size, the thermophysical properties of the specimen, and the location of sensors in contact with the specimen which act as heat sink. Other important considerations in the selection of the heating equipment are the accuracy of temperature control and the thermal inertia of the system. Compatibility with the vacuum environment furthermore imposes the use of a solution offering low heat content and low degassing rates.

2.5. Sensors and transducers

In order to control the mechanical test and to correlate the response of the specimen with the imposed loading, a number of testing parameters, such as temperature, strain, displacement, load, etc., need to be measured. These measurements require the use of suitable sensors and transducers mounted at appropriate locations on the specimen. They have to be sensitive and accurate, should not interfere with the property to be measured, and not be susceptible to external disturbing influences.

2.6. Interaction

The solution adopted for any of the items listed above has an explicit impact on the others. The factors that require specific attention in this respect are shown schematically in figure 1. They affect the performance of the setup mainly in three areas: alignment, temperature distribution and location and mode of failure. The final concept depends on the relative importance attributed to these factors.

2.6.1. Specimen alignment is extremely important in the uniaxial testing of ceramics /2/. When determining a given mechanical property, it is absolutely necessary to indicate the level of alignment at which it is obtained /2, 3/. A procedure has been developed in house to quantify the alignment performance of any uniaxial load train. It is based on the on-line determination of the magnitude and angular position of the maximum bend vector along the gauge length of the specimen. Their determination requires the measurement of bending strains in at least two sections along the specimen's gauge length. Investigations have shown that the maximum bend vector is not influenced by the applied load level when the specimen is mounted in a rigid load train /2, 3/. The bending strains are easily measured at room temperature by means of strain gauges and subsequently remain unaffected during heating and testing when the cold grip concept is adopted.

2.6.2. The selection of a cold gripping system implies that heating is applied locally to the gauge length part of the specimen. This impairs the uniformity of the axial temperature gradient which may thus require improvement by additional measures, such as better thermal insulation, multiple zone heating, etc.

2.6.3. To ensure failure in the gauge section away from the grips, waisting of the specimens can be used. This may be necessary even for tabbed specimens since the use of bonded tabs to cushion the specimen in the grips does not prevent shear failure caused by the geometry of the clamping ends. Waisting may be unavoidable when testing composites which show an increase in strength with increasing temperature. If waisted specimens are used, temperature gradients in the region of the specimen between the grips and the end of the reduced section may cause local stresses larger than those in the gauge section, and give raise to failure outside the gauge length. Therefore the geometric and thermal transition zones should not coincide.

3. EXPERIMENTAL REALISATION AND PERFORMANCE

3.1. General description

An ultrahigh vacuum chamber is mounted onto a four column electromechanical universal testing machine. The vacuum level attained during the tests is better than $1*10^{-6}$ mbar, over the whole temperature range. The oxygen partial pressure is below $5*10^{-10}$ mbar. The specimen heating and gripping assembly, as well as the watercooled contacting extensometer are installed inside the vacuum chamber. The allowable specimen length ranges from 120 to 200 mm. Using appropriate grip inserts, specimen thicknesses between 2.2 and 3.8 mm can be accommodated. The load cell is also kept under vacuum

(10^{-2} mbar) in order to avoid errors caused by ambient pressure variations. During bake-out and testing its temperature is kept constant by separate watercooling.

3.2. Vacuum
The watercooled UHV chamber is designed to provide maximum flexibility towards specimen geometry and specimen mounting procedure. Low vacuum is obtained using a rotary pump, while high vacuum is realised by turbomolecular pumping. Bake-out of the chamber walls using internally placed quartz lamps and external heating elements is necessary in order to remove excess water vapour in the system. The presence of vacuum-incompatible materials within the hydraulic gripping system inside the chamber imposes the use of rather mild bake-out conditions, resulting in a partial water vapour pressure of about $1*10^{-7}$ mbar. A simple mass spectrometer with a resolution of $1*10^{-10}$ mbar is connected to the chamber and a mass spectrum ranging from 2 to 50 amu is taken before every test.

3.3. Specimen clamping and resulting alignment
The samples are rigidly clamped in a pair of watercooled hydraulic grips that allow a repeatable mounting of the specimen as well as a fine adjustment of the clamping pressure applied to the specimen ends. The magnitude of the maximum bend vector obtained on an accurately machined reference specimen, is $4*10^{-5}$. Although this value is too high for testing monolithic ceramics, it is adequate for uniaxial testing of CFRCMCs where the inherent curvature of the specimens after processing in most cases induces a higher level of misalignment than that caused by the load train and gripping assembly.

3.4. Specimen heating
The specimens are heated indirectly by induction over a 60 mm length, via a high-purity graphite susceptor centered about the sample. In order to avoid heat losses by radiation towards the inner chamber walls, the susceptor is surrounded by a zirconia insulation tube with high open porosity. The maximum achievable temperature on the specimen is 1600 °C. This maximum can be easily increased by using more power, but at the cost of a loss in vacuum quality of about one decade. The temperature is measured by a dual wavelength infrared pyrometer and by three shielded B-type thermocouples along the gauge length. Shielding is necessary in order to avoid carbon contamination of the noble thermocouples. However, the ceramic insulation behaves as a heat sink and therefore the readings of the shielded thermocouples have to be calibrated in a one-shot experiment against those of new bare B-type thermocouples mounted in holes drilled in a dummy specimen. The middle shielded thermocouple is used for controlling the furnace temperature. Figure 2 shows temperature differences from the setpoint against location along the gauge length. At 1200 °C the temperature gradient is less than 15 °C over a 10 mm distance both above and below the specimen centre. This reduces to less than 5 °C over a central distance of 10 mm. Figure 3 shows the variation with time of the temperature measured by the shielded thermocouples during a creep test. The maximum recorded differences are ±3 °C. The short-term perturbations are explained by inaccuracies in the data acquisition chain while the long-term variations are taken to be temperature variations of the thermocouple.

3.5. Mechanical contact extensometry
During the tests, the deformation is measured continuously by a low contact force mechanical extensometer mounted sideways on the specimen. It is placed inside a thermostatically controlled watercooled Faraday cage to avoid thermal drifts and electromagnetic interferences. The strain resolution is better than 10^{-5}. The nominal gauge length is 10 mm.

ACKNOWLEDGEMENTS

The authors thank M. Bourgeon and J.-M. Parenteau (SEP, Etablissement de Bordeaux) for providing the material and for valuable discussions. They are also grateful for the experimental assistance offered by P. Young and G. von Birgelen. The work has been performed within the Research and Development Programme of the CEC.

REFERENCES

1. M. Steen and J. Bressers, Fortschrittsberichte der Deutschen Keramischen Gesellschaft, 7 (1992) 205-227
2. J.-N. Adami, D. Bolsch, J. Bressers, E. Fenske and M. Steen, Journal of the European Ceramic Society, 7 (1991) 227-236
3. J. Bressers, M. Steen and J.-N. Adami, in "Harmonisation of Testing Practice for High Temperature Materials" (1992) 255-272, Elsevier Applied Science

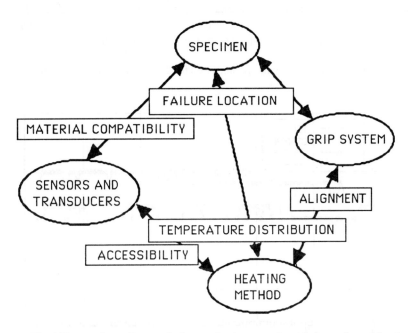

Fig. 1 Factors depending on the interaction between the solutions adopted for the separate items (schematic)

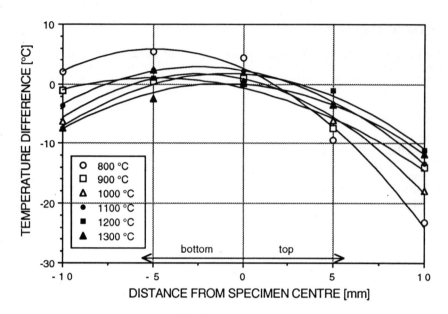

Fig. 2 Temperature gradient along the specimen axis for different temperature setpoints

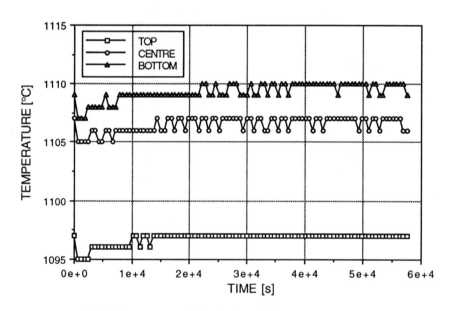

Fig. 3 Temperature stability in time during a creep test

REDUCTION OF SCATTER IN MECHANICAL PROPERTIES OF COMPRESSION MOULDED LONG GLASS FIBRE REINFORCED POLYPROPYLENE COMPOSITE

H. HAMADA, G. SHONAIKE, T. SATO*, S. YAMAGUCHI, Z. MAEKAWA, M. KOSHIMOTO**

Kyoto Institute of Technology - Matsugasaki - Sakyo-Ku 606 - Japan
*Sumitomo Precision Prod. Ltd - 1-1d Fuso-Cho
Amagasaki Hyogo - Japan
**Idemitsu Petrochem Co. - 1-1 Anesaki - Kaigan
Ichibara, Chiba 299 01 - Japan

ABSTRACT

The effect of fibre length on mechanical properties of three-ply long glass fibre reinforced polypropylene composite was investigated. Samples with various glass fibre lengths (i.e. 13, 25, 50mm and continuous fibre) in the laminates were prepared by stamping. The results of bending properties of shorter glass fibre lengths in the laminate exhibited a large scatter. On substitution of the middle layer with a continuous fibre, the scatter was reduced. The results suggest that orientation and volume fraction of glass fibre are responsible for the scatter.

INTRODUCTION

Compression moulding of long glass fibre reinforced thermoplastic composites by stamping is receiving increasing attention from both the scientific and industrial communities. The products are characterized by good mechanical properties (1-3) and short moulding cycle (4). However, the mechanical properties are known to exhibit a large scatter and the difference as large as 200 percent are common (5). Various investigations have shown that the change in mechanical properties is due to the nonuniform distribution of glass fibre in the composite.

The purpose of the present studies is to investigate the effect of material combinations on bending properties of three-layer glass fibre reinforced polypropylene composite. The mechanical scatter in the moulded panel was checked.

1. EXPERIMENTAL DETAILS

1.1. Materials

In this work, glass fibre mat reinforced polypropylene was used. The blank materials (semi-moulded) consist of 40wt% glass fibre and 60wt% polypropylene. The following notations were used to identify the samples: 1, 2 and 5 correspond to 13, 25 and 50mm glass fibre lengths and M = continuous glass fibre.

1.2. Stamping technique

Compression moulding of the samples was carried out by stamping. The blanks were heated to approximately 200 C in an infrared oven. The mould geometry (Fig.1) was 300 x 300mm and the charge area was 173 x 173mm. The charge area defined as the area of the blank to that of the die was 33%. The die temperature was 50 C, moulding pressure was 14.7MPa and the closing speed was 7.5mm/sec. In this work three plies were used.

1.3. Bending test

Bending test samples were machined out of the laminated composites. Three-point bending test was carried out at room temperature using Instron Tensile Testing Machine. The cross-head speed was 3mm/min with span length of 63mm.

RESULTS AND DISCUSSION

The results of the bending properties of three-ply long glass fibre reinforced polypropylene composite were found to depend on the fibre length. Table 1 shows the composition and glass fibre lengths of the three-layer composite. Figs. 2-4 are the bending strengths of all samples determined from various locations. Locations 1-4 are the blank or flow area (a) and above these locations is the charge area (b). Location 1 was close to the edge of the mould whilst location 10 was the centre of the charge area. The vertical line indicates the boundary between the two areas. During stamping, flow of the material occurred from the charge to the blank area. In Fig.2, bending strengths of F13, F25, F50 and FM initially reduced but increased on approaching the boundary. In Fig.3 where F151 and F252 samples are shown, the middle layer glass fibre was 50mm long and the behaviours were similar to those in Fig.2. In both Figs 1 and 2, scatter of bending properties was noted. Bending strengths of F1M1, F2M2 and F5M5 samples are presented in Fig.4. In this case, the middle layer fibre was substituted with continuous glass fibre. As can be seen in the figure, bending strengths are almost the same in both the blank and charge areas. This indicates that the scatter of mechanical properties may be reduced in a three-layer glass fibre reinforced polypropylene composite by introducing a continuous fibre in the middle-layer. Bending moduli of all samples are shown in Figs.5-7. The trend is similar to that of bending strength but the scatter of bending modulus was higher than that of bending strength. Table 2 shows the representative coefficient of variation of bending modulus and strength. The coefficient of variation of F2M2 sample was the lowest, i.e. the scatter was reduced. The volume ratio of glass fibres measured at various locations are depicted in Fig.8. As can be seen in the figure, the volume fraction in the blank region was lower than in the charge region. Based on the above results, the volume fraction of glass fibre may not be the only parameter responsible for the scatter. Therefore, it is suggested that slippage flow which occurred during stamping is partially responsible for the scatter.

During moulding , the middle layer resin was able to flow whilst flow of resin in both top and bottom layers was restricted due to material solidification. This unbalance leads to slippage flow. Flow models for the representative samples are shown in Fig.9. In Fig.9 (a) high slippage flow occurred by shear force at the interlamina and this represents the situation where the material displayed a large scatter of mechanical properties. In Fig.9(b), slippage flow was eliminated and no significant scatter of mechanical properties was noted.

CONCLUSION

Bending properties of three-layer glass fibre reinforced polypropylene composites have been investigated. The results conclusively show that the scatter of mechanical properties may be reduced by substituting the short middle layer fibre with continuous fibre. Interlamina slippage flow during stamping was found to be responsible for the scatter. Large scatter occurred in samples with high slippage flow whilst in samples without slippage flow, little or no scatter was observed.

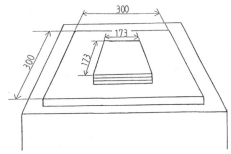

Fig.1 Schematic representation of charge pattern.

Table 1 Composition of mouldings.

	Fibre Length		
	Top Layer	Mid-Layer	Bottom Layer
F13	13	13	13
F25	25	25	25
F50	50	50	50
FM	Mat	Mat	Mat
F151	13	50	13
F252	25	50	25
F1M1	13	Mat	13
F2M2	25	Mat	25
F5M5	50	Mat	50

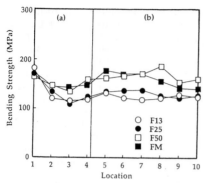

Fig.2 Bending strengths of
F13, F25, F50 and FM
samples.

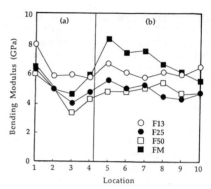

Fig.5 Bending modulus of
F13, F25, F50 and
FM samples.

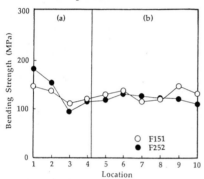

Fig.3 Bending strengths of
F151 and F252 samples.

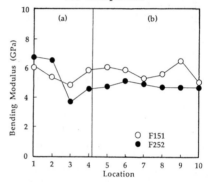

Fig.6 Bending modulus of
F151 and F252 samples.

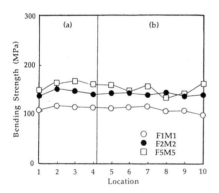

Fig.4 Bending strengths of
F1M1, F2M2 and F5M5
samples.

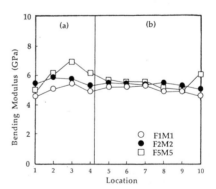

Fig.7 Bending modulus of
F1M1, F2M2 and F5M5
samples.

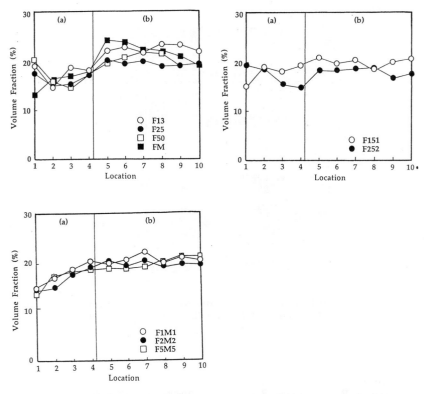

Fig.8 Volume fraction:(a) F13, F25, F50 and FM samples, (b) F151 and F252 samples (c) F1M1, F2M2 and F5M5 samples.

Table 2 *Coefficient of variation of bending modulus and strength.

	Bending Modulus	Bending Strength
F50	16.43	14.05
F151	14.43	13.40
F2M2	5.63	7.97

* representative

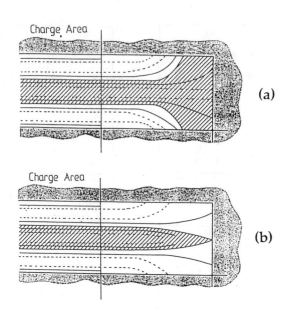

Charge Area

(a)

Charge Area

(b)

Fig.9 Representative Flow models:(a) F13, F25, F50 and FM samples (b) F1M1, F2M2 and F5M5 samples.

References
1. Gupta V.B., Mittal R.K. and Sharma.K. Compos. Sci. Tecnol. Vol.37 No4 (1990) p 353
2. Maekawa Z., Fujii T., Ikeda T. and Yamada T. Proc. 27th Japan   Cong. on Material Research (1983) p 222
3. Denault J., Vu-Khanh T. and Foster B. Polym. Compos. Vol.10 No5 (1989) p 313
4. Mallick P.K. Fiber Reinforced Composites, Marcel Dekker Pub. New York (1988) p 335
5. Kenig S. Polym. Compos. Vol.7 No1 (1986) p 50

MECHANICAL PROPERTIES AND NOTCH SENSITIVITY OF SHORT CARBON FIBRE-EPOXY COMPOSITES

S. HITCHEN, S. OGIN*, P. SMITH*

*DRA - Building R50 - Materials and Structures
GU14 6TD Farnborough Hants - UK*

ABSTRACT

The effect of fibre length on the mechanical properties and notch sensitivity of a short carbon fibre-epoxy composite is investigated. Reducing the fibre length from 15 mm to 1 mm increased the toughness whilst decreasing the modulus and it appears that maximum toughness and maximum modulus cannot be obtained simultaneously for short fibre composites. The notched strength of the three laminates studied showed a sensitivity to notch root radius with the tensile strength decreasing at smaller radii of curvature.

1.0 INTRODUCTION

An area of practical importance for components made from short fibre composites is their ability to tolerate the presence of stress raisers such as damage, a notch or cut out in a moulded structure. Currently there are comparatively few studies in the literature concerned with the effects in short fibre composites compared with the extensive work in this area for continuous fibre composites (for example [1,2]).

The present work is part of a wider study with the overall aim of investigating and modelling damage tolerance of short carbon fibre epoxy composites. In this study the effect of fibre length on the Young's modulus and toughness of a random two dimensional short carbon fibre reinforced epoxy has been investigated together with the effect of radius of curvature on the notched strength.

2.0 EXPERIMENTAL

Toray T300 PAN based carbon fibre was cut into 1, 5 or 15 mm lengths and random short fibre laminates, approximately 1.5 mm thick, were produced using a pre-preg technique. The fibre volume fraction of each laminate was measured using a matrix 'burn-off' method. The modulus and unnotched tensile strengths were determined from quasi-static tests on parallel sided specimens, 20 mm wide and 160 mm long, using an Instron 1175 tensile testing machine. Specimens were loaded to failure at a cross-head speed of 1 mm/min. and 10 specimens were tested at each fibre length. The fracture toughness was measured using single-edge-notch specimens, 20 mm wide and 160 mm long, containing cracks of length 1 to 11 mm. The fracture toughness was calculated from the expression

$$K_c = \frac{P}{BW} Y \sqrt{\pi a} \tag{1}$$

where P is the load at fracture and a is the crack length. B,W are the specimen width and thickness respectively and Y is a geometrical correction factor which was determined using a compliance calibration technique /3/ for each of the three laminate types. The fracture toughness calculated using the experimentally determined correction factors was approximately constant over the range of crack lengths and these values were used to determined the toughness, G, using the expression;

$$G_c = \frac{(K_c)^2 (1 - (\upsilon_{12})^2)}{E} \tag{2}$$

where E is the Young's modulus and υ_{12} is the Poisson's ratio.

The effect of radius of curvature on the tensile strength of laminates containing 1, 5 and 15 mm, length fibres was investigated using specimens, 25 mm wide and 160 mm long, which contained two holes connected by a cut made with a jewellers saw. The total notch width was kept constant at 10 mm and the tip radius of curvature was varied in the range 0.5 mm to 5 mm by using holes of different diameters. At each radius of curvature, 6 specimens were tested for each of the three different fibre length materials.

3.0 RESULTS AND DISCUSSION

The results from the tensile tests on the laminates containing the three fibre lengths, 1, 5 and 15 mm are listed in Table 1 together with the fibre volume fraction determined using a matrix 'burn-off' technique. Since each laminate type contained a slightly different volume fraction ranging from 14% for laminates with 1 mm length fibres to 22% for laminates with 5 mm length fibres, it was not possible to determine the effect of fibre length on the modulus and toughness from the measured

416

experimental data. Therefore, theoretical analyses /4/ have been used to normalise the data to a laminate containing a fibre volume fraction of 20%.

The normalised data are shown in Figure 1. The effect of increasing the fibre length (1 mm to 15 mm) is to reduce the toughness whilst increasing the modulus. Cooper /5/ developed a model for toughness of composites based on the work required to pull broken fibre from the matrix, taking into account the distribution of strength reducing defects in the fibres. The important conclusion, as far as reinforcement with discontinuous fibres is concerned, is that the toughness is greatest when the fibre length equals the critical fibre transfer length. This model may help to explain the observed trends in toughness with fibre length. From the results described here it appears that maximum toughness and maximum modulus cannot be achieved simultaneously and short fibre composites must be designed for optimum combination of desired mechanical properties.

The effect of radius of curvature on the tensile strength of laminates containing 1, 5 and 15 mm length fibres was investigated using specimens which contained four different hole configurations. The effect of radius of curvature on the tensile strength of a laminate containing 5 mm length fibres is shown in Figure 2 and shows that the tensile strength decreases at sharper radii of curvature. Similar trends were observed for laminates containing 1 mm and 15 mm length fibres. In order to compare the trends at the three fibre lengths, the experimental data were normalised with respect to the tensile strength measured from the static tests on unnotched specimens (Table 1).

In figure 3 the effect of radius of curvature on the mean normalised strength is shown. The standard deviations are not included, but are listed in Table 2 together with the mean values. The behaviour observed at each of the three fibre lengths is similar with the experimental data appproximately superimposing at each radius of curvature. Hence the tensile strength of laminates containing 1, 5 and 15 mm length fibre is affected by the radius of curvature, but the observed trend is independent of fibre length. Work is currently in progress to model the observed experimental trends.

4.0 CONCLUSIONS

During the current work the effect of fibre length on the mechanical properties of a random short carbon fibre epoxy composite has been studied. The modulus and toughness data for laminates containing 15, 5 and 1 mm length fibres were normalised to a laminate having a fibre volume fraction of 20%. The effect of reducing the fibre length (15 to 1 mm) was to increase the toughness whilst decreasing the modulus and it appears that maximum toughness and modulus cannot be obtained simultaneously for short fibre composites. All three laminates showed a radius of curvature effect with the normalised strength (notched strength/unnotched strength) decreasing at sharper radii of curvature. This trend was independent of fibre length.

ACKNOWLEDGEMENTS

The authors gratefully acknowledge Kobe Steel for funding this work and would like to thank Dr. D.C. Phillips and Dr.G. Wells for their helpful comments.

REFERENCES

1. M.T. Kortschot and P.W.R. Beaumont, ASTM STP 1110, ed. T.K. O'Brien (1991) p596

2. S.M. Spearing, P.W.R. Beaumont and M.F. Ashby, ASTM STP1110 ed. T.K. O'Brien (1991) p617

3. J.T. Barnby and B. Spencer, J. Mat. Sci. 11(1976) p78

4. A. Kelly and W.R. Tyson, J. Mech. Phys. Sol. 13(1965) p114

5. G.A. Cooper, J. Mat. Sci 5(1970) p645

Fibre Length mm	Fibre Volume Fraction %	Modulus GPa	Tensile Strength MPa
15	19.8+/-1.3	14.7+/-0.6	151+/- 9
5	22.5+/-1.4	14.3+/-0.6	157+/-10
1	14.4+/-2.0	10.2+/-0.5	127+/-13

Table 1: tensile properties of laminates containing 1,5 and 15mm length fibres

Fibre Length mm	Radius of Curvature mm	Normalised Strength
15	5	0.434+/-0.04
15	2	0.385+/-0.02
15	1	0.295+/-0.02
15	0.5	0.275+/-0.02
5	5	0.390+/-0.03
5	2	0.360+/-0.03
5	1	0.304+/-0.03
5	0.5	0.305+/-0.02
1	5	0.460+/-0.04
1	2	0.378+/-0.03
1	1	0.323+/-0.02
1	0.5	0.310+/-0.02

Table 2: The effect of radius of curvature on the normalised tensile strength of laminates containing 1, 5 and 15mm length fibres

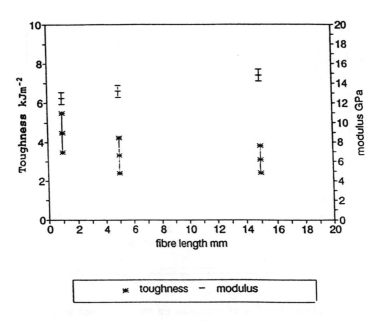

Figure 1: Normalised modulus and toughness (V_f = 20%) vs fibre length

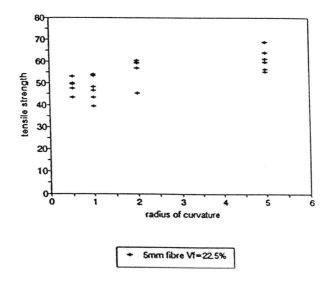

Figure 2: The effect of radius of curvature on the tensile strength of a laminate containing 5mm length fibres (notch width constant at 10mm)

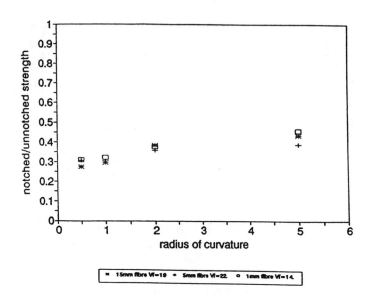

Figure 3: The effect of radius of curvature on the normalised tensile strength for laminates containing a range of fibre volume fractions (notch width constant at 10mm)

STUDY OF MAGNESIUM ALLOY REINFORCED WITH LOW DENSITY ALUMINA FIBRES

K.P. HICKS, V.D. SCOTT, R.L. TRUMPER*

*School of Materials Science - University of Bath - Claverton Down
BA2 7AY Bath - UK
DRA (Maritime Division) - Are Holton Heath - UK

ABSTRACT.

Composites based upon magnesium alloy reinforced with three types of alumina fibre and manufactured by liquid metal infiltration have been investigated. The composite containing alumina fibres of standard density showed reaction between fibre and matrix to a depth of ~200nm. A composite containing low density alumina fibres exhibited more severe reaction during processing such that the whole of the fibre was converted to magnesium oxide during manufacture; the aluminium thereby released was taken into solution in the melt to change substantially its constitution and increase the amount of $Mg_{17}Al_{12}$ in the composite matrix. The use of a coating of yttria on low density fibres was apparently ineffective since the fibres were again extensively reacted.

1. INTRODUCTION.

The low density of magnesium, $1.74g.cm^{-3}$ as compared with $2.7g.cm^{-3}$ for aluminium, makes it an attractive proposition for use in lightweight metal matrix composites (MMC). In order, however, to realise the potential of magnesium for such an application, it is desirable to utilise lightweight reinforcing fibres so as to maximise the specific properties. Thus investigations have been carried out to develop low density fibres for incorporation into MMC based upon magnesium [1,2]. There is, however, a basic difficulty with this metal due to its high reactivity in certain environments and, consequently, problems of compatibility with reinforcements, such as alumina, may arise. The chemical reactions which may occur at the fibre/matrix interface can take place during manufacture or subsequent heat treatment, either of which stage of processing will undoubtedly affect the interface microstructure and, in turn, the integrity of the mechanical bond. Here, as part of an ongoing programme of research aimed at the development of magnesium alloys reinforced with continuous fibres, data are reported on the microstructural properties of three types of alumina fibre reinforced composite which have been manufactured by liquid metal infiltration.

2. EXPERIMENTAL.

The matrix metal used for the composites was magnesium alloy AZ31 (3wt.% Al, 1wt.% Zn with 0.5wt.% Mn added for corrosion resistance). Three types of alumina fibre (Safimax: ICI Ltd.) were used - low density (2.2 g.cm^{-3}) termed LD, low density with a coating of yttria (YLD) and standard density (3.3 g.cm^{-3}) termed SD.

The composites were manufactured using a liquid metal infiltration (LMI) technique. Each type of fibre was placed lengthways in its own steel tube and the three tubes placed together in a die heated to ~450°C. The molten metal was held at ~950°C. in the crucible and then allowed to infiltrate the fibres, the liquid front moving perpendicular to the length of the fibres. This large degree of melt superheat was to aid infiltration and accentuate any fibre/matrix reactions that may occur. The final composites consisted of rods, ~100mm long and 10mm diameter, with the fibres aligned along the axis, and giving nominal fibre volume fractions of 10%.

Samples for optical microscopy and scanning electron microscopy (SEM) were prepared by metallographic polishing on successively finer diamond slurries, taking special care to avoid relief polishing and any etching of matrix second phases. SEM studies were carried out in a JEOL 35C instrument to which was attached an energy-dispersive x-ray spectrometer (EDS) for compositional analysis.

Thin electron-transparent specimens for transmission electron microscopy (TEM) were prepared by a combination of dimpling and ion beam milling techniques and examined in a JEOL 2000FX instrument fitted with EDS.

3. RESULTS.

3.1. Reinforced with Standard Density (SD) Alumina Fibres.

An optical micrograph, Fig. 1, shows good infiltration of the fibres with only limited porosity where fibres were closely packed. The distribution of fibres was not uniform and was up to 20% by volume in some regions of the composite. A few second phases in the matrix were found, using SEM/EDS, to contain aluminium and manganese. EDS analysis of the fibres revealed a small amount of magnesium in outermost regions of the fibre, but this was difficult to quantify because of the small size of the reaction zone compared with the volume of material (~ μm^3) excited by the electron probe.

TEM study showed the reaction zone was ~200nm thick, Fig. 2a. A corresponding selected area diffraction (SAD) pattern indicated the presence of an f.c.c. structure with a lattice parameter of 4.21Å, in accord with crystalline MgO. EDS analysis from this region, Fig. 2b, confirmed the presence of magnesium. The crystals in the reaction zone are noticeably larger than those in the rest of the fibre. Here, SAD, Fig. 2c, showed the presence of δ-Al$_2$O$_3$, a tetragonal structure with a = 7.94Å.

3.2. Reinforced with Low Density (LD) alumina fibres.

The distribution of fibres was similar to the SD composite. The matrix contained, however, many more second phases with many of them having a Mg:Al ratio of ~3:2. EDS of the fibres showed that the aluminium was replaced by magnesium, Fig. 3.

Fig. 4a is a TEM picture from a fibre and adjoining region of matrix. The crystals within the fibre are much larger, often greater than 50nm, than those in unreacted fibre (~5nm). EDS analysis, Fig. 4b, reveals no evidence of aluminium, whilst SAD, Fig. 4c, confirms that it consists of magnesium oxide; no evidence of the spinel, $MgAl_2O_4$, was found. The matrix Mg-Al second phases were identified by SAD as $Mg_{17}Al_{12}$, see also Fig. 6.

3.3. Reinforced with Yttria-Coated Low Density (YLD) Alumina Fibres.

Fig. 5, an optical micrograph, reveals a darker outer ring to fibres which corresponds to the yttria coating. As well as $Mg_{17}Al_{12}$ phases in the matrix, a coarse particle is visible which EDS found contained substantial quantities of yttrium. EDS analysis of the fibres revealed that only magnesium and oxygen was present. A small amount of yttrium was present at the fibre/matrix interface.

TEM studies revealed a relatively coarse-grained MgO microstructure to the fibre, Fig. 6a, much the same as found with the uncoated LD fibre material. The major difference was the presence of particles at the fibre/matrix interface which EDS showed contained yttrium and oxygen; some magnesium was detected but this may have been interference from the matrix. Figs. 6b, 6c and 6d were taken from the matrix particle, labled S, and identify it as a single crystal of $Mg_{17}Al_{12}$ oriented such that the electron beam is along $[3\bar{5}1]$.

4. DISCUSSION.

Reaction occurred between standard density (SD) alumina fibres and magnesium during manufacture of the composite by liquid metal infiltration /3,4/. The reaction extended ~200nm into the fibre to form magnesium oxide with a grain structure somewhat greater than that of the original alumina, ~50nm compared with ~20nm. When low density (LD) alumina fibres were used as reinforcement, the reaction spread throughout the fibre converting it entirely to magnesium oxide during the time taken for infiltrating the fibre preform and cooling to room temperature. The aluminium that was released was dissolved in the magnesium alloy whilst in the liquid state and subsequently precipitated out during cooling as the equilibrium phase $Mg_{17}Al_{12}$. Assuming that the aluminium has been totally rejected from the alumina fibres (volume fraction 0.1), then the amount released is given by 53 wt.% x 2.2/3.3 x 0.1. This quantity when dissolved in the matrix alloy would produce an increase of approximately 3.5 wt.% in the aluminium level, to ~6.5 wt.%, a figure which is not inconsistent with the amount of second phase seen in the composite matrix.

In the case of the composite reinforced with SD alumina fibre, the rate of reaction would be controlled by diffusion of magnesium ions through the aluminium oxide lattice and their interaction with the cations, with a substantial contribution resulting from the ready diffusion paths provided by the large number of grain boundaries encountered in this fine-grained material. With LD fibres, on the other hand, the reaction zone is some four orders of magnitude greater under identical thermal conditions, which implies that the movement of magnesium has been facilitated by the large surface area provided by a network of interconnected micropores in this structure.

The coating of yttria which was applied to the LD fibre to inhibit the diffusion reaction was unsuccessful. Results showed that the coated fibre was just as severely reacted with magnesium as the uncoated product. Indeed, it would appear that the yttria coating on the fibre may not have been continuous and, further, may have suffered damage during composite processing, both of which allowed contact to occur between the alumina and magnesium components.

ACKNOWLEDGEMENTS.

SERC and MOD support is gratefully acknowledged. British Crown Copyright 1993/DRA. Published with permission of the Controller of her Brittanic Majesty's Stationery Office.

REFERENCES.

1.   Dinwoodie J. and Horsfall I., ICCM6/ECCM2 , Vol. 2, eds.F.L.Matthews, N.C.R.Buskell, J.M.Hodgkinson and J.Morton, (Elsevier: London and New York) 1987, p390-401
2.   Fox S., Flower H.M. and West D.R.F., ECCM4, eds. J.Fuller,G. Gruninger, K.Schulte, A.R.Bunsell and A.Massiah, (Elsevier: London and New York) 1990, p.243-248 .
3.   Pfeifer M., Rigsbee J. and Chawla K., J. Mater. Sci., **25,** (1990), 1563.
4.   Page R.A., Hack J.E., Sherman R. and Leverant G., Metall. Trans. **15A,** (1984), 1397.

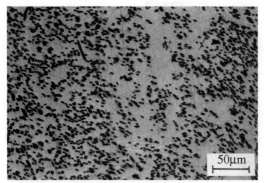

Fig. 1. SD alumina reinfroced
AZ31 alloy,
optical micrograph.

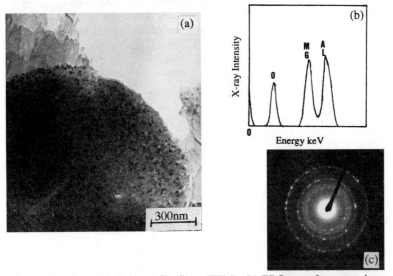

Fig. 2. As Fig. 1. (a) Thinned SD fibre (TEM). (b) EDS trace from reaction zone.
(c) SAD taken from centre of fibre showing δ-Al_2O_3 structure.

Fig. 3. Backscattered electron
image (SEM) LD
reinforced AZ31. Shows
fibres and matrix phases.

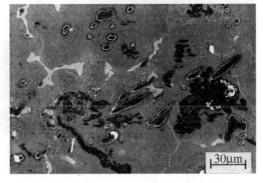

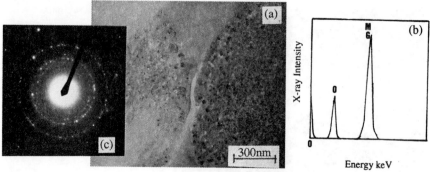

Fig. 4. As Fig. 3. (a) Thinned LD fibre (TEM). (b) EDS trace from fibre.
(c) SAD taken from centre of fibre showing MgO structure.

Fig. 5. YLD alumina reinfroced
AZ31 alloy,
optical micrograph.

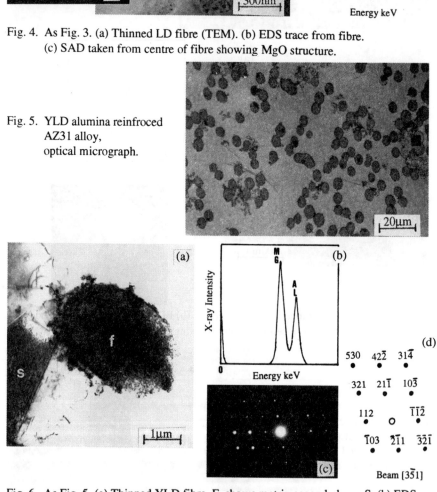

Fig. 6. As Fig. 5. (a) Thinned YLD fibre, F, shows matrix second phase, S. (b) EDS
trace from S. (c) SAD taken from S showing $Mg_{17}Al_{12}$ structure. (d) Analysis
of SAD pattern.

MECHANICAL BEHAVIOR OF TWO TYPES OF 2D SiC-SiC COMPOSITES : COMPARISON OF THE FRACTURE TOUGHNESS

M. GOMINA, M.-H. ROUILLON, M. DRISSI-HABTI, T. DESPIERRES

LERMAT - URA CNRS N°1317 - ISMRA - 6 Bd du Maréchal Juin 14050 Caen Cedex - France

ABSTRACT

The fracture toughness of two types of 2D SiC-SiC ceramic composite materials is investigated using large dimensions of compact tension specimens. The behaviors of the crack growth resistance curves are shown to be strongly dependent on the damage mechanisms pertinent to each type of materials. At the contrary of the classical approach of the behavior of CMCs, it appears that a higher increase in the fracture toughness can be obtained even with a relatively strong fiber - matrix bonding, by promoting intensive matrix microcracking.

INTRODUCTION

The fiber-matrix interface is of prime importance for the mechanical performance of a fiber reinforced composite. Today, it is accepted that the damage tolerant capability of ceramic-matrix composites (CMCs) reinforced with continuous fibers is due to a weak bonding between the fibers and the matrix, which prevents the matrix microcracks propagating across the reinforcement once they are initiated. Thus, the increase in the fracture toughness of CMCs is attributed to the matrix crack bridging by intact and fractured fibers /1,2/, sliding and pull-out of fibers in the crack wake /3/. As a consequence, the approach which is commonly adopted in developing CMCs proceeds by control of the fiber-matrix bonding to achieve long pull-out lengths. While this has led to successful materials /4/, an alternative mean of improving the fracture toughness is to promote elastic energy consumption by the brittle matrix.

In the present paper, the crack growth resistance behaviors of two types of 2D SiC/C/SiC composite materials are investigated and compared, using both energy- and stress intensity - analyses.

1. Experimental

1.1. Materials

Two different types of 2D SiC/C/SiC composite materials have been elaborated by the Société Européenne de Propulsion, using bidirectional woven cloths of Nicalon fibers as reinforcement (40 vol %). Both the pyrolytic carbon (the material controlling the fiber-matrix bonding, of thickness $\simeq 0.1$ μm) and the SiC matrix (ultimate strain $\varepsilon_r^m \simeq$ 0.035 %, Young's modulus $E_m \simeq 350$ GPa) were obtained by mean of a chemical vapor infiltration process. The difference between the two types of the composites (termed type A and type B) lies in the texture of the interphase /5/. Type B materials have been confered a more efficient load transfer capability from the fibers on the matrix, their ultimate strain is higher than 0.40 %, while type A materials possess an ultimate strain value $\varepsilon_r \simeq$ 0.20 %. Typical uniaxial tensile loading curves are shown in Fig. 1 for the two types of materials whose longitudinal Young's modulus is $E_c = 230$ GPa.

1.2. Specimens and mechanical tests

Compact tension specimens of thickness B = 3 mm, similar to those suggested by the ASTM Standard E399-72, were employed for frac-ture testing, with a characteristic dimension W = 40 mm. One large surface of the specimen was polished using different grades of dia-mond paste and the notch plane was machined in the edge-wise direction. The relative notch depth was changed from $a_o /W = 0.2$ to 0.5.

Fracture mechanics tests were conducted on a displacement controlled testing machine using a cross-head speed of 50 μm/min. During the quasistatic crack extension, the crack length was directly measured on the polished surface of the specimen using a travelling microscope. The crack resistance curves were worked out either in the form of the total fracture energy, R, /6/, or in the form of the stress intensity factor, K_R /7/. The crack growth resistance R can be divided into a nonlinear energy fracture toughness GR1 related to the variation in the specimen compliance as the crack grows (noted GR* under linear elastic assumptions) and a contribution GR2 linked to the extensive damaged zones responsible of the permanent displacements observed when unloading the specimen.

2. Results

2.1. Type A materials

The interrelationship of the nonlinear fracture toughness parameters R, GR1, GR2 and the matrix crack length is plotted in fig. 2. The quasi-elastic term GR1 increases to a maximum value of 4200 J/m^2 and then decreases as the crack extends. The total fracture energy R is essentially dominated by the non-elastic term GR2 : it reaches a value as high as 22 kJ/m^2 by a crack extension of 20mm and then decreases rapidly.

The resistance curves of this type of materials are illustrated in Fig. 3 for three values of the relative notch depth a_o/W in the form of K_R versus crack length. It appears a notable difference between the curves associated to the notch depths $a_o/W = 0.3$ and 0.4 and the one associated to $a_o/W = 0.5$, although the matrix crack ini-tiation value is unchanged ($K_R^{ini} = 3.5$ MPam$^{1/2}$).

2.2 Type B materials

When utilizing the loading-unloading procedure to this type of the composites for checking the amount of permanent displacements, one can notice that all the local compliance lines meet at the same point. The local compliance line is defined as the tangent local compliance line to reloading loop drawn from the preceding unloading point. That means type B materials behave linear elastic. The energy release rate GR* calculated assuming elastic fracture mechanics applicability is shown in Fig. 4 as a function of the crack exten-sion. A steady state level of 23 kJ/m^2 is attained after a crack growth increment $\Delta a \simeq 12$ mm.

Fig. 5 illustrates the K_R versus Δa relationship for the same specimen as in Fig. 4. The K_R curve increases up to a plateau value of 67 MPa m$^{1/2}$ and then rised.

3. Discussion

All the resistance curves were derived using experimentally mea-sured crack lengths directly on the polished surface of the specimens under load. These crack lengths measured in the brittle matrix are identical to the one obtained on the surfaces of rupture by using a dye penetrant technique /8/. For the type A specimens notched at a_o/W = 0.3 and 0.4, the variation of fracture toughness K_R with crack growth can be written as $K_R = B \Delta a^{1/2}$ respectively for $a \leq 22$ to 24 mm and $a \leq 24$ mm (Fig. 3). Then the frontal process zone is totally contained within the remaining ligament. This behavior is characte-ristic of small scale brigding effects while the parabolic shape of the K_R curves for higher crack lengths and for the specimen notched at $a_o/W = 0.5$ is representative of large scale bridging effects /2/. When the frontal process zone of the crack tip enters in the compres-sion zone of the ligament, the remaining fibers bridging the crack fail due to the bending stress. This results in a decrease of the total fracture energy R (Fig. 2). Thus, the surfaces of rupture of this type of the composites show large pull-out lengths which testify the presence of a relatively weak fiber-matrix interface (Fig.6a).

On the contrary, the surfaces of rupture of the specimens of type B materials show very short pull-out lengths and the fracture in the matrix around the fibers is not fully brittle. As a consequence of the high density of microcracks in this specimens which testify a good load transfer from the fibers on the matrix and hence a strong enough interface (Fig.6b), different planes of rupture of the matrix are also observed (Fig. 8b). Hence the high ultimate strain and the linear elastic behavior of this type of the composites is due to the high density of microcracks in the matrix. Another point supporting the linear elastic hypothesis to describe the mechanical behavior of this type of 2D SiC-SiC composites is the applicability of the Irwin relationship when using the plateau values of the K_R and GR* curves :

$K_R=(EGR*)^{1/2}$ leads to E=195 GPa which is consistent with E_c=230 GPa.

It should be noticed that for both types of materials, the CVI-SiC matrix is of good mechanical quality since the crack initiation toughness K_R^{ini} = 3.5 MPa m$^{1/2}$ is quite similar to the values commonly reported for polycristalline SiC.

4. Conclusion

So far, it was accepted that an increase in the fracture toughness of CMCs, is associated to fibers debonding, crack bridging and long pull-out lengths, all processes requiring necessarily a weak bond between the fibers and the matrix. The direct comparison of the mechanical behaviors of two types of 2D SiC-SiC composites clearly show that promoting intensive matrix cracking by a judicious process route leading to a relatively strong interface can toughen as well.

Acknowledgments

This work is part of our contribution at the joint research groups created by CNRS (GS 4C and GS thermostructuraux) and SEP and Aéro-spatiale for investigations on CMCs. We wish to thank SEP for diligence in machining and delivery of the specimens.

REFERENCES
1. Thouless M.D. and Evans A.G., Acta Metall., 36 N°3(1988) p 517-522.
2. Zok F. and Horn C.L., Acta Metall. Mater., 38.N°10(1990) p1895-1904.
3. Miyajima T. and Sakai M., in "Fracture Mechanics of Ceramics", Vol. 9, (R.C. Bradt et al., eds), Plenum Press, New York(1992) p 83-95.
4. Lowden R.A., Stinton D.P., Ceramic Engineering and Science Procee-dings. 9. N° 7-8 (1988) 705-722.
5. Monthioux M., Cojean D., in "Developments in the science and technology of composite materials" (1992) p 729-734, ECCM 5, Bordeaux.
6. Sakai M., Urashima K. and Inagaki M.J., Am. Ceram. Soc., 66, N°12 (1983) p 868-874.
7. Srawley J.E., Int. J. Fracture (1976) p 475-476.
8. Rouillon M.H., Thèse de Docteur de l'Université de Caen (1993).

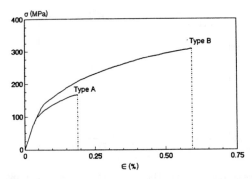

Fig. 1 : Behaviors of the two types of materials under uniaxial tensile loading.

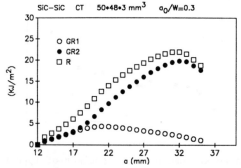

Fig. 2 : Nonlinear fracture mechanics parameters for a CT specimen of type A materials.

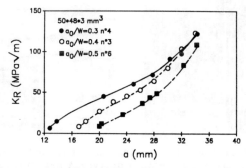

Fig. 3 : K_R-a curves for three values of notch length for CT specimens of type A materials(W = 40 mm).

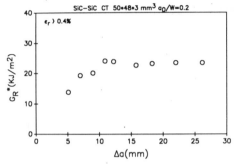

Fig. 4 : Representation of the elastic energy release rate G_R^* as a function of the matrix crack growth increment.

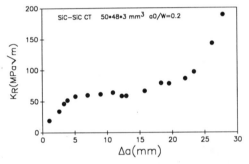

Fig. 5 : K_R- Δa relationship for a specimen of type B materials.

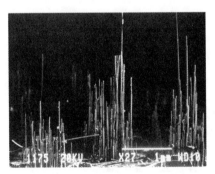

Fig. 6 : SEM photographs showing long pull-out lengths on the surface of rupture of a specimen of type A materials (a); intense matrix microcracking in type B materials (b).

432

PHOTOELASTIC ANALYSIS OF FIBRE FRACTURE AND BUCKLING IN COMPOSITE MATERIALS

M. ANDRIEUX, M.-H. AUVRAY*, P. SIXOU

Laboratoire de Physique de la Matière Condensée
UA 190 CNRS - Parc Valrose - 06108 Nice 02 - France
*Onera - BP 72 - 92322 Chatillon Cedex - France

ABSTRACT

Non destructive photoelastic observation of the stress field around a single carbon fibre in a model composite material has been carried out. Two types of applied stress are considered, the traction mode parallel to the fibre axis leading to a fibre multifragmentation process and the compression mode causing buckling of the fibre. The isochromatic fringes pattern and stress trajectories pattern are derived.

INTRODUCTION

The studies of the interface fibre-matrix constitute one of the areas of composite materials development. Namely the load transfer at this interface is known to influence composites properties. Some degree of knowledge of the effects of fibre geometry and orientation on the local stress distribution within the material certainly would be prerequisite to any attempts to improve composite material design .

The object of this work is to study the resin-carbon load transfer at the interface between a carbon fibre and a polymer matrix, using the photoelastic response of the polymer. We are concerned with the traction mode (fracture appearance, multifracture) as well as the compression mode where buckling of the fibre appears. One of the future aim is to test fibre/matrix load transfer by comparing the calculated photoelastic response of the composite material to the observations.

EXPERIMENTAL WORK

A carbon epoxy single-fibre model composite has been studied in the compression mode and in the traction mode with a load parallel to the fibre axis.

In the traction configuration, the multifragmentation process (see figure 1) is characterized by the critical length Lc (typically 350 μm for a carbon fibre of 7 μm diameter) determined through the average length of the final pieces at saturation. The fragments approach a lower limit in lenght which is dependent on the fibre fracture strength σ_f , the fibre diameter d_f and the interfacial shear strength τ , according to the Kelly-Tyson expression (1) .

$$\tau = \frac{\sigma_f . d_f}{2 . L_c} \qquad (1)$$

The undeformed epoxy resin is optically isotropic. However when reinforced with the carbon fibre, the resin adjacent to the fibre becomes optically anisotropic (birefringent) due to anisotropic stress field . For the compression mode, the buckling and the birefringence arise either from the shrinkage of resin on the fibre that occurs during cooling from the resin cure temperature or from the applied external stress . For the traction mode, a notable photoelastic effect is experimentally observed at the interface in the vicinity of fibre ruptures.

The optical observations are made using polarized light in the optical configuration given schematically in figure 2 . On entering a birefringent region of resin, polarized light is resolved into two components normal to the direction z of propagation, the fast axis and the low axis . Their relative amplitudes remain the same but a phase difference between them exist that is proportional to the thickness of material traversed and the difference in magnitude of the refractive index for the fast and low axes.

In the general case the direction of optical axis changes all along the path of light through the composite and the image observed in photoelasticity has no simple interpretation.
In a two dimensional configuration the birefringence (h.Δn) and stress σ are linked by the following equation (2), where $\varepsilon_{i,j}$ and $\sigma_{i,j}$ are respectively the principal strain and stress, k and C are the optical constants of medium, h is the thickness of medium.

$$\Delta n_{ij} = k(\varepsilon_i - \varepsilon_j) = C(\sigma_i - \sigma_j) \qquad (2)$$

In spite of the three-dimensional character of the stress field that is more pronounced in the traction mode than in the compression (buckling) mode, the isochromatic fringes pattern (see figures 3.a and 4.a) and stress trajectories pattern (see figures 3.b and 4.b) were observed and worked out. Assuming a two dimensional configuration for the compression mode we can deduce the numerical values of the principal stress and of the maximum shear stress. These patterns will be compared to patterns simulated from a mechanical modelization of the single filament composite.

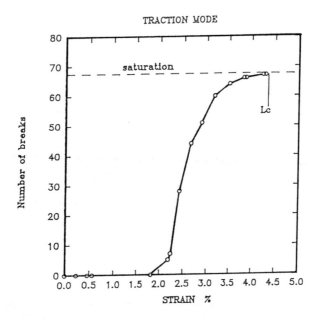

figure 1 : Number N of fibre breaks against strain applied (ε%).
The critical length Lc is obtained after saturation

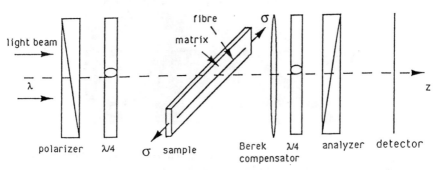

figure 2 : Schematic arrangement of the optical setup

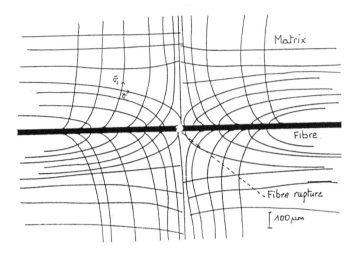

figure 3.a : Stress trajectories derived from the isoclinic fringe pattern
for the traction mode in the vicinity of a fibre rupture

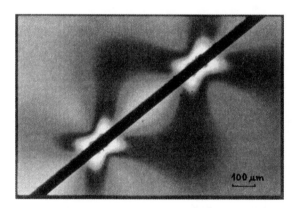

figure 3.b : Experimental fringe pattern in the traction mode. We can see
ruptures of the fibre (monochromatic light).

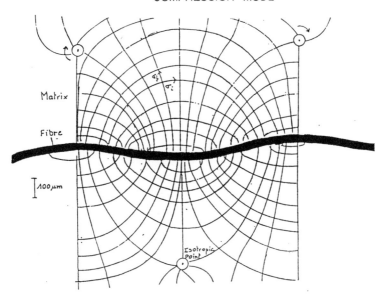

figure 4.a : Stress trajectories derived from the isoclinic fringe pattern for the compression mode with fibre buckling in the absence of fibre rupture

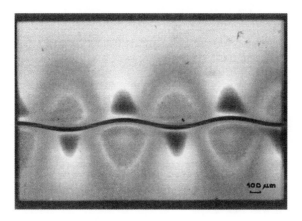

figure 4 b : Experimental fringe pattern in the compression mode with fibre buckling (white light).

MODELLING

DETECTION AND LOCALIZATION OF PUNCTUAL DEFECTS IN SOME COMPOSITE MATERIALS USING CATHODOLUMINESCENCE

R. CHAPOULIE, S. DUBERNET, F. BECHTEL, M. SCHVOERER

CRIAA - Université Michel de Montaigne Bordeaux III - CNRS URA 1515 - Maison de l'Archéologie - Esplanade des Antilles 33405 Talence Cedex - France

ABSTRACT

Cathodoluminescence (CL) properties of a SiC-SiC composite material are reported. Two different types of emission are detected at room temperature depending on the physico-chemical conditions of processing the materials. A first emission in the red part of the spectrum is observed from the as-received samples and is located at 615, 642 and 670 nm. A second emission situated mainly in the blue part is also revealed from the most highly oxidized samples. This specific emission with a maximum pointing out at 460 nm, is due to the presence of silica grown inside the composite.

INTRODUCTION

Nowadays composite materials find many applications in various fields such as aerospace, medicine or sporting equipment. But their properties and performances might be disrupted because of physical or chemical defects. Detection and characterization of such defects seem essential and are possible thanks to their properties of luminescence. Previous analysis using cathodoluminescence (CL) with Al_2O_3 synthetic crystals /1/ led us to carry out investigation into Al_2O_3-ZrO_2 and SiC-SiC composites /2, 3/. Here will be recalled the main results concerning the synthetic alumina and the Al_2O_3-ZrO_2 composite while a discussion about the SiC-SiC composite will be developed. Indeed punctual defects were revealed as well as silica grown at high temperature for the SiC-SiC composite. It is known to be difficult to demonstrate the presence of crystalline or amorphous silica using X-ray diffraction, scanning electron microscopy or transmission microscopy /4/.

I. CATHODOLUMINESCENCE AND ITS EXPERIMENTAL ASPECT

Cathodoluminescence is a well known phenomenon which has been largely utilized for cathode-ray-tubes /5, 6/. The emission of luminescence due to electron bombardment is associated to excitation/de-excitation processes of valence electrons from atoms and to electron-hole radiative recombinations. These occur in the vicinity of atoms or atom aggregates or physical defects which constitute some electronic anomalies of the crystal field. Most common defects detected using CL are atom vacancies (electron or hole centres) and/or chemical impurities such as rare earths or transition metals.

Three sets of equipment allowing CL detection have been used : a Nuclide system (cold cathode) to take colour pictures, another cold cathode system (laboratory-made) working at liquid nitrogen temperature or room temperature allowing spectral analysis of bulk materials, and a CL detector linked to a SEM (JEOL 820) to carry out spectral analysis (photomultiplier tube RTC 56 TUVP) and to take black and white photographs from very small surfaces (a few square micrometres). The latter system has been widely developed in the last few years for the study of semiconductors /7/ and more generally of inorganic solids /8/.

II. ANALYSED SAMPLES

II.1. Synthetic alumina

The study of synthetic alumina which was carried out previously is reported else where /1/. Here are only given the final outcomes.

Figure 1 shows a typical cathodoluminescence spectrum of α–Al_2O_3 prepared under reducing atmosphere (Ar/H_2) from which three sets of emission can be observed centred at 330 and 510 nm for the two wide bands and 694 nm for the main line in the red part of the spectrum. The interpretations were the following /1/ :

1. the band at 330 nm is due to an oxygen vacancy which has trapped an electron and acts as a luminescent centre,

2. the band at 510 nm has been interpreted as due to the presence of an impurity atom, titanium. In fact several doping processes with increasing Ti concentrations have shown the appearance of a blue band at 440 nm certainly due to Ti^{4+} ions /9/, while the former band at 510 nm was disappearing, which was probably due to titanium in a previous different electronic state,

3. the lines located in the red part of the spectrum among which the main line is pointing at 694 nm, find their origin in the presence of Cr^{3+} ions.

Futhermore the same types of emissions were also detected from α–Al_2O_3 prepared under oxidizing atmosphere but with a lower general intensity. Thus concerning the alumina samples, the cathodoluminescence investigation carried out with synthetic crystals grown in specific atmosphere and doped with relevant impurity atoms, has shown the influence of each parametre on the CL behaviour i. e. shape and intensity /1, 2/. This type of approach, also largely developed in the field of solid state and geological materials /7, 8, 10, 11/, led us to investigate into Al_2O_3-ZrO_2 and SiC-SiC composites.

II.2. Composite materials Al_2O_3-ZrO_2 and SiC-SiC

The samples analysed were all prepared by SEP (Société Européenne de Propulsion, Bordeaux). A first experimentation with Al_2O_3-ZrO_2 has been carried out with the as-received samples. It showed chromatic contrasts within the zirconia matrix as well as in the fibres made of alumina /2/. No futher investigation was attempted because of the importance given to the study of the SiC-SiC composite. This composite is made of Si-C-O microcrystallized fibres surrounded by a crystalline SiC matrix /12/. Samples to be studied (15x10x15mm) are embedded in an epoxy resin ; the surface the electrons impinge upon is carefully polished to increase the luminescence radio. The results presented here cannot include colour pictures. Therefore only observations can be quoted. A SiC-SiC sample under artificial light presents the SiC fibres gathered in packs and some holes of porosity inside the SiC matrix. A CL view of this as-received sample reveals red emitting areas around the holes and others coming from little pieces of SiC trapped in the blanks. The emission spectrum (fig. 2) presents 3 maxima at 615, 642 and 670 nm. A very accurate localization has been performed usy the SEM while filtering the wavelength at 670 nm (fig. 3).

Then thermal treatments were carried out. Annealing at 900°C under vacuum conditions involves a homogeneous brownish emission from the matrix and still a red emission around the holes. Another sample of the same nature annealed at 900°C in ambient atmosphere did not show any red luminescence, it only showed a brown emission from the matrix apparently corresponding to the different steps of crystal growth. A very bright emission is also detected around the holes; it has been attributed to the presence of silica grown by oxidation. The last sample of the same nature annealed at 1,200° C in ambient atmosphere, showed a red CL, a brown emission from the matrix and a very bright blue emission due to the silica formed around the holes. From these observations new samples were analyzed which were made using a pyrocarbon phase located in between the fibre and the matrix. This phase is oxidized around 600°C involving oxygen diffusion inside the material followed by silica growth. The localization and determination of the origin of the silica was of primordial importance.

As shown in figure 4, the strong CL emission of silica enables to localize it around the fibres. The width path is less than 1 μm.

The CL spectrum analysis given in figure 5, is a typical case of silicate emission. The maximum points out at 460 nm in the blue part of the spectrum.

This very intense blue emission was finally selected using a blue colour filtre, to give a cartography of the CL emission due to the presense of silica as it can be done in X-ray fluorescence mapping to determine the element distribution onto the surface of a sample. But in this case it was punctual defects which were under investigation.

III. DISCUSSION

About the SiC-SiC composite, in both cases i. e. for the red and the blue emissions, CL has enabled to determine a precise localization of different types of defects in SiC-SiC composites. Concerning the red CL origin, several hypotheses are mentioned : silica or carbon vacancies /13/, impurities such as nitrogen or oxygen atoms /14, 15/ simultaneous existence of α-SiC and β-SiC which can both show specific emissions /16/. For the blue CL due to silica many hypotheses may also be quoted resulting from the study of crystalline and/or amorphous SiO_2. Oxygen vacancies are most probably involved in connection with interstitiel oxygen atoms to form radiative recombination centres /17, 18/. But we must evoke the possibility of the presence of impurity atoms although in this case this hypothesis is not our favourite one. Aluminium and germanium notably are mentioned to be responsible for the blue emission in some silicates /19, 20/. A definite interpretation would of course need a far more developed investigation.

Finally cathodoluminescence appears as a relatively simple method to reveal punctual defects in composite materials. Their identification can be achieved by processing some material in very well defined conditions as it was done with the synthetic alumina : atmosphere, temperature, annealing time and eventually doping with luminescent elements.

Aknowledgement
This research was carried out within the framework of a research convention with the Société Européenne de Propulsion (SEP), Division Propulsion à Poudre et Composites and the Laboratoire des Composites Thermostructuraux (LCTS). It was held thanks to the financial support our laboratory receive from the DRED (Enseignement Supérieur), the CNRS (Département SHS) and the Région Aquitaine. We would like to thank also some of the CNRS members working in our laboratory: NEY C. (drawing), SELVA P. (maintenance) and GESS J.P. (photography).

REFERENCES

1. Schvoerer M., Guibert P., Piponnier D. et Bechtel F., PACT 15, Ed. Conseil de l'Europe, Strasbourg, (1988) pp 93-109.
2. Chapoulie R., Doctorat Univ. Bordeaux 3, (1988) p 214.
3. Chapoulie R., Schvoerer M. et Müller P., C. R. Acad. Sci. Paris, t. 313, Série II, (1991) pp 1105-1110.
4. Filipuzzi L., Doctorat Univ. Bordeaux 1, (1991) p 181.
5. Garlick G. F., Advances in Electronics, Vol.2, (1950) pp 151-184.
6. Leverenz H. W., An Introduction to Luminescence of Solids, New-York, J.Wiley Inc., (1950) p 569.
7. Yacobi B. G. et Holt D. B., J. Appl. Phys. 59 (4), (1986) pp R1-R24.

8. Yacobi B. G. et Holt D. B, Cathodoluminescence Microscopy of Inorganic Solids, Plenum Press, New-York, (1990) p 292.
9. Blasse G. et Verweij J. W. M., Mat. Chem. Phys., 26, (1990) pp 131-137.
10. Remond G., Cesbron F., Chapoulie R., Ohnenstetter D., Roques-Carmes C. et Schvoerer M., Scanning Microscopy, Vol. 6, N°1, (1992) p 23-68.
11. Marshall D. J., Cathodoluminescence of Geological Materials, Unwin Hyman, London, (1988) p 146.
12. Naslain R., Rossignol J. Y., Hagenmuller P., Christin F., Heraud L. et Choury J. J., Rev. Chim. Min., 18, (1981) pp 544-564.
13. Geiczy I. I., Nesterov A. A. et Smirnov L. S., Sov. Phys. Semicond., 5 (3), (1971) pp 439-442.
14. Geiczy I. I. et Nesterov A. A., 3rd Int. Conf. Silicon Carbide, Miami Beach, Flo., (1973) pp 213-219.
15. Kholuyanov G. F., Vodakov Yu. A., Violin E. E., Lomakina G. A. et Mokhov E. N., Sov. Phys. Semicond. , 5, (1971) pp 32-37.
16. Gmelin Handbook of Inorganic Chemistry, Si Suppl., B2, Silicon Carbide, Part 1, (1984) pp 206-229.
17. Itoh C., Tanimura K., Itoh N. et Georgiev M., Rev. Solid State Sci., 4 (2-3), (1990) pp 679-690.
18. Hayes W., Kane M. J., Salminen O., Wood R. L. et Doherty S. P., J. Phys. C: Solid State Phys., 17 (16), (1984) pp 2943-2951.
19. Solntsev V. P. et Lysakov, J. Appl. Spectros., 22, (1975) pp 339-341.
20. Luff B. J. et Townsend P. D., J. Phys.: Condens. Matter, 2 (40), (1990) pp 8089-8097.

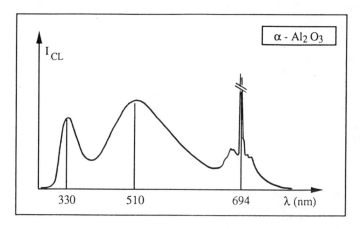

Fig 1 : Cathodoluminescence (CL) spectrum of synthetic α–Al_2O_3 prepared under reducing atmosphere (Ar/H$_2$), after /1/.

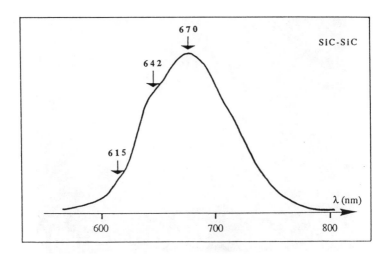

Fig. 2 : CL emission spectrum of some red emitting areas visually detected from the as-received SiC-SiC sample.

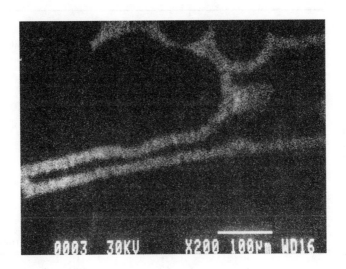

Fig.3 : Monochromatic CL picture (λ=670 nm) using a SEM showing the localization of a red emitting zone from the as-received SiC-SiC sample.

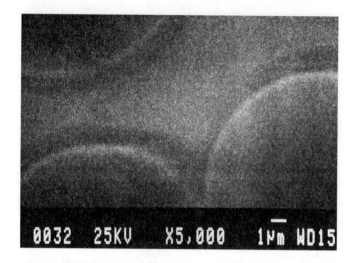

Fig. 4 : CL image without wavelength filtering and using a SEM, of a SiC-SiC sample annealed at
900°C during 300 hours, showing the localization of the silica grown around the fibres.

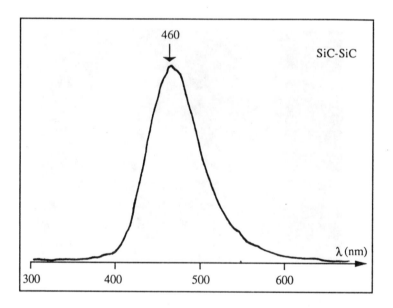

Fig. 5 : CL spectrum using the SEM of SiO_2 in a highly oxidized SiC-SiC composite (area under
investigation 110 μm X 90 μm).

SCALINGS IN FRACTURE STATISTICS OF A UNIDIRECTIONAL COMPOSITES WITH A RANDOM DISTRIBUTION OF DEFECTS

C. BAXEVANAKIS, D. JEULIN, J. RENARD

Ecole des Mines de Paris - Centre des Matériaux P.M. Fourt
BP 87 - 91003 Evry Cedex - France

ABSTRACT: The present work deals with the prediction of the fracture stress of a unidirectional composite T300/914; it is based on a multi-scale approach. The fracture behaviour of a representative volume element (meso scale) of the material is studied using finite element simulations with associated statistics. The model requires the distribution of the defects along the fibres which has been estimated through two different tests: a) Single Fibre Test b) Multifragmentation Test. The theoretical results are compared with experimental data obtained with unidirectional specimens of T300/914.

1. INTRODUCTION

The statistical aspect of the fracture stress of a unidirectional composite is due to imperfections in the volume of the material and specially to the random distribution of the defects along the fibres. When the local stress overcomes the critical stress of a defect, the fibre is broken. Whenever one or more fibres are broken the load must be transferred through the matrix to adjacent fibres in order to restore the equilibrium. The local stress concentration creates damaged zones, the accumulation of those induces the failure of the material. The fracture mechanism is well understood but far as to be well modelled.

Assuming that a composite material can withstand a number of isolated fibres and taking into account that a composite specimen tested in the laboratory contains a few thousand fibres, the numerical simulation of the failure of a complete composite seems overwhelming. To overcame this problem a multi-scale approach is proposed. We consider firstly in a meso scale a small part as representative of the whole composite and then on a macro-scale the material is approximated as an assembly of representative volume elements. On each scale the different involved parameters have to be understood, estimated through adequate experiments and then introduced to the model. This approach presents several advantages such as: a) The scale effects and the influence of the involved micromechanical parameters in the failure of the material can be estimated b) The strength of a specimen(scale of the laboratory) can be extended to real components or structures.

2. STATISTICAL MODELLING

2.1 Statistical Formulation

The statistical formulation is based on statistical models developed by D.JEULIN [1]. We assume that the defects are ponctual and distributed according to a spatial Poisson point process:

i) The defects arise at separate sites (there are no overlaps)

ii) The number of defects occurrences in separate volumes are independent random variables

iii) The number of defects occurrences is independent of the position in the volume.

The above assumptions define a spatial point Poisson process and it can be easily proven that the number of defects in any volume V is a Poisson random variable with parameter θV. θ is a positive real number, the mean number of defects occurrences per unit volume or the intensity of the process. For fracture statistics, we consider a material with an infinite critical stress $\sigma_c(x)$ everywhere, except at the points of the Poisson point process. More precisely, the average number of points with $\sigma_c < \sigma$ contained in the volume V is equal to $\theta(\sigma)V$. The probability of finding exactly an activated flaws in a volume V is given by:

$$P\{N = n\} = \frac{(\theta(\sigma)V)^n}{n!} \exp(-\theta(\sigma)V) \qquad 1.$$

As a first criterion of fracture, we assume a critical density of defects θc; we allow in the volume V, the presence of Nc (Nc=θcV) critical defects (with $\sigma_c(x) < \sigma(x)$) before the unstable propagation of a crack which induces the ruin of the material. The non fracture probability is equal to:

$$P\{\text{no fracture}\} = P\{N \leq Nc\} = p_o + p_1 + ... + p_N \qquad 2.$$

From the above equation, we can easily deduce the cumulative fracture probability.

$$P\{N > N_c\} = 1 - \sum_{n=0}^{Nc} \frac{(\theta(\sigma)V)^n}{n!} \exp(-\theta(\sigma)V) \qquad 3.$$

Another type of fracture criterion without damage of the material is the brittle fracture. In this case, it is very common to make "the weakest link assumption", so that the ruin of the material is provided when the fracture stress of the most critical defect is reached. According to the previous assumption, the cumulative fracture probability is calculated from the equation 3 for Nc=0.

$$\Pr\{\sigma_R < \sigma\} = 1 - e^{-\theta(\sigma)V} \qquad\qquad 4.$$

σ_R: Fracture stress of the material

σ: Applied stress

From the knowledge of the function $\theta(\sigma)$ and of the parameter θc, it is easy to calculate the fracture probability of a specimen with a volume V by applying equations 3. and 4. For application to finite element simulations, these probability distributions are used to generate random variables reproducing the fracture stress at the scale of the calculation. Indeed, we will use a special mesh which corresponds to the volume V, containing a single defect using an appropriate critical stress function.

2.2 Estimation of the Statistical Parameters
2.2.1 Single Fibre Test
Single fibre test is used to estimate the tensile strength distribution of the fibres. The fibre is broken when the local stress overcame the fracture stress of the most critical defects. Thirty single fibres T300 were tested with a gauge length of 50 mm and the two parameters Weibull distribution is used to describe the experimental data. The Weibull model is special case of the above model (equation no 4) if the density function is replaced by a power law function (5).

$$\theta(\sigma) = \left(\frac{\sigma}{\sigma_0}\right)^m \qquad\qquad 5.$$

Based on 30 single fibre tests and using a linear least square method the parameter m, and σ_0 are estimated (m=6.7 and σ_0=4734). For simulations the fracture stress as a function of the fracture probability is given by:

$$\sigma = \sigma_0 \left[-\frac{\ln(1-\Pr)}{L}\right]^{(1/m)} \qquad\qquad 6.$$

2.2.2 Multifragmentation Test
Another method to estimate the density of flaws along the fibres is the multifragmentation test. A high stiffness fibre T300 with 50 mm gauge length is embedded in an epoxy matrix, and the single fibre specimen is pulled in tension. The fibre breaks on defects as the tensile load is increasing. Each break provokes a decohesion between the fibre and the matrix along a length Ls. This may be viewed as "a forbidden zone" for the next failure of a defect. From the number of fibre breaks as a function of the applied stress, we can obtain a complete estimation of the flaws population along the fibre.

The density of defects along a carbon fibre as a function of the applied stress is modelled by a sigmoïdal function [2]:

$$\theta(\sigma) = A\left(1 - e^{-\left(\frac{\sigma-\sigma_s}{\sigma_0}\right)}\right)^m \qquad\qquad 7..$$

where:

A: 8.66 defects per unit length (maximum number of defects per unit length)

σ: applied stress on the fibre

σ_0: 6318 (scale parameter)

σ_s: Threshold stress below which the fracture probability is zero.

m: 7.26 (shape parameter)

The figure 1 presents the correlation between the experimental data and the theoretical results. Assuming that the fibre breaks according to weakest link assumption, from the equation 6 and 7 we can simulate the fracture stress of a fibre length Ls, as a function of the fracture probability. This is useful for simulation of the random variable σ_R from a random number generator Pr.

$$\sigma_R = \left(-\ln\left[1 + \frac{\ln(1-Pr)}{ALs} \right] \right)^{1/m} \qquad \qquad 8.$$

With a condition to respect:

$$Pr < 1 - e^{-(ALs)} \qquad \qquad 9.$$

Therefore, the distribution function is valid for a fibre length Ls≥0.5mm for the T300 fibre. This length corresponds to the mean fragment length at the saturation limit of the single fibre specimen due to the decohesion. Consequently the distribution function cannot be extrapolated for fibre lengths lower than 0.5mm.

Comparing the two above experimental technics of sampling defects along the fibres we can deduce that using single fibre test the most critical defects can be only sampled in contrast using the multifragmentation test the whole population of defects is estimated. We have to point out also that the density function of the defects is not intrinsic parameter of the material because it depends of the sampling method. Therefore it is very interesting to study the influence of the different density functions on the fracture stress of the unidirectional composite.

3. FINITE ELEMENT SIMULATION

The proposed model is based on finite element analysis. Comparing with previous approaches where analytical solution are adopted, the finite element method gives an exact approximation of the stress field around the broken fibres. Statistical information concerning the distribution of defects along the fibres obtained with the above experimental methods are introduced to the finite elements simulations. Each finite element is firstly characterised by its elastic properties, Young and Poisson moduli, which can be changed during calculation. A critical stress corresponding to the fracture stress determined by the appropriate statistics, is associated to each element of the mesh. Each time that in one element, at one increment of calculation, the stress at any Gauss point of the considered element becomes greater than its critical stress, the elastic characteristics (Young modulus and Poisson coefficient) of this element are replaced by those of the matrix for the next calculation increment. The elastic matrix is evaluated, and therefore an additional iteration is necessary to reach the convergence.

In more details, we consider a representative element of the volume of the material which contains N_f fibres (Fig.2). Each fibres is divided in N_l links of a length L_s ($L_s \geq 0.5$mm). Each of them is associated with a random fracture stress simulated from the equations 6. and 8. The statistical links are divided in N finite element in order to reach the convergence of the calculation.

The fracture behaviour of the material is simulated as follows: when the average stress at the Gauss points of an element overcomes the associated fracture stress, the stiffness of the element is weaken. The fracture occurrences provokes a local drop of the stiffness. The fracture of the element can be viewed as an important drop of the global stiffness of the element.

The global stiffness of the REV is estimated using a homogenization approach by effective moduli.

$$< \sigma >= C^{\textit{eff}} < \varepsilon > \qquad\qquad 10.$$

The average macro stress $<\sigma>$ and the average macro strain $<\varepsilon>$ are defined as the average spatial values of the local stresses and strains.

$$< \sigma >= \frac{1}{V} \int_V \sigma dv \qquad\qquad < \varepsilon >= \frac{1}{V} \int_V \varepsilon dv \qquad\qquad 11.$$

All the results are obtained assuming 2-D finite elements in plane stress with periodic boundary conditions.

4. RESULTS AND DISCUSSION

We consider a representative volume element (Fig. 2) of the material (REV), the fracture of which induces the ruin of the material. The REV has a length of 6mm and it contains 6 fibres in parallel with a Vf=60%. Each fibre is divided in Nm links of a length of 6mm, the fracture stress of each link is simulated from the equations 6. and 8. The theoretical results are compared to the experimental data. The experiments were carried out [4] on unidirectional T300/914 specimens loaded in tension.

The figure 3. presents the average macroscopical stress as a function of the average macroscopical strain. We point out that the material behaves linearly before an important drop of the stiffness occurs, which deals with the fracture of the element. This reproduces the observed behaviour in experiment. This type of fracture behaviour reflects a local damage of the material and can be considered as "brittle" because it is with a low damage accumulation.

A comparison between the experimental and the estimated fracture stress probability is shown in figure 4. Thirty simulations were carried out, each simulation being associated with a different fracture stress distribution along the fibres using a sigmoidal density function (equation 8). More precisely the sigmoïdal function represents the density of the defects along the fibre as a function of the applied stress. However the single fibre embedded in the epoxy matrix sustains a residual compression stress field (905 MPa) due to different thermal expansion coefficients between fibre and matrix. As a result of the residual stresses the new value of σ_0 is 5432 MPa instead of 6318 MPa so that the density function is translated towards lower values of stress. The curves 1 and 2 in the figure 4 illustrate the estimated fracture stresses obtained with the above density functions. The curve No 2 where residual stresses are taken into account presents a good correlation with the experimental values. However the model overestimates specially the low values of the fracture stress. This fact is due to: a) various micromechanical parameters which are neglected, such as: plasticity of the matrix, debonding, etc... b) scale effects; the model operates on a meso scale in comparison with the real composite.

The effect of the statistics of the defects through the choice of $\theta(\sigma)$ in equations 1-4 is illustrated by the comparison of the results of simulations obtained for the following functions :
- the sigmoïdal function given in equation 7 fit from the multifragmentation test.
- the approximation of the sigmoïdal function by a power law obtained for the lower values of the stress σ.
-a power law function obtained by fitting experimental data from single fibre tests to a Weibull model (equation 5).

The sigmoïdal function used in the simulations differs from the others functions on the tail concerning the defects with the highest critical stress. It gives also the best agreement between simulations and experiments (Fig.5). This means that the tail of the distribution of the defects is of major importance in our simulations; therefore it cannot be accessed from single fibre tests where the weakest defects are sampled. This conclusion is in contrast to the assumption shared by different authors that the fracture of a unidirectional composites is due to the fracture of the most critical defects along the fibres. The physical explanation is that the fibres are broken firstly to the most critical defects and then damaged zones are created around to the first breaks due to the stress concentration, which induces the ruin of the material. The same type of damage is obtained during the simulations.

To estimate the size effect, calculations were carried out for different lengths (3mm, 6mm, and 12mm) and for different numbers of fibres. No difference was observed in the results.

The distribution of fibres in the material is an other random parameter involved in the fracture behaviour which has to be investigated. One hundred simulations were carried out where the fibres are positioned randomly. The estimated strength for a periodic and a random configuration of fibres is shown in figure 6 and the two curves are superposed. According to the model it turns out that the random distribution of fibres plays an important role in the fracture behaviour of the material only for low volume fraction of the fibres. (40%-55%). For high volume fraction no difference is observed between a periodic and a random distribution of fibres.

5 CONCLUSIONS AND PERSPECTIVES

A multi scale approach based on finite elements analysis is proposed to estimate the fracture behaviour of a unidirectional composite. On each scale the involved random parameters have to be estimated through appropriate tests (as the multifragmentation test) and then to be introduced to the model. A representative volume element is used to describe the fracture behaviour of the studied material and we point out that:
- The model gives a good correlation with the experimental results.
- The distribution of defects along the fibres is of major importance in the fracture of unidirectional composite. The whole population of defects has to be estimated and as an adequate test the multifragmentation test is suggested.
- The distribution of fibres in the material for high values of Vf does not play an important role in the fracture of the material.

The above results describing the fracture behaviour of a representative volume element will be used as input in a statistical mesh of the next scale, the distribution of the local fibre volume fraction will be taken into account in order to estimate in a reliable way the scale effects.

ACKNOWLEDGEMENT

The authors are grateful to the following sponsors:
- DRET(contract 89-149-00-470-75-01) and CNES (contract n° 89/365)

6. REFERENCES

1. D. JEULIN "Modèles Morphologiques de Structures Aleatoires et de Changement d'Echelle" Thèse d'Etat soutenue le 25 Avril 1991, Université de Caen.

2. C. BAXEVANAKIS, D. JEULIN, and D. VALENTIN "Fracture Statistics in Single Fibre Composites" Communication to the "International Conference on Microphenomena in Advanced Composites" Herzlia, Israel June 28 - July 1, 1992, accepted for publication in a special issue of Composites Science and Technology (1993).

3. A. SOMMER "Caractérisation Mécanique des Fibres de Carbone Unitaires" DEA de l'Institut Nationale des Sciences et Techniques Nucleaires (INSTN); option : Métallurgie Spéciale et Matériaux; June1984.

4. E. PETITPAS "A Micromechanical Approach to Increased Confidence in the Design of a Bearing Composite Structure " Convention Euram/Armines 12 Month Report 1989

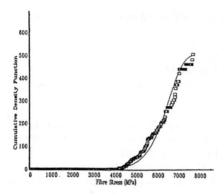

Figure 1: Theoretical and experimental cumulative density function.
(single fibre specimen)

Figure 2: Used mesh of the representative volume element

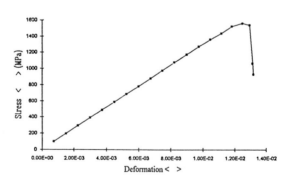

Figure 3: Theoretical damage behaviour of the representative volume element

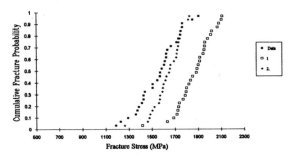

Figure 4: Comparison between experimental and theoretical values. Residual sresses in a single fibre specimen are taken into account (2).

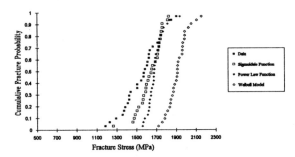

Figure 5. Comparison between experimental data and theoretical results obtained using different density functions of defects along the fibres.

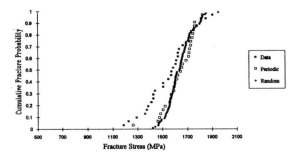

Figure 6: Comparison between experimental and theoretical results obtained with a periodic and a random distribution of fibres in the representative volume element

COMPARISON OF DIFFERENT METHODS FOR COMPUTER SIMULATION OF THE ULTRASONIC WAVE PROPAGATION IN MULTILAYERS

P. DELSANTO, G. KANIADAKIS, D. IORDACHE*

Dipartimento di Fisica - Politecnico di Torino - Corso Duca Degli Abruzzi 24 - 10129 Torino - Italy
*Physical Dept Polytecnic Bucharest Splaint Independentei 313 - Bucarest - Romania

ABSTRACT

The formalism of finite difference equations provides an excellent tool for the computer simulation of the ultrasonic wave propagation in materials, in which the physical properties are homogeneous or vary continuously. A FDE treatment becomes, however, questionable in the presence of sharp interfaces between layers of different physical properties. In the present contribution we discuss several treatments of the interface region and show that an incorrect method can lead to large numerical errors.

INTRODUCTION

A detailed understanding of the mechanism of ultrasonic wave propagation is essential for a quantitative ultrasonic characterization of any kind of materials. Composite materials can often be treated, at least as a first approximation, as multilayers. If there are no irregularities or inhomogeneities within each layer, the problem of wave propagation can be easily solved analytically or numerically. If, however, defects or non elementary boundary conditions are present, the problem can become hopelessly complex or time consuming.

The recent advent of massively parallel computers /1/ suggests a new approach, in which the specimen is divided into a very large number of elementary parts, which we shall call "cells". Each cell, which is assumed to have constant physical properties in its inside, is put into a one-to-one correspondence with the processors of a parallel computer. Thus a lattice is created in which the interaction of each cell with its neighbours is simulated by the interaction of the corresponding processors. This local interaction simulation approach (LISA) has been applied to study a variety of propagation /2/, diffusion /3/, absorption /4/ and desorption /5/ problems. In all cases once the detailed physical properties and initial configurations are specified, an iteration equation predicts the evolution of the system for as many time steps as needed.

LISA has three basic advantages. First, it is extremely time efficient; in fact, since many processors work simultaneously, the computer time is reduced of a corresponding factor (equal to the number of processors if the "speed up" is perfect). Second, since each processor is independent of each other, all the cells can be independent. Therefore the interactions among them may vary arbitrarily. Thus very complex cases may be easily treated. Finally, since the problem is reduced to iteration equations, which are always the same, except for the coefficients, which are inputted as initial data, the same computer code can be used to treat a variety of very different problems.

There are, however, different ways of obtaining iteration equations. A simple method consists in deriving them by means of a transformation of the wave equation into a finite difference equation (FDE). This procedure works well in elementary situations, such as in homogeneous media, but it leads to serious ambiguities in the presence of interfaces separating two media of different physical properties. The purpose of the present contribution is to show that these ambiguities can cause severe errors, which can be eliminated (or reduced) if a more physical approach is adopted.

A correct simulation of the wave propagation is, of course, even more important in two or three dimensions. In fact, the problem of a correct treatment of the crosspoints, i.e. nodepoints separating four different media, is very difficult /6/ and will be treated in a following paper /7/. However, the most relevant difficulties of a correct FDE treatment can already be appreciated and are easier to visualize in the one-dimensional case, to which the present contribution is devoted.

FDE METHODS

Let us assume that an ultrasonic longitudinal pulse is transmitted through a material plate, normal to its surface. Then the elastodynamic wave equation can be written as one-dimensional:

$$\partial_x[S(x)\partial_x w(x,t)] = \rho(x)\ddot{w}(x,t) \quad , \tag{1}$$

where S is the stiffness, ρ the density and w the longitudinal particle displacement. We then define an elementary time δ, so that the continuous time t becomes $t\delta$, with $t = 0, 1, 2...$, and divide the propagation path in "cells" of length ε. Heuristically or by means of the usual FDE formalism /2/ Eq.(1) becomes

$$w_{i,t+1} = w_{i+1,t} + w_{i-1,t} - w_{i,t-1} \quad , \tag{2}$$

for each nodepoint i and time t, provided that the "optimal" condition for the longitudinal propagation velocity

$$v = \sqrt{S/\rho} = \varepsilon/\delta \tag{3}$$

is satisfied. It is important to note that Eq.(2) is particularly suitable for parallel computing, since all the $w_{i,t+1}$ can be computed independently and simultaneously.

Let us now assume that the plate is a bilayer, with a sharp interface separating two media of physical properties S, ρ, v and S', ρ', v', resp. Let us assume for simplicity:

$$v' = \varepsilon'/\delta = v = \varepsilon/\delta \quad . \tag{4}$$

Then Eq.(2) is valid everywhere except around the interface. The treatment of the interface is not obvious, since a rigorous FDE treatment would require that $S(x)$ and $\rho(x)$ be continuous functions. Since the interface is assumed to be sharp, $S(x)$ and $\rho(x)$ must be "smoothed out" over a certain Δx, which is, of course, arbitrary. If the parameters of the two layers are not too different, the procedure is acceptable and the results are independent of Δx, provided that ε is sufficiently small. Otherwise the procedure becomes very questionable and the results will depend on the model used to smooth out $S(x)$ and $\rho(x)$.

We consider in the following several FDE methods corresponding to different models of the interface. The first three methods correspond to the three models represented in Fig.s 1-3, resp., with similar plots for $\rho(x)$. The fourth method is similar to the first one except that additional smoothing is obtained by partly writing:

$$S_i \approx (S_{i+1} + S_{i-1})/2 \quad . \tag{5}$$

Other methods or models are, of course, possible but, in our opinion, less plausible or repetitious. The four methods considered yield around the interface the iteration equation:

$$w_{i,t+1} = a_i w_{i+1,t} + b_i w_{i-1,t} + c_i w_{i,t} - w_{i,t-1} \quad , \tag{6}$$

where

$$c_i = 2 - a_i - b_i \quad , \tag{7}$$

and a_i and b_i are given in Table 1, as functions of

$$t = \frac{2S}{S + S'} \quad , \qquad t' = \frac{2S'}{S + S'} = 2 - t \quad , \tag{8}$$

which are the transmission coefficients in the forwards and backwards direction, resp. From Table 1 it can be seen that $c_i = 0$ for the first three methods, but not for the fourth.

Table 1: Coefficients a_i and b_i in the interface region for the four FDE methods considered.

method	a_{I-1}	b_{I-1}	a_I	b_I	a_{I+1}	b_{I+1}
1st	$(3t+1)/4t$	$(5t-1)/4t$	$(t'+1)/2$	$(t+1)/2$	$(5t'-1)/4t'$	$(3t'+1)/4t'$
2nd	1	1	$(t+1)/2t$	$(3t-1)/2t$	$(3t'-1)/2t'$	$(t'+1)/2t'$
3rd	1	1	t'	t	1	1
4th	$(t+1)/2t$	1	$(t'+1)/2$	$(t+1)/2$	1	$(t'+1)/2t'$

RESULTS AND DISCUSSION

The methods discussed in the previous Section and their generalizations to two and three dimensions are extensively used in the literature. E.g. the first and fourth methods are applied in refs. 8 and 9, resp. The third method leads to the same results as the more physical sharp interface method (SIM) /2/, in which the iteration equation at the interface is provided by the local matching of both the displacements and stresses. The reliability of SIM (and therefore of the third FDE method) can be immediately inferred by the values of the coefficients a_I and b_I, which are equal to the transmission coefficients t' and t, resp.

Therefore we will compare the results of the other three methods with those of the third one. For the comparison we consider the propagation of an ultrasonic pulse of gaussian shape, through an interface separating two media of different stiffness and density. Since we have assumed that S and ρ are proportional (Eq. 3 and 4), we limit ourselves to vary one parameter: S or, equivalently, t (see Eq.8).

Fig.4 shows the results obtained with the four methods when t is not too different from unit: $t = 0.6$. This value corresponds to a stiffness ratio $S'/S = 7/3$. In this case the differences among the various methods are not large and at least the second and fourth methods can be considered satisfactory. In fact the results of the third and fourth method almost coincide.

The situation becomes very different for larger stiffness ratios. Fig.5 shows the transmitted pulses for the case $t = 0.4$, i.e. $S'/S \approx 4$. Here the second and, even more, the first method are completely off target. The fourth method, on the contrary, shows at first an excellent agreement with the third method. Unfortunately, however, at larger times the iteration procedure breaks down and is no longer applicable. This is a consequence of the nonvanishing $c_{i\pm1}$ coefficients, which can be easily explained within the framework of the FDE theory.

To conclude, of the four FDE methods considered, the third is the only one which can provide a correct treatment of sharp interfaces. The other methods are satisfactory for small values of the stiffness ratio (up to about 2). The best one is the fourth method, which, however, is not stable. The least accurate method seems to be the first one, which shows a poor agreement already for $S' = 7/3S$.

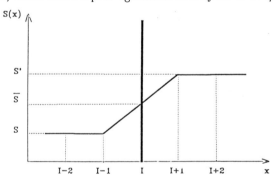

Figure 1: Interface model corresponding to the first method.

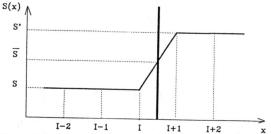

Figure 2: Interface model corresponding to the second method.

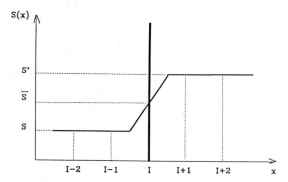

Figure 3: Interface model corresponding to the third method.

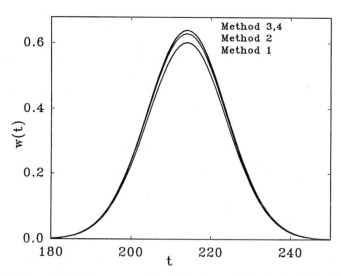

Figure 4: Transmitted pulse, as computed with the four methods for $t = 0.6$, i.e. $S'/S = 7/3$.

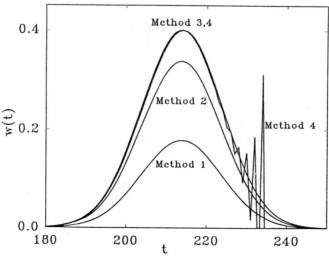

Figure 5: Transmitted pulse, as computed with the four methods, for $t = 0.4$, i.e. $S' = 4S$.

References

[1] P.J. Denning, W.F. Tichy: *Highly Parallel Computation*, Science V.250, (1990) p.1222

[2] P.P. Delsanto, T. Whitcombe, H.H. Chaskelis, R.B. Mignogna: *Connection Machine Simulation of Ultrasonic Wave Propagation in Materials. I: the One-Dimensional Case* , Wave Motion, V.16,(1992) p.65

[3] G. Kaniadakis, P.P. Delsanto, C.A. Condat: *A Local Interaction Simulation Approach to the Solution of Diffusion Problems*, Math. Comp. Modelling, in press

[4] G. Kaniadakis, P.P. Delsanto, C.A. Condat: *Random Walk Simulation of Absorption Processes*, submitted to Nuovo Cimento

[5] G. Kaniadakis, P.P. Delsanto: *Simulation of Desorption Effects in Vacuum by Parallel Processing*, submitted to J.Phys.Chem.Sol.

[6] A. Ilan: *Stability of Finite Difference Schemes for the Problem of Elastic Wave Propagation in a Quarter Plane*, J.Comp.Phys., V.29 (1978) p.389

[7] P.P. Delsanto, R.S. Schechter, H.H. Chaskelis, R.B. Mignogna, R. Kline: to be submitted to Wave Motion

[8] R.A. Kline, Y.Q. Wang, R.B. Mignogna, P.P. Delsanto: *A Finite Difference Approach to Acoustic Tomography in Anisotropic Media*, Oklahoma Univ. preprint (1993)

[9] A.H. Harker, J.A. Ogilvy: *Coherent Wave Propagation in Inhomogeneous Materials: a Comparison of Theoretical Models*, Ultrasonics, V.29 (1991) p.235

THE WEDGE TEST APPROACH TO MODE I
INTERLAMINAR FRACTURE TOUGHNESS FOR C.F.R.P.

T. SCHJELDERUP, C.G. GUSTAFSON*

Sintef Production Engineering - 7034 Trondheim - Norway
*University of Trondheim - 7034 Trondheim - Norway

ABSTRACT

This paper describes the use of the ASTM D3762-79 Standard Test Method for Adhesive-Bonded Surface Durability of Aluminum on different unidirectional CFRP laminates in different hot environments. The experiments show that the strain energy release rate G_{Ic} decreases rapidly after a very short exposure time, and flattens out after some 8 hours. Relatively tough materials at room temperature behave very differently at moderately elevated temperatures and in aggressive media, and can give poorer fracture properties than less tough materials. The Wedge Test is an inexpensive and quick test to get a qualitative measure of the Mode I fracture toughness of CFRP laminates in different environments.

INTRODUCTION

A lot of work has been carried out to find suitable testing methods for the Mode I interlaminar fracture toughness G_{Ic} of polymer composites. The most common method is with Double Cantilever Beam (DCB) specimens. Practice has shown that this is a relatively time consuming method, and that the results are not very consistent. However, different approaches are being taken to develop a test method which is consistent enough to give reliable data that can be used in design.

The DCB Mode I test is normally done at room temperature and in a standard atmosphere, i.e. not in a very aggressive environment for the composite. This paper describes the use of the ASTM Wedge Test on different CFRP laminates in different hot environments, and to study the fracture growth on a long time basis. This is not an alternative to the above mentioned DCB Mode I test, but a supplementary method to indicate the resistance to fracture growth in different (hot) environments.

1. EXPERIMENTAL

1.1 Materials

The DCB specimens was made from UD filament wound plates of about 3 mm thickness. The materials were Ciba-Geigy LY556/HY917/DY070 (100/90/1 by weight) and CY221/HY2954 (100/35 by weight), and fibers from Toray Industries and Enka AG. The combinations used can be found in table 1.

Table 1. Materials.

Plate no.	Fiber	Epoxy system
3	Torayca T700	LY556/HY917/DY070
4	"	CY221/HY2954
7	Tenax HTA	LY556/HY917/DY070
8	"	CY221/HY2954
9	Torayca T700	LY556/HY917/DY070

1.2 The winding process

The plates were filament wound on a modified lathe, without tension control and without a temperature controlled resin bath. The same procedure was used for each plate, and they were made by the same personnel. The lack of tension and temperature control can lead to some differences in the final properties, but for this study they are assumed to have minor effect.

The LY556 based epoxy system was preheated to 45 °C before winding, the tougher CY221 system was not preheated. After the winding, plates from both systems were pressed at 0,4 MPa (4 bar) and 80 °C for 4 hours. Postcuring for both was at 140 °C for 8 hours.

The Protocol for Interlaminar Fracture Testing of Composites by ESIS was followed as far as possible in making the DCB test specimens. A thin PVDF film was arranged in the midthickness of the plate during the winding process. The geometry is presented in figure 1.

1.3 Test procedure for traditional G_{Ic} testing

The specimens were tested in an Instron universal testing machine and at 0.5 mm/min. We used the corrected beam theory, Berry's method (experimental compliance method), and the modified experimental method, in calculating the strain energy release rate G_{Ic}. All methods are well described in the ESIS protocols. This testing gave us the value of 'n', the slope of the curve in the log a - log C plot in Berry's method (figure 2).

1.4 The Wedge Test procedure (modified)

The specimen is opened carefully, the PVDF film is removed, and the wedge is forced into this 'crack' until a new crack propagates approximately 1 mm. The stressed specimen is exposed to a liquid environment at elevated temperatures, and the crack growth with time is evaluated.

The wedges are made from aluminum with nominal dimensions width 20 mm, thickness 3 mm, and length 60 mm.

A short feasability study was performed to see the usefulness of the method. TenaxHTA/LY556 specimens were used for this study. Two specimens (number 6H and 6K) were painted with white ink on both edges and the wedge was forced into the specimens until the crack propagated. The point at which the wedge was in contact with the specimen, was marked (on both sides), and the distance from these points to the crack tip was measured under a microscope.

Specimen 6H was then immersed in distilled water at 90 °C, and specimen 6K in air at room temperature and approximately 50% RH. The crack growth for both specimens were meassured after 17 and 24 hours. No crack growth was observed on specimen 6K, but the other had a crack propagation of 0.97 mm. It was therefore decided to carry out a larger test series. The experimental setup can be found in table 2.

Table 2. Experimental setup.

Environment	Material - specimen(s)
50 °C, acetone	T700/LY556 - 3E T700/CY221 - 4G
50 °C, 5% salt solution	T700/LY556 - 3J T700/CY221 - 4H
50 °C, water based hydraulic oil	T700/LY556 - 3L T700/CY221 - 4J
50 °C, distilled water	T700/LY556 - 3K
50 °C, air	T700/CY221 - 4K
90 °C, distilled water	T700/LY556 - 3I, 9D, 9E, 9F T700/CY221 - 4E TenaxHTA/LY556 - 7D TenaxHTA/CY221 - 8C
90 °C, air	T700/CY221 - 4B

The crack growth was measured after 1, 4, 8, 24, 168, and 1000 hours.

2. THEORY

We wanted to see the effect of different environments on the Mode I interlaminar fracture toughness of CFRP laminates. The experimental compliance method, also known as Berry's method, gives the strain energy release rate G_{Ic} by first plotting the compliance C vs crack length a on a log-log plot. The slope of the curve in this plot, n, can then be used to give G_{Ic} as follows:

$$G_{Ic} = \frac{nP\delta}{2Ba} \tag{1}$$

where n is a constant (the slope), P is the force, δ is total crack opening, B is the specimen width, and a is crack length.

The force P is an unknown parameter in our experiment, but can be calculated from the formula for deflection of a cantilever beam with length L, flexural modulus E, and moment of inertia I:

$$f = \frac{PL^3}{3EI} \tag{2}$$

By rearranging equation (2), the force P can be expressed:

$$P = \frac{3EIf}{L^3} \tag{3}$$

The moment of inertia I for a rectangular cross section with width b and hight h is:

$$I = \frac{bh^3}{12} \tag{4}$$

Putting (4) into (3), gives P:

$$P = \frac{Ebh^3f}{4L^3} \tag{5}$$

Since B = b, L = a og δ = 2f, the Mode I fracture toughness G_{Ic} can be expressed:

$$G_{Ic} = \frac{Eh^3n\delta^2}{16a^4} \tag{6}$$

3. RESULTS

The specimens for the wedge test were taken from the same filament wound plates as used in earlier, traditional Mode I tests. From these tests, the slope n and modulus E were known. Putting these values into equation (6), the G_{Ic} could be calculated for each specimen at each time interval, see figures 3 and 4.

3.1 Results from 50°C test

The specimens immersed in acetone, cracked immediately some mm (visual observation). None of the other environments led to such observations at the start of the experiments.

Acetone had the biggest effect on the specimens, as expected. The crack growth, and thereby the reduction in G_{Ic} was dramatic. Crack growth of up to 38% from original length was observed on the CY221 system after just 1 hour. This leads to a reduction in G_{Ic} down to only 27% of original value. The LY556 system had a reduction in G_{Ic} down to 50% of original value. This means that an epoxy system which is considered as a tough epoxy at room temperature and air, totally collapses in more aggressive environments.

Both air and distilled water had a minor effect at this temperature. Salt solution and hydraulic oil had a moderate effect on the LY556 system, but a significant effect on the 'tougher' CY221 system (figure 3).

3.2 Results from 90 °C test

At this temperature, the specimens were only tested in distilled water. And again, as figure 4 shows, the 'tough' CY221 system totally loses its good Mode I fracture properties. After only minutes, this system shows much lower toughness than the LY556 system.

4. DISCUSSION

The bending modulus is a fiber dominated property. The validity of Equation (6) with a constant E and n is therefore assumed to hold for the different environments. Only when the shear modulus of the matrix is decreased by passing above the T_g, will E and n in equation (6) be changed. In case of matrix softening, E and n will decrease, and the calculated G_{Ic} value is somewhat high. The T_g for the LY556 system is 110-120 °C, and for the CY221 system somewhat lower than 90 °C. Therefore the G_{Ic} values for 90 °C and the CY221 system might even be lower than calculated.

5. CONCLUSIONS

The ASTM Wedge Test is, slightly modified, well suited for environmental testing of interlaminar fracture properties in Mode I for CFRP laminates. It is very simple to use in order to rank materials, and gives a clear indication if a matrix system is unsuitable for use in a specific environment.

The testing can be stopped after only one day as the G_{Ic} curves flatten out early.

Tough systems in a standard atmosphere can exhibit very poor properties in liquid and hot environments.

ACKNOWLEDGEMENTS

The authors are indebted to Raufoss A/S for financial support.

REFERENCES

1. European Structural Integrity Society Polymers & Composites Task Group: Protocols for Interlaminar Fracture Testing of Composites.

2.   P. Davies: Development of a standard for interlaminar fracture testing of composites. Proceedings of 1st International Conference on Deformation and Fracture of Composites at UMIST, Manchester, UK 25-27 March 1991.

3.   ASTM D3762-79: Standard Test Method for Adhesive-Bonded Surface Durability of Aluminum (Wedge Test).

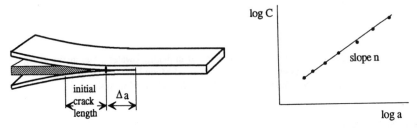

Fig 1 Specimen with inserted wedge.   Fig 2 Berry's method.

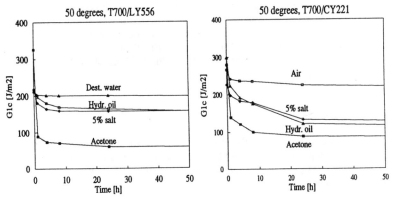

Fig 3 Results from testing in different environments at 50 degrees Celsius.

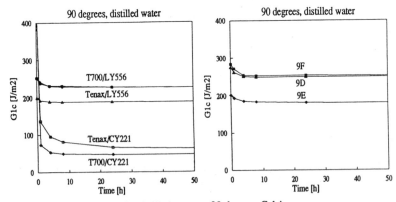

Fig 4 Results from testing in distilled water at 90 degrees Celsius.

468

APPLICATIONS

CFRP : A SOLUTION FOR THE DESIGN AND THE CONTROL OF FLEXIBLE ROBOTIC MANIPULATORS

B. BLUTEAU, M. NOUILLANT, A. OUSTALOUP

Equipe Crone - LAP-ENSERB - Université Bordeaux I - 351 Cours de la Libération - 33405 Talence - France

ABSTRACT

This paper deals with the identification and the control of flexible robotic arms with one degree of freedom. A comparison is made between the dynamic performances of four robot arms, one of which is made of aluminum, the others of composite. In order to test the robustness performances of the control law, several pay-loads are taken into account. The control robustness performances are evaluated and compared by using two controllers, a classic PD controller and a CRONE controller, through step responses of the system. The best performances was obtained with the CRONE control and the composite arm, so giving evidence of the coupling interest between design and control in advanced robotic applications.

INTRODUCTION

The achievement of high dynamic performances in modern robotic applications has triggered a vigorous research thrust in various multi-disciplinary areas such as design and control. Some solutions based on light weight, flexible robotic arms have been investigated these past few years. It is well known now that the best materials for optimizing the stiffness-weight ratio are the composite materials. Moreover, these materials exhibit larger structural damping than usual metallic materials ; so the vibration amplitude can be substantially reduced. The increased damping due to passive effect (i.e. structural damping) is not sufficient by itself to eliminate structural deflection. Therefore, the design of control algorithms taking into account the flexible dynamics is needed. In order to determine experimentally the natural frequency and the damping factor of each mode for the aluminum and composite arms, a sinusoidal input torque was applied to the hub. Finally, the operational identification of the co-located transfer function obtained is carried out through a method based on the Viete's functions [1, 2]. The existing control strategies such as optimal control [3, 4] or output feedback control

[5] can reveal inadequate for they do not take into account the reparametration of the plant and the uncertainties of the model. For the various arms, the control robustness performances are evaluated and compared by using two controllers, a classic PD controller and a CRONE controller, through step responses of the system under different pay-loads. CRONE is a French abbreviation of "Commande Robuste d'Ordre Non Entier" [6]. This control technique is based on the non integer derivative operator. The robustness performances of the CRONE control have often been verified in the case of linear plants, or non linear plants with static non-linearities [7] and with high non-linearities. Although flexible modes are situated at relatively high frequencies, the control dynamic rapidity has been chosen so that the regulator synthesis takes into account the effect of these modes. If the choice of such a control strategy leads to relatively high solicitations of the driver, it is shown in the experiments that the composite arm has a smaller input torque and a better settling-time than the arm made of aluminum. Moreover, with the CRONE control we obtain the robustness of stability degree which is expressed by a constant first overshoot of the step response whatever the pay-load in a specified range.

1 - EXPERIMENTAL APPARATUS

The experimental flexible manipulator testbed under study (figure 1) consists of a tubular beam. It is about 1.2 mm thick and 2 m long with an inner diameter of 20 mm. Table 1 gives the caracteristics of the four materials used for comparison. The experimental arm is mounted directly on the vertical drive shaft of the DC servo-motor and its extremity is on a cushion of air so as to avoid friction at the free end. The aluminum arm used for comparison have the same length and the same flexural rigidity as the composite arms.

The DC servo-motor featuring a permanent magnetic field, model MC17H, manufactured by AXEM is activated by a current generator, model A648, manufactured by PRODERA, whose transfer factor is 4 A/V in a 7 kHz bandwidth.

The input variable of the plant is the tension u(t) of the current generator. The ouput variable is the angular position $\theta_m(t)$ of the DC servo-motor. The angular position $\theta_m(t)$ is measured by a circular potentiometer whose cursor is coupled with the drive shaft of the actuator. These variables determine a transfer function of the plant such as :

$$G(s) = \frac{\theta_m(s)}{U(s)} .$$

(1)

2 - IDENTIFICATION

An operational identification of the so defined transfer function based on frequency analysis has been performed by means of a method using the Viete's functions. For each harmonic test, there are 70 measure frequencies distributed unevenly. The quicker the phase variations, the higher the local density is. Full details of this technique may be found in [1] and [2]. By taking into account only two flexible modes, the transfer function obtained is the following one :

$$G(s) = \frac{g}{s^2} \prod_{i=1}^{2} \frac{\omega_i'^2 + 2\zeta_i' \omega_i' s + s^2}{\omega_i^2 + 2\zeta_i \omega_i s + s^2} .$$

(2)

Table 2 gives the resonance frequencies, the anti-resonance frequencies and in the same way their damping factors. Figures 2 give the Black loci of $G(j\omega)$ for the various materials. $G(j\omega)$ presents an alternate set of resonances and of anti-resonances, the argument of $G(j\omega)$ varying between 0 and -180°.

3 - CONTROL DESIGN

3.1 - Study conditions

In order to fix a frequency ω_u which should be (for example) superior to the first two local resonance frequencies of the plant, it is convenient to provide a gain which localizes the open loop Black locus spires around the axis OdB. This frequency configuration then leads to a set of unit gain frequencies, the lowest one of which determines the rapidity of the closed loop dynamics. This amounts to saying that (contrary to the general case in control) the initially chosen frequency ω_u is not indicative of the rapidity of the control dynamics.

In order to ensure convenient resonance factors, the observation of figure 2 shows that a solution consists in shifting the spires to the right. Such a shifting can be obtained through a phase advance and particularly through a constant phase advance in order to insure the robustness of the control versus the variation of the payload at the free end. A constant phase CRONE controller can therefore meet this requirement.

In order to point up the interest of such a controller, the performance tests also turn on a proportional-differential controller.

So, a PD controller and a constant phase CRONE controller successively provide a same phase advance of 50° at the same frequency ω_u fixed at 10 rd/s. By insuring the same phase margin at the same frequency ω_u, such study conditions ensure the same dynamics to the control. This makes it possible to carry out a comparative study which is indicative of the robustness and sensitivity performances obtained respectively with the PD controller and the CRONE controller.

For a given frequency ω_u, the input sensitivity being exclusively linked with the controller gain beyond ω_u, the frequency placements of the controllers are such as the placement of the PD is symmetrical in relation with ω_u whereas that of the CRONE is asymmetrical (figure 3).

3.2 - Controllers

The PD controller admits a transmittance of the form :

$$C(s) = C_o \frac{1 + s/\omega'_1}{1 + s/\omega_1} \, ,$$

(3)

with : $C_o = 32,38$;
$\omega'_1 = 3,640$ rd/s and $\omega_1 = 27,475$ rd/s.

The ideal version of the CRONE controller is defined by the order m' transmittance :

$$C_{m'} = C_o \left(\frac{1 + s/\omega_b}{1 + s/\omega_h} \right)^{m'} \, ,$$

(4)

in which : $m' = 0,611$; $C_o = 3,924$;
$\omega_b = 0,05$ rd/s and $\omega_h = 50$ rd/s.

(5)

The corresponding real version is defined by the order N transmittance :

$$C_N (s) = C_o \prod_{i=1}^{N} \frac{1 + s/\omega'_i}{1 + s/\omega'_i} \, ,$$

(6)

in which : $N = 4$; $C_o = 3,924$;

$$\alpha = \omega_i/\omega'_i = 2,872 \; ; \; \eta = \omega'_{i+1}/\omega_i = 1,957 \; ;$$

(7)

$$\begin{array}{ll}
\omega'_1 = 0,0699 \text{ rd/s} & \omega_1 = 0,2009 \text{ rd/s} \\
\omega'_2 = 0,3934 \text{ rd/s} & \omega_2 = 1,1300 \text{ rd/s} \\
\omega'_3 = 2,2123 \text{ rd/s} & \omega_3 = 6,3548 \text{ rd/s} \\
\omega'_4 = 12,440 \text{ rd/s} & \omega_4 = 35,736 \text{ rd/s} \; .
\end{array}$$

(8)

4.4 - Performances

4.1 - Comparison for the four materials with the PD controller

Figures 4 and 5 illustrate respectively the frequency performances in open loop and the times performances. Figure 4 presents the Black loci of the open loop frequency response, $\beta(j\omega)$, such as :

$$\beta(j\omega) = C(j\omega)\, G(j\omega) \text{ for the PD} \qquad (9)$$
and $\qquad \beta(j\omega) = C_N(j\omega)\, G(j\omega) \text{ for the CRONE.} \qquad (10)$

Figure 5 gives the step responses of the control to a step of amplitude 0,05 rd, such an amplitude causing the free end to be shifted tangentialy by 10 cm at the final time.

These Black loci reveal that the last spires are not practically shifted to the right. It is true that it is not necessary to shift them, even if the resonance factors that they define are greater than the others. Given that the spectrum of a step prompting varies as $1/\omega$, it is indeed the first resonance factor which assesses stability degree, as illustrated by the step responses in figure 5. The highest the structural damping capacity the quicker the superfluous oscillations decrease and the best performances are obtain for the carbon-epoxy (1) composite arm.

4.2 - Robustness performances

In order to test robustness performances, two different payloads have been employed. The first, $m_1 = 160g$, is the mass of the air cushion device ; the second, $m_2 = 2.3$ kg is obtained by an additional mass of 2.14 kg. By taking into account the inertia of the link and the inertia of the rotor, the total inertia I on the motor axis is $I_1 = 0.95$ kgm^2 for the mass m_1 and $I_2 = 9.5$ kgm^2 for the mass m_2. This allowed us to have a variation by a factor of 10 of the inertia between the two configurations.

Figures 6 illustrate the comparative time performances in closed loop for the best composite arm with the PD and the CRONE controller.

Through the relative constance of the first overshoot (despite an inertia variation by a factor of 10), the responses in figure 6.b indeed express the robustness of the stability degree of the control in the case of the CRONE controller.

Figure 7 presents the corresponding time variations of the input of the plant, the physical magnitude of which is a torque expressed in Newton meters, while time is expressed in seconds. The curves clearly reveal that PD and CRONE controllers ensure the same input immunity.

CONCLUSION

The experimental results clearly indicate that the composite arm (carbon-epoxy (1)) exhibit performances superior to that obtained with the aluminum arm and the others composite arms. The best performances was obtained with the CRONE control and the composite arm, so giving evidence of the coupling interest between design and control in advanced robotic applications.

REFERENCES

[1] Oustaloup, A., Lanusse, P. and Elyagoubi, A., "Synthesis of a Wide Band Template Based on the VIETE's functions", *Session "The CRONE Control" - IMACS Symposium on "Modelling and Control of Technological Systems"*, Lille, France, May 1990.
[2] Oustaloup, A., Lanusse, P., "Operational Identification based on the VIETE's functions", *ACRO workshop, Automatic research group, Non Linear Pole, CNRS*, Bordeaux, France, April 1990.

[3] Cannon, R.H., Schmits, E., "Initial Experiments on the Endpoint Control of Flexible One-link Robot," *The International Journal of Robotics Research*, Vol. 3, 1984, pp. 61-75.

[4] Hastings, G. and Book, W.J., "Experiments in the Optimal Control of a Flexible Manipulator", *Proc. of American Control Conference*, pp. 728-729, Boston, 1985.

[5] Wang, W.J., Lu, S.S., and Hsu, C.F., "Output Feedback Control of a Flexible Robot Arm," *Proceedings of 25th CDC*, Athens, Greece, Dec. 1986, pp. 91-95.

[6] Oustaloup, A., "The CRONE Control", European Control Conference, ECC-91-CNRS-Grenoble, July - 2 - 5,1991.

[7] Oustaloup, A., "CRONE Control, Principle, Synthesis, Performances with Non-linearities', *9th Conference " Analysis and Optimization of Systems"*, INRIA, Antibes, France, June 1990.

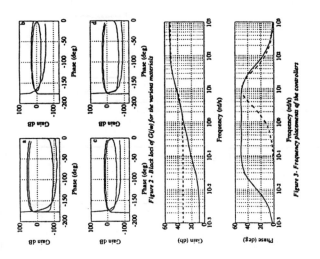

Figure 2 - Black loci of G(jω) for the various materials

Figure 3- Frequency placements of the controllers

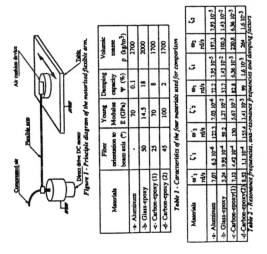

Figure 1 - Principle diagram of the motorized flexible arm.

Table 1 - Characteristics of the four materials used for comparison

Materials	Fiber orientation to beam axis (°)	Young Modulus E (GPa)	Damping capacity Ψ (%)	Volumic masse ρ (kg/m³)
-+- Aluminium	-	70	0.1	2700
-b- Glass-epoxy	50	14.5	18	2000
-◊- Carbon-epoxy (1)	25	70	8	1700
-◊- Carbon-epoxy (2)	45	100	2	1700

Table 2 - Resonance frequencies, anti-resonance frequencies and damping factors

Materials	ω'₁ rd/s	ζ'₁	ω'₂ rd/s	ζ'₂	ω₁ rd/s	ζ₁	ω₂ rd/s	ζ₂
-+- Aluminium	7.07	6.5 10⁻⁴	127.1	7.03 10⁻⁴	7.01	7.85 10⁻⁵	197.1	7.95 10⁻³
-b- Glass-epoxy	3.24	9.95 10⁴	59.2	1.27 10²	32.7	1.43 10²	100.5	1.43 10²
-c- Carbon-epoxy(1)	7.12	4.42 10⁴	130	5.67 10²	12.8	6.36 10²	220.8	6.36 10³
-d- Carbon-epoxy(2)	1.92	1.1 10³	155.4	1.41 10³	99	1.6 10³	264	1.6 10³

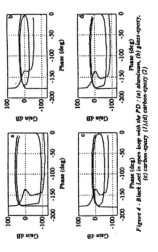

Figure 4 - Black Loci in open loop with the PD : (a) aluminum, (b) glass-epoxy, (c) carbon-epoxy (1),(d) carbon-epoxy (2)

475

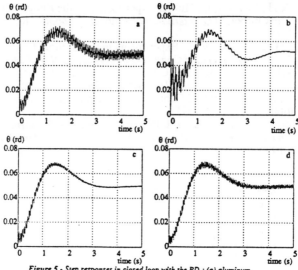

Figure 5 - Step responses in closed loop with the PD : (a) aluminum, (b) glass-epoxy, (c) carbon-epoxy (1),(d) carbon-epoxy (2)

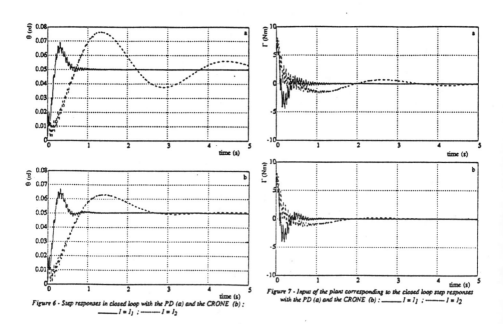

Figure 6 - Step responses in closed loop with the PD (a) and the CRONE (b) : ———— $l = l_1$; ---------- $l = l_2$

Figure 7 - Input of the plant corresponding to the closed loop step responses with the PD (a) and the CRONE (b) : ———— $l = l_1$; ---------- $l = l_2$

THE INFLUENCE OF PROCESSING CONDITIONS ON THE ADHESION STRENGTH OF CARBON FIBRE COMPOSITES

C. BAILLIE, H. VIRAY

University of Sydney - Dept Mechanical Engineering
2006 Sydney - Australia

ABSTRACT

The processing and the testing conditions of composites strongly influences the mechanism and efficiency of stress transfer across the interface. In this paper the influence of processing is tested using the fibre fragmentation test as an indicator of changes to the interface in two carbon fibre/polymer systems after various processing conditions. Carbon in epoxy resin is tested after varying curing time and temperature and carbon in PMMA is tested after varying moulding temperature, pressure and cooling rate. In both cases where chemical bonding is unlikely to occur, the most significant changes are those affecting the resin shrinkage onto the fibre. A greater shrinkage will induce a more efficient stress transfer. The usefulness of this information on consideration of bulk composites is then discussed.

INTRODUCTION

It is now well established that the transfer of stress across the interface in composite materials to a large extent determines the strength and toughness of the bulk composite. In order to improve composite properties many studies have been undertaken /1/ to endeavour to understand the nature of the stress transfer and develop the means to alter interfacial adhesion. There are two ways of approaching the problem, firstly by using bulk composite tests of the interfacial reaction or secondly by using model composite tests often employing single fibre techniques /1/. The former gives an answer which would appear to be more useful as it predicts large scale behaviour in service, however, often it is difficult to interpret the results and many assumptions are required as to the actual mechanisms acting. The single fibre techniques are more satisfactory in that it is easier to directly measure changes at the level of the interface but complicated analytical models are necessary in order to determine materials parameters from the data. There is also a problem concerning conditions of processing or testing which do not mimic bulk manufacturing or service conditions. Examples of this might be the influence of processing pressure and temperature, cooling rate, fibre tension, test temperature and humidity, rate and method of loading, presence of neighbouring fibres etc. Various attempts have been made to investigate some of these influences on bulk composites /2,3,4/ and to a certain extent on the interface /5,6/ but no comprehensive investigation has been made.

It is the intention of the authors to study each of the above phenomena using single fibre techniques. In the present paper the influence of processing conditions on the results of fragmentation tests of a carbon/epoxy system is investigated. Additionally results from a carbon/PMMA (polymethylmethacrylate) system will be presented. For the latter system, a thermoplastic composite, there is usually little chemical reaction at the interface. Also the viscoelasticity of the material will influence stress transfer. This is an area which has been to a large extent neglected.

1.EXPERIMENTAL

1.1.Materials

The fibre selected for the work was PAN based high strength carbon fibre tested untreated and treated by an electrolytic oxidation treatment at $25Cm^{-2}$.
DGEBA epoxy resin with a tetrafunctional amine hardener of 100/12 ratio was the basic thermosetting resin system used with the following variations in processing route

E1- cure 3 hours 60 C, post cure 2 hours 120 C
E2- cure 1 hour 60 C, post cure 2 hours 120 C
E3- cure 3 hours 60 C, post cure 2 hours 155 C
E4- cure 3 hours 60 C, post cure 1 hour 120 C

PMMA was used as the basic amorphous thermoplastic material with the following variations in processing route

P1- 2.64MPa, 230 C, 35 C/min
P2- mould pressure 2.64MPa, mould temperature 230 C, cooling rate 65 C/min
P3- 1.76MPa, 230 C, 30 C/min
P4- 2.64MPa, 170 C, 30 C/min
These variations were only tested with $25Cm^{-2}$ treated fibres.

Manufacturing methods have been well established for the epoxy system /7/ but had to be developed for the PMMA system as there were problems of excessive shrinkage of the polymer on cooling which caused buckling of the fibre. Therefore a

478

method of keeping the fibres taught in the resin whilst being heated was developed using a special frame to which the fibres could be cemented before placing them in the mould.

1.2.Testing

Fragmentation tests were performed using a Polymer labs Minimat microtensometer on a microscope stage at 0.5mm/min for the epoxy system and 0.01mm/min for the PMMA system. These rates were principally chosen for experimental reasons and will be investigated further in a subsequent publication. The breaks in the epoxy sample were identified with the aid of polarised light as the material is birefringent. The sample could be scanned for breaks and then straining continued. The PMMA proved more difficult as it is not birefringent and behaves viscoelastically. Only with a magnification of 1600X could any failures be seen. At this magnification many other features could also be observed including debonding. This will have an important influence on the use of analytical models to process data as described in reference /8/. However, as this meant that a large amount of time was required to scan the fibre length for breaks, the PMMA suffered from stress relaxation during this time. A method was developed whereby the breaks could be identified from a read out of the stress at all times on the PMMA sample throughout the test. A load drop was noted each time a fragmentation event occurred. Other features, for example crazing and void formation could also be monitored in this way by associated load drops. Details of the method used to obtain data from the PMMA tests is described in a separate publication /9/.

2.RESULTS

The results for this work are summarised in Table 1. The processing variation is given along with the median saturation fragment length after fragmentation tests have been carried out. Fragment lengths are used in this paper to represent changes to the interface and no attempt is made to calculate materials properties from the raw data as reliable analytical models are only in their infancy /8/. Only results for untreated carbon fibres in epoxy resin are shown as the treated fibre tests showed no change with varied cure conditions. By comparing E1 with E2,E3 and E4 in turn it is evident that for the epoxy system the only resin processing variation which has an influence is that of increasing the post cure temperature with an untreated fibre. It shows a reduction in saturation fragment length, an indicator of effective stress transfer, with increased post cure temperature. Figure 1 gives a cumulative frequency plot of saturation fragmentation lengths for untreated carbon fibre in E1 and E3 showing the influence of the post cure temperature.

If P1 is compared with P2,P3 and P4 in turn there appears to be several processing conditions influencing the stress transfer in the PMMA/treated carbon fibre system. The first of these is cooling rate, at least down to 35 C/min, where a higher cooling rate increases the saturation fragment length or reduces the stress transfer. For the following two comparisons of mould pressure and temperature there were slight differences in the cooling rates but subsequent tests have shown that cooling at different rates below 35 C/min do not alter the results significantly. A larger moulding pressure appears to reduce the saturation fragment length or improve stress transfer and finally a higher moulding temperature reduces the fragment lengths and increases the stress transfer efficiency. Figure 2 gives a cumulative frequency plot of saturated fragmentation lengths for treated carbon fibre in P1 and P2 showing the influence of cooling rate.

3.DISCUSSION

It is evident that processing conditions influence the stress transfer capability very strongly and it is important to understand why in order to control the variations and produce the optimum bulk properties. Firstly the epoxy resin post cure temperature only affects the untreated fibre and no other conditions have any influence. The post cure seems to have more influence than the cure temperature and the temperature is more dominant than time. To investigate this further the coefficient of thermal expansion of the materials were measured using a dilatometer /1/ and the resin which was postcured at 155 C/min had a much higher value. Cooling from a higher temperature as well as a higher coefficient of thermal expansion will cause the shrinkage stress to be greater. This may have affected the untreated fibre and not the treated fibre because in the latter case the interface is dominated by chemical bonds /10/.

The temperature, pressure and cooling rate affect the stress transfer in PMMA/carbon. The higher cooling rate may reduce shrinkage of the polymer onto the fibre. Although the fibres were treated, the interface will depend only on mechanical and physical bonds due to the nature of the PMMA. In other words the shrinkage of the resin onto the fibre would not be masked by chemical bonds as is possible in the above case. Looking at figure 2 the fact that changed cooling rate alters the values of fragment lengths is very apparent except that there is not only a shift along the x-axis but a different dispersion of values as well. No attempt is made to explain this behaviour until further work has been carried out on the viscoelastic behaviour of the PMMA.

The higher temperature of moulding would certainly increase the shrinkage onto the fibre as would higher pressure and both of these increase the stress transfer efficiency via improved mechanical bonding. Also physical bonding may be improved due to better wettability at higher pressures as well as lower viscosity at higher temperatures. Again further work is necessary before definite conclusions can be drawn. Also with a single fibre composite system the influence of increased resin shrinkage is to increase the interfacial strength whereas in a bulk composite often the reverse occurs with tensile forces being induced at the interface. In a real composite there will be a combination of both types of microstress and the situation is very complex. However, it is necessary to first study the direct influence of changing conditions on the basic properties before complex models generating realistic predictions can be made from any results.

4.CONCLUSIONS

The results of this paper have confirmed the influence of changing selected processing conditions on the fibre/matrix interface which would have an important effect on the bulk properties of the composite material. Both a thermosetting and a thermoplastic resin system, epoxy and PMMA were selected and fragmentation tests carried out after manufacturing using different temperatures, pressures or cooling rates. It was discovered that only untreated fibre/epoxy resin interfaces and not treated fibre/epoxy resin interfaces were strongly influenced by temperature during post cure. It was considered that this must be due to shrinkage pressure increasing the mechanical adhesion which would be masked by the chemical component in the case of treated fibres. For the PMMA the treated fibres were affected by temperature, pressure and cooling rate. Those conditions which appeared to be conducive to greater shrinkage stress thus increasing mechanical bonding or wetting thus increasing physical bonding improved the stress transfer efficiency.

5.REFERENCES

1.C.A.Baillie, to appear proc "ICCM 9", Madrid, (1993)
2.N.K.Sung, T.J.Jones, N.P.Suh, J.Mat.Sci. 12 (1977), 239-250
3.I.Low, Y.W.Mai, S.Bandyopadhayay, Comp. Sci. Tech. 43 (1992), 3-12
4.B.Bouette, C.Cazeneuve, C.Oytana, Comp. Sci. Tech. 45 (1992) 313-321
5.H.D.Wagner, H.E.Gallis, H.Wiesel private communication
6.A.Turgut, E.Sancaktar, 14th An Mtg.Adhesion Soc., Clearwater, Florida(H.M.Clearfield, Ed) (1991), 24
7.L.M.Zhou, C Baillie, Y.W.Mai, submitted to Proc Roy Soc
8.C.A.Baillie and M.G.Bader to appear J.Mat Sci, 1993
9.C.A.Baillie and H.B.Viray submitted for publication, (1993)
10.C.A.Baillie, J.F.Watts, J.E.Castle, J.Mat.Chem 2(9) (1992), 939-944

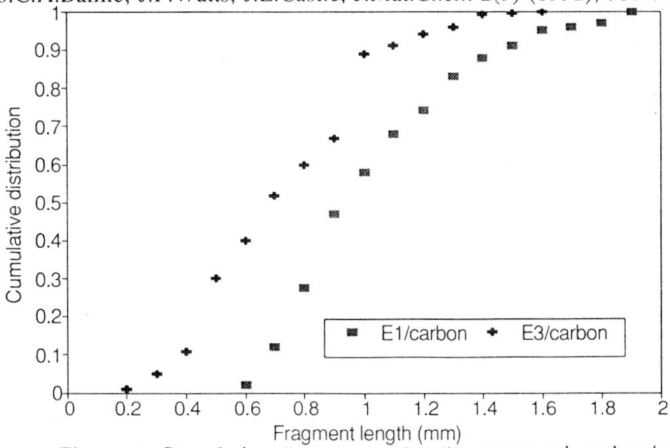

Figure 1 Cumulative frequency plot for untreated carbon/epoxy of different post cure temperatures

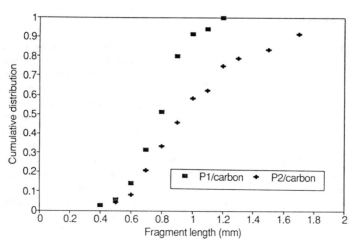

Figure 2 Cumulative frequency plot for treated carbon/ PMMA samples of different cooling rates

Polymer system	Processing conditions			Interface tests
Epoxy/amine	Cure time (hours)	Post cure time (hours)	Post cure temperature (C)	Median fragmentation length (mm)
E1	3	2	120	0.9
E2	1	2	120	0.9
E3	3	2	155	0.7
E4	3	1	120	0.9
PMMA	Pressure (MPa)	Temperature (C)	Cooling rate (C/min)	Median fragmentation length (mm)
P1	2.64	230	35	0.8
P2	2.64	230	65	1.0
P3	1.76	230	30	1.3
P4	2.64	170	30	1.7

Table 1

THERMOPLASTIC COMPOSITES REINFORCED WITH WOOD FIBRES

F. CARRASCO, P. PAGES*, J. SAURINA**, J. ARNAU**

Dept of Chemical Engineering - Universitat de Girona - Spain
*Dept of Material Science
Universitat Politecnica de Catalunya - Spain
**Dept of Industrial Engineering - Universitat de Girona - Spain

ABSTRACT

Mechanical properties (tensile strength, tensile modulus and impact energy) of high density polyethylene (HDPE)-wood fiber (aspen) composites have been studied in this work. The effect of coupling agents (Epolene C-18 and Silane A-174) on composite properties were discussed by taking into consideration the hydrolytic stability at the fiber-matrix interface. The results showed a good retention in tensile strength and impact energy with the use of silane coupling agent as compared to maleated ethylene and control (no coupling agent). Tensile modulus was less affected by the fiber-matrix interface than did tensile strength and impact energy.

0. INTRODUCTION

Wood fibers offer many advantages such as lower cost, desirable aspect ratio, low density, and they are a renewable resource. However, there are limitations on their use in high volume thermoplastics: difficulty in compounding in thermoplastic matrices (poor dispersion), weak adhesion between the hydrophilic fiber and the hydrophobic polymer, thermal degradation at higher processing temperature and poor moisture resistance /1/. Most of the above mentioned problems can be overcome to a large extent by the modification of fiber surface with suitable additives or coupling agents. The nature of interface determines the degree of interaction between the filler and polymer. The ultimate properties of the composite are greatly influenced by the extent to which the stress can be transferred form the matrix to the fiber, at the interface. With proper selection of additive/coupling agent, the fiber-matrix adhesion and hence the mechanical performance of short fiber reinforced thermoplastics can be improved /2-6/. In our study, we examined the effect of low temperatures (from -28 to 7°C) and moisture on mechanical properties of high density polyethylene-wood fiber (aspen) composites (pretreated or not with coupling agents).

1. EXPERIMENTAL

1.1. Materials
High density polyethylene (HDPE 2909, DuPont Canada) (melt index = 1.35 g/min and density = 0.960 g/cm^3) and wood fiber (aspen, L/D=8.7) were used for the fabrication of composites. Composites were made with 40% by weight of fiber. Two coupling agents with different organo-functional groups were used to improve the hydrolytic stability at the fiber-matrix interface: maleated ethylene (Epolene C-18, Eastman Kodak) and γ-methacryloxypropyltrimethoxy silane (Silane A-174, Union Carbide).

1.2. Fabrication of composites and mechanical properties evaluation
Wood fibers were pretreated with a coupling agent in a roll mill (C.W. Brabender No. 065) before compounding with the polymer. The following procedure was used for the fiber pretreatment with epolene: wood fibers, after dried in an oven at 55°C for 14 h, Epolene C-18 (2% by weight of fiber) and a small amount of polymer (HDPE 2909, 5% by weight of fiber) were tumble mixed at room temperature and added to the heated rolls (160°C). After mixing thoroughly for 10 min, the fibers were ground to mesh size 60. The mixing of polymer and pretreated fiber was done at 160°C in a C.W. Brabender roll mill. Samples containing 40% by weight of fiber in the composites were prepared. The polymer-fiber mixture was ground to mesh size 20 prior to molding. The procedure for pretreatment of wood fiber with Silane A-174 was the following: To a solution of 150 g of carbon tetrachloride were added 20 g of oven-dried fibers, 0.2 g of an initiator (benzoyl chloride) and 0.4 g of Silane A-174. The mixture was reflux-heated for 3 h. After 30 min cooling, solvent was evaporated and the fibers were oven-dried at 55°C for 24 h. Dog-bone shaped tensile specimens (ASTM D-638, type V) were obtained by compression molding in a Carver Laboratory Press. Samples were tested according to the ASTM D-638 procedure. The test results were calculated by a HP86B computing system using Instron 2412005 General Tensile Test Program. Izod-impact tests (un-notched) were performed in a TMI 43-01 Impact Tester.

2. RESULTS AND DISCUSSION

2.1. Moisture absorption (see Figure 1)
During the initial period of exposure, the rate of water absorption is high and tends towards an assymptotic value at long exposure time. The unfilled HDPE sample was inert to water due to its hydrophobic nature. However, all the composites absorbed different amounts of water depending on the type of pretreatment applied. The control

sample (no coupling agent) absorbed the highest amount of water (1.8%) after 90 days of exposure at out-door conditions. The use of coupling agents resulted in a reduction in water absorption: epolene-pretreated samples absorbed 1.2% of water and silane-pretreated samples, 0.7%. The results show that Silane A-174 is a good pretreatment agent leading to a reduction in moisture absorption and thus, to an improved hydrolytic stability of composites.

2.2. Tensile strength (see Figure 2)

Silane A-174 treated fiber composites showed higher tensile strength than that of unfilled HDPE. Epolene C-18 treated fiber composites produced slightly better tensile strength than control, but the strength values were inferior than unfilled and Silane A-174 treated fiber composites. This finding is very encouraging because it means that it is possible to produce composite materials at lower cost and presenting higher mechanical properties by using the appropriate coupling agent. Earlier studies have shown that tensile strength of short fiber reinforced polymer depends strongly on the polymer-fiber bonding at the interface /7,8/. Therefore, a key indicator of the performance of the coupling agent is its ability to provide good bonding at the interface. In our study, Silane A-174 treated fiber composites presented a higher tensile strength and a good durability after exposure to severe environmental conditions. This can be attributed to the improved hydrolytic stability due to the formation of oxane linkages (Si-O-fiber) at the fiber-matrix interface. In the case of Epolene C-18, the hydrophilic nature of the carboxylic acid group (C-O-fiber) is particularly sensitive to moisture absorption.

2.3. Tensile modulus (see Figure 3)

There was little difference between the performance of Silane A-174 and Epolene C-18 coupling agents, in terms of tensile modulus. This finding confirms the well known fact that modulus is a bulk property which depends primarily on the geometry and proportion of fibers in the composite. In the case of fiber reinforced polymers, tensile modulus is relatively less sensitive to the degree of polymer-fiber interaction /2,7/. Our experimental data also confirm that tensile modulus is less influenced by the type of coupling agent (i.e. fiber-matrix interaction) than did tensile strength. The results suggest that the decrease in tensile modulus of the composites is due to a higher water absorption by the fiber.

2.4. Impact energy (see Figure 4)

As expected, the HDPE-wood fiber composites showed poor impact energy when compared to unfilled polymer. However, it shoud be noted that the unfilled HDPE itself showed a drastic drop in impactenergy with the increase in out-door exposure, whereas only a slight decrease in impact energy was observed HDPE-wood fiber composites. After 90 days exposure time, the impact energy of degraded HDPE is similar to that of silane-pretreated material. Generally speaking, strong linkages between the fiber and

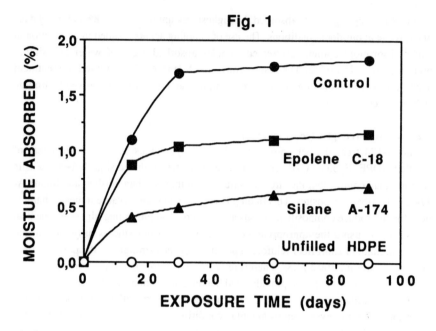

Fig. 1

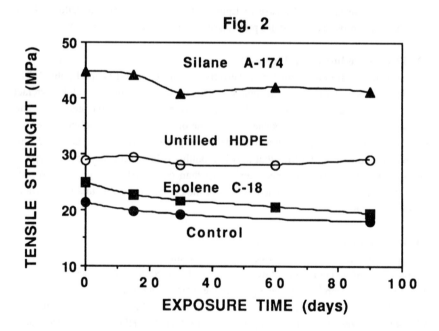

Fig. 2

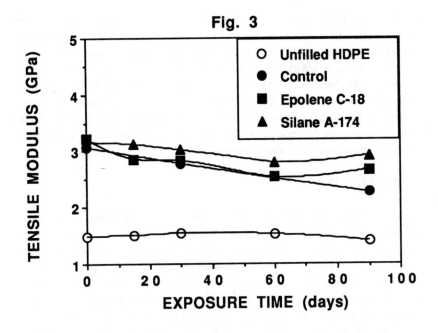

Fig. 3

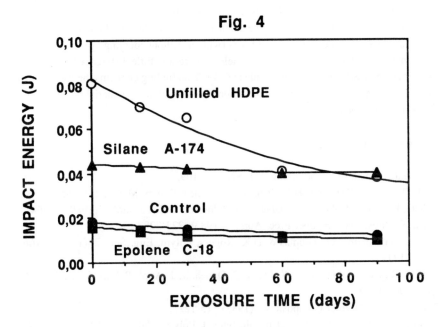

Fig. 4

matrix produces poor impact strength. This is because a crack can propagate from the matrix through a fiber and on the matrix again if the interface between the fiber and matrix resists separation. While in the case of weak bonding between the fiber and matrix, the fiber easily separates from the matrix, thus diverting the crack and absorbing the energy. The decrease in impact energy of the composite after low-temperature exposure is dependent on the reduction in the crosslinking of the polymer, with water as plasticizer, as a result of chain scission in the polymer backbone (note that the reduction for the silane-pretreated composite is less than that of epolene-pretreated or no coupled composites, because the latter absorb more water).

3. CONCLUSIONS

All the composites absorbed different amounts of water depending on the type of pretreatment applied. The control sample (no coupling agent) absorbed the highest amount of water (1.8%) after 90 days of exposure at out-door conditions. The use of coupling agents resulted in a reduction in water absorption: 1.2% (epolene) and 0.7% (silane). Therefore, Silane A-174 is a good pretreatment agent allowing a significant reduction in moisture absorption and thus, favoring the hydrolytic stability of composites. Composites produced with only wood fibers or epolene-pretreated wood fibres showed less tensile strength than that of pure polymer. However, the pretreatment of wood fibers with Silane A-174 led to a significant improvement of composite strength (even higher than that of HDPE). It means that silane provide good bonding properties at the interface. The addition of wood fibers (with or without coupling agent) always led to an increase in tensile modulus. Nevertheless, there was little difference between the performance of Silane A-174 and Epolene C-18. This finding confirms the well known fact that tensile modulus depends on the proportion of reinforcing material rather than on fiber-matrix interaction.

REFERENCES

1. E. Galli, Plastics Compounding, 5(1982), 105-109.
2. H. Dalvag, C. Klason & H.E. Stromvall, Intern. J. Polym. Mater., 11(1985), 9-38.
3. R.G. Raj, B.V. Kokta & C. Daneault, J. Adhesion Sci. Technol., 3(1989), 55-64.
4. M. Xanthos, Plast. Rubber Proc. Appl., 3(1983), 223-228.
5. R.T. Woodhams, G. Thomas & D.K. Rogers, Polym. Eng. Sci., 24(1984), 1166-1171.
6. R.G. Raj, D. Maldas, B.V. Kokta & C. Daneault, J. Appl. Polym. Sci., 37(1989), 1089-1103.
7. D.M. Bigg, Polymer Composites, 8(1987) 115-122.
8. R.G. Raj, B. V. Kokta & C. Daneault, Inter. J. Polym. Mater., 14(1990), 223-234.

ADVANCED COMPOSITES

REPAIR METHODS FOR CFRP SANDWICH STRUCTURES

K. WOLF, R. SCHINDLER

*Eurocopter Deutschland GmbH - Postfach 80 11 40
81663 München - Germany*

ABSTRACT

This paper reports on an investigation of repair techniques for honeycomb sandwich with thin-walled CFRP skins. Based on the external patch concept two repair procedures that are adaptable to field level maintenance operations were developed. These techniques were evaluated through a static test program and compared with a depot level repair. The investigated repair methods as well as the results of the test program are presented.

1. INTRODUCTION

Sandwich construction is one of the basic design concepts for structural components of helicopters (fuselage shells, stabilizers, rotorblades, etc.). Owing to features such as high strength-to-weight ratios as well as high resistance to fatigue loading and corrosion especially honeycomb sandwich with skins made of carbon fibre reinforced plastics (CFRP) are extensively used on advanced helicopter structures. Due to the thin skins sandwich structures are susceptible to low velocity impacts (dropped tools, small stones, etc.). Therefore, damages resulting in serious strength or stiffness degradation may occur more frequently than in the case of thick-walled monolithic composite structures. As the application of sandwich construction normally results in large integrated components a replacement of damaged parts would be expensive and time consuming. Consequently, a growing demand exists for repair methods that can be accomplished cost-effectively on the aircraft.

In order to study the applicability of such repairs to helicopter airframe structures a research program was initiated as part of the Brite/Euram "DESIR" project /1/. The sandwich repair program consists of two phases: In Phase 1, which has been already completed, field-level repair procedures were developed and evaluated by static testing. In Phase 2, which is currently in progress, mechanical testing is continued to investigate the quality of the considered repairs under hot/wet conditions and fatigue loading. In the following the Phase 1 activities will be described.

2. REPAIR CONCEPTS

A variety of repair concepts for sandwich type structures have been considered in several previous programs (e.g. /2/, /3/). In the current program, therefore, the effort is focused on modifications of known basic concepts in order to simplify the procedures and to make use of new materials such as low temperature curing adhesives and repair prepregs. For the repair evaluation a sandwich configuration was chosen which is typical for flat and shallow curved fuselage panels, i.e., a honeycomb core with thin skins made of CFRP fabric. Such panels normally are designed to sustain compression loads which cause structural failure by local instabilities (e.g. face wrinkling). For this type of structure skin and core damages caused by low velocity impacts (e.g. dropped tools) are considered to be critical.

As a basis for the selection of appropriate techniques following requirements were defined:
- Restoration of original strength/stiffness and durability for all design environments
- Minimum expenditure of time, money and effort
- No special tools necessary
- Capability for repair on the aircraft
- Low level of personnel skill.

Based upon these criteria the external patch concept was chosen as the most promising approach for the sandwich configuration considered. From this basic concept two field-level repair techniques designated RS2 and RS3 were derived. The configurations of these repairs are shown in Figure 1. RS2 is a cure-in-place patch repair which makes use of low temperature curing repair prepregs. The concept of RS3 is to bond a pre-cured patch made of the parent laminate material to the skin using low temperature curing adhesives. For setting the standard for the evaluation procedure a third method called RS1 was defined. RS1 is a typical depot-level repair using materials and processes equivalent to those used for manufacturing the original sandwich component. For all repairs patches having the same lay-up as the parent laminate skins were used.

3. EXPERIMENTAL PROGRAM

To evaluate the processibility and the structural integrity of the selected repair methods a total of 24 sandwich specimens were used. The basic specimen configuration was a 200 by 350 mm sandwich plate made of a 15mm thick 48kg/m³ NOMEX core and 2-ply fabric skins with [±45,0/90] orientation as shown in Figure 2. The specimen ends were reinforced by fillers and tabs. Two different CFRP skin materials were investigated: a 125°C curing fabric prepreg (Material "1") and a 175°C system (Material "2"). All specimens were damaged by low velocity impacts resulting in punctured faces and partially crushed cores. The average diameter of the damage zones was 35mm. Although the backside skins remained undamaged the compression strength of the sandwich plates was seriously affected by the impact. A strength reduction of about 40% was determined by preliminary testing /1/.

Each repair technique was applied to 4 specimens of the same skin material. Details of the used procedures are listed in Table 1. Repair prepregs and adhesives that can be processed with vacuum pressure only and at curing temperatures lower than those originally used to manufacture the parent skins were chosen. Based on data of preliminary studies (DMA and lap shear strength tests) suitable curing cycle parameters were determined to get an acceptable bonding strength at elevated temperatures and humidity. The repairs were processed without problems and non-destructive inspection revealed no bonding defects.

A total of 18 specimens (3 per repair method and skin material; the remaining specimens were used for NDT evaluation) were tested to failure under static uniaxial compression loading. All tests were performed at room temperature and ambient conditions.

4. RESULTS

The results of the compression tests are summarized in Figure 3. All failure loads are normalized by the theoretical strength of the undamaged sandwich specimens which was determined using the analytical methods described in /4/. The predicted failure mode is face wrinkling. Figure 3 shows that all three types of repair are capable to recover the full compression strength of the undamaged sandwich which was reduced to nearly 60% by the impact. All specimens failed by skin rupture and no patch delamination could be observed. Only the Material "1"-specimens repaired using RS2 failed in the repaired section. This might be explained by the fact that after curing the patches showed a very low resin content in the damage area resulting in a local strength reduction.

In Table 2 main performance parameters and processing variables of the three evaluated repair techniques are compared. Regarding handling, equipment and curing effort RS3 seems to be the most effective technique for field-level repair. RS3 also produces a very smooth surface because no peel-ply has to be used for curing.

5. CONCLUSIONS

Cure-in-place and pre-cured patch repair procedures were developed for field-level repair af damaged honeycomb sandwich structures with thin-walled CFRP skins. These techniques as well as a standard depot-level repair were evaluated in an experimental program. The results of the first phase of the evaluation indicate that:

o Both techniques utilizing unaugmented vacuum pressure proved to be applicable to repair CFRP sandwich structures under field-level limitations or on-aircraft.

o All repair methods investigated restore the RT compression strength and stiffness of the original sandwich part.

However, the final conclusion concerning the performance of the techniques cannot be drawn before the results of the second phase of the program will be available.

6. ACKNOWLEDGEMENTS

The financial support given by the CEC (Contract No. BREU-0085) is gratefully acknowledged. The authors would like also to thank our partners in the DESIR project - Agusta, CIRA, RISØ, and Westland Helicopters - for contributing some results.

7. REFERENCES

1. DESIR Consortium, " Simulation, Detection and Repair of Defects in Polymeric Composite Materials - An European Research Program", ECCM-6 (1993), Bordeaux

2. M.Torres and B.Plissonneau, in "The Repair of Aircraft Structures Involving Composite Materials", AGARD CP402 (1986)

3. R.E. Trabocco, T.M.Donnellan and J.G.Williams, in "Bonded Repair of Aircraft Structures", (A.A.Baker/R.Jones, ed.) (1988) 175-209, Nijhoff, Dordrecht

4. G.Dreher, in Proc. of 2nd Int. Conf. on Sandwich Construction (1992), Gainesville

RS1	RS2	RS3
Cut out damaged skin	Cut out damaged skin	Clean hole
Remove damaged core	-	-
Prepare skin surfaces for bonding (abrading, cleaning)	Prepare skin surfaces for bonding (abrading, cleaning)	Prepare skin surfaces for bonding (abrading, cleaning)
Apply adhesives to bottom and wall of core cavity and put core-plug into place	Fill the damaged part of the core with filler paste	Fill the damaged part of the core with filler paste
Position uncured prepreg over the hole using film adhesive	Position uncured prepreg over the hole	Position pre-cured patch over the hole using paste adhesive
Prepare assembly for curing (peel ply, vacuum bag)	Prepare assembly for curing (peel ply, vacuum bag)	Prepare assembly for curing (vacuum bag)
Cure (autoclave): 1h at 135°C * / 2h at 175°C **; vaccum and 3*/4** bar pressure	Cure (oven or heater blanket): 4h at 100°C * / 2h at 120°C **; vaccum pressure	Cure (oven or heater blanket): 2h at 65°C * / 1h at 90°C **; vaccum pressure

Table 1: Main steps of repair procedures (* Material "1" / ** Material "2")

	RS1	RS2	RS3
Equipment	Autoclave, vacuum	Heater lamp or blanket, vacuum	Heater lamp or blanket, vacuum
Personnel skill	100%	~80%	~60%
Repair Time (without curing cycle)	100%	~70%	~50%
Curing Time	100* % / 100** %	400* % / 100** %	200* % / 50** %
Curing Temperature	100* % / 100** %	74* % / 68** %	48* % / 51** %
Curing Pressure	3* / 4** bar	Vacuum	Vacuum
Restoration of Compression Strength (RTA)	100.2* % / 98.5** %	96.4* % / 99.5**%	99.4* % / 105.7** %

Table 2: Comparison of evaluated repair methods (* Material "1" / ** Material "2")

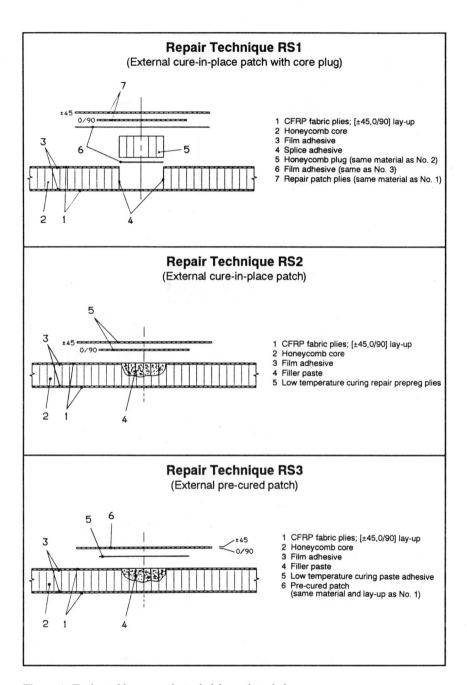

Figure 1: Evaluated honeycomb sandwich repair techniques

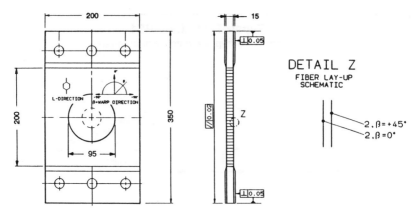

Figure 2: Specimen configuration

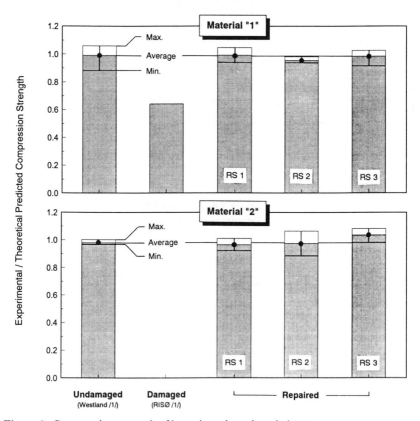

Figure 3: Compression strength of investigated repair techniques

QUANTITATIVE ASSESSMENT OF THE EFFECTS OF MICROCRACKING ON THE PROPERTIES OF A CROSSPLY CMC

J. LORD, N. McCARTNEY

National Physical Laboratory - Queens Road
Teddington Middlesex TW11 0LW - UK

ABSTRACT

Uniaxial tension and compression tests have been carried out at room temperature on a crossplied SiC fibre reinforced barium osumilite glass ceramic composite. The practical issues of the test methods are considered in some detail, and results show that the measured stress-strain behaviour is very reproducible. Attempts have been made to interpret the data using high quality micromechanical models of stress transfer and transverse cracking developed at NPL. While the modelling of the effects of damage formation at low applied stresses is reasonably successful, damage growth at higher loads is complex and needs detailed microstructural investigation.

1 INTRODUCTION

Ceramic materials are of great interest in engineering applications because they retain their properties at higher temperatures than metals, and can be resistant to environmental attack at such temperatures. Their exploitation is, however, inhibited by their brittle behaviour, and their sudden catastrophic mode of failure. In an attempt to overcome these problems, fibre reinforced ceramic composites are now receiving increasing interest. Although the major driving force is the high temperature applications of ceramic matrix composites there is significant interest also in their mechanical behaviour and properties at ambient temperatures. While most work has been focused on the behaviour of unidirectional composites, more recently work has been reported on CAS laminated composites /1-4/. The objective of this paper is to summarise briefly some results on the measurement of the mechanical properties of a cross-plied barium osumilite ceramic matrix composite, and compare these with recently developed micromechanical models of cracking in such materials /5-10/. A review describing the characteristics of the stress transfer models is given in reference /11/.

2 EXPERIMENTAL PROCEDURES

2.1 Material

The material tested in this work was purchased from AEA Technology, Harwell, in the form of a number of identical tiles each approximately 210 mm square and 3.2 mm thick. The composite consists of a barium osumilite glass ceramic matrix reinforced with *Ube* SiC fibres, with a nominal fibre volume fraction of 0.45 in a 16 ply, 0/90 balanced fibre layup.

2.2 Testpiece Preparation

Two basic testpiece designs were examined for the tensile tests - a parallel sided specimen and a profiled specimen with a narrow gauge length. The specimens were machined from separate tiles - designated B, G and I - and are coded according to the fibre orientation in the outer plies relative to the loading direction. Uniaxial compression tests have been carried out in an IITRI-compression rig, using parallel-sided specimens taken from Tile F, with the fibres in the outer plies perpendicular to the loading direction. For both the tensile and compression tests, aluminium tabs were bonded to the ends of the testpiece to protect the composite from damage and crushing in the grips during testing.

The testpieces were cut from the tile using a diamond slitting wheel, and the edges finished by grinding. The profile of the tapered tensile specimens was machined by surface grinding, and the top and bottom surfaces were prepared for strain gauging by degreasing followed by mild abrasion with 320-grade SiC paper.

2.3 Test Procedures

The tensile tests were carried out at room temperature on an Instron servoelectric testing machine equipped with wedge-action grips at a constant crosshead displacement of 0.5 mm min $^{-1}$. Most of the tests were carried out to failure, but some unloading-reloading tests were performed to monitor the stress-strain response with different levels of matrix microcracking. The compression specimens were tested in the IITRI rig under the same loading conditions.

Strain measurements were made with a combination of strain gauges and extensometers, using a computer controlled multichannel data acquisition system developed by NPL. A comparison of strain measurements was made using strain gauges of different sizes on the same testpiece, from a number of suppliers, and transverse gauges were applied to some specimens to monitor the lateral strains and Poisson's ratio during the test. In all cases, strain was measured on both sides of the specimen to check misalignment and minimise the effects of out of plane bending. For some of the tensile specimens, up to three pairs of gauges were used at different positions on the specimen to check the consistency of the measurements.

3 RESULTS

Table 1 shows the data from the tensile and compression tests, grouped according to the tile and the type of specimen. Details of the type and location of the strain gauges are also given. Fig.1 is a typical stress-strain curve from a tensile test to failure (Specimen B6) plotted as the mean output from a pair of strain gauges and Fig.2 shows a plot of the tangent and secant moduli for the same test.

Three distinct linear regions can be identified on the stress-strain curve in Fig.1, but the first linear region of the curve exists only at very low strains (below ~ 0.06%). Beyond this strain limit there is a progressive loss of stiffness followed by a second, longer, linear portion from 0.15 - 0.3 % strain. Above ~ 0.8% strain the curve is linear to failure. This behaviour is typical of the tensile stress-strain curve for all the tests on this material. The initial value for E_A, the axial Young's modulus, is quoted in Table 1 and the mean value from all the tests is 114.0 GN/m^2. The early linear part of the curve is short and it is difficult to measure the initial modulus accurately, but there does not appear to be a significant variation in modulus between the tiles, nor does the modulus vary greatly with testpiece orientation. The variation in properties between tests is probably a function of the variability of the material both within a single tile and between tiles, rather than that attributable to the test method. Typical values for the tangent modulus of the second and third linear regions of the curve are 60 and 30 GN/m^2 respectively. The three linear regions of the stress-strain curve can also be identified by the flat regions on the tangent modulus - stress plot in Fig.2.

Specimens B6 and B7 were tested with multiple strain gauges, in pairs one on either side of the testpiece and located at different positions in the gauge length. Results from test B6, with three different sizes of gauges at different locations, show excellent agreement. For specimen B7, there is some difference in the values obtained for the initial modulus for the different sets of gauges, but the stress-strain curves are almost identical.

The average tensile strength from all tests was 516 MPa, but the range of values measured varied from 351 - 615 MPa. The highest values were obtained with the tapered specimen geometry and these also gave the highest strains to failure, typically 1.4 - 1.5%. Parallel sided specimens from Tile G, which had the fibres in the outer plies perpendicular to the loading direction gave slightly lower values of strength and strain to failure. Specimen G2 was the only testpiece to fail prematurely in the tabs at a low stress, possibly due to a defect in the material. Generally, for both types of specimen geometry, failure originated from a point, 10-15 mm outside the tab/gripping region. Failure was sudden and catastrophic with the specimen disintegrating into a number of fragments.

Poisson's ratio was measured in a number of tests using combined longitudinal and transverse strain gauges. Fig.3 shows the output from the transverse gauges from the early part of the test on Specimen B9. The mean value for Poisson's ratio, calculated from the initial linear part of the curve, for all the tests is 0.200. Beyond a stress of 0.275 GN/m^2 the strain rate reverses and the composite begins to expand in the transverse direction. This coincides with the point in Fig.1, where the transition from the second to the third linear portion of the stress-strain curve occurs. This behaviour is consistent with that observed in earlier work on a similar crossply ceramic composite /12/.

Results from the compression tests in the IITRI rig gave a mean value for the compressive strength of 463 MPa, and analysis of the stress-strain data showed that the curve remained linear almost to failure. Values measured for the axial modulus ranged from 116.8 - 138.0 GN/m^2, with a mean value of 126.2 GN/m^2, and the typical failure strain was 0.38%

4 MODELLING

The approach used for modelling the behaviour of the crossply laminates during loading is to apply recently developed micromechanical models /5-10/ and to use best estimates for the thermoelastic constants of the fibres and matrix material, and for the fracture energy of the matrix. In predictions the fibre and matrix are taken to be isotropic solids having the following properties:

$$E_f = 171 \text{ GN/m}^2 \qquad v_f = 0.2 \qquad \alpha_f = 3.1 \times 10^{-6} \text{ / °C}$$
$$E_m = 85 \text{ GN/m}^2 \qquad v_m = 0.25 \qquad \alpha_m = 2.8 \times 10^{-6} \text{ / °C}.$$

Knowing the volume fraction of fibres (0.45) in the composite it is possible to use a micromechanical model /8/ to predict the following thermoelastic constants for single plies of the laminate, assuming that the plies are undamaged:

$$E_A = 123.8 \text{ GN/m}^2 \qquad E_T = 114.3 \text{ GN/m}^2 \qquad v_A = 0.226 \qquad v_T = 0.241$$
$$\mu_A = 46.9 \text{ GN/m}^2 \qquad \alpha_A = 2.983 \times 10^{-6}/°C \quad \alpha_T = 2.934 \times 10^{-6}/°C .$$

The above thermoelastic constants for single plies are then used to predict the thermoelastic constants for an uncracked crossply laminate where all plies are assumed to have the same thickness (0.2 mm). The resulting values are:

$$E_A = 119.1 \text{ GN/m}^2 \qquad E_T = 119.1 \text{ GN/m}^2 \qquad v_A = 0.217$$
$$\alpha_A = 2.960 \times 10^{-6}/°C \quad \alpha_T = 2.960 \times 10^{-6}/°C .$$

The axial modulus of the matrix $E_m = 85$ GN/m^2 was in fact selected so that the axial modulus of an uncracked laminate had a value $E_A = 119.1$ GN/m^2 that was close to experimental values. The selected value was measured from stress-strain curves for a single sample tested in tension and compression for which the initial modulus in tension and compression has a unique value of 119.1 GN/m^2, indicating that the laminate was initially uncracked. The predicted axial Poisson's ratio has the value 0.217 which is slightly higher than the values measured experimentally which are in the range 0.191 - 0.209 (see Table 1).

The above thermoelastic constants were then used to predict, using recently developed methods /6/, the first cracking stress and strain of a laminate that is subject to transverse cracking in the 90° plies. Account is taken of thermal residual stresses by assuming that the stress-free temperature is 1000°C. The stress at which any cracks in the 90° plies just close is predicted to be 0.00247 GN/m^2 which is very small. One approach /6/ uses energy balance methods to calculate the first ply cracking stress as a function of crack density assuming that all cracks form simultaneously. The value used for the matrix fracture energy $2\gamma_m$ is 20 J/m^2. When using the generalised plane strain model of stress transfer /5,6/ there is a single minimum value of cracking stress corresponding to a single value of crack density that characterises the damage state as soon as transverse cracking has been initiated. The first ply cracking stress is 0.072 GN/m^2 and the corresponding first ply cracking strain is 0.088%. The initial crack density is predicted to be 0.96/mm. Such a crack density reduces the axial modulus from the uncracked value of 119.1 GN/m² to 80.8 GN/m². Such a modulus reduction is not observed experimentally indicating that transverse cracking is occurring progressively in a manner that is determined by the statistical variability; an issue that is beyond the scope of this paper, but discussed in /6,7/. In Fig.1 the point A denotes the predicted value of transverse crack initiation which corresponds reasonably well with the experimentally observed point of damage initiation, although it should be noted that the initial modulus of the data shown in Fig.1 is 107.6 GN/m² which is slightly less than the value of 119.1 GN/m² used in the predictions.

Neglecting the statistical variability of the material, additional transverse cracking would occur when the stress is large enough for transverse cracks to form at the mid-points between the existing cracks. Energy balance principles predict that the stress at which the crack density would double to the value 1.923/mm is 0.234 GN/m², occurring at a strain of 0.349%. The modulus is then predicted to reduce from 80.8 GN/m² to 66.75 GN/m². This prediction is shown as point B in Fig.1 and clearly corresponds to the onset of more pronounced damage growth observed experimentally. Assuming that only transverse cracks can form, it is predicted that the crack density would double again to a value of 3.85/mm when the stress is 1.105 GN/m² and the strain is 1.733%. It is clear from the experimental data in Fig.1 that additional damage modes must be occurring in the material. The portion of the stress strain curve approaching point B in Fig.1 seems to be linear, although from Fig.2 the tangent modulus measurements indicate that there is a point of inflection. A question that arises is whether any transverse cracking is occurring in this region. The results of unloading-reloading experiments (to be published) indicate that the unloading and reloading curves are almost coincident, that they are linear and they almost extrapolate to a single point at which transverse cracks would just close. The gradients of the unloading curves decrease as one progresses to point B. The conclusion is that transverse cracking is indeed occurring and that it is unlikely that cracks are growing significantly into the 0° plies at this stage.

Subsequent damage growth could occur in a variety of different ways. First of all transverse cracks could propagate into the 0° plies leaving fibres intact /2/ and such cracking would be accompanied by fibre/matrix debonding. Such debonding would lead to a further loss of stiffness and straining, and introduce the prospect of stress transfer by friction acting at the debonded interfaces. Matrix cracking within the 0° plies, independent of any cracking in the transverse plies could also possibly occur leading to a further loss of stiffness and additional straining. If such events occurred then modelling experience suggests that the crack densities formed would be much more dense than that found in the transverse plies as crack densities scale with respect to the fibre radius in UD composites, whereas they scale with the ply thickness in laminates. An additional damage mode that could occur is delamination between plies where the delaminations are initiated at the transverse cracks in the 90° plies. Stress transfer between plies may occur at the locations of delaminations by friction or roughness effects. Again this type of damage mode would lead to a loss of stiffness and additional straining. Clearly the accurate modelling of the behaviour of the laminate in the region of the stress-strain curve having strains in the range 0.4 - 0.8% is very difficult and discussion beyond the scope of this paper.

Following the onset of significant damage at a stress of approximately 0.27 GN/m², it is seen from Fig.2 that the tangent modulus falls to a minimum when the stress is 0.3 GN/m² and then rises to a steady value of approximately 30 GN/m² which is maintained to the point of laminate failure. It is useful to attempt to predict this limiting value from micromechanical models. If one assumes that the 90° plies have cracked to the point of saturation such that the 90° plies are no longer supporting any load, and if one also assumes that matrix cracks have formed in the 0° plies such that the matrix in these plies is no longer supporting load, then the fibres in the 0° plies are clearly supporting all the load and one can then estimate their stress-strain behaviour. On neglecting completely the presence of the matrix, and of any stress transfer by friction or roughness, it follows that the stress-strain behaviour should be a straight line. The gradient of this line would have the value $V_f E_A{}^f/2 = 38.475 \text{ GN/m}^2$ which is larger than that measured by experiment. The linear portion of the stress-strain curve should also extrapolate to point close to the origin as thermal residual stresses are small. Clearly from Fig.1 such an extrapolation of the experimental curve would lead to a very large offset. It is concluded that the discrepancy between measured and predicted values of limiting modulus and the large extrapolated offset are due to some form of stress transfer between fibre and matrix. This conclusion is supported by unloading-reloading experiments (to be published) that indicate hysteresis effects.

Fig.3 shows the dependence of stress on transverse strain indicating an initial reduction in Poisson's ratio of the laminate resulting from transverse cracking in the 90° plies. When the stress is approximately 0.27 GN/m² the transverse strain rapidly begins to increase from negative values to zero and then to a small positive value. Continued loading to failure leads to larger positive values and apparently negative effective Poisson's ratios. Clearly the matrix in the external 0° plies is relaxing, presumably because of fibre/matrix debonding (initiating from matrix cracks in the 0° plies) and/or orthogonal cracking in the 0° plies. The larger positive lateral strains may be caused by matrix and/or fibre debris in delaminations, orthogonal cracks, and/or in the interfacial regions between fibre and matrix. It is not possible to model these type of effects with accuracy unless the details are known of the damage mechanisms that are operating. Detailed microstructural examination of the composites is currently in progress and will be reported in a future publication.

CONCLUSIONS

A major conclusion derived from the experiments on barium osumilite glass ceramic crossplied CMC is that the measured stress-strain behaviour is very reproducible, exhibiting some orderly characteristics. The second conclusion is that damage initiation and growth in these composites is very complex and requires the use of sophisticated models of stress transfer and cracking in order to interpret the results accurately. Thirdly, microstructural examination of damage in the composites (currently in progress) is required to assist in the interpretation of stress-strain behaviour for regions of the stress-strain curve where significant matrix damage in the 0° plies is encountered.

ACKNOWLEDGEMENT

The authors would like to acknowledge Dr Jonathon Davies, Harwell Laboratories, for providing information on the fibre and matrix properties that have been used for modelling purposes. The research in this paper was carried out as part of the 'Materials Measurement Programme', a programme of underpinning research financed by the UK Department of Trade and Industry.

REFERENCES

1   PRYCE A W and SMITH P A, J. Mater. Sci. **27** (1992) 2695.
2   BEYERLE D S, SPEARING S M and EVANS A G, J. Am. Ceram. Soc. **75** (1992) 3321.
3   KARANDIKAR P and CHOU T-W, Comp. Sci. Tech. **46** (1993) 253.
4   DAVIES C M A, HARRIS B and COOKE R G, Composites **24** (1993) 141.
5   McCARTNEY L.N., J. Mech. Phys. Solids **40** (1992) 27.
6   McCARTNEY L.N., Composites **24** (1993) 84.
7   McCARTNEY L.N., (1993) to be published.
8   McCARTNEY L.N., NPL Report DMM(A)57, March 1992.
9   McCARTNEY L.N., J. Comp. Tech & Res. **14** (1992) 133.
10   McCARTNEY L.N., J. Comp. Tech & Res. **14** (1992) 147.
11   McCARTNEY L.N., Proceedings of EUROMAT '93, Paris, June 1993, in press.
12   COOKE R G, Proceedings Inst. of Ceramics Convention, St. Catherines College, Oxford, April 1991, in press.

Spec code (& type)	Strain ▪ m'ment (g length)	E (GN/m²)	Prop limit (MN/m²)	Strength (MN/m²)	ν	εf (%)	Comment
Tensile Data							
B3 - 0 (tapered)	XY (3)	116.7	49	534	0.209	1.42	
B6 - 0 (tapered)	XY (3)	107.6	65	568	0.198	1.52	+ 80 mm†
▪ ▪	LY (1.5)	107.1	64	568		1.48	centre
▪ ▪	LY (6)	107.7	58	568		1.48	- 80 mm†
B7 - 0 (tapered)	LY (3)	115.9	57	615		1.53	centre
▪ ▪	LY (20)	120.5	75	615		1.54	- 80 mm†
B8 - 0 (tapered)	XY (3)	107.7	67	586	0.191	1.51	
B9 - 0 (tapered)	XY (3)	119.3	35	592	0.203	1.53	
I1 - 0 (parallel)	LY (6)	111.8	42	495		1.36	
I2 - 0 (parallel)	LY (6)	110.3		502		1.37	
I3 - 0 (parallel)	LY (3)	123.4	51	514		1.47	
G1 - 90 (parallel)	XY (3)	112.4	32	447		1.11	
G2 - 90 (parallel)	LY (6)	116.4	31	351		0.83	Failed in tabs
G3 - 90 (parallel)	LY (10)	118.3	25	489		1.26	
	LY (6)	115.1	34	495		1.33	
Compression Data							
F1 - 90 (IITRI)	LY (3)	116.8	480	486		0.42	
F2 - 90 (IITRI)	LY (3)	122.9	463	552		0.46	
F4 - 90 (IITRI)	LY (3)	129.9	340	445		0.34	
F13 - 90 (IITRI)	LY (3)	138.0	86	452		0.36	
F14 - 90 (IITRI)	LY (3)	123.5	191	381		0.32	

▪   XY - Combined longitudinal and transverse gauges
     LY - Longitudinal gauges only

†   Position of gauges from centre of specimen gauge length

TABLE 1

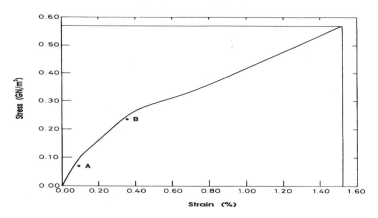

Fig. 1   Tensile stress-strain curve

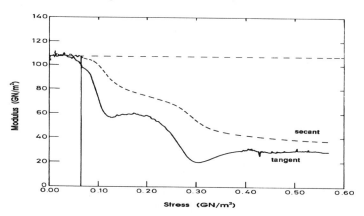

Fig. 2   Tangent and secant modulus plot

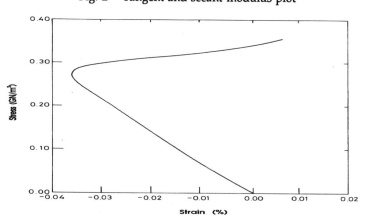

Fig. 3   Output from the transverse strain gauges

505

FIBRE REINFORCED CERAMIC COMPOSITES AS LOW ACTIVATION MATERIAL FOR FUSION REACTOR APPLICATIONS

P. FENICI, H.W. SCHOLZ, L. FILIPUZZI*

*Commission of the European Communities - Joint Research Centre
Institute for Advanced Materials - 21020 Ispra (Varese) - Italy
Sep - BP 37 - 33165 St Médard en Jalles Cedex - France

ABSTRACT

The most serious safety and environmental concern for thermonuclear fusion reactors involves the induced radioactivity of the first wall and structural components. The development of low activation materials will simplify the maintenance problems of components and will improve the possibility to dispose or to recycle structural materials.
Ceramic fibre reinforced SiC materials offer highly appreciable low activation characteristics in combination with good thermomechanical properties. This class of materials is under investigation in the frame of the European Fusion Technology Programme as a structural material for application in Fusion Reactors. The scope is to put into practice the enormous potential of inherent safety of a fusion reactor.

INTRODUCTION
The development of fusion technology is a long-term objective of the European Community. This development foresees the construction of an Experimental Reactor in the near future, based on the magnetic confinement of a Deuterium-Tritium plasma, which would be followed by the design and construction of a Demonstration Reactor.
The operative conditions of various components differ considerably according to their position with respect to the plasma. The first wall and divertor are subjected to the most severe interactions, a high neutron flux with energies up to 14 MeV, a pulsed thermal flux (~ 0.4 MW m^{-2}) and a bombardment of charged particles of up to 1 KeV. The blanket and the shielding are exposed to neutrons and gamma fluxes which diminish with the distance from the plasma.
Presently fusion reactors are still at the stage of Conceptual Design studies and the Tokamak concept is the one most studied. This is especially so in Europe where the NET (Next European Torus) is the main fusion device /1/ which according to the

European fusion energy development strategy, will follow JET (Joint European Torus) and provide the basis for designing a DEMO (Demonstration Reactor). Due to the considerable advances made in nuclear fusion research over many years, the stage was reached for international collaboration between the European Communities, Japan, the Russian Federation and the USA, under the auspices of the International Energy Agency. The aims of the collaboration are to develop an experimental reactor to demonstrate the scientific and technological feasibility of fusion power. The next step will then be ITER, the International Thermonuclear Experimental Reactor /2/.

In fusion reactors thermonuclear reactions do not generate radioactive products but induce radioactivity in the structural and blanket material. In order to minimize the latter effect materials exhibiting a residual activity below that of conventional steels are studied. The term "low activation" is used to define them. Three routes are followed: elemental substitution in existing materials such as austenitic and martensitic steels, the development of novel alloys and ceramic based materials, and the use of isotopically separated elements, so-called isotopiclaly tailored materials.

Ceramic matrix composites (CMC) with continuous ceramic fibre reinforcement offer significant advantages in comparison to classical ceramic materials in terms of overcoming brittleness and improving thermal cycling resistance. They are now under study as potential low activation structural material for fusion reactors. A co-operative research programme on this subject has been established between CEA, ENEA, SEP and JRC. The potential use of existing industrial material in a fusion environment is hereafter reported, with a special emphasis on those properties which are more relevant for fusion.

1. THE LOW ACTIVATION CHALLENGE

The definition of "Low Activation Material" has not always been univocal. It will mainly depend on what type of radiological quantity should be optimised, i.e., what radiological aspect (Security, Maintenance, long-term Disposal) is given priority. Integral definitions which take into account limits derived from all these aspects are more advanced. They are still based, however, on differing assumptions on tolerable decay respectively cooling times. Given limitations for radioactivity, surface dose rates and afterheat generation of a radioactivated material should also include information on its exposure spectrum and time. Moreoever, the Low Activation Challenge goes beyond classical radiation protection limits and evokes consideration of inherent safety, recycling and also a greater flexibility in future reactor design, e.g., less critical requirements for shielding and shut-down cooling.

So far, using advanced computational methods, we know the base elements which allow the development of low activation materials. They are C, O, Cr, V,Mn, Fe, Si.

Commercial SiC/SiC, manufactured by the Société Européenne de Propulsion (SEP), were evaluated by dose-rate calculations (FISPACT code and EAF2-library of nuclear reactions). Before, this class of materials, though not elementally tailored, was already shown to potentially fulfil all current low-activation limits of interest in Europe /3/. At that point, the analysis of the impurities, however, was based only on chemical/spectroscopic methods. The study was then extended on quantitative analysis of scarce but critical trace-elements by charged particle activation analysis (CPAA) on different types of materials /4/. The results have shown clearly that a higher grade of purity is obtained in materials produced by Chemical Vapour Infiltration (CVI). Moreover, after irradiation of CVI-type material in the HFR (High Flux Reactor) Petten, there were gained complementary results from neutron activation analysis (NAA). After the quantitative analysis of the content of 32 different crucial trace elements, the conclusion is that european industrial 2D reinforced SiC/SiC materials will not cause any problem in terms of low activation criteria. This is of particular importance after the considerations by Cierjacks /5/, where the low-activation advantages of Cr- and/or V-based alloys were critically reviewed under the aspect of fusion spectrum induced sequential nuclear reactions. These sophistications in calculating the neutron induced radioactivity of future fusion reactor materials are not expected to cause comparable restrictions for SiC based materials. The important sequential reactions [(p,n); (d,n); (α,n)] are not leading to critical or long-lived radioisotopes.

2. THE MATERIAL

Fiber Reinforced High Temperature Composites are widely used for the aerospace industry. Carbon-, metal- or ceramic-matrix composites have already found their application in components which work under extreme conditions; new components with improved performances are continuously put into operation as new fibres and new manufacturing processes are developed. Among the different possibilities in combining matrix and fibre, the class of SiC fibre/SiC matrix is of specific interest for the long term fusion thermonuclear application. SEP has developed this class of composites, using Chemical Vapour Infiltration (CVI), and has the industrial capability to produce large parts (size around 2.5 x 3 m). The measured properties of these ceramic matrix composites are reported in Table I. These emerging materials present characteristics which are likely to be adapted to some extend in order to meet particular specifications necessary for fusion applications. SEP is present in thermonuclear fusion programmes not only in the long term research, but is also developing and supplying industrial materials, in particular C/C, adapted to the needs of the near term experimental machines.

Due to the good high temperature properties of CMC, the application in energy generating systems will increase their thermodynamical efficiency. The advantages of using SiC/SiC, instead of metals, in a fusion environment can be summarized as follows:
- rapid decrease of induced radioactivity after reactor shut-down,
- low after-heat especially useful in the case of a loss-of-coolant accident,
- high temperature capabilities and thermal shock-resistance,
- if used as first wall material, its sputtered atoms will lead to lower plasma radiation energy losses than those of higher Z elements,
- low specific weight
- very abundant availability of basic material.

3. RADIATION STABILITY

It is important to consider that the primary advantage of Ceramic Matrix Composites (CMC) materials over bulk ceramics lies in the ability to obstruct catastrophic crack propagation. The fracture thoughness of CMCs does not rely on the thoughness of the matrix but on the interaction of matrix cracks with fibres. The effect of possible microcracking of the matrix due to formation of neutron-radiation induced gas bubbles should not constitute failure of CMCs.

The response of SiC composites to neutron irradiation is not known at present. Anyhow encouraging results were found in a limited number of low fluence neutron irradiation experiments. Both fibres and matrix materials (β-SiC) have proven to stay practically unaffected in mechanical properties up to a neutron fluence of 9.10^{24}n m^{-2} with E $>$ 1 MeV at 923 K /6/. The Nicalon SiC fibres exhibited no change in tensile strength and Young's modulus in irradiations with 14 MeV neutrons at the RTNS-II /7/.

Nevertheless, as all low Z materials, SiC will develop substantial quantities of He under high energy neutron irradiation. The He generation rate is about 10 times that of type 316 stainless steel. The formation of He in SiC/SiC by transmutation reactions due to fast neutrons is currently simulated by irradiations at the Ispra cyclotron. Under a helium jet cooling, 39 MeV α-particles were implanted in CMC specimens up to high fluences (2500 appm). First results show no swelling of specimen irradiated at 973 K. Analysis whether the inherent residual porosity of this type of materials is an advantage are ongoing. Ion induced radioactivity was identified by decay analysis to be originated by the ^{32}P isotope via the reaction $^{30}Si(\alpha,pn)^{32}P$, $E\alpha > 14.08$ MeV. This will permit bending tests and microscopic analysis of the implanted specimens within the next months.

Meanwhile, 2D SiC/SiC specimens were sent to HFR Petten for neutron irradiation of different fluences at 973 K. Mechanical tests after irradiation are foreseen and will be eased by radiation protection techniques developed for the ion irradiated samples.

4. COOLANT LEAK RATE

Gas leak tightness is an important issue for fusion reactor structural materials. If CMCs are to function as a pressure boundary, the coolant leak rate into the plasma and the tritium leak rate out of the blanket will be a major design criteria. Helium is considered as coolant in conceptual design studies of Fusion Reactors using ceramic composites /8/. Chemical Vapour Deposited (CVD) SiC has a very low hydrogen diffusion coefficient of $\sim 10^{-10}$ m^2 s^{-1} even at at 1273 K /9/. Such a low hydrogen diffusion coefficient lets appear CVD SiC quite impermeable to helium.

Measurement of leak rate have been made on SiC/SiC tubes, with an internal diameter of 15 mm and 2 mm wall thickness. Pressure was applied from an external reservoir, and measured inside the testing sample in function of time, in order to evaluate the leak rate. The preliminary results show that at room temperature and with an internal pressure of 6 bar Nitrogen the leak rate is of the order of 0.5 l h^{-1} max. Using Helium, the leak rate increases of a factor from 2 to 7, while if the temperature is increasing from room temperature to 300°C a reduction of a factor 2 in leak rate is observed. The preliminary results of permeability in tubes are presented in Fig.1.

5. HIGH TEMPERATURE COMPATIBILITY WITH BREEDER MATERIALS

The breeders are liquid or solid lithium compounds. The NET option for the liquid breeder is Pb-17Li water cooled or self-cooled. The Pb-17Li eutectic alloy melts at 518 K and does not show vigorous chemical reactions with air or water.

Specimens of SiC/SiC in the form of squares (1 x 1 cm) were introduced in a Mo container together with molten Pb-17Li. The containers were then sealed by electron beam welding. The as sealed Mo crucibles were placed into AISI 316 containers which were also subsequently electron beam welded. All operations were carried out in a stainless steel glove box which can be maintained at a vacuum of 1.3 x 10^{-4}Pa or under pure Argon and is equipped with an electron beam welding device.The crucibles containing SiC were heat treated in an Al$_2$O$_3$ tube, under vacuum, at 1073 K for 1500 h.

After the tests the crucibles were opened in the glove box, the Pb-17Li melted and the samples extracted. The specimen were then sectioned, metallographically mounted, polished and examined by Scanning Electron Microscopy. When necessary X-ray diffraction analyses were carried out. The SiC/SiC composites stay inert with Pb-17Li at 1073 K: no trace of a reaction product can be observed, there is only a penetration of Pb-17Li into the open pore volume as the density of the material is only about 85 -90 % of the theoretical value.

From simple thermodynamic consideration it could be expected that if a reaction would occur between carbide-ceramics and the Pb-17Li alloy, the active agent should be Li. The reaction involving SiC is reported hereafter together with the ΔG of reaction /10/. The term "Li" is used to signify lithium in the form of Pb-17Li:

$$SiC + "Li" \rightarrow Li_2C_2 + Si$$

Temperature	800 K	1000 K	1200 K
ΔG (reaction) KJ mol^{-1}	98.7	98.9	99.00

It can be seen that the reaction given above is unlikely to occur, indicating that SiC is stable in Pb-17Li. /11/.

CONCLUSION

Like every other advanced technology, fusion requires a major research effort in the area of materials. While for the experimental reactors (next step) research is directed towards the optimisation of existing materials, for commercial reactors completely new possibilities are emerging. The objective pursued is the development of materials which will reduce the radioactivity induced by neutrons to minimum levels through the control of the composition.

For the first time it has been proven experimentally that in an industrial product, as the SiC/SiC, all low activation criteria valid in Europe are fulfilled.

Radiation stability is an area where further research is necessary. Although the neutron irradiation data base is small, low fluence and He implantation experiments do not show any drastically adverse response to radiation damage.

Allowable leak rates have not yet been quantified for a Fusion Reactor. Preliminary results show that the Chemical Vapour Infiltration is particularly well suited for adjusting the matrix composition to improve leak tightness.

From experimental as well as thermodynamic considerations it clearly appears that SiC based materials could be used at high temperature as low activation material to contain a Pb-17Li blanket, .

REFERENCES

1. The Net Team, Fus. Eng. and Design, 21 (1993) 1-358
2. P.H. Rebut et al., Fus. Eng. and Design, 22 (1993)
3. P. Rocco, H. Scholz and M. Zucchetti, J. Nucl. Mater., 191-194 (1992) 1474-1479
4. K. Casteleyn and H. Scholz, J. of Trace and Microprobe Techn. (1993), in press
5. S.Cierjacks, Fus. Engin. and Design, 13 (1990) 229-238
6. K.Okamura et al., J. Nucl. Mat., 155-157 (1988), 329-333
7. K. Okamura et al., J. Nucl. Mat. 133-134 (1985), 705-708
8. The ARIES-I-Tokamak Reactor Study, (1991), UCL-PPG-1323
9. R.A. Causey et al., J. Am. Ceram. Soc., 61 (1978) 221
10. T. Sample, Nottingham University, personal communication
11. V. Coen et al. in "High Temperature Corrosion of Technical Ceramics", (R.J Fordham, ed.) (1990) 169-179, Elsevier Science Publisher, Barking

Property	Units	Temperature		
		296 K	1273 K	1673K
Fibre content	%	40	40	40
Specific Gravity	kg m-3	2500	2500	2500
Porosity	%	10	10	10
Tensile Strength	MPa	285	285	270
Elongation (Tensile)	%	0.65	0.75	0.8
Young's Modulus (Tensile)	GPa	230	200	170
Flexural Strength	MPa	300	400	280
Compressive Strength				
In Plane	MPa	580	400	300
Through the Thickness	MPa	420	380	250
Shear Strength (interlaminar)	MPa	40	35	25
Thermal Diffusivity				
In Plane	10-6m-2 s-1	12	5	5
Through the Thickness	10-6m-2 s-1	6	2	2
Coefficient of Thermal Expansion				
In Plane	10-6K-1	3	3	
Through the Thickness	10-6K-1	2.5	2.5	
Fracture Toughness	MPa√m	30	30	30
Specific Heat	Jkg-1 K-1	620	1200	
Total Emissivity		0.8	0.8	0.8

Tab. 1: Characteristics of SEP's SiC/SiC material, average values.

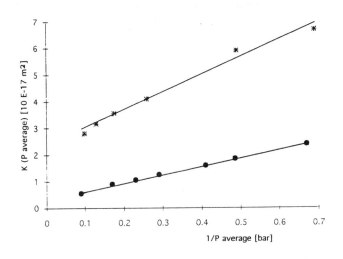

Fig. 1: Room temperature permeability of N_2 (●) and He (*) in SiC/SiC tubes.

THERMAL PROPERTIES OF Ti-SiC AND Ti-TiB2 PARTICULATE REINFORCED COMPOSITES

S.P. TURNER, R. TAYLOR, F.M. GORDON*, T.W. CLYNE*

*Manchester Materials Science Centre - University of Manchester/UMIST
Grosvenor Street - M1 4MS Manchester - UK
*Dept of Materials Science and Metallurgy - University of Cambridge
Pembroke Street - Cambridge CB2 3QZ - UK*

ABSTRACT

Titanium based composites are particularly prone to interfacial chemical reaction during processing and in high temperature service. Such reactions can impair mechanical properties and will affect the thermal conductivity. In this investigation, composites of commercial purity titanium reinforced with 10% and 20% by volume of SiC and TiB_2 were produced by powder blending, and extrusion. Thermal diffusivities of these composites were measured as a function of temperature using the laser flash technique. Thermal conductivities were derived from these data using a rule of mixtures assumption for reference book specific heat data and measured densities.

1. INTRODUCTION

It is being increasingly recognised that transport properties such as thermal and electrical conductivity may be of considerable significance for advanced composites in various potential applications[1]. Thermal conduction also plays an important role in determining thermal shock behaviour. For most metal composites the reinforcement generally has a lower thermal conductivity than the matrix. However, SiC and TiB_2 both have higher thermal conductivity than titanium. Hence, an enhancement of thermal conductivity is expected to occur from the presence of both types of reinforcement. However interfacial chemical reaction can be significant and has been shown to effect the thermal conductivity[2].

In the present paper, experimental data are presented for Ti-SiC and Ti-TiB_2 particulate composites, with and without substantial heat treatments. the main objective is to clarify the role of interfacial heat transfer during heat conduction in these materials and to establish correlation with microstructural information.

2. EXPERIMENTAL

2.1 Thermal Diffusivity Measurements

Thermal diffusivity measurements were made using the laser flash technique. The technique uses a small disc-shaped specimen, the front face of which is subjected to an instantaneous, uniform energy pulse. From the recorded temperature history of the opposite face thermal diffusivity may be calculated[3].

In the UMIST apparatus the heat pulse is supplied by a 100J Nd/glass laser with a wavelength of $1.067\mu m$, and a pulse dissipation time is 0.6ms. The specimen, in a pyrolytic graphite sample holder, is heated to the measurement temperature inside a graphite susceptor located inside an induction coil. This assembly is located within a vessel that can contain either vacuum or inert gas. Radiation from the specimen rear face is collected by a calcium fluoride lens and mirror system and focused onto an InSb infrared detector. The biased output from this detector is amplified and fed into a microcomputer via an analogue to digital convertor. The computer is programmed to calculate the thermal diffusivity, α, using equation (1).

$$\alpha = \frac{\omega/\pi^2 \; L^2}{t_{1/2}} \qquad (1)$$

where ω/π^2 is a dimensionless heat term, L is the sample length and $t_{1/2}$ is the time period required for the rear face of the specimen to reach half of the maximum value. The boundary conditions for the flash method assume a homogeneous sample uniformly irradiated, negligible pulse duration time and no heat losses. Corrections for finite pulse time are made using the method due to L.M. Clark and R.E. Taylor[4] and the term ω/π^2 is corrected for heat losses using the analysis due to R.D.Cowan[5]. Corrections for sample length change during heating were made using the thermal expansion coefficient of titanium. The thermal conductivity, λ, is the product of thermal diffusivity, α, density, ϱ and specific heat, C_p.

$$\lambda = \alpha.\rho.C_p \qquad (2)$$

The conversion was carried out using reference book volume specific heat values for the titanium, SiC and TiB_2[6], which were fitted to third order polynomials over the measurement temperature range. The volume specific heat of the composites were taken as a weighted mean between the values of the constituents.

2.2 Specimen Manufacture

Composites containing 10% and 20% by volume of reinforcement were made by dry blending of Ti powder (99.8% pure, 1400ppm oxygen, 50-150μm particle size) and either SiC or TiB_2 particles (10-30μm). The powder mixture was cold pressed into a copper can which was evacuated and heat sealed by electron beam welding prior to extrusion. The can was soaked for 2 hours in a furnace at 900°C and then extruded. Specimens were machined in the form of circular discs, 3mm thick and 8mm in diameter. These were machined from both composite and unreinforced extrudates so that the axis of the disc was parallel to the extrusion axis. Measurements were made during heating and cooling between 100°C and 700°C. The two types of composite were heat treated by annealing for times up to 300 minutes at 950°C.

3. RESULTS AND DISCUSSION

It can be seen in figure 1 that while an enhancement of the thermal diffusivity/conductivity was expected from the presence of both particulate reinforcements, this behaviour is observed only with the Ti-TiB$_2$ composites. The thermal conductivities of extruded Ti-TiB$_2$ composites are significantly greater than that of the unreinforced matrix and are enhanced with increasing volume fractions of reinforcement. In contrast the thermal conductivity of the Ti-SiC composites are considerably lower than the unreinforced titanium and moreover decrease with increasing volume fraction of SiC reinforcement.

These results have been interpreted in terms of the thermal resistance of the reaction layers that exist between the matrix and two types of particulate reinforcements. The greater reaction kinetics between SiC and Ti produces a thicker reaction layer than that produced between Ti and TiB$_2$ and is also accompanied by a much larger volume change (-4.6%). It is this volume change, causing interfacial microcracks, and reaction layer chemistry/thickness that is believed to increase interfacial thermal resistance, thus reducing the Ti-SiC composite thermal conductivity.

Previous work/7-9/ investigating the interfacial reaction between Ti and SiC particles has revealed the formation of complex reaction products at the interface, TiC, Ti$_5$Si$_3$. This chemical reaction is accompanied by a volume decrease of ~4.6% and hence tends to cause tensile hoop stresses at the interface causing radial cracks. It is these reacted/cracked interfaces within the Ti-SiC composites that are responsible for a high interfacial thermal resistance, thus lowering the thermal diffusivity/conductivity below that recorded for pure titanium. Furthermore, the nature of the radial cracks round the particulates may explain why the effective decrease in thermal conductivity is reduced at higher temperatures, since differential thermal expansion stresses may act to close cracks and reduce the interfacial thermal resistance.

3.1 EFFECT OF HEAT TREATMENT

In figure 2 ratio plots of the thermal conductivity of Ti to that of Ti reinforced with TiB$_2$ and SiC before and after heat treatment at 950°C are shown. For the Ti-SiC composite the already low thermal conductivity is further decreased by only 80 minutes heat treatment at 950°C. This is believed to be due to an increase in reaction layer thickness and a change in the network of radial cracks emanating from the interface in the heat treated composite /7-9/. Even after 300 minutes at 950°C the thermal conductivity of the Ti-10vol%TiB$_2$ is still greater than that of the unreinforced titanium. However the reaction between TiB$_2$ and matrix and its effect on thermal conductivity is more complex, as illustrated by the effect of heat treatment time on Ti + 10vol%TiB$_2$, figure 3. Heat treatments were conducted at 950°C for 40, 80 and 300mins. When compared with the as extruded Ti-10vol%TiB$_2$ composite a heat treatment of 40mins at 950°C results in an increase in thermal conductivity over the entire measurement temperature range and is most apparent at high temperatures.

This enhancement of thermal conductivity could be accounted for by two effects i) the interfacial thermal resistance between the TiB$_2$ and the titanium may have been reduced as a result of the heat treatment. ii) the volume fraction of TiB$_2$ plus reaction products increases as the reaction proceeds during heat treatment. This improved thermal conductivity at high temperatures suggests that the Ti-TiB$_2$ system is prone to cracking in the interfacial reaction zone or between this and the matrix or reinforcement. This could explain the much smaller increase in thermal conductivity at low temperatures where cracks offer high thermal resistance. At higher temperatures where crack closure occurs the interfacial thermal resistance is reduced resulting in a much greater increase in thermal conductivity over that of the as extruded Ti-10vol%TiB$_2$ composite. Heat treatments for 80 and 300mins at 950°C results in a decrease in thermal conductivity at low temperatures. This is despite the increase in volume fraction of TiB$_2$ plus reaction product that would have accompanied the heat treatment. In these materials the breakdown of the reaction zone around the reinforcing particulates is more pronounced. As with the material heat treated for 40mins, crack closure at high temperatures produces an increase in thermal conductivity over that for as extruded Ti-10vol%TiB$_2$. This increase however is reduced with increasing time at 950°C.

4. CONCLUSIONS

(a) The thermal diffusivity/conductivity of extruded Ti-TiB$_2$ composites are significantly greater than that of the unreinforced matrix and is enhanced with increasing volume fractions of reinforcement.

(b) The thermal diffusivity/conductivity of the Ti-SiC composites are considerably lower than the unreinforced titanium and decrease with increasing volume fraction of SiC reinforcement.

(c) In the Ti-TiB$_2$ system, heat treatment causes and initial increase in the thermal conductivity of the composite. Further heat treatment reduces the thermal conduction. The observed behaviour is explained in terms of interfacial cracking and damage in the matrix-particle reaction layers.

(d) In the Ti-SiC system any heat treatment decreases the composite thermal conductance. This is believed to be caused by the higher reaction rate and radial cracks resulting from the accompanying volume changes taking place at the matrix-particle interface.

5. ACKNOWLEDGEMENTS

This work forms part of a project supported by SERC. One of the authors (FHG) is supported by BP plc.

REFERENCES

1. A. Kelly, 'Composites for the 1990s', Phil. Trans. Roy. Soc., A322, (1987) 409-423
2. A.J. Reeves, R. Taylor. and T.W. Clyne, 'The Effect of Interfacial Reaction on Thermal Properties of Titanium Reinforced with particulate SiC', Mat. Sci & Eng., 141, (1991) 129-138
3. W.J. Parker, R.J. Jenkins, C.P. Butler and G.L. Abbott, 'Flash Method of Determining Thermal Diffusivity, Heat Capacity and Thermal Conductivity', J. Appl. Phys., 32, (1961) 1679-1684
4. L.M. Clark and R.E. Taylor, 'Finite Pulse time effects in the Flash Diffusivity Method', J. Appl. Phys., 46, (1975) 714-719
5. R.D. Cowan, 'Pulse Method of Measuring Thermal Diffusivity at High Temperature', J. Appl. Phys., 34, (1962) 1679-80
6. Y.S. Touloukian, 'Thermophysical Properties of Matter Vol 4 Specific Heat' (E.H.Buyco ed) (1970) 598-600
7. A.J. Reeves, H. Dunlop, and T.W. Clyne, 'The Effect of Interfacial Reaction Layer Thickness on Fracture of Ti-SiC$_p$ Composites', Met.Trans., 23, (1991) 970-981
8. P.B. Pragnell, A.J. Reeves, T.W. Clyne. and W.M. Stobbs, 'A Comparison of Interfacial Reaction Mechanisms in the Al-SiC and Ti-TiB$_2$ MMC Systems', in '2nd European Conference on Advanced Materials and Processes, Euromat 91', (1992), (T.W. Clyne and P.J. Withers eds.), Cambridge, UK, Institute of Metals, 215-229
9. A.J. Reeves, W.M. Stobbs, and T.W. Clyne, 'The Effect of Interfacial Reaction on the Mechanical Behaviour of Ti Reinforced with SiC and TiB$_2$ Particles', in '12th Riso Symposium - Processing Microstructure & Properties of Metal Matrix Composites', (N.Hansen et.al. eds.) (1991) Riso National Laboratory, Denmark.

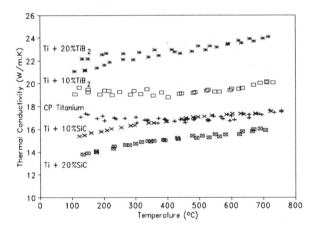

Figure 1.   *Thermal conductivity data for unreinforced titanium and the Ti-SiC and Ti-TiB$_2$ composites.*

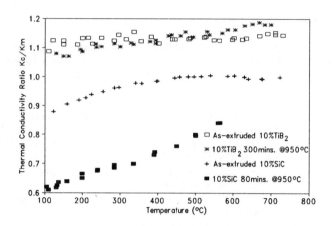

Figure 2.   Thermal conductivity ratio plots for Ti-10%TiB2 and Ti-10%SiC composites, with and without heat treatments.

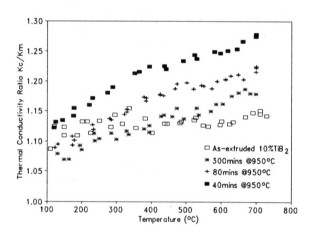

Figure 3.   Thermal conductivity ratio plots for Ti-10%TiB₂ composites, with and without a heat treatment.

518

CERAMIC COATINGS ON CERAMIC
SUBSTRATES BY RCVD

C. VINCENT, H. VINCENT, T. PIQUERO, J. BOUIX

Laboratoire de Physicochimie Minérale I - URA 116 - Université Claude Bernard-Lyon I - 43 Boulevard du 11 Novembre 1918 69622 Villeurbanne - France

ABSTRACT

Double layer-coatings were performed on planar graphite substrates by a succession of two reactive CVD. The process consists of forming a carbide layer by means of a reaction between the substrate and a gas phase, then in converting part of this carbide into a new ceramic (carbide, boride) by a second RCVD. For illustration of this technique, a double layer TiC/TiB_2 has been characterized. Then results have been useful to modify the surface of every single filament of a carbon fibre.

INTRODUCTION

Over recent years, there has been a growing interest in the deposition of multiple layers of ceramics on reinforcing fibres to resolve the problems on interfacial compatibilities such as the fibre degradation caused by chemical reactions and by strong interfacial bonding between the fibre and the matrix or to improve the wettability by liquid metals or to increase the fibre-matrix adherence /see for example 1-5/. The production of these coatings by a conventional CVD process is not perfect because of the difficulty of making an homogeneneous coating on all the fibres in a tow. In a reactive CVD (RCVD) process, the coating is formed by reaction between the fibre surface and a flowing vapour. The coating formation is thus controlled by a diffusion mechanism, and its homogeneity is much better. However, as yet RCVD technique allows only the coating of a single-element substrate by a ceramic compound containing this element. For instance, by surface reaction between a carbon fibre and gaseous $SiCl_4$, BCl_3 and $TiCl_4$ diluted in hydrogen at atmospheric pressure, one obtains a thin layer of SiC, B_4C or TiC respectively /1-5/. The purpose of this work was to adapt the RCVD technique to form a double layer of either two carbides or one carbide and one boride on carbon fibres. The process consists of forming a first layer of a carbide on a carbon substrate, then in converting part of this layer into a new layer.

In this publication, preliminary results concerning the formation of a double-layer coating on bulk graphite substrate are given, the first layer is either TiC or B_4C, the second layer is SiC, B_4C or TiB_2. We show that the process can be transposed to the carbon fibre treatment. All experiments were carried at atmospheric pressure.

I. RESULTS ON PLANAR GRAPHITE SUBSTRATE

The RCVD conditions have been optimized on graphite disks (8 mm in diameter and 1 mm thick) and after this study, some experiments were performed on high strength carbon fibres, T300 supplied by Torayca. The gas precursor ($TiCl_4$ or $SiCl_4$) is liquid at room temperature, therefore its vapour was carried in the reactor by H_2 or by Ar. The flowrates of BCl_3 and the carrier gas were monitored by gas mass flowmeters. The first layer of TiC and B_4C was obtained by RCVD in the conditions described previously /6-10/. The conditions of the second RCVD, chosen after preliminary experiments are summarized in table 1. In order to characterize the coating and the influence of the RCVD parameters on the morphology texture of the ceramic layer, several complementary techniques were used. The layer nature was identified by a X-ray diffractometer and by an XPS spectrometer. The surface and the cross-section of the coating were observed by SEM and analysed by a WDX spectrometer. The thickness measurements of the layers were made by optical microscopy with an image analyser on a polished cross-section of treated substrate embedded in resin.

In all the cases studied in this work, we have always observed a reaction between the carbide and the gas phase with formation of a double layer-coating. SEM micrographs of representative specimens are shown in figure 1: the presence of two adherent and continuous layers are clear.

1.1. Formation of TiC or B_4C on a planar graphite substrate
TiC coating was obtained thanks to the reaction :
$$TiCl_{4(g)} + 2 H_{2(g)} + C_{(s)} \rightarrow TiC_{(s)} + 4 HCl_{(g)}$$
The effect of the temperature of the RCVD of TiC was examined in the temperature range from 1200 to 1800°C. The thickness of the layers slightly increased by increasing the reaction temperature. The growth rates were 2.2 and 2.8 $mg.h^{-1}$ for temperature of 1200 and 1800°C, respectively. The TiC phase was crystallized and it has a columnar growth. EMPA showed a gradient of the Ti content in the layer : the carbide composition in contact with the graphit was close to stochiometric TiC, the one on the layer surface was $TiC_{0.6}$.

B_4C coating has been formed according to the following reaction :
$$4 BCl_{3(g)} + 6 H_{2(g)} + C_{(s)} \rightarrow B_4C_{(s)} + 12 HCl_{(g)}$$

The conditions of the RCVD were T=1500°C and t=30 min. The thickness of the coating was about 70 μm.

1.2. Formation of a double layer-coating of carbides
1.2.1. TiC-SiC double layer
The SiC coating was formed by reacting $SiCl_4$-H_2 with TiC at T>1200°C.

$$SiCl_{4(g)} + TiC_{(s)} \rightarrow SiC_{(s)} + 4 TiCl_{4(g)}$$

The conversion of TiC into SiC was observed by means of the (111) XDR peak position of SiC ($2\theta = 35.7°$) and TiC ($2\theta = 36.1°$). SEM shows the presence of two adherent and continuous layers. The upper layer was identified by XPS as SiC containing silicium oxycarbide and silica (BE's 100.7, 101.5 and 103.2 eV respectively for the Si_{2p} peak). After sputtering the surface by a Ar ion beam, Si and C are the major constituents in the coating. Examination of the cross-section confirms that the SiC layer is dense and exhibits an excellent contact between the TiC layer and the SiC one. However, notice that the TiC layer exhibits a porous structure. The hole formation can be explained by diffusion phenomena, and their presence reduces the adherence between the carbide layers and the graphite. The nature of the carrier gas of $TiCl_4$ is very important (Table 1). When argon is used, EMPA reveals the presence of C in the SiC layer, and when H_2 is used as carrier gas, SiC is the only identified phase. These results are in agreement with a previously thermodynamic forecast /11/. Moreover, the use of Ar instead of H_2 requires a higher temperature. EMPA shows the Si presence in the TiC layer, such presence can occur by grain-boundary or lattice diffusion.

1.2.2. B₄C-SiC double layer
$SiCl_4$-H_2 was also reacted with B_4C for 30 to 60 min at 1030-1230°C. Continuous coatings were formed /12/. XPS analysis revealed that SiC was indeed formed, and possibly some silicon oxide. The presence of a double layer B_4C-SiC was confirmed by XDR, XPS and EMPA. An important observation was the fact that the conversion of the B_4C layer into SiC can be complete. The RCVD temperature is an important parameter, the growth rate and SiC morphology depend on temperature. 1130°C is the optimal temperature, and with this condition, the SiC layer is dense and adherent. At higher temperature, the SiC coating is porous and its growth is dentritic, a such layer is not consistent with an effective diffusion barrier.

1.3. Formation of TiC-TiB₂ double layer-coating
The reaction between TiC and BCl_3-H_2 was investigated with the aim to obtain a TiB_2 coating by RCVD on a carbon substrate. Some preliminary experiments have showed that TiC reacted really with the reactant gases to give a continuous and adherent coating of TiB_2, but this boride was not always the alone condensed phase : elementary boron was detected by XDR in different coatings , B formation was depending on H_2/BCl_3 ratio (table 1). A thermodynamic analysis was made to examine situations involving RCVD conversion of TiC in TiB_2 and to define the optimal conditions of deposition. A difficulty of these calculations comes from the definition of the initial system because the nature of the substrate is evolving during the RCVD treatment

and the amount of TiC in contact with the initial gas is continually decreasing. An approach consists of envisaging different sytems TiC-BCl$_3$-H$_2$, each one being characterized by a different TiC mole number n°_{TiC}. The RCVD initial step is defined by n°_{TiC} in excess, and the final step by n°_{TiC} equal to 1 The phase diagram (figure 2) resumes the equilibrium calculations, at 1300K. It shows the formation of TiB$_2$ whatever n°_{H2} and n°_{TiC} but also the possible codeposition of TiB$_2$-B$_4$C, TiB$_2$-B$_4$C-C, TiB$_2$-B$_4$C-B and B$_4$C-B. The different domains are given as a function of n°_{H2}, on the basis of 1 mole of BCl$_3$.

The formation of B$_4$C expected from the calculations did not correspond to the experimental results. This carbide was never identified experimentally by XDR and by EMPA. Besides, the B formation was only observed for an initial reactant gas rich in H$_2$ (H$_2$/BCl$_3$>20). This result was again not conform to calculations. In RCVD reactor, equilibrium was not reached and it is known that such mixtures lead to B in absence of C. Complementary experiments have shown that B formation was also favorised at high temperature (T>1300°C).

II. CARBON FIBRE COATING

PAN-based T300 fibres have been coated with TiC-SiC, B$_4$C-SiC and TiC-TiB$_2$ double layers. The duration of each RCVD was 1 minute. The fibre surfaces were characterized by XPS and the presence of SiC and TiB$_2$ were confirmed by this technique. For instance, the photopeaks C$_{1s}$, B$_{1s}$ and Ti$_{3p/2}$ (figure 3) correspond to a TiC/TiB$_2$ coated fibre. Notice the presence of BN in the coating but it is a characteristic of T300 fibre treated in BCl$_3$-H$_2$ mixture.

The tensile strengths, determined by a single filament method of the coated fibres, are shown in table 2. The σ_R values indicate that the titanium carbide or the boron carbide coating results in a marked strength loss. The bonding between the carbide and the fibre is strong and when a tensile strength is applied to a such fibre, cracks are formed in the brittle layer and the degradation in strength is due to the crack propagation into the fibres. The fibres coated by a double layer-coating exhibit higher strength. The partial recovery of the properties of the initial fibre seems to indicate that a lower bonding between the first carbide layer and the core carbon fibre. This phenomenon is not yet clear, it can be connected with the observations mentioned above of the effect of the second RCVD on the texture of TiC deposited on planar graphite substrate.

III. CONCLUSION

RCVD technique is a suitable technique for obtaining one or two continuous coatings of a ceramic material on a ceramic substrate. Experiments have been performed on bulky graphite substrate and on carbon fibres. They show that the process permits to obtain carbide coating but also boride coating by means of two successive processes. In a first step, a deposit of titanium carbide was performed using TiCl$_4$-H$_2$ gas mixture ; in a second step, double coating TiC/TiB$_2$ has been formed by attack of TiC by BCl$_3$-H$_2$ gas mixture.The conversion of the first layer can be partial or complete.

Acknowledgements : We thank the French Direction Recherches et Etudes Techniques for its financial support.

REFERENCES
1. A. Shindo, K. Honjo, "Composites 86 : Recent Advances in Japan and US" (K. Kawata, S. Umekawa, A. Kobayashi, ed.) Proc. Japan-US. CCM-III, Tokyo (1986) 767
2. M.Sung Kim, J. S. Chun, Thin Solid Films, 107 (1983) 129
3. C. Colombier, B. Lux, J. Mater. Sci. 24 (1989) 462
4. T.E. Strangman, R.J. Keiser, US Pat. 4,668,579 (26 May 1987)
5. T.M. Besmann, J. Am. Ceram. Soc. 69(1) (1986) 69
6. H. Vincent, J.L. Ponthenier, J. Bouix, J. Cryst. Growth, 92 (1988) 553
7. C. Vincent, J. Dazord, H. Vincent, J. Bouix, L. Porte, J. Cryst. Growth 96 (1989) 871
8. H. Vincent, H. Mourichoux, J.P. Scharff, C. Vincent, J. Bouix, Thermochimica Acta 182 (1991) 253
9. H. Vincent, C. Vincent, J.L. Ponthenier, H. Mourichoux, J. Bouix, 3ème Confér. Europ. Matériaux Composites (ECCM-3), Bordeaux, 20-23 Mars 1989 "Development Science and Technology of Composite Materials" (A.R. Bunsell, P. Lamicq, A. Massiah, ed., Elsevier Appl. Science : London and New-York) (1989), 257
10. J. Bouix, C. Vincent, H. Vincent, R. Favre, Mat. Res. Soc., Symposium Proceedings " C. V. D. Refractory Metals and Ceramics", 29 Nov - 1 Déc.1991, (T.M. Besman, B.M. Gallois, Pittsburgh, Pennsylvania, ed.) Vol. 168, 305
11. H. Vincent, J.P. Scharff, C. Vincent, J. Bouix, Mat. at High Temperatures 10 (1) (1992) 2
12. T. Piquero, J. Bouix, J.P. Scharff, C. Vincent, H. Vincent, J. Alloys and Compounds 185 (1992) 121

		First layer	Second layer			
Sample	TiC	B_4C	SiC	SiC	SiC	TiB_2
Substrate	Graphite	Graphite	TiC/C_{gr}	TiC/C_{gr}	B_4C/C_{gr}	TiC/C_{gr}
Carrier gas	H_2	H_2	H_2	Ar	H_2	H_2
Precursor	$TiCl_4$	BCl_3	$SiCl_4$	$SiCl_4$	$TiCl_4$	BCl_3
T (°C)	1200		1200	1400	1230	1030
t (min)	120	30	60	60	30	15
TiC	70 (µm)		50 (µm)	70 (µm)		40 (µm)
B_4C		70 (µm)			50 (µm)	
SiC			20 (µm)	<1 (µm)	20 (µm)	
TiB_2						15 (µm)

Table 1 . RCVD conditions for the formation of a second layer of SiC, B_4C or TiB_2 on a planar graphite substrate coated by a first layer of a carbide.

	T300	Coating			Coating	
		TiC	TiC/SiC	TiC/TiB_2	B_4C	B_4C/SiC
σ_R (MPa)	3100	1600	2100	2200	2200	2700

Table 2 . Tensile strength of coated T300 fibre

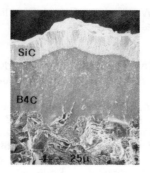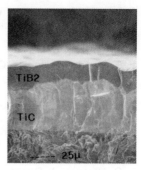

Fig 1 : SEM micrographs of cross-sections of double layer-coatings

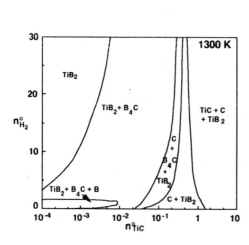

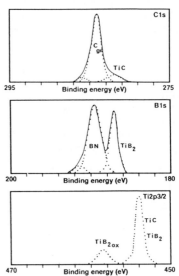

Fig 2 : calculated RCVD phase diagram

Fig 3 : XPS analysis of a T300 fibre coated by a double layer TiC-TiB$_2$

CREEP DAMAGE OF Al2O3/Al ALLOY COMPOSITES

X.C. LIU, C. BATHIAS

ITMA - CNAM - 2 Rue Conté - 75003 Paris - France

Abstract

Creep damage at 300°C for two cast aluminium alloy reinforced by Al_2O_3 short fibres randomly oriented in the matrices have been studied. Creep damage began by dislocation creep in matrix, followed by grain slipping, then fibre rupture and F-M interface debonding and sliding, and finally matrix free deformation and fracture. Fibres traversing grain boundaries can reinforce grain boundary and resist its sliding at elevated temperature.

Introduction

Metallic matrix composite (MMC) is a class of potential materials that can be used at elevated temperature. The incorporation of a higher melting point reinforcement can substantially increase the temperature capability relative to that of the matrix alone, for example, commercial aluminium alloys are restricted to structural applications at temperatures below about 200°C, however, the incorporation of 10-20% by volume of SiC fibre in aluminium matrix allows similar short-term strength to be retained to maximum temperatures of about 400°C , even more elevated temperature [1]. Most of the structural materials are hoped to use for long time, shape changes arising from creep generally are undesirable and can be the limiting factor in the life of a part. For example, blades on the spinning rotors in turbine engines slowly grow in length during operation and must be replaced before they touch the housing, thus, it is important to estimate the long-term stability of such materials under stress at elevated temperatures.

Some models for the creep of unidirectional fibre MMCs, with short or long fibres, and loaded in or off the fibre directions, were described without considering the effect of fibre matrix (F-M) interface [2-7]. Considered the effects of F-M interfaces on creep deformation, Goto and Mcleam [8,9] described a creep model for unidirectional fibre-reinforced MMC. They concluded that weak interfaces had a very large effect on the creep behaviors of aligned short fibre composites, but no significant effect on creep performance of continuous fibre composites. Nieh[10] and Lilholt [11] studied creep behavior of SiCw/2024 and SiCw/6061 composites respectively, in which the whiskers were aligned. It was concluded that minimum creep rate for the composites was strongly dependent on the applied stress.

Evidently, orientation and dimension of fibres in composite are important factors for the creep characteristics. However, few published papers on the creep of random orientation short fibre MMC can be found. The present paper deals with the creep damage of two random orientation alumina short fibre cast aluminium alloy composites, most of our attention are paid to roles of randomly oriented fibres during the course of creep.

Experimental

Two random short fibre composite materials with a fibre volume fraction of 20%, Al_2O_3/Al-7Si-0.6Mg and Al_2O_3/Al-5Si-3Cu-1Mg (in wt%), were used. The structure and properties about the fibre are presented in table 1.

The composites were manufactured by the method of squeeze casting. After casting, they were subsequently subjected to the following heat treatment:

For Al_2O_3/Al-5Si-3Cu-1Mg: 6 hrs. at 510°C + water quench (70°C) + 6 hrs. at 160°C

For Al_2O_3/Al-7Si-0.6Mg: 6 hrs. at 540°C + water quench (25°C) + 6 hrs. at 160°C

Typical fibre orientation and microstructure in the composites are shown in Fig.1. From it, it is clear that fibre orientation in the matrix is almost random in tridimensional space, and the dispersion of intermetallic phases or precipitates in the matrix is homogeneous.

Table 1. information about SAFFIL Al2O3 fibers

composition	Al2O3: 97%, SiO2: 3%
crystal structure	δ-Al2O3 (polycrystalline)
density	3.3
thermal expansion coefficient	$8 \times 10^{-6}/°C$
mean length	150 μm
mean diameter	3 μm
service temperature	<1600°C
strength	2000 MPa
strain to failure	0.67%
elastical modules	300 GPa

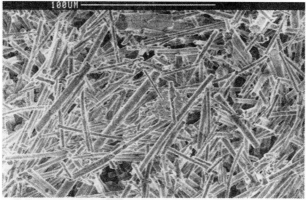

Fig.1.Fibre orientation in the composites (deeply etched sample) x 100

For a conventional metallic material, if the creep test is effected at constant external tensile load rather than at constant stress, then the stress will constantly increase as the creep strain reduces the cross-sectional area of the specimen and this can cause creep to accelerate. However, reduction of the cross-section area of the composite specimens used in our study during the creep tests was very small, hence, even the creep tests were carried out at a constant external tensile load, it can be considered to be done at a constant stress. The main difficulty to do creep test for this kind of materials comes from their small deformation. In order to precisely detect creep deformation of the materials, an Instron 8501 machine was used, which can determine the displacement with a precision of 1 micron. An induction heating instrument equipped a precise temperature control system was used. The creep deformation-time curves were continually and in time recorded through the whole creep test course with a graphic recorder. The tests were conducted on dog-bone specimens at 300°C and several stress levels. In order to know the process of creep damage, some polished specimens before the testing were examined by optical or electronic microscopes before and after creep testing for a certain time.

Results and Discussion

I. Typical creep curve

Typical deformation-time curve is presented in Fig.2., from which it can be seen a dominant primary-creep, i.e., the deformation is mainly concentrated in the primary-creep, and a stable secondary-creep. Also can be seen from it that after the deformation in primary-creep had arrived at a maximum value, the specimen began to contract for a certain time, then, the deformation restarted to increase, the phenomenon is defined as short-term negative creep. Most of the curves show this kind of negative creep for the composites used in the study.

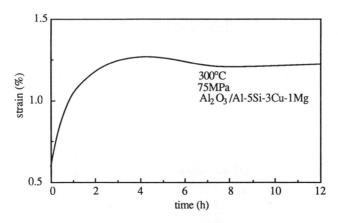

Fig.2. Typical creep curves showing a noticeable negative creep

For the specimen with a band free of fibres, and the band of which traverse most of the specimen cross-section as shown by the photo in the right of Fig.3, even the load was not high,

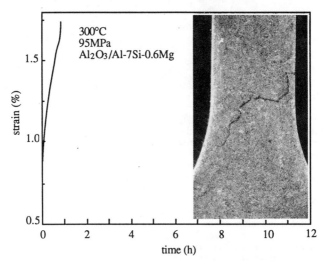

Fig.3.Creep curve of the specimen with a band free of fibres
(the black line in the photo is a band free of fibres x 6)

there were neither negative creep nor secondary-creep, and the deformation in ternary-creep and total strain to failure were still very small (the left of Fig.3.).

During creep, dislocations are created and forced to move through the material. This leads to work hardening as the dislocation density increases and the dislocations encounter barriers to their motion. At low temperature, an ever-diminishing creep rate results; however, if the temperature is sufficiently high, dislocations rearrange and annihilate through recovery events. The combined action of hardening and recovery processes during primary-creep can lead to negative creep in the case that the hardening action is dominant for a certain time. This short-

528

term negative creep was also observed in tests on quenched and tempered 2.25Cr-1Mo steel at 482°C and 538°C for the same steel in the normalized and tempered condition [12]. When both hardening and recovery processes during primary-creep are stable and in part compensated each other, a stable secondary-creep arrives.

Finally, it should be pointed out that there is no significant difference between the creep curves of the two composite materials. However, negative creep for Al_2O_3/Al-7Si-0.6Mg composite was much smaller with respect to Al_2O_3/Al-Si-3Cu-1Mg composite.

II. Creep damage evolution

It is well known that dislocation creep, grain boundary slip and diffusion creep are three possible and main creep mechanisms in metallic materials. Since the creep of ceramic fibre at 300°C can be neglected, creep deformation of the MMC mainly consists of that of the matrix, in other words, the above three creep mechanisms are also the main creep mechanisms to the composites. However, it is necessary to consider the role of F-M interface in MMC.

Creep deformation began by dislocation creep in matrix. First all, plastic deformation band appeared(Fig.4a.). However, the size is over that of gains, therefore, it can not been observed in a gain. After creep of the first several hours, an obvious grain slip appeared, especially, in the region poor in fibres (Fig.4b.). Afterwards, some weak fibre-matrix interfaces began to slid, as shown by the traces near the slid fibres (Fig.4c). When the total deformation was over the deformation to fail of the fibre, some fibres broke at the most weak point (Fig.4d). When the cracks run across F-M interfaces, the interfaces debonded. Without the restriction from fibres, the matrix can freely deform, the ruptured fibres were deeply inserted in the stretched hollows, so that it is difficult to find fibre in the fracture surface (Fig.4e.), and finally, the specimen is ruptured(Fig.4f).

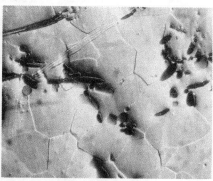

(a.) Matrix plastic deforming
(b.) Grain boundary slipping

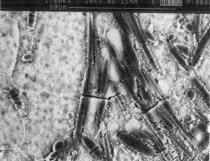

(c.) F/M interfaces sliding
(d.) Fibre cracking

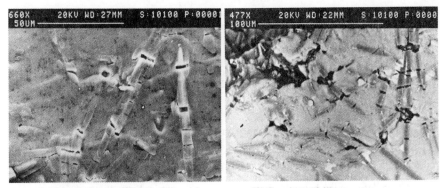

(e.) Matrix free stretching   (f.) Specimen failing

Fig.4.Creep damage development 1100x

III. Role of fibres during creep

Even though like metallic materials, dislocation creep, grain boundary slip and diffusion creep are the main damage mechanisms for matrix composite materials, the fibres in MMC will strongly affect matrix creep deformation in the case of strong F-M interfaces.

Firstly, the fibres traversing grain boundaries efficiently resist grain boundary slips, therefore, it is difficult to observe noticeable grain boundary slips in the region rich in fibres as shown in Fig.5. By contrast, in the region poor in fibres for the same specimen, grain boundary slips can be clearly seen as shown by Fig.4b., in which the grain boundaries have been displayed by creep rather than by etchant, and the lines labelled by A were obviously displaced a little at a grain boundary. It follows that a small creep rate of the composites is related to this role of fibres.

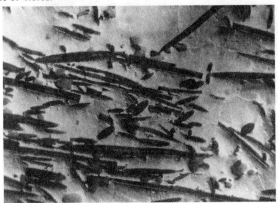

Fig.5.Intergrainal slip during creep in region rich in fibres: x1100

Secondly, it has been well demonstrated that ceramic fibres in MMCs strongly resist dislocation glide, and a number of dislocations pile-up in the front of the fibres. Evidently, this role still exists even at elevated temperature, especially when the fibres are randomly distributed in the matrix. In this case, liberal glide of dislocation, no matter in which direction, is restricted in a very limited space, leading to so many dislocations piled-up in the fronts of the fibres that dislocation glide will be more and more difficult. This is a substantial creep strengthening mechanism, and its strengthening efficient is much more significant than these in both particulate and unidirectional fibre MMC. That is the main reason why a noticeable short-term negative creep occurred in the tests for the composites. Of cause, when external load is high enough to overcome the resistance from stress field induced by the piled-up dislocations in the fronts of fibres, the piled-up dislocations can move again by formation of dislocation loops around the fibres, at that monument, the negative creep commences reducing.

Finally, fibres in MMC are the main constituent carrying load, the load supported by matrix is smaller than the applied normal stress, thus the creep is conducted at smaller real stress , the creep contributed by grain boundary slip is relatively increased, because the smaller the load, the more important the portion contributed to creep by grain slip.

All the above points are favorable to reduce creep rate of the composites. Hence, generally speaking, the creep rate of a fibre MMC can be smaller one order of magnitude relative to their matrix alloy alone. On the other hand, since the creep damage often begins in a region poor in fibre. A band without fibres, will be the most weak part, where the material will rupture, as shown by the specimen in the right of Fig.3.

Conclusions

Creep damage began by dislocation creep in matrix, followed by the slip of both grain boundary and weak F/M interfaces, then fibre rupture and F-M interface debonding, and finally matrix free deformation and fracture. The presence of short-term negative creep in primary-creep is an important feature for the composites. Fibres traversing grain boundaries can reinforce and hence resist grain boundary sliding at elevated temperature.

References

(1). S.Nishide, High Temperature Metal Matrix Composites, Mechanical Properties and Applications of MMC, Proceedings of the Japanese-French workshop, Paris, 19-21 February 1992, Ed. S.Nishijima and C. Bathias, PP.151-155.
(2). S.T.Mileiko, Steady State Creep of a Composite Material with Short Fibres, J. Mater. Sci., 1970, 5, PP.254-269.
(3). D. Mclean, Viscous Flow of Aligned Composites, J. Mater. Sci., 1972,7, 98-104.
(4). A. Kelly and K.N.Street, Creep of Discontinuous Fibre Composites II. Theory for the Steady State, Proc. Roy. Soc., 1972, A328, PP.283-293.
(5). H. Lilholt, Relations Between Matrix and Composite Creep Behavior, Fatigue and Creep of Composite Materials, Eds. H. Lilholt, R. Talreja, Riso National Laboratory, Roskilde, 1982, pp.63-76.
(6). H. Lilholt, Creep of Fibrous Composites Materials, Comp.Sci. and Techn., 1985, 22, pp. 277-294.
(7). M. Mclean, Modelling of Creep Deformation in Metal Matrix Composites, Proc. of the Fifth International Conference on Composite Materials, Eds. W.C.Harrigan, J. Strife, A.K. Dhingra, The Metallurgical Society, AIME, Warrendale, 1985, pp.37-51.
(8). S.Goto and M. Mcleam, Role of Interfaces in Creep of Fibre reinforced Metal-Matrix Composites-I, continuous fibres, Acta Metall. Mater. Vol. 39, No.2 pp.153-164, (1991)
(9). S.Goto and M. Mcleam, Role of Interfaces in Creep of Fibre reinforced Metal-Matrix Composites-II, short fibres, Acta Metall. Mater. Vol. 39, No.2 pp. 153- 164, (1991)
(10). T.G.Nieh, Metal Trans. 15A (1984) 139-146.
(11). H. Lilholt and M. Taya, Creep Behavior of the Metal Matrix Composite Al 2124 With SiC Fibres, Proc. of the International and European Conference on Composites Materials, Vol.2, Eds. F.L.Matthews, N.C.R. Buskell, J.M.Hodgkinson, J.Morton, 1987, Elsevier Applied Science, London and New York, PP.2.234-2.244.
(12). H.R. Voorhees, Assessment and Use of Creep-Rupture Properties, Metals Handbook Ninth Edition Vol.8 Mechanical Testing, American Society for Metals, Ed. by John R. Newby et al. 1985 pp. 329-342.

HIGH STRAIN RATE SUPERPLASTICITY OF AlN AND TiC PARTICULATE REINFORCED ALUMINIUM ALLOYS

T. IMAI, G. L'ESPERANCE*, B.D. HONG*, J. KUSUI**

Government Industrial Research Institute Nagoya
1 Hirate-Cho - Kitaku - Nagoya 462 - Japan
*Ecole Polytechnique de Montreal - PO Box 6079 - Station "A"
Montreal (Quebec) H3C 3A7 - Canada
**Toyo Aluminium Company - 341-14 Higashiyama
Otani Hino-cho - Gamou-gun Shiga 529-16 - Japan

ABSTRACT

AlN/6061 and TiC/2014 Al alloy composites fabricated by a P/M route were hot-rolled after extrusion. AlN/6061 Al composites exhibited a m value of 0.5 and a total elongation of 300-350 % at strain rates from 0.1 up to 1.0 (1/sec) at 873 K. TiC/2014 Al composite exhibited a total elongation of 200-300 % at strain rates from 0.08 up to 1.3 (1/sec) at 818 K, although its m value was only 0.23. Flow stress-true strain curves of TiC/2014 Al composite indicated that softening occurs during superplastic deformation. Interfacial segregation of elements Mg and Cu in both composites was detected by nanoprobe microanalysis in TEM. Filaments which may be associated with viscous flow were also observed on fracture surfaces of the two composites.

INTRODUCTION

Ceramic whisker or particulate reinforced aluminum alloy composites have great potential for automobile engineering components and aerospace structure since the composites have a high specific elastic modulus and specific tensile strength, excellent wear resistance and heat resistance, low thermal expansion and good dimensional stability.

Ceramic whisker or particulate reinforced aluminium alloy composites, however are limited by their low tensile ductility at room temperature, low fracture toughness, low formability in conventional forming techniques. They are also difficult to machine with ordinary tools.

Recently, however, it has been found that aluminium alloy composites reinforced by SiC or Si_3N_4 whiskers or particulates can exhibit high-strain-rate-superplasticity (HSRS) at strain rates in the range from 0.1 to 10 (1/sec) /1-7/. Although the superplastic deformation behavior was attributed to grain boundary sliding of fine grains /8/, liquidus interface sliding between the matrix and reinforcement /8,9/ and dynamic recrystallization, a number of papers showed that the strain rate sensitivity of the flow stress (m value) of metal matrix composites (MMC) /3, 4, 5, 6/ was about 0.5, indicating that the grain boundary sliding is a dominant deformation mechanism.

In addition to SiC and Si_3N_4 whiskers or particulates, some investigations showed that AlN and TiC /10/ particulates are also prospective reinforcements for composite materials. Therefore, the purpose of this study is to investigate the superplastic behaviour of AlN and TiC particulate reinforced aluminium alloy composites.

1. EXPERIMENTAL PROCEDURES

Reinforcement materials used were two kinds of aluminium nitride particles (AlN1, AlN2) and titanium carbide particles (TiC). AlN1 (average particle size of 1.42 μm) was fabricated by a deoxidation nitrogenous method and AlN2 (average particle size of 1.78 μm) by a direct nitrogenous method. The chemical compositions of AlN1 and AlN2 particles and of TiC particles (average particle size of 1.0 μm) are shown in Table 1.

6061 and 2014 aluminum alloy powders (particle size under 45 μm) were mixed with AlN and TiC particles respectively in a alcoholic solvent for 24 hours. The mixed powders were sintered at 773 K in vacuum under a pressure of 200 MPa for 20 minutes. As-hot-pressed billets were reforged at 773 K in air and at a pressure of 495 MPa for 20 minutes.

Thermomechanical processing used to produce superplasticity in these composites was hot rolling after extrusion. Hot extrusion was performed with an extrusion ratio of 44 at 773 K. Hot rolling at 818 K and reheating for about 5 minutes between passes of composites covered by steel plates were repeated until a final thickness of the composites became about 0.7 mm.

Tensile testing specimens, 4 mm wide and 5 mm long, were tested at 873 K for AlN/6061 Al composites and at 818 K for TiC/2014 Al composites. The initial testing strain rate varied from 0.01 to about 2 (1/sec). The microstructure and fracture surface were characterized by TEM and SEM, both equipped with a ultra-thin window X ray detector allowing the detection of light elements down to boron.

2. EXPERIMENTAL RESULTS AND DISCUSSION

2.1 AlN/6061 Al composite

Fig. 1 shows the relationship between flow stress and strain rate of 6061 aluminum alloy composites reinforced by AlN1 and AlN2 particles. Testing temperature is 873 K. The flow stress can be related to true strain rate ($\dot{e}$) via the equation $\sigma = K\dot{e}^m$, where K is a constant and m = δ (logσ)/δ (log$\dot{e}$), is the strain rate sensitivity of the flow stress. It is necessary that the m value is larger than 0.3 to consider the material superplastic. The m value of the two composites is about 0.5 as shown in Fig. 1 indicating that the predominant deformation mechanism of AlN1/6061 and AlN2/6061 Al composites is grain boundary sliding.

Total elongations of both composites are shown in Fig. 2. The total elongation of AlN$_1$/6061 Al composite was less than 200 % for strain rates in the range from 0.1 to 0.9 (1/sec), while that of AlN$_2$/6061 Al composite exhibited a value larger than 300-350 % in the same strain rate range. The AlN$_2$/6061 Al composite exhibited a total elongation larger than 250 % at the very high strain rate of 2 (1/sec). Since the testing temperature of 873 K is higher than that of the solidus temperature (855 K) of 6061 aluminium matrix, the presence of a liquid phase is expected, particularly at the matrix/reinforcement interface where segregation can occur/9/.

2.2 TiC/2014 Al composite

The total elongation of TiC/2014 Al composite is shown in Fig. 3 as a function of strain rate. It exhibited a total elongation larger than 200 % for strain rates in the range from 0.08 to 1.3 (1/sec) and a total elongation larger than 300 % at a strain rate of 1.3 (1/sec). It becomes obvious that the TiC/2014 Al composite behaves as a high strain rate superplastic composite at 818 K, although its strain rate sensitivity of the flow stress (m value) is only about 0.23. This composite was also deformed in the partially liquid condition since the deformation was performed at a temperature higher than the solidus (780 K) of 2014 Al.

The flow stress of TiC/2014 Al composite as a function of true strain exhibited a maximum value at a true strain of about 0.5 and then decreased as shown in Fig.4. This behavior indicates that dynamic recrystallization occurred during the superplastic deformation.

2.3 Microstructural characterization

TEM observations were performed for AlN$_2$/6061 Al and TiC/2014 Al composites before and after superplastic deformation. The grain shape in the two composites and in the two conditions is largely equiaxed with an average diameter of about 2~3 μm. Some elongated grains with a length to width ratio of about 2~3 were occasionally observed in the deformed condition. The AlN particle size distribution is relatively narrow with values between 1.0~2.5 μm in diameter while that of TiC particles is wider with a diameter from 0.5 μm to 3.0μm.

High magnification TEM observation and X-ray microanalysis (by energy dispersive spectrometry, EDS) showed that both AlN and TiC particles were chemically stable during the fabrication of the composites and their high temperature testing. There was no chemical reaction product observed at the matrix/reinforcement interfaces of AlN/6061 Al and TiC/2014 Al composites. As shown in Table 2, however, significant interfacial segregation of O, Mg and Cu in the AlN$_2$/6061 Al composite and of Mg and Cu in the TiC/2014 Al composite was measured by EDS. This segregation locally reduces the melting point of the matrix material near the matrix/reinforcement interfaces and therefore promotes interfacial sliding during superplastical deformation. Similar results were obtained in an investigation of Si$_3$N$_4$w/2124 Al composites /9/.

2.4 Fracture surfaces

Figs 5(a) and (b) show SEM micrographs of fracture surfaces of AN2/6061 and TiC/2014 Al composites hot-rolled after extrusion, respectively. The rounded grains indicate that the fracture surfaces of both composites were covered by partially melted matrix material at the moment of the rupture. This is consistent with the fact that both composites were deformed above the solidus temperatures of the 6061 and 2014 Al matrices.

It is also possible to see filaments on the fracture surfaces of both composites. Similar filaments were observed on the fracture surfaces of superplastically deformed - $Si_3N_4w/2124$ /9/ and -$Si_3N_4w/7064$ Al /11/ composites, in which Mg and Cu segregation at interfaces between the matrix and reinforcements was also observed. The presence of a liquid phase expected at the matrix/reinforcement interface for reasons explained above is thought to be related to the formation of these filaments.

3. CONCLUSIONS

AlN1/6061, AlN2/6061 and TiC/2014 Al alloy composites fabricated by a P/M route were hot-rolled after extrusion and their superplastic characteristics investigated. The main results can be summarized as follows.

(1) Both the AlN1/6061 and AlN2/6061 Al composites hot-rolled after extrusion exhibited a strain rate sensitivity of the flow stress of 0.5 at 873 K. The AlN2/6061 Al composite had a total elongation of 300~350 % at strain rates in the range from 0.1 to 1.0 (1/sec) while the AN1/6061 Al composite had an elongation of less than 200 % in the same strain rate range. Further microstructural observations are underway to investigate the causes for this difference.
(2) The TiC/2014 Al composite hot-rolled after extrusion exhibited a total elongation of 200-300 % at strain rates from 0.08 up to 1.3 (1/sec) at 818 K, although the m value was only 0.23.
(3) Flow stress-true strain curves of the TiC/2014 Al composite indicated that softening may have occurred as a result of dynamic recrystallization.
(4) A fine grain matrix structure with an average size of 2~3 μm was observed in all the composites investigated. No chemical reaction was observed between Al_2N and 6061 Al, or between TiC and 2014 Al during fabrication of the composites and subsequent testing. Significant solute segregation was observed at AlN/6061 Al and TiC/2014 Al interfaces.
(5) The appearance of fracture surfaces was consistent with the presence of a liquid phase during deformation of the composites.

REFERENCES

1. T.G. Nieh, C.A. Henshall and J. Wadsworth, Stripta Metall., 18-12 (1984) 1405
2. M.W. Mahoney and A.K. Ghosh, Metall. Trans., 18A (1987) 6533.
3. T. Imai, M. Mabuchi, Y. Tozawa and M. Yamada, J. Materials Science letters, 9 (1990) 255.
4. M. Mabuchi, T. Imai, K. Kubo, H. Higashi, Y. Obada and T. Tanimura, Materials Letters, 11-10, 11, 12 (1991) 339.
5. M. Mabuchi, K. Higashi, Y. Okada, S. Tanimura, T. Imai and K. Kubo, Scripta Metall. Mater., 25 (1991) 2517.
6. T. Imai, M. Mabuchi and T. Tozawa, Superplasticity in advanced materials, edited by S. Hori, M. Tokizane and N. Furushiro (A Publication of JSRS, 1991) 373.

7. T. Tsuzuki and A. Takahashi, Proc. of 1st Japan International SAMPE Symposium (Nov. 28-Dec 1, 1989) 243.
8. T.G. Nieh and J. Wadsworth, Superplasticity in Advanced Materials edited by S. Hori, M. Tokizane and N. Furushiro (A Publication of JSRS, 1991) 339.
9. G. L'Esperance, T. Imai and B.D. Hong, ibid, 379.
10. A.K. Kuruvilla, V.V. Bhanuprasad, K.S. Prasad and Y.R. Mahajan, Bull. Mater. Sci., (printed in India) Vol 12 (1989), No 5, 495.
11. T. Imai, G. L'Esperance, B.D. Hong, M. Mabuchi and Y. Tozawa, Advances in Powder Metallurgy to Particulate Materials - 1992, compiled by J.M. Caus and R.M. German, Vol 9 (1992) 181.

Table 1 Chemical compositions of AlN and TiC particles used

materials	Fe	Si	Cu	Mg	Ni	Cr	O	N	C	Ca
AlN1	ppm 10	ppm 50	--	--	--	--	wt% 0.90	wt% 33	ppm 230	ppm 170
AlN2	ppm 69	ppm 102	ppm 5	ppm 14	ppm 5	ppm 6	wt% 1.29	wt% 32.9	wt% 0.03	--
TiC	wt% 0.28						wt% 0.60	wt% 0.29	wt% 19.41	

Table 2 Chemical composition (wt%) at interfaces and in matrices of the two composites investigated

	AlN2/6061 Al			TiC/2014 Al	
	O	Mg	Cu	Mg	Cu
interface	0.7*	2.5	0.5	2.1	12.1**
matrix	0.2*	0.5	0.1	0.7	1.9

* The values listed for oxygen are net X-ray intensity ratios of oxygen to aluminum, i.e. I_O/I_{Al}.
** θ phase ($CuAl_2$) precipitated as small particles on the TiC particle surface.

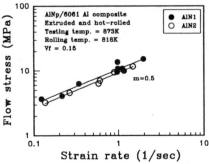

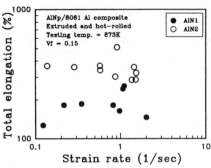

Fig. 1 Relationship between flow stress and strain rate of AlN/6061 Al composite hot-rolled after extrusion.

Fig. 2 Relationship between total elongation and strain rate of AlN/6061 Al composite hot-rolled after extrusion.

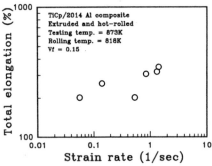

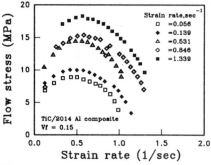

Fig. 3 Relationship between total elongation and strain rate of TiC/2014 Al composite hot-rolled after extrusion.

Fig. 4 True stress-true strain curves of TiC/2014 Al composite hot-rolled after extrusion at various rates.

Fig. 5(a) Fracture surface of the AlN2/6061 Al composite. 873 K, 0.56 (1/sec).

Fig. 5(b) Fracture surface of the TiC/2014 Al composite. 818 K, 0.85 (1/sec).

538

FABRICATION OF CARBON FIBRE REINFORCED ALUMINIUM COMPOSITE BY POWDER METALLURGY

A. VESELA, J. JERZ, P. SEBO

Institute of Materials and Machine Mechanics - Slovak Academy of Sciences - Racianska 75 - 836 06 Bratislava - Slovakia

ABSTRACT

Influence of pressing (733-773 K) and rolling (753-803 K) temperature on the strength of Al 6061-chopped carbon (T300, M40) fibres composite prepared by powder metallurgy were studied. In both cases only a slight increase in strength for low fiber volume fraction was observed. In these cases noncumulative mode of fraction results and for higher fiber volume fraction fiber clusters are forming decreasing the strength. Rolling increases the strength of the composites to the same level as the matrix.

INTRODUCTION

Composite materials have created considerable interest as structural engineering material.Carbon fibre reinforced aluminium alloy composites have, theoretically and practically, the potential for very high specific strength and specific stiffness.

Carbon fibres for unidirectional metal matrix composites have the advantage of lower price in comparison with other ceramic fibres. However,the wide application of carbon fibre-aluminium composites are still not realized due to several problems (poor wettability and strong chemical reaction with many metals) connected with the fabrication difficulty of the composites.

The wettability of aluminium against carbon fibre has been shown to improve if the temperature is raised over approximately 1273 K /1/.Interfacial chemical reaction is known to take place between fibre and matrix at above 773

K. Therefore, carbon fibres are usually coated with
metallic or ceramics layers or matrix-alloying methods to
improve the compatibility and wettability of carbon fibres
and aluminium are generally involved /2,3/. All these
methods of carbon fibre-aluminium composites used the
matrix in a liquid state. To avoid the problems with
wetting and interaction at the material processing
temperature (above the matrix melting point) the present
work is an effort to study the development and the
behaviour of high strenth type chopped carbon fiber
reinforced 6061 powder aluminium alloy matrix composites
prepared by powder metallurgy technique. The aim was to
find the parameters od pressing,volume fraction and
following optimal thermal forming from point of view the
best mechanical properties.

I.EXPERIMENTAL

The aluminium alloy 6061 powder and high strength
Torayca T 300 and high modulus M 40 chopped carbon fibres
were used in the present investigation. Aluminium alloy
powder and given amount of discontinuous carbon fibres 2
mm long were wet mixed and homogenized using ultrasound.
The slurry was poured into the pressing form with the
effort to achieve maximum directional alignment of the
fibres. The samples of 30 x 12 x 5 mm dimensions were
prepared by pressing the mixture at the temperatures of
733, 753 and 773 K, pressure of 30,40 and 50 MPa and time
of 900 s in a vacuum of $1x10^{-3}$ Pa. The samples were
prepared with 0, 3, 6, 8 and 10 vol.% of both types of
carbon fibres. Some of them were rolled and/or heat
treated. Thus we prepared three types of specimens:
- as pressed
- pressed and rolled at 753 and/or 803 K with reduction
rate of about 4 % per pass and with the total reduction of
75 %,
- pressed, rolled and T 6 heat treated: 803 K, 3 hours
followed by an immediate water quenching for solution
treatment and 443 K and 8 hours for artificial aging.The
time of 3 hours for heat treatment (803K) included where
appropriate the time when the samples were in furnace for
rolling.
From all types of these composites "dog bone" shaped
longitudinal tensile specimens were prepared. Tensile
testing was performed at room temperature and at
temperature of 573 K. Fracture surface from the tensile
tests were examined under a JEOL 100 C scanning electron
microscope.

II.RESULTS AND DISCUSSION

The typical properties of raw materials used in this
experiment are in Table 1.

540

To achieve good exploitation of powder metallurgy composites containing fibres or particles the relative density of the material after the manufacturing process should be higher than 95 %. The manufacturing process should result in modest damage of fibres, good bonding and the inicial blends should be homogeneous.

The theoretical, experimental and relative density of composites prepared by pressing at the temperature of 753 K,pressure of 50 MPa, time of 900 s in dependence on fibre volume fraction is in Fig.1. The relative densities of pressed and rolled composites vary from 99 % of the theoretical density for pure Al 6061 material to 94,4 % for materials containing 10 vol.% of fibres.The lower densities for 10 and higher vo.% fibres are due to the formation of fiber clusters (Fig.2).These clusters preclude the plastic flow of matrix material between the fibres lowering thus the number of strengthening fibres which results in decreasing the mechanical properties. For higher pressing temperature the relative density is higher.

Tensile strength of the matrix and the composites strengthened by 6 vol.% of both types of carbon fibres (CF) in dependence on the pressing temperature is in Fig.3. The strength of the matrix for all pressing temperatures is practically the same. The strength of composite reinforced by T 300 CF is slightly decreasing with increasing pressing temperature which can be due to the reaction occuring at the interface between fiber and matrix. M 40 CF (high modulus) due to their better developed structure have lower ability to react with aluminium (weaker bonding strength) and the dependence of composite strength is similar to the curve for matrix.

Fig.4 summarizes the influence of fabrication of composite parameters for 6 vol.% of carbon fibres.Number 1 corresponds to the matrix Al 6061, 2 is for composite with T 300 CF and 3 is for composites with M 40 CF.Rolling temperature of 803 K leads to slightly higher strength both of the matrix and composites (compare a, b). Rolling temperature (in studied range) does not effect the relative (composite/matrix) strength increasing. Strength increasing is the highest for 6 vol. % of T 300 CF only in the case of nonaged matrix and also there is no strengthening for M 40 CF (comp.a,b and c). Strength of Al 6061-6 vol. % M 40 CF is the same as that of nonaged matrix.

Strength of the matrix and of both types of composite after aging is increasing, however no one type of CF increases the strength of the composite relatively to the matrix (comp.b,c).

The tensile strength of composites in dependence on fibre volume fraction is shown in Fig.5. Three various treatments give three different matrices (with different strength).

The strength of as pressed specimens is highest for

about 3 vol.% of CF and mildly is decreasing with increasing fiber volume fraction. Strength increase is due to the strong bond between fiber and matrix which can be seen from noncumulative mode of fraction in Fig.6 and that each fiber is surrounded by the matrix. With increasing the fiber volume fraction there is more fibers which are not fully surrounded by the matrix and the clusters of fiber are arising. The dependence of the strength on fiber volume fraction of as pressed and rolled CM with T 300 CF is similar to that of as pressed CM. There is a better diffusion bonding of the matrix powder and the maximum reinforcement is shifted to about 6 vol.% of CF. For high modulus (M 40) CF the fracture mode is cumulative already for 3 vol.%. The fibres are pulling out because of weak bonding strength to the matrix.

For the composites with highest matrix strength (pressed,rolled and T 6 heat treated) the fracture mode is noncumulative. Although the bond strength is high the fibres do not contribute to the composite strength. This can be explained by lowering the fiber strength because of their reaction with aluminium /4/ and occuring aluminium carbide. Testing as pressed and as pressed and rolled and T 6 heat treated composites at 523 K showed the decrease of composite strength in both cases.

III. CONCLUSIONS

1. Manufacturing processes of Al 6061-chopped T 300 and/or M 40 CF composite material moderately damage the carbon fibres (pressing,rolling).
2. Careful blends pouring, pressing and rolling make most fibres aligned.
3. Higher pressing temperature (733-773 K) has slight effect on the strength of CM:decreases th strength of CM with T 300 CF and increases with M 40 CF.
4 Higher rolling temperature (753-803 K) increases the strength of the matrix and composites. It has no effect on the reinforcement by CF.
5. The strength increasing of as pressed CM is limited to low fiber volume fraction due to the forming fiber clusters for higher volume fraction.
6. T6 heat treatment of Al 6061-CF composite increases its strength in comparing with no aged specimens.Due to interaction between alumium and CF there is no increasing in composite/matrix strength.

Acknowledgements

The authors are grateful to the Slovak Grant agency for science (grant No 2/999286/93) for partial supporting of this work.

References

1. S.Rhee,J.Amer.Ceram.Soc.,53(1970),386
2. L.Aggour,E.Fitzer,M.Heyman and E.Ignatowitz,Thin Solid Films, 40(1977),97
3. Y.Kimura, Y.Mishima, S.Umekawa and T.Suzuki,J.Mat.Sci., 19(1984),3107
4. H.M.Cheng,S.Akiyama,A.Kitahara, K.Kobayashi and B.L. Zhou,Scripta Met.et Mater.,26(1992),1475

TABLE 1. Typical properties of carbon fibres and Al alloy matrix used

Material	Tens.strength [MPa]	Tens.modulus [GPa]	Density [kgm^{-3}]
T 300	3530	221	1760
M 40	2860	390	1810
Al6061 (as pressed)	148*		2719
Al6061+rolled	213*		
Al6061 + rolled + T6	341*		

*Values measured in this work

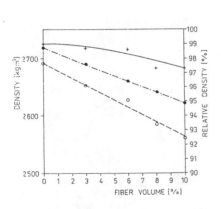

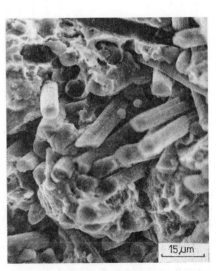

Fig.1.Theoretical (•),experimental (o) and relative (+) density of comp.materials.

Fig.2. Cluster of CF in Al 6061-10 vol.% T 300 CF.

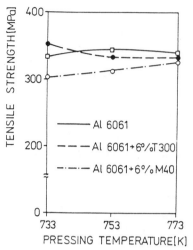

Fig.3.Influence of pressing
temperature on the strength
of Al 6061 matrix and composites Al 6061-6 % T 300 CF.

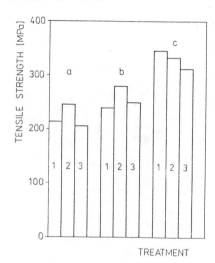

Fig.4.Tensile strength of the
matrix and CM with 6 vol.% CF
in dependence on the fabrication parameters:T_P=753 K,P=50
MPa,t=900 s;rolling temp.T_R=
a)753 K,b)803 K,c)803 K+T6.
1-Al 6061;2-Al 6061+6% T 300;
3-Al 6061+6% M 40 CF.

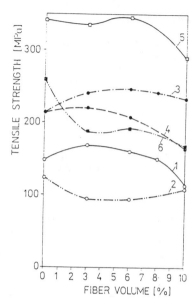

Fig.5.Tensile strength ofCM
in dependence on fiber volume
fraction for:as pressed CM at
RT (1) and 523 K (2),rolled
(at 803 K) CM for T 300 (3)
and M 40 (4) at RT and rolled
and T 6 heat treated for room
(5) and 523 K temperature (6).

Fig.6.Noncumulative fracture
mode of Al 6061-3 % T 300 CF
As pressed specimen.

ELECTRON-BEAM AND FRICTION WELDING
OF METAL-MATRIX COMPOSITES

W. B. BUSCH, B. SUTHOFF

Fraunhofer Institut für angewandte Materialforschung
Lesumer Heerstrasse 36 - 2820 Bremen (Lesum) - Germany

1. INTRODUCTION

An economic way for energy savings can be the use of light-weight structures. These can be either reinforced plastics or light metals if strength and stiffness are high enough. The stiffness can be enhanced in both systems by an addition of fibres or especially in light metals also with particles. Adhesive bonding is one way to produce a complex structure, but it is only useful for temperatures which differ relatively little from room temperature. For applications at elevated temperatures neither the polymer matrix nor the adhesives are stable. Therefore other joining techniques for reinforced materials on the basis of metal-matrices have to be found [1-3]. The industrial use of new materials such as MMC's depends not only on the availability, but for a high degree on the workability too. This includes all aspects of machining and joining.

With two very different fusion techniques the principal possibilities of these materials for welding shall be demonstrated. The limits of the fusion welding as well as first results of mechanical testing of MMC's will be demonstrated.

2. TEST MATERIALS

For our tests we have chosen two different materials with principle modifications of the reinforcements: a short fibre reinforced magnesium alloy with about 20 Vol.-% Saffil-fibres and an aluminium alloy, which contains about 20 Vol.-% particles of SiC.

The chemical composition of the magnesium alloy AZ91 is given in <u>Table 1</u>. Characteristic for this alloy are aluminium contents of nearly 10 Vol.-% and small additions of zinc and manganese. The compound was produced by squeeze-casting with preforms of Saffil fibres in dimensions of 180x100x20 mm.. The aluminium fibres, containing up to 4 % SiO_2 have diameters of 3 μm were commercially produced in preforms. The infiltration happens at typical melting temperatures of magnesium and with a maximum pressure during infiltration of about 100 MPa. The mechanical properties of this material are given in Table 2.

The micrograph of this magnesium-base material shows, that the fibres are nearly randomly distributed. This was expected, because of the use of platelets, Figure 1. The distances between the single fibres are more or less overall constant, but in single specimens there are areas of several square millimetres nearly without any reinforcement. These irregularities are typical for squeeze cast materials with higher material thicknesses and relative high amounts of fibres. For our investigations we ignored such irregular samples, for the welding tests as well as for the mechanical tests, when we saw these irregularities before testing.

The other test material is an aluminium based material, reinforced with particles of silicon carbide SiC. The production is done conventionally by blending, consolidating and extrusion. In comparison to the fibre reinforced material, the particles are much coarser with diameters of 15 to 20 μm. The chemical composition is given in Table 1 and the mechanical properties in Table 2. The microstructure loks very similar in all directions and the particles are equally distributed, Fig.2.

3. WELDING TECHNIQUES

The use of conventional welding technologies may be in some cases possible for these materials, but we expect that only automatic welding techniques are usable for this material because of the constant and reproducable heat input. For our tests we used two very different joining technologies: the electron-beam technology with its very high temperature gradients and the friction welding with a low temperature gradient. Our welding tests are not completed yet, but we can present first results.

3.1 Electron-beam welding

Electron-beam welding technology is characterised by a very concentrated energy input. The energy densities of beam welding is typically in the regime of 10^4 bis 10^6 W/mm^2. This energy density is high enough to melt or even to decompose the ceramik particles or fibres. On the other hand a minimum energy is needed to melt the base material. The art of beam welding is to melt this material without decomposing the ceramic phase. For our material we saw no other way than melting an unreinforced intermediate layer, which indirectly melts the edges of the reinforced materials. This sounds complicated but seams in practice verx clear: Between the two material ends that have to be joint a narrow foil of nearly the same material is mounted. Only this unreinforced material of about a quarter millimeter width will be melded directly by the electron-beam. This molten pool will melt the two borders of the reinforced materials and some particle reinforcements will be floated into the melt. When the beam has passed through, the melt will solidify. As the result, the strength of the welding seam will be a litthe higher than the plain material, but it will still apear weak compared with the reinforced parts. If there is still an interaction between the beam and the ceramic particles some small voids will apear in the welding seam, Fig 3. If the beam does not precisely match well with the narrow welding path, there will be numerous even coarse voids. The result is a welding seam with bad mechanical properties.

3.2 Friction welding

The problem with the high energy input existists not for friction welding. But especially for fibre reinforced materials the deviation of the fibre direction may cause problems. Particle reinforced materials are in principle easy joinable by friction welding. But this technique is restricted to more rotationally symmetric matrials. Because of this restrictions this technique can be used only for round or highly symmetric rodmaterials, but not for sheet materials. An advantage besides the low energy density of this technology is the consolidation under an external pressure, which prevents the formation of voids.

The result of a friction weldment is a fair welding seam with fine grain sizes of the matrix and particles with unchanged diameters, Fig.4.

Characteristic for this process is the formation of friction bulges, which may play an inportand role in fibre reinforced materials: the alignement of the fibres is disturbed and the strength is dramatically lower. In particle containing tubes or rods no influence is measurable. For this reason in the monent we join only particle containing materials with the friction joining process, but later on also fibre reinforced materials shall be joined by friction welding.

4. CHARACTERISATION OF THE WELDS

It could be shown, that possilbe interactions between the welding process an the ceramic phase are of fundamental interest. If especially for the electron-beam technology the welding parameters are wrong, the appearence of the welding seam differs very much from that, shown in Fig.3 as the material is nearly cut, Fig.5: In a SEM it can be observed, that a beam-material interaction does not always evaporate the ceramik, but in some cases only melts it. This viev shows an electron-beam cut in a fibre reinforced magnesium alloy. The free ends of the fibres that stuck out of the metal show rounded tips.

Changes in the metal-ceramic interfaces are also possible by the welding process. Therefore an analytical TEM shall be used. In this very monent these investigations have not yet begun.

With both materials we carried out some mechanical testings after welding. As this re-search program is still in progress we can only demonstrate first reslts. These are room-temperature tensile tests and fatigue tests of the fibre reinforced magnesium-base alloy.

The tensilt tests for the welded magnesium fibre reinforced specimens show for both modifications (both parts reinforced or reinforced/plain) nearly identical values, Fig.6. This result was espectes, because the interlayer in the reinforced modification was plain.

Both results are significant lower than for the reinforced base material. A change in the Youngs's modulus was not yet investigated. The large scatter of the tensile stregth of the specimens can be a result of insufficient welding. Observations of the fracture surfaces show, that the specimens fail either on defects in the base-material or on voids or inclusions in the weld. If these defects result from beam interactions with the ceramic fibres has not to be investigated.

The results of the fatigue tests show the same tendency: the fatigue strength of the welded material are remarcably lower than for the base material, Fig 7. This is naturaly a result of the weak interlayer. Changes in the welding parameters and in the thickness of the interlayer shall rise these unsufficient mechanical values.

For aluminium alloys the tendency for the void farmation is less pronounced. But as we have to use for electron-beam weldments an interlayer too, this layer must be as narrow as possible to achieve sufficient high mechanical values. First tests show fair tensil strength. The fatigue tests are foreseen for the end of this year.

For friction weldings no mechanical values are available in the monent. First tests showed however, that there are no void in the welding seam. So we expect nerly the same mechanical properties as for the base material.

5. SUMMARY

Welding of MMC's is not as simple as the joining of plain materials. Difficulties can be caused either by any phase reaction between the electron-beam and the reinforcements or by deviating the fibres perpendicular to the main loading axis. But we colud show that the joining of this class of material is possible. During the beam welding the interaction between beam and ceramic can decompose the particles or fibres. This decomposition leeds to voids in the solifying welding seam. Since we use small interlayers and are able to produce very narrow beam geometries by measuring the energy density of the electron-beam, these failures are nearly demished.

Friction welding does not decompose the ceramic or change the particles in their shape. The joining problem for particle reinforced material is a deviation of the fibres and resulting from this a mechanically low strength connection area. But for particle containing MMC's, we expect excellent mechanical properties.

[1] Arun Junai,A.,H.Botter and C.A.Brak
    Diffusion welding of Al-SiC Metal-Matrix-Composites
    Lastechnik (57 (1991),(1),20-23
[2] Utsunomiya,S.,Y.Kagava and Y.Kogo
    Laser Induced Phenomena in Metal-Martix-Composites
    Proc.Conf.: Composites '86 - Recent Advances in Japan and the United States
    Tokyo, Japan, June 23-25, 1986
[3] Cola,M.J.,Lienert,T.J.,Gould,J.E. and J.P.Hurley
    Laser welding of a SiC particulate reinforced aluminium metal-matrix-composite
    Proc.ASM-International Conf.:Weldability of Materials,Detroit 1990
    ASM-International, Materials Park,Ohio, USA

Alloy Typ		chemical composition in weight-%							
		Mg	Al	Zn	Mn	Si	Fe	Cu	Cr
AZ 91		Bal.	9.9	0.85	0.15	---	---	---	---
A6061	1.19	1.19	Bal.	0,10	0.07	0.52	0.29	0.18	0.07

Table 1: Chemical composition of the test materials

	AZ 91		A 6061	
	plain	fibres	plain	particles
Yield Strength	138 MPa	225 MPa	260 MPa	450 MPa
Tensile Strength	65 MPa	275 MPa	470 MPa	520 MPa
Hardness	84 HB	164 HB	120 HB	154 HB

Table 2: Some mechanical properties of the test materials

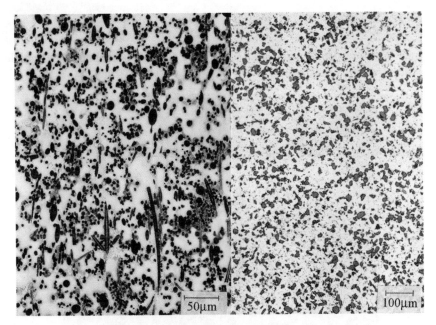

Fig 1: Microstructure of the magnesium alloy AZ91 reinforced with 20 vol-% Saffil-fibres

Fig.2: Microstructure of the aluminium alloy A6061 reinforced with 20 vol.-% SiC particles

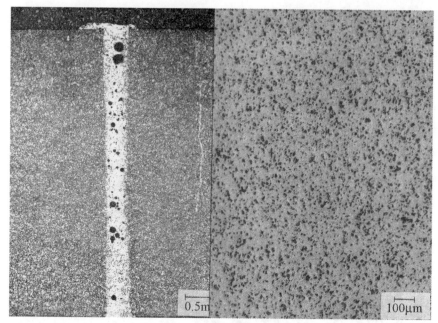

Fig 3: Elecron-beam welding seam of a
reinforced magnesium AZ91with
only a few small welding defects

Fig.4: Friction welding of the particle
reinforced aluminium alloy
A6061

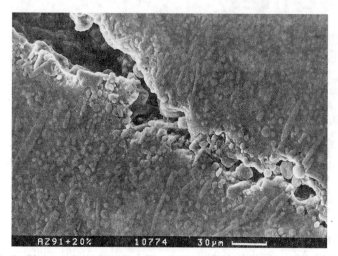

Fig 5: Electron-beam cut in a fibre reinforced magnesium alloy AZ91
caused by a decomposition of the ceramic fibres

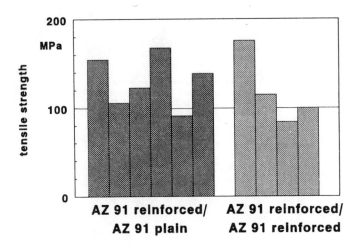

<u>Fig 6</u>: Tensile tests of fibre reinforced magnesium alloy AZ 91
in two different welding conditions

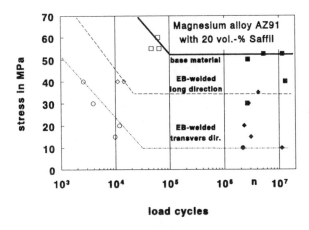

load cycles

<u>Fig 7</u>: Fatigue results of the fibre reinforced magnesiumbase alloy
AZ91+20 Vol.-% Saffil-fibres in the welded and unwelded
condition, hollow symbols: broken; filled symbols: stopped

EFFECT OF RECYCLED MATERIAL ON THE STRUCTURE AND PROPERTIES OF LOW DENSITY CARBON-CARBON COMPOSITES

I. DAVIES, R. RAWLINGS

Imperial College of Science, Technology and Medicine
Dept of Materials - Prince Consort Road South Kensington
SW7 2BP London - UK

ABSTRACT

Flexural and compression tests have been carried out on low density 2-D planar-random carbon-carbon composites, containing recycled (reworked) fibrous material fractions (compared to the total fibrous content) ranging from 0.0 to 1.0. Tests were carried out on two orientations of the composites in order to investigate the mechanical anisotropy. The mechanical property values so obtained, namely flexural and compressive strengths, initial modulus and mechanical anisotropy, have been correlated with the changes in density and microstructure associated with the additions of rework to the composite.

1. INTRODUCTION

Most of the literature related to carbon-carbon (C-C) composites is concerned with those that have densities in excess of $1.0 Mgm^{-3}$, eg /1-3/, but there also exists a group of C-C composites with a density range of approximately $0.10-0.50 Mgm^{-3}$ that is used primarily for thermal insulation /4-9/. The microstructure of these low density C-C composites generally exhibit a 2-D planar random structure as illustrated in Figure 1 /10,11/, and can be thought of as consisting of dense (xy) planes of fibrous material (intralayer regions) stacked on top of one another, with each plane or layer being only weakly bonded to its nearest neighbours via a region sparsely populated with fibrous material (interlayer regions). The mechanical and physical properties of these composites have been found to be isotropic within the xy plane, though different to those found in the zx (or the equivalent zy) planes /12,13/.
These low density C-C composites contain three components, namely matrix, virgin fibre, and 'rework' fibre, where the 'rework' fibre is milled excess material produced from machining operations that has been recycled /10,11/. The effect of the addition of reworked fibre on the microstructural anisotropy of the composites has been investigated utilising image analysis /11/, and it was found that the addition of

reworked material reduced the microstructural anisotropy. This phenomenon is the consequence of the differing aspect ratios of the virgin and reworked fibrous material; the reworked material has an aspect ratio of nearly two orders of magnitude lower than that of the virgin fibre /10,11/, and thus the former has a greater tendency to randomly align itself than the latter, resulting in a less microstructurally anisotropic material /10,11/.

As the composites mentioned in this report are manufactured on a commercial basis, it would be useful to know what proportion of reworked fibre could be utilised in the production of the composites without adversely affecting the properties to a significant extent, as an increased proportion of reworked fibre in the composite would contribute to a reduction in production costs. The aim of this paper was thus to investigate the effect of the addition of reworked material on the mechanical properties (flexural and compressive) of a series of low density C-C composites, ranging from a composite containing no reworked fibre to a composite whose total fibrous content was reworked fibre.

2. EXPERIMENTAL PROCEDURE

Seven samples that contained different proportions of virgin fibre, reworked fibre, and matrix, were manufactured by Calcarb Ltd in a process summarised in detail elsewhere /10,11/. The proportions of materials were altered so as to give reworked fibre fractions (as compared to the total fibrous material content) of 0.00, 0.16, 0.35, 0.50, 0.65, 0.85, and 1.00. Sample sizes were of the order of 300 x 300 x 300mm, and the bulk density of each sample was evaluated by measuring the mass and volume.

Specimens for 3-point flexural testing were cut to an approximate size of 5 x 10 x 60mm and a span of 40mm. Compression specimens had an approximate cross-section of 10 x 10mm and a length of 15mm. A crosshead speed of 1mm/min was employed for all tests. Schematic designations of flexural and compressive specimens are given in Figure 2 and Figure 3 respectively.

3. RESULTS AND DISCUSSION
3.1. DENSITY

Figure 4 shows that there is a general increase in density with increasing rework content, with a fourfold increase in density from the 0.00 to the 1.00 rework fraction samples. The sample containing 0.16 rework fraction had a slightly lower density than the 0.00 rework fraction sample (0.12Mgm^{-3} compared to 0.13Mgm^{-3}), and this anomalous behaviour was attributed to a slight deviation of the manufacturing process for these samples /10/. The increase in density with rework fraction is due to the reworked fibrous material being able to pack more closely together, when compared to the larger aspect ratio virgin fibres. However, it is suggested that even a small proportion of virgin fibre has a large disrupting influence on the packing of the rework material within the composite, and thus efficient packing of the rework fibrous material can only take place when the proportion of virgin fibre within the composite is either zero or near-zero hence the marked increase in density at high (>0.85) rework fractions.

3.2. MECHANICAL PROPERTIES
3.2.1. FLEXURAL AND COMPRESSIVE STRENGTH

Figure 5 and Figure 6 illustrate the flexural and compressive strengths respectively, as a function of rework content. Points of note are :- (i) the strengths are fairly consistent with those found by earlier researchers /4-8,12,13/, (ii) properties measured in the xy plane (termed 'x/y') are generally greater than those measured in the zx or zy planes (termed 'z'), (iii) x/y and z compressive strengths are more similar than the corresponding flexural strengths, (iv) the strengths of the 1.00 rework fraction sample are much greater than those of other samples, (v) the 0.16 rework fraction samples have slightly lower values when compared to the 0.00 rework samples, and (vi) both strength values remain fairly constant up to 0.85 rework fraction.

The difference in x/y and z strengths is due to the different failure mechanisms in x/y and z samples /10,12,13/. In an x/y sample, the flexural and compressive strengths are determined by the tensile failure and compressive buckling failure respectively, of the intralayer fibres, whereas for z samples, the flexural strength is limited by the tensile strength of the interlayer fibres. As on average there are more intralayer than interlayer fibres /10,11/, it is not surprising that x/y flexural values are greater than the respective z values.

The failure in compression of the z samples is complex and has been described in detail elsewhere /10,13/, it is sufficient for present purposes to note that the compressive strength value corresponds to failure of intralayer fibres as was the case for x/y compression samples. This accounts for the relatively small difference between the x/y and z strengths.

The dependence of strength, both flexural and compressive, on rework fraction is similar to that observed for density, with little effect of rework on strengths up to 0.85 rework fraction and then significant improvement in strengths at higher rework fractions. It should be noted however, that specific values (mechanical property value/density) decrease within the rework range 0.16-0.85 due to the slight increase in density with increased rework content.

3.2.2. COMPRESSIVE MODULUS

Although x/y compressive samples exhibited a single elastic modulus region up until failure (Figure 7 /10,13/), it can be seen from Figure 8 /10,13/ that z compressive samples exhibit three seperate moduli regions (OA, AB and BC) up to the point when the maximum strength is reached, though only the first modulus region was found to be elastic. The reason for there being three distinct z compressive moduli values has been discussed elsewhere /10,13/ and only data for the initial modulus will be discussed here.

The initial compressive modulus as a function of rework content is given in Figure 9 and shows similar features to the flexural strength in that (i) x/y values are greater than z values, (ii) the 1.00 rework fraction values are significantly greater than values for the other samples, and (iii) there is a plateau between the 0.16 and 0.85 rework fraction values.

The reason for the x/y values being greater than the z values is that the initial compressive modulus of the x/y samples is determined by the initial compressive elastic modulus of the intralayer region, whereas for the case of the z samples, it has been reasoned /10,13/ that the initial compressive modulus is determined by the initial compressive elastic modulus of the interlayer region. As the intralayer region has a greater density of fibres than the interlayer region, it follows that the x/y initial compressive modulus values would be greater than the z values. The constancy of the moduli up to rework fractions of 0.85 is consistent with the relatively small change in density whereas the large increase in density on going from 0.85 to 1.00 rework fraction accounts for the marked rise in moduli. It can thus be said that the addition of reworked material does not degrade the initial compressive modulus of the samples.

3.2.3. MECHANICAL ANISOTROPY

The mechanical anisotropy is defined as the ratio of the x/y value to the z value. As stated earlier, it was found using image analysis that the microstructural anisotropy of the samples decreased with an increasing rework content, and from Figure 10 it can be seen that the flexural strength and initial compressive modulus anisotropies also decrease with increasing rework fibrous content. In both cases the maximum anisotropy values are approximately eight for a sample containing no rework material, and they both decrease to a value of approximately two for a sample containing all of its fibrous material as rework.

In simple terms the x/y modulus and flexural strength values are determined by intralayer properties and the corresponding z values by interlayer behaviour. The reason for the decrease in mechanical anisotropy with increasing rework content is as a result of a relatively greater increase in the z direction property values with increasing rework content, as can be seen from Figure 5 and Figure 9. The reason for this is that, as rework content is added, it reduces the alignment of the fibrous material in the xy planes and disproportionately increases the density of material in the interlayer regions /10,13/, and thus z values increase more rapidly with rework content than do the x/y values.

In contrast to the modulus and flexural strengths, the compressive strength in both the x/y and z samples is determined largely by interlayer properties. It would therefore be expected that both the x/y and z compressive strength values would exhibit similar rework content dependences. This is indeed the case and the anisotropy of the compressive strength, although showing a slight positive dependence on rework content, remains close to unity for all rework contents.

4. CONCLUSIONS

The following conclusions can be drawn:-
(i)   Increasing the proportion of reworked fibrous material in a composite increases the density of the composite, due to the easier packing of the lower aspect ratio (compared to that of the virgin fibre) of the reworked material.
(ii)   With the exception of compressive strength, the in-plane x/y mechanical property values tend to be greater when compared to the out-of-plane z values.
(iii)   Mechanical property values remain fairly constant for rework fractions up to 0.85, which correlates with a relatively small change in density. In contrast the large increase in density on increasing the rework content from 0.85 to 1.00 is accompanied by a marked improvement in strength and stiffness.
(iv)   As the rework fraction increased, the flexural strength and initial compressive modulus anisotropy values both decreased from approximately eight to two, as a consequence of the decreasing microstructural anisotropy within this range. The anisotropy of the compressive strength values was around unity for all rework contents.

ACKNOWLEDGEMENTS

The authors wish to express their thanks to the Science and Engineering Research Council for the provision of CASE funds to support this project, and also to Chris Gee and Steve Ellacott of Calcarb Ltd. (12, North Road, Bellshill, Strathclyde, ML4 1EN) for the supply of all materials used in this project, their additional funding and encouragement and assistance.

REFERENCES

1. Fitzer, E., et al, Carbon 25(2) (1987) 163-190
2. Manocha, Latit M., et al, Carbon 25(3) (1988) 333-337
3. Kowbel, K., Shan, C. H., Carbon 28(2/3) (1990) 287-299
4. Reynolds, C. D., Ardary, Z. L., Report No. Y/DA-6925, Union Carbide Corporation, Oak Ridge Y-12 Plant, Oak Ridge, Tennessee, USA, October 5th, 1976
5. Kureha F. R., Technical information F-4003, 84.01, Kureha Chemical Industry Co. Ltd., Tokyo, Japan
6. Fiber Materials, Inc., Product Data (1987), Biddeford, Maine, USA
7. Ellacott, S. D., Report No. 9/004/02, February 1988, Calcarb Ltd., Bellshill, Scotland
8. Matcon, Report No. 1407, 13 Wolverton Avenue, Kingston-upon-Thames, Surrey, UK
9. Wei, G. C., Robbins, J. M., Am. Ceram. Soc. Bull. 64(5) (1985) 691-699
10. Davies, I. J., PhD Thesis, University of London, 1992
11. Davies, I. J., Rawlings, R. D., To be published in J. Mat. Sci.
12. Davies, I. J., Rawlings, R. D., Paper submitted to Carbon
13. Davies, I. J., Rawlings, R. D., Paper submitted to Composites

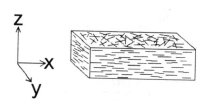

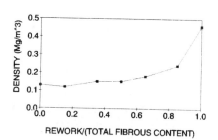

FIGURE 1 :- SCHEMATIC
REPRESENTATION OF A BLOCK
OF 2-D C-C COMPOSITE.

FIGURE 2 :- SCHEMATIC
DESIGNATION OF FLEXURAL
SPECIMENS.

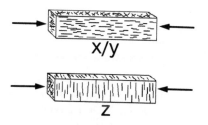

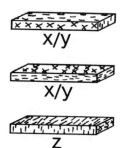

FIGURE 3 :- SCHEMATIC
DESIGNATION OF COMPRESSIVE
SPECIMENS.

FIGURE 4 :- DENSITY AS A
FUNCTION OF REWORK
CONTENT.

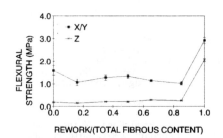

FIGURE 5 :- FLEXURAL
STRENGTH AS A FUNCTION OF
REWORK CONTENT.

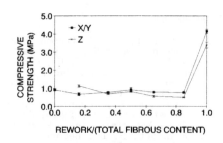

FIGURE 6 :- COMPRESSIVE
STRENGTH AS A FUNCTION OF
REWORK CONTENT.

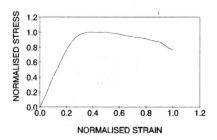

FIGURE 7 :- COMPRESSIVE
STRESS-STRAIN CURVE FOR AN
X/Y SAMPLE.

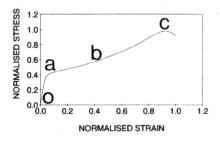

FIGURE 8 :- COMPRESSIVE
STRESS-STRAIN CURVE FOR A
LOW DENSITY Z SAMPLE.

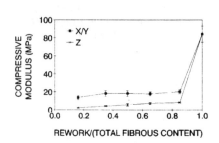

FIGURE 9 :- INITIAL
COMPRESSIVE MODULUS AS A
FUNCTION OF REWORK
CONTENT.

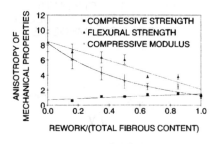

FIGURE 10 :- ANISOTROPY OF
MECHANICAL PROPERTY AS A
FUNCTION OF REWORK
CONTENT.

INTEGRATING VIBRO-ACOUSTIC CONTROL FUNCTIONS INTO COMPOSITES AND ADVANCED MATERIALS

B. GARNIER, R. de MONTIGNY

METRAVIB R.D.S. - BP 182 - 69132 Ecully Cedex - France

1. ABSTRACT AND INTRODUCTION

The integration of composites as a solution for an industrial structural application results most of the time from the following needs:
- offer the same strength with a reduced weight
- resist to a harsher environment.

Both contexts bring to an increased difficulty to satisfy directly vibro-acoustic requirements; however, adding damping and cladding materials is contradictory with the weight reduction objective.

Fortunately, by taking into account some vibro-acoustic requirements from the preliminary design, it appears possible to integrate directly within the composite some "added value functions". In particular, either active or passive vibro-acoustic functions appear relatively easy to integrate into the structural composite.

We will provide here several demonstrative examples:
- the design and test of "speaking trim-panels" to act directly as "anti-noise" sources in commercial aircraft to reduce the propeller noise in the cabin;
- the design and manufacturing of a tuned "acoustic barrier" to be integrated in advanced sonar arrays;
- the integration of vibration or pressure sensitive sensors (piezopolymers or silicon crystals made micro sensors) into multilayered materials to build completely integrated detectors.

1. FIRST EXAMPLE: CONCEPT OF "SPEAKING TRIM PANELS" TO INTEGRATE AN ACTIVE NOISE CONTROL INTO COMMERCIAL AIRCRAFT

The propeller noise is a major source of discomfort in the small commercial aircraft driven by turboprop's. Because the source is relatively stationary, active control may offer a powerful solution. A first approach could be to introduce active mounts between the motors and the wings, but most of the noise in the cabin is generated by the wall vibration induced by the aerodynamic cyclic pressure load (cf. blade passing frequency) which creates a distributed noise source unsolvable by this active mount approach, more relevant for business jets.

The adequate alternative is to create directly in the cabin the adequate "anti-noise" by a set of loudspeakers controlled to minimise the cabin noise, at the passengers head level. It was the aim of an E.C. funded BRITE-EURAM contract called ASANCA, federating several Aircraft Manufacturers and Contract Research Organisations (cf. reference [1].

A big problem of applying active noise and/or vibration control to aircraft may be the weight of the actuators (loudspeaker and/or shaker systems), necessary for generating the secondary noise and/or vibration field. The required performance of loudspeakers is for example, a high acoustic output at frequencies below 500 Hz. Currently, the smallest conventional loudspeakers which will meet this requirement have a weight of more than 1.3 kg. Since a larger number of such sources may be needed for aircraft to obtain a global noise reduction in the cabin, the resulting total weight of the control system could be very large.

A basic idea was to develop actuators which are more or less an integrated part of the interior trim and which are restricted to the required low frequency range so that significant weight and space can be saved.

The actuators developed by METRAVIB R.D.S. in this E.C. funded program and presented in fig. 1 are based on the piezo-electric principle - here a piezopolymer PVf_2.

The advantages of the piezo-electric principle are the simple and robust design, the high forces, and the low weight. The major disadvantage is the small displacement. This disadvantage may however be overcome by utilising the bimorph principle. Here the piezo-electric material is combined to an elastic material, which will amplify the displacement. By applying a thin piezo-electric polymer film (PVf_2 to a specially designed trim panel the whole panel may be made to radiate sound, thus creating a simple and robust loud-speaker with nearly no additional weight (Fig. 2).

The typical efficiency of such robust loudspeakers may be expressed in terms of acoustic level versus electric power. We measured on the prototypes a typical 0.2 Pa/W, which brought us to control internal noise levels up to 95 dB with this very simple "speaking trim panel".

3. SECOND EXAMPLE: DESIGN AND MANUFACTURING OF A GRP TUNED "ACOUSTIC BARRIER"

Some concept of acoustic absorbent coatings are presented in the companion paper; We now focus on an even more specific design, based on decimetric "inclusions" embedded in a damped polyurethane.

The scope is now to offer a "narrow-band" sound reflection based on the definition of a resonant cavity: at the resonance frequency, every incoming sound wave is totally

reflected (Fig. 3). At lower frequencies, and in particular for the static pressure resistance when the submarine is diving, the inclusion behaves as rigid. Of course, the inclusion resonance is adjusted (= tuned) to the sonar frequency of major interest. The difficulty was to design the inclusion. It was made from a Glass Reinforced Epoxy tube optimised to offer:
- the required transversal resonant frequency,
- the maximum static stiffness (to minimise the volume change while diving),
- the ultimate diving depth pressure resistance,
- the long term resistance to cyclic pressure fatigue.

The typical cross section is shown in Fig. 4. We used FEM techniques to optimise the cross-sectional fibre distribution. At the final production stage, the natural frequency dispersion of any tube was less than 3 %. Several hundred square meters of two-layers barriers (2 tubes of different shapes tuned to two operational frequencies) were produced, tested and integrated successfully, to back the sonar arrays of our most recent submarine.

4. THIRD EXAMPLE: FULLY INTEGRATED ACOUSTIC ARRAYS

The previous system was designed as a specific component to be integrated in a complex "flank array" system (Fig. 5). It is again evident that such design does not use all the integration opportunities of composites; as a consequence, we push now the development of more advanced concepts of multi-layered media integrating all the features required in such system:
- the masking layer against hull vibration and structure-born noise,
- the detection layer: hydrophones + reflector,
- the external coating: protection + flow noise reduction.

Instead of the previous bolts + suspension - expensive, weight consuming - (cf. shock loads over dimensioning) and subjected to corrosion (sea-water), the alternative concept appears externally as a solid block of polyurethane (Fig. 6). It could be manufactured through our E.E.I.G. METOD. The detection is made either by a network of "classic" piezoceramic hydrophones over moulded in the intermediate layer, or preferable by a patchwork of piezopolymer thin sheets offering directly quasi the same acoustic impedance compared to the surrounding medium. The only strong impedance mismatch required by principle is the reflector + masking + decoupling layer immediately backed to the pressure hull, with a relatively low sound speed/high air content (Fig. 6).

Any precise project of this kind is of course classified, but we will present an intermediate prototype as tested on our specific test body "Limande" operated now by the French D.C.N. in a lake of the Southern Alps (Fig. 7). The 2 x 3 test panels of 1 x 1 meter were composite structures combining GRP, solid polyurethane over mouldings, classic and PVf_2 hydrophones, and helped to define the optimum configurations.

5. CONCLUSION

The integration of sensors or actuators is not only the opportunity to mention the concept of "smart structures", but offers real opportunities to introduce built-in metrological functions. In a recent survey, METRAVIB R.D.S. stated the immediate possibility to develop:
- aircraft composite parts integrating a cyclic fatigue counter: from stress

sensitive fibres, a small chip could count the number of occurrences of stress levels above a given threshold, offering the possibility to introduce a fatigue monitoring really relevant to the precise plane history, or similar applications,
- metallic or concrete floors integrating an intrusion detection network,
- real time shock detectors integrated in the security barriers of motorways,
- tire pressure control embedded on the pavement of motorway toll stations,
etc.,
more and more applications being now arising in our everyday environment.

REFERENCES

1. I.U. Borchers, U. Emborg, A. Sollo, E.H. Waterman, J. PAILLARD, P.N. Larsen, G. VENET, 1 Al. "Advanced study for active noise control in aircraft (ASANCA). DGLR/AIAA 14th Aeroacoustics Conference, 14-14 May 1992, Aachen, Germany, DGLR/AIAA-92-02-092

piezo-electric polymer film

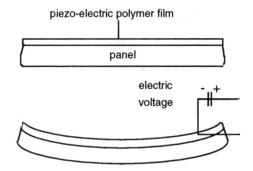

Figure 1: Principle of piezo-electric Polymer film attached to a panel

Figure 2: Details of PVf_2 actuators integrated into a DORNIER 228 trim panel

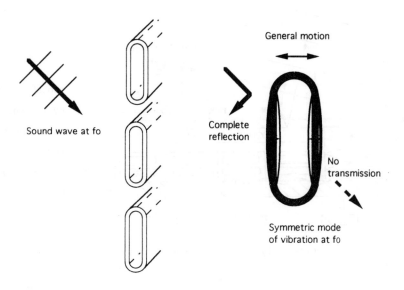

Sound wave at fo

Complete
reflection

General motion

No
transmission

Symmetric mode
of vibration at fo

Figure 3: Resonant acoustic barrier

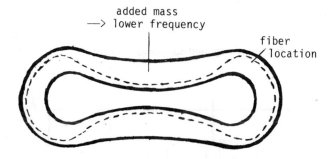

added mass
⟶ lower frequency

fiber
location

Figure 4: Typical cross-section

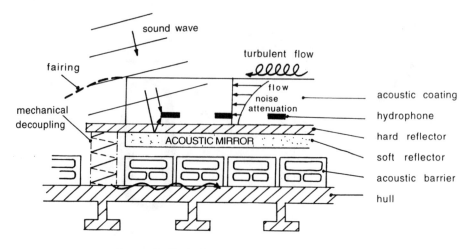

Figure 5: Typical set-up of a "flank-array"

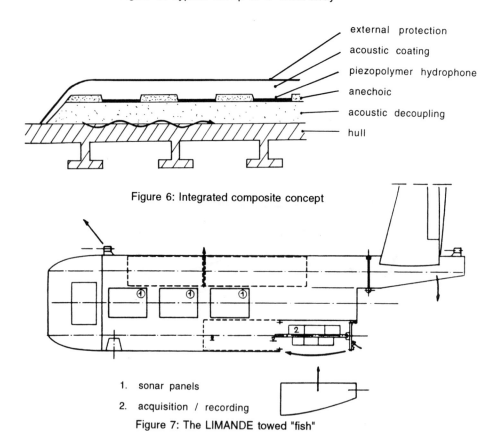

Figure 6: Integrated composite concept

1. sonar panels
2. acquisition / recording

Figure 7: The LIMANDE towed "fish"

FABRICATION TECHNOLOGIES, PROCESS MODELLING, NON DESTRUCTIVE TESTING

FIBER-OPTIC CURE MONITORING OF COMPOSITE RESINS

R.E. LYON, S.M. ANGEL*, K.E. CHIKE**

Federal Aviation Administration - Fire Research Branch
FAA Technical Center - 08405 Atlantic City Intl Airport - USA
*Lawrence Livermore National Laboratory - Livermore CA 94551 - USA
** University of South Carolina - Dept of Chemistry
Columbia SC 29210 - USA

ABSTRACT

Fiber-optic Raman spectroscopy was used to monitor the curing of epoxy resins *in-situ* for eventual application to polymer composite processing. A 200-μm diameter quartz fiber-optic sensor immersed in liquid resin was used to obtain Raman spectra for a concentration series of diglicidylether of bisphenol-A in its own reaction product with diethylamine and to monitor the room-temperature cure of phenylglicidylether with N-aminoethylpiperazine. The concentration of epoxide groups calculated from the Raman spectra were in excellent agreement with Fourier-transform near-infrared absorbance measurements made under the same conditions indicating the viability of the Raman technique for near-real-time cure monitoring of polymer composites.

INTRODUCTION

Non-flammable polymers and composites for aircraft interiors are the focus of a new FAA materials research effort. Low flammability is a characteristic of thermally-stable polymers such as phenolics, polyimides, and bismaleimides, but these materials can be expensive and difficult to process. Consequently, parallel development of an economical sensor-based processing technology which measures both extent of chemical reaction and temperature *in-situ* will help to offset the potentially higher cost of new fire-resistant materials and facilitate their acceptance by airframe and cabin manufacturers.

In previous publications (1-3) we described how changes in the fiber-optic Raman spectrum of polyethertriamine-cured diglicidylether of bisphenol-A (DGEBA) epoxy could be used to follow the degree of cure *in-situ* by assuming a linear relationship between the scattering intensity of the 1240 cm^{-1} vibrational band of the oxirane ring and the concentration of epoxide groups in the reaction mixture. Raman peaks at 1112 cm^{-1} and 1186 cm^{-1} associated with bisphenol-A backbone vibrations of the phenyl ring and gem-dimethyl groups, respectively, were used in a ratio with the 1240 cm^{-1} epoxide band to normalize the Raman spectrum for density fluctuations and instrumental variations. This normalization method is based on the assumption that, unlike the 1240 cm[-1] epoxide band, the scattering intensity of the backbone vibrations would not change during the cure reaction and so could be used as internal standards. The objective of the present study was to calibrate the proposed Raman scattering technique for epoxy cure monitoring using conventional near-infrared absorbance spectroscopy, and to demonstrate quantitative fiber-optic Raman spectroscopic monitoring of model epoxy-amine curing reactions for eventual application to other thermoset resins and their fiber composites.

EXPERIMENTAL

A series of epoxide concentrations was prepared for Raman and near-infrared spectral measurements by diluting a high-purity diglicidylether of bisphenol-A with a stoichiometric DGEBA-diethylamine liquid reaction product in which all of the epoxide groups had been consumed in an epoxide-secondary amine reaction. The diethylamine-endcapped DGEBA adduct was prepared by reacting 50.0 grams of diglicidylether of bisphenol-A (99%, Dow Chemical DER 332, epoxide equivalent weight 173 g/eq) with 42.3 grams of diethylamine (99%, Aldrich Chemical, amine hydrogen equivalent weight 73.14 g/eq) under argon in a sealed 250-ml roundbottom flask maintained at 50°C. Diethylamine (DEA), a secondary aliphatic amine with only one active hydrogen was chosen for endcapping the DGEBA to preclude crosslinking reactions leading to gelation and ensure a pourable liquid resin for mixing with pure DGEBA. The epoxide groups in the DGEBA-DEA reaction mixture were completely consumed after 24 hours at 50°C as determined by the disappearance of the 4525 cm[-1] epoxide overtone band in the near-infrared absorbance spectrum, according the method of Dannenburg (4). The DGEBA-DEA adduct was a glassy liquid at 23°C with a density of 1.035 g/cm^3.

The DGEBA-DEA adduct was warmed to 60°C to reduce the viscosity and mixed in 10% w/w increments with pure DGEBA to provide a series of known molar concentrations of epoxide ranging from 0.0 epoxide equivalents/liter for the DGEBA-DEA adduct to 6.7 epoxide equivalents/liter for the pure diglicidylether of bisphenol-A. This approach allowed us to systematically vary the epoxide concentration while maintaining a nearly constant molar concentration of the bisphenol-A backbone which is used as internal standard in the Raman technique. Moreover the DGEBA/adduct blend provided a neat-liquid mixture representative of practical amine-cured epoxies. The chemical structures of DGEBA, DEA, and DGEBA-DEA adduct reaction product are shown in Figure 1.

Epoxide-amine reaction kinetics at room-temperature were investigated using both Raman and near-infrared spectral measurements of a stoichiometric mixture of ±1,2-epoxy-3-phenoxypropane (phenylglicidylether, 99%, Aldrich Chemical, epoxide equivalent weight 150.0 g/eq) and N-aminoethylpiperazine (AEP, 99%, Aldrich Chemical, amine hydrogen equivalent weight 32.25 g/eq). Reasonable room temperature reaction rates were achieved with the cycloaliphatic primary amine, AEP, in combination with the monofunctional epoxide resin, phenylglicidylether– the latter selected to preclude network formation. Reaction of the epoxide group with hydroxyl groups formed during the epoxide-amine reaction is negligible at temperatures less than about 100°C. The phenyl group of phenylglicidylether (PGE) provided the 1112 cm[-1] Raman band used as an internal reference for normalizing the spectra. The chemical structures and stoichiometric reaction product of PGE and AEP are shown in Figure 2.

The optical components used for making Raman scattering measurements are shown in Figure 3. Raman spectra were obtained at ambient temperature (23±1°C) by immersing a dual-fiber optical probe directly into the liquid epoxy-amine mixtures contained in 30-ml polyethylene beakers. The probe consisted of two 250-μm, step-index, silica-on-silica fiber-optics (PolyMicro) which had been stripped of their polyimide jacket and adhesively bonded side-by-side into a SMA connector. Fibers were prepared by mechanical polishing of the ends with 3M Imperial lapping film, using a 0.3-μm grade film for the final polish. A stainless steel tube was bonded over the parallel fibers and SMA connector to provide strain relief for handling and a convenient pencil-probe geometry for measurements. A GaAlAs diode laser (Spectra Diode Labs Model SDL-5412) operating at 50 mW and a wavelength of 820-nm was filtered using a 10-nm bandpass filter to remove spontaneous emission before being focused into one fiber of the probe to illuminate the sample at the probe tip. Scattered light is collected by the second fiber in the probe and travels to the spectrometer where it is collimated as it emerges from the fiber using an f/2 lens before passing through an 819-nm holographic long-pass cut-off filter (Physical Optics, Inc.) to block Rayleigh scattered light. The Stokes-scattered light is then focused with an f/4 lens onto the slit of an image-corrected spectrograph (Chromex) containing a 600 groove/mm grating blazed at 750 nm and a back-illuminated charge-coupled device (CCD) imaging detector (Princeton Instruments). All Raman spectra were obtained with 60 second exposures at a nominal 8-12 cm^{-1} resolution. Near-infrared and mid-infrared spectra were obtained using 64 scans at 4 cm^{-1} resolution on a Digilab FTS-40 FTIR/NIR (BioRad) instrument at 23±1°C. A 1-mm pathlength, disposable Pyrex cell was used for all of the NIR measurements.

RESULTS AND DISCUSSION

Mid-infrared and Raman spectra for the 600-1400 cm^{-1} spectral region of the diglicidylether of bisphenol-A are plotted in Figure 4. This "fingerprint" region of the vibrational spectrum consists of skeletal bending vibrations of polyatomic molecules and stretching of single-bonds linking a substituent group to the remainder of the molecule. Some of the same group frequencies are active in both the mid-IR and Raman spectra of DGEBA even though the selection rules are different for these two vibrational spectroscopies. Bands which are common to both spectra include the epoxide ring stretch at 1240 cm^{-1}, the in-plane deformation of the gem-dimethyl group of bisphenol-A at 1186 cm^{-1}, and the C-H out-of-plane bending of the para-disubstituted phenyl group at 830 cm^{-1}. In a previous publication (1) we reported the use of the 830 cm^{-1} Raman band for *in-situ* temperature determination using the ratio of the Stokes and anti-Stokes intensities. A strong para-disubstituted phenyl band at 1112 cm^{-1} is observed only in the Raman spectrum and is used interchangeably with the 1186 cm^{-1} band as an internal reference.

In Figure 5 are plotted Raman spectra for three epoxide concentrations corresponding to 100%, 40%, and 0% DGEBA dissolved in the stoichiometric DGEBA-DEA adduct (see Figure 1). Only the 1170–1290 cm^{-1} spectral region used for cure monitoring is displayed in Figure 5 and the spectra have been vertically shifted for clarity. Raman scattering intensity was normalized to that of pure DGEBA using the 1186 cm^{-1} gem-dimethyl reference band as an internal standard. It is observed that the 1240 cm^{-1} epoxide ring vibration decreases in intensity but does not disappear as the epoxide concentration in the solutions goes to zero for the pure DGEBA-DEA adduct. The small residual peak at 1245 cm^{-1} observed at zero epoxide concentration and the Raman peak at 1220 cm^{-1} are thought to be due to the Ø-O-C aromatic ether stretch and would not be expected to change in intensity during the cure reaction. The band at about 1267 cm^{-1} attributed to C-O and C–C stretching increases in intensity with decreasing epoxide concentration. NIR absorbance spectra for the DGEBA dilution series showed the complete disappearance of the 4535 cm^{-1} epoxide ring overtone combination band in the DGEBA-DEA adduct (0% DGEBA) confirming complete reaction of the epoxide groups at the calculated 1:1 stoichiometry.

Calibration curves for Raman scattering and NIR absorbance versus epoxide concentration are shown in Figure 6. Raman response is expressed as the ratio of the 1240 cm^{-1}/1112 cm^{-1} peak intensities, while NIR response is the corrected absorbance of a C-H overtone band of the epoxide group at 4535 cm^{-1} according to the method of Dannenburg (4) measured in a one millimeter pathlength cell. Abscissa values for epoxide concentration in units of equivalents/liter were calculated from the experimentally determined epoxide equivalent weight of DGEBA (173 g/mol) and the individual weight fractions and measured densities of DGEBA (ρ=1.155 g/cm^3) and the DGEBA-DEA adduct (ρ= 1.035 g/cm^3) by assuming volume additivity in the mixtures.

Raman spectra for the PGE-AEP stoichiometric mixture at room temperature immediately after mixing and after 29 hours are shown in Figure 7. Prominent in the Raman spectra of the resin/hardener mixture are the strong, symmetric, aromatic ring stretching vibration near 1000 cm^{-1} and the in-plane hydrogen bending vibrations at about 1025 cm^{-1} and 615 cm^{-1} characteristic of the monosubstituted aromatic phenylglicidyl ether. The instantaneous degree-of-cure versus time for the PGE-AEP reaction at 23°C was calculated from the ratio of the 1240 cm^{-1} epoxide band to the 1000 cm^{-1} reference band of the phenylglicidyl ether. The linear relationship between normalized Raman scattering intensity and epoxide concentration demonstrated in Figure 6 permits the degree-of-cure during the chemical reaction of phenylglicidyl ether (PGE) with n-aminoethylpiperazine (AEP) at room temperature to be evaluated simply from the initial, R(0), instantaneous, R(t), and final, R(∞), ratios of the 1240 cm^{-1} to the 1112 cm^{-1} peak intensity according to the relationship– $\alpha(t) = [R(0)–R(t)] / [R(0)-R(\infty)]$, where $\alpha(t)$, the degree-of-cure at time, t , ranges from 0.0-1.0 during the cure process. The degree-of-cure during the PGE-AEP reaction at room-temperature was also measured by NIR absorbance spectroscopy for comparison to the Raman method with the corrected 4535 cm^{-1} NIR absorbance values in place of the Raman peak ratios. Results for degree-of-cure versus time for the PGE-AEP reaction measured by the fiber-optic Raman and near-infrared absorbance methods are compared in Figure 8 and show excellent agreement.

CONCLUSIONS

Fiber-optic Raman spectroscopy utilizing a near-visible diode laser excitation source and a pair of 200-μm diameter quartz optical fibers is a useful technique for remote, *in-situ* , measurement of the concentration of epoxide groups in neat resin mixtures. Normalized intensity of the 1240 cm^{-1} Raman peak was linearly related to the molar concentration of epoxide groups providing a simple relationship for calculating the degree-of-cure in epoxies. Fiber-optic Raman measurements were in close agreement with near-infrared absorbance data for epoxide concentration versus time during the chemical reaction of an epoxy resin with an aliphatic amine curing agent. Somewhat lower precision is indicated for the fiber-optic configuration used to obtain the Raman spectra in the present study compared to the near-infrared absorbance technique, although the sensitivity of the two spectroscopic methods was found to be comparable. Recent FT-Raman studies of epoxy resins, however, indicate that significant improvement in quantitative capability of the fiber-optic Raman technique is expected with relatively simple modifications to the optical configuration. Previous publications have demonstrated that temperature can also be determined *in-situ* from simultaneous Stokes and anti-Stokes Raman spectra measured over the same fiber-optics used to measure cure chemistry in the Stokes region.

ACKNOWLEDGMENTS

This work was performed under the auspices of the U.S. Department of Energy by the Lawrence Livermore National Laboratory under contract number W-7405-ENG-48. The authors would like to acknowledge the assistance of Dr. Thomas Vess in performing Raman measurements and Prof. M.L. Myrick for useful discussions.

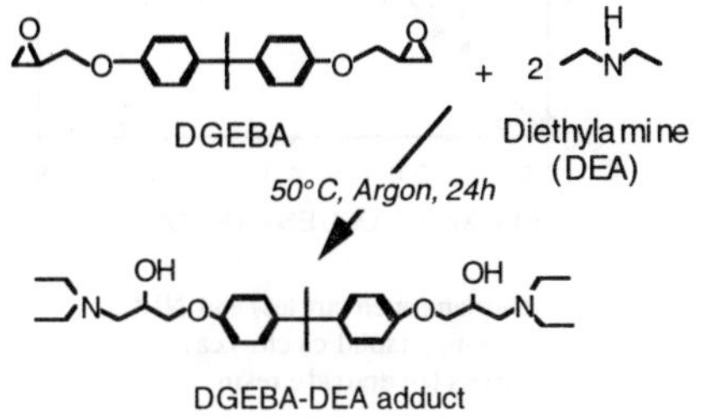

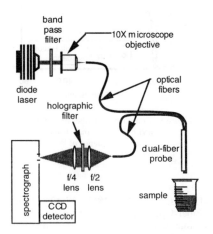

Figure 1. Reaction of diglicidylether of bisphenol-A (DBEGA) and diethylamine (DEA)

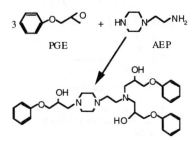

Figure 2. Reaction of phenylglicidylether (PGE) and N-aminoethyl piperazine(AEP).

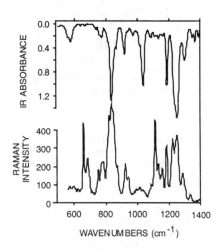

Figure 3. Experimental arrangment for fiber-optic Raman measurements.

Figure 4. Raman and mid-infrared spectra of diglicidylether of bisphenol-A..

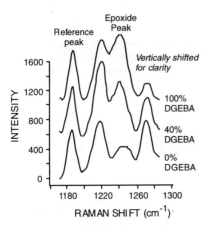

Figure 5. Fiber-optic Raman spectral for 0,40, and 100% DGEBA in DGEBA-DEA adduct..

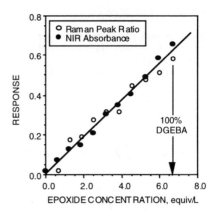

Figure 6. Raman (open circles) and NIR absorbance (solid circles) calibration curves for epoxide resin.

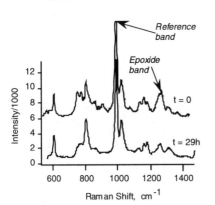

Figure 7. Raman spectra of PGE-AEP mixture at t = 0, and at, t = 29 hours at 23°C.

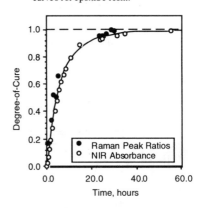

Figure 8. Comparison of fiber-optic Raman and NIR absorbance methods of measuring kinetics of PGE-AEP cure reaction.

REFERENCES

1. M.L. Myrick, S.M. Angel, R.E. Lyon, T.M. Vess, SAMPE J, **28**(4), 37-42 (1992)
2. R.E. Lyon, M.L. Myrick, and S.M. Angel,"Fiber-Optic Raman Spectroscopy for Epoxy Adhesive Cure Monitoring," Proc. 6th International Symposium on Structural Adhesives Bonding, Morristown, NJ, May 4-7, 1992
3. R.E. Lyon, M.L. Myrick, S.M. Angel, and T.M. Vess,"Fiber-Optic Raman Spectroscopy for Cure Monitoring of Advanced Polymer Composites," Proc. SPE Antec 92, Detroit , MI, May 3-7, 1992
4. H. Dannenberg, SPE Trans., **3**(1), 78-88 (1963)

THERMOPLASTIC COMPOSITES IN VEHICLE APPLICATIONS

C. BARON, R. MEHN*

HÜLS AG - GB 6.4 FEA 3 - Bau 1227 PB 16 - 45764 Marl - Germany
*BMW AG - Abt EW-1 - 80788 München - Germany

ABSTRACT

New thermoplastic composites with PA 12 matrix are introduced in this paper. They offer mechanical properties which are comparable with thermoset composites and additionally very short manufacturing cycles. Other advantages of PA 12 composites are the good drawing, the low moisture pick up and high energy absorption beside the low weight and good recycling prospects. For automotive structural components efforts in design, moulding and joining were made to substitute conventional lightweight material in highly loaded structural components. Beside the FEM-calculation with anisotropic finite shell elements and material datas of thermoplastic composites with PA 12 matrix an optimisation of the process parameters were investigated.

INTRODUCTION

The processing technology for continuous fibre reinforced composites with thermoset matrices is well developed and production costs are comparable with metallic materials. However, for medium and large productions the processing cycle time is too long. With thermoplastic composites, very short manufacturing cycles can be achieved besides some other advantages. At the moment thermoplastic composites can be divided into three classes. A low cost and in the automotive industry well established material is GMT (Glassmat Reinforced Thermoplastics), a PP-matrix mainly reinforced with long fibres.
Under the name "Advanced Composites", continuous fibre reinforced high temperature (HT) matrices like PSU, PES, PEI, PEKK and PEEK are offered. They find their applications in the aerospace industry /1/.
Between these two extremes thermoplastic composites with so called engineering plastics become more and more commercially available. These

composites retain mechanical properties that exploit the strength of the used fibres when the matrix can support the mechanical forces from one fibre to the other and has a sufficient bonding to the fibre itself. PA 12 is such an engineering plastic that is appropriate.
In order to manufacture automotive structural components made of these thermoplastic composites in an economically viable manner, numerous detailed problems concerning the design of the highly loaded parts as well as a suitable series-mature manufacturing process still have to be investigated. Especially if advanced materials introduced in load bearing structure components, a range of various structural as well as design aspects have to be clarified during the concept phase.

CHARACTERISTICS OF PA 12 COMPOSITES

Prepregs out of glass-, carbon- or aramid-fabrics were produced under the tradename VESTOPREG by using the powder impregnation technique /2,3/. The low viscosity of the used PA 12 matrix enables a good penetration between the fibre bundles and therefore a good impregnation of the fabrics. A full impregnation of laminates with more than two layers of the prepreg must be done by consolidation in a hot press or double-belt-press.
In comparison with thermoset laminates the following advantages were found:
- lower specific weight
- laminates are formable in reshaping processes
- more economic production of components
- good impact behaviour
- normal thermoplastic joining techniques can be used
- recycling to a short fibre reinforced thermoplastic compound is possible

Thermoplastic composites out of PA 12 prepregs can be fabricated to laminates at temperatures of $250^{\circ}C$ and low pressures of about 5 to 10 bar. This can be achieved in a hot or cold press, but also with vacuum forming. Recently, PA 12 laminate plates were offered commercially by Krempel/Vaihingen and Ten Cate/Netherlands.
Tapes with unidirectional fibres were produced from Baycomp / Canada by using of PA 12 powder in a fluidised bed.
For special applications, where a connection of two or more parts with damping behaviour is necessery, a co-vulcanisation with rubber without any adhesion promoters is possible (so called K&K-Technique).

MECHANICAL PROPERTIES

Mechanical properties of composites with PA 12 presented here are obtained on laminates out of VESTOPREG. Currently two types of glass fibre fabrics were offered as prepregs: a satin 1/7 woven fabric with nearly 50% of the fibres in both directions and an unidirectional woven fabric with 90% of the fibres aligned in the main direction.
As shown in fig.1, the flexural strength of glass fibre reinforced PA 12 is comparable with the values of the high temperature composites. The strength behaviour of the used E-glass fabrics is almost fully exploited. Glass mat thermoplastics (GMT/PP) with not aligned and general not continuous fibres can not reach these high mechanical properties. Also the adhesion between the glass fibres and the used PP matrix is not the best.

574

The interlaminar shear strength (ILSS) is an established test method (EN 2377) to determine the adhesion between fibres and matrices. The test equipment is similar to the three point bending test but with shorter specimen. ILSS values can not be normalised to the same fibre volume content. For that reason minimum and maximum values for some relevant composite laminates are shown in fig.2. The ILSS values of the PA 12 laminates seem moderate compared to typical values for glass/epoxy and are high compared to PP or PF (Phenolic resin) reinforced by glass fibres. The higher values for epoxy are a result of the higher young's modulus of the resin itself. On the other hand, thermoplastic composites are tougher than the thermoset epoxy and this improves the fracture mechanism like crack growing and debonding. Because of the simple test specimen and test equipment the ILSS test is suitable for quality control of the prepreg or laminate production.

One important advantage of composites with thermoplastic matrices is the higher energy storage capacity compared to metals. Fig.3 shows schematic the stress/strain behaviour of steel and of the unidirectional glass-fabric/PA 12 composite laminate. With the equation:

$$U \cong \epsilon^2 * E$$

where U is the capacity of energy storage, ϵ is the fracture strain and E the young's modulus of the material, we get values of 1.68 MPa for steel and 5.06 MPa for the PA 12 composite. The enormous advantage of composites against steel can be seen when the capacity of energy storage is divided by the density of the material. This is an important fact for applications like in cars.

MANUFACTURING OF AUTOMOTIVE STRUCTURAL COMPONENTS

Due to the good material properties of thermoplastic composites with PA 12 matrix, efforts in design, moulding and joining techniques were achieved to substitute conventional lightweight material in highly loaded structural components for a safety seat with integrated belt system.
Fig. 4 shows a preliminary result of structural investigations for determining the stiffness behaviour of the entire load bearing structure of the seat under crash condition. The presented FEM-calculation-model, based on generally anisotropic finite shell elements and material datas of VESTOPREG laminates, sufficiently describes the global stiffness relations and allows beyond it to estimate the intermediate matrix failure or fibre failure in component areas where stress concentrations are observed. Subsequently the calculation results are exploited to support the optimization work for a number of design features.
In competition with other manufacturing techniques the moulding/pressing technique offers a number of advantages related to the demands for series-mature processes /4/.
The basic idea behind the moulding of continously glass fibre reinforced thermoplastic laminates into structural components is the usage of flat, consolidated semifinished products, the subsequent melting of these laminate plates and the moulding procedure into the final shape.
The semifinished products, consisting of numerous single prepregs, have to be composed and adapted to the component geometry and the specific loading conditions as well as to the weight demands. Subsequent to the consolidation of the stacking sequence of prepregs into a compact semi-finished laminate, the moulding into the desired shape will be carried out. After melting the thermoplastic matrix of the laminate, approxima-

575

telly at 240 °C, in the heating station and the fast transporting into the press station, the moulding and consolidation to the final component shape will be achieved in a heatable closed die.
Depending on the required surface quality of the components, the matrix content of the laminate, the wall thickness, the consolidation time, the moulding velocity and the die temperature can vary between room temperatur and closely to 150 °C.
Apart from the optimization of the process parameters the die technique is a very relevant factor in optimizing the moulding results. Especially the spring-forward phenomena, typically for composite components with curved regions, must be considered very carefully during the design of the die. An other important topic is the appropriate design of sufficient curvatures in areas of the shells where structural joints exist. For each case a compromise must be found in order to support formability during moulding and structural requirements and also to fulfil the demands concerning the joining techniques.
Fig. 5 illustrates some preliminary samples of moulding investigations for the manufacturing of complex shells for structural components of a safety seat and principal parts of other automotive components.

CONCLUSION

The advantage of new thermoplastic composite materials with PA 12 matrix was described. They show mechanical properties that exploit the potentional of the used glass-fabrics with advantages in the processing technique. These composites can be competitive with metals and other existing composites, especially when they will applied for thin load bearing structures.
A considerable number of further investigations is still required before a series-mature and reproductible production based on thermoplastic composites with PA 12 matrix can be employed in the automotive industry. One of the main hinderances beside the design and manufacturing problems is the lack of continously reinforcing fibre systems, orientated by the structural and geometrical requirements and fabricated in an advanced textile technique. The mentioned glass fibre systems, impregnated by PA 12 powder in an economically manner, would help to shorten the immensly long time for the fabrication of the semi-finished products and to avoid the waste currently generated during the contour cutting processes.
Despite the currently existing problems discussed above, the new VESTOPREG composite materials offers the potential to be used for applications in high-loaded automotive structural components with good recycling prospects.

REFERENCES

1.   G.Kempe, H.Krauss, AIAA 86, Sep. 7-12, 1986, London.
2.   S.R.Iyer, L.T.Drzal, Journal of Thermoplastic Composite Materials, Vol.3, Oct.1990, pp. 325-355.
3.   Chr.Baron, Techtextil Symposium 1993, June 7-9, 1993, Frankfurt.
4.   R.Mehn, F.Seidl, R.Peis, D.Heinzmann, Verbundwerk 92, July 1-3, 1992, Wiesbaden.

Property [1]	Test method	Unit	G 101 Satin 1/7	G 201 crowfoot UD
Weight		g/cm²	412	402
Density 23°C	DIN 53479	g/cm³	1,62	1,68
Fibre content		%	50	54
Tensile test	DIN 29971	N/mm²		
Tensile strength 0°Fibre orientation			350	710
90°Fibre orientation				90
Elongation at break 0°		%	1,6	1,5
90°				2,2
Tensile modulus 0°		N/mm²	26000	44000
90°				7200
Flexural test	DIN 29971			
Flexural strength 0°		N/mm²	480	840
90°				140
Flexural modulus 0°		N/mm²	25000	41000
90°				9200
ILSS	DIN EN 2377	N/mm²	42	44
HDT Method A	DIN 53736	°C	177	178

1) Preliminary values, no guaranteed minimal or maximal values

Fig.1 Properties of laminates from VESTOPREG

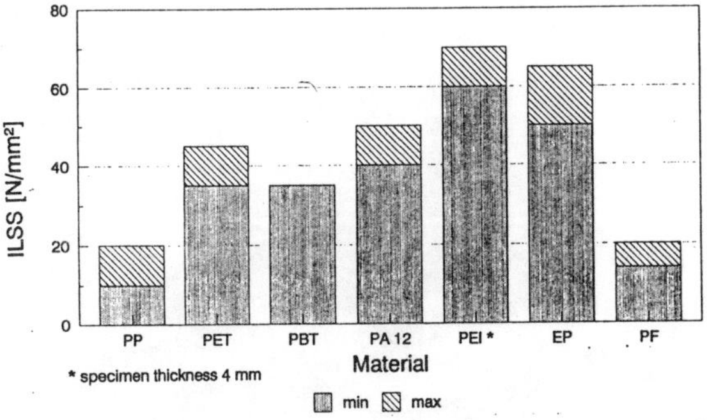

Fig.2 Comparison of interlaminar shear strength (ILSS) of composite laminates with different matrices

* specimen thickness 4 mm

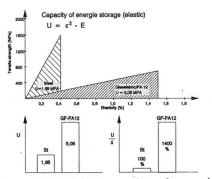

Fig.3 Comparison of the capacity of energie storage

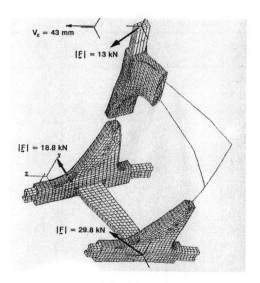

Fig.4 FEM-calculation model; deformation of the seat frame structure under-crash load condition

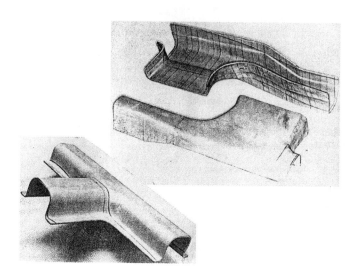

Fig.5 Moulded automotive structural components made of glass fibre rein-forced PA 12

CONTACT PRESSURE AND TENSILE STRESS DISTRIBUTION IN CROSSING FIBRES IN COMPOSITE MATERIALS PREFORMING

C.-H. ANDERSSON, M. KARLSSON, O. MANSSON

Lund Inst. of Technology - Dept of Production and Materials Eng.
Box 118 - 221 00 Lund - Sweden

Abstract

The stress concentration in the zone of contact between crossing fibres subjected to transversal and longitudinal loads is analyzed. A model for calculation of the statical contact stress in linear elastic isotropic reinforcement fibres is outlined. The mathematical treatment is based on the analysis by Timoshenko and Goodier. Hertzian contact stress between crossing fibres is simulated and related to textile and ceramic fibre handling.

1. BACKGROUND AND INTRODUCTION

The high productivity and the state of development of modern textile techniques and machinery offers a great potential in the handling of continuos fibres for composite material preforms. Short fibres are in principle handled by modified traditional ceramic techniques, originally developed for the wet handling of clays and powders before sintering. Consolidation of preforms in order to increase the volume fraction is usually done by compression of the still wet green bodies.

Textile structures are defined as integrated fibrous structures produced by fibre entanglement or yarn interlacing, interloping, intertwining or multiaxial placement. For example in weaving and braiding, the integrity of the structure is given by the crimp. The point of these structural features of fabrics is the arrangement of some kind of fibre to fibre contact for the subsequent build up of frictional stresses and plastic deformation, /2/3/.

The textile techniques and thinking were however developed from experience of natural polymeric viscoelastic fibres exhibiting extremely ductile behaviour. Most load bearing fibres for composites exhibit almost linear elastic behaviour, i.e. absence of viscous deformation and thus virtually no relaxation of applied stress during processing. In order to utilise the mechanical properties of the reinforcement, the choice of preforming technique, structure and infiltration process parameters is thus limited by the different modes of local stress build up, such as:

- Stress build up due to bending. The difference between the outer and inner radii of the fibre in bending gives a strain difference, /2/3/.

- Frictional stress build up due to sliding over bent surfaces. The stress increases exponentially with the contact angle and inversely to the bending radius, /9/.

- Hertzian contact stress due to transverse compression resulting in radial tensile stress at the edge of the fibre - fibre contact, /4/.

- Stress build up due to infiltration of liquid matrix material, /5/.

2. THEORY

The fibres are modelled as homogeneous cylinders of linear elastic isotrop material. The angle between the longitudinal axis of the cylinders in contact is ψ, see 1. The radii R_{11} and R_{12} are the main radii of cylinder 1, while R_{21} and R_{22} are the main radii for cylinder 2. The radii R_{11} and R_{21} are the main curvature radii, and the radii R_{12} and R_{22} are the main cross section radii of the cylinders. Thus the relations between the radii are $R_{11} > R_{12}$ and $R_{21} > R_{22}$. If two arbitrary points at the cylinder surfaces are approaching each other, the distance between the points will decrease until they are in contact. The decreasing distance, $z_1 + z_2$, between these two arbitrary points can be expressed as follows /1/:

$$z_1 + z_2 = Ax^2 + By^2 \tag{1}$$

Where A and B are constants that are depending of the curvature of the surfaces. Equation (1) can be rewritten with the relations presented by Timoshenko /1/ as:

$$2(A + B) = \frac{1}{R_{11}} + \frac{1}{R_{12}} + \frac{1}{R_{21}} + \frac{1}{R_{22}} \tag{2}$$

$$2(B - A) = \{(\frac{1}{R_{12}} - \frac{1}{R_{11}})^2 + (\frac{1}{R_{22}} - \frac{1}{R_{21}})^2 + 2(\frac{1}{R_{12}} - \frac{1}{R_{11}})(\frac{1}{R_{22}} - \frac{1}{R_{21}})\cos 2\psi\}^{\frac{1}{2}} \tag{3}$$

For the theoretical model with two cylinders in contact, the curvature radius can be assumed to be very large compared to the cross sectional radius, thus $R_{11} = R_{21} = \infty$ and

$R_{12} = R_{22} = R$. The value of the constants A and B are positive, since the expression $z_1 + z_2$ need to be positive. It is therefore possible to conclude that all points with an equal sum of distances $z_1 + z_2$ have to be on an ellipse. Two cylindrical bodies (fibres) that are in contact due to transverse pressure, will form a contact zone with an elliptical boundary. The approximation of the main radii put into equations (2) and (3) makes it possible to determine the constants A and B:

$$A = \frac{1}{2R}(1 - \cos \psi) \qquad B = \frac{1}{2R}(1 + \cos \psi)$$

3. FIBRE - FIBRE CONTACT

Introduce the constants k_1 and k_2, as below:

$$k_1 = \frac{1 - \nu_1{}^2}{\pi E_1} \qquad k_2 = \frac{1 - \nu_2{}^2}{\pi E_2}$$

Where E is the Young modulus and ν is Poissons number for the fibres. The sum of the deformations $u_1 + u_2 = C - z_1 - z_2$ at two points in the contact zone, where C is a constant, can be calculated as below.

$$C - z_1 - z_2 = C - Ax^2 - By^2 = (k_1 + k_2) \int \int \frac{qdA}{r}$$

The force qdA is acting on a finite element in the contact zone, where r is the distance to this element from the centre of the ellipse. The integral must be calculated all over the zone of contact. The problem is now to find a function of the pressure q to put in the expression above. Hertz has shown with sufficient accuracy that it is possible to replace the pressure function q with the constant maximum pressure in the zone of contact q_0, /10/. This maximum pressure occurs in the centre of contact.

$$F = \int \int qdA = \frac{2}{3}\pi abq_0$$

The constants a and b are the half axis of the elliptical contact zone, see 4, and q_0 is the maximum pressure in the centre of the contact zone ($x = 0, y = 0$). The maximum pressure of the contact zone is consequently:

$$q_0 = \frac{3F}{2\pi ab} \tag{4}$$

The maximum pressure of the contact zone is 1.5 times larger than the average pressure in the zone. The radii of the elliptical contact zone have to be calculated before the pressure can be determined:

$$a = \alpha(\psi)\{\frac{3\pi}{4}F\frac{k_1 + k_2}{A + B}\}^{\frac{1}{3}} \tag{5}$$

581

$$b = \beta(\psi)\{\frac{3\pi}{4}F\frac{k_1 + k_2}{A + B}\}^{\frac{1}{3}} \tag{6}$$

Where α and β are coefficients depending of the angle of contact between the fibres ψ, according to Timoshenko /1/. In the edge of the contact zone at the half axis a and b, the stress will be, $\sigma_x = -\sigma_y$ and $\tau_{xy} = 0$. The study of a finite element situated at the boundary of the elliptical contact zone, illustrated by 2, allows the following expressions to be set up, by using polar co-ordinates, $\tau = |\sigma_\theta| = |\sigma_r|$. The shear stress in the boundary at the half axis can be expressed:

$$\tau_1(x = \pm a, y = 0) = (1 - 2\nu)q_0\frac{\gamma}{\delta^2}(\frac{1}{\delta^2}\text{arctanh}\delta - 1) \tag{7}$$

$$\tau_2(x = 0, y = \pm b) = (1 - 2\nu)q_0\frac{\gamma}{\delta^2}(1 - \frac{\gamma}{\delta}\arctan\frac{\delta}{\gamma}) \tag{8}$$

Where the coefficients γ and δ are:

$$\gamma = \frac{b}{a} \qquad\qquad \delta = \sqrt{1 - \gamma^2}$$

When b is approaching a, in other terms, when the zone of contact becomes circular, equation (7) and (8) transforms to the simple expression:

$$\tau = \frac{1 - 2\nu}{3}q_0$$

Generally the shear stress equals the tensile stress at the half axis, $\tau_1 = \sigma_{ra}$ and $\tau_2 = \sigma_{rb}$. Equations (7) and (8) with the maximum pressure inserted in eq. (4) results in the expressions below:

$$\sigma_{ra} = (1 - 2\nu)\frac{3F}{2\pi ab}\frac{\gamma}{\delta^2}(\frac{1}{\delta^2}\text{arctanh}\delta - 1) \tag{9}$$

$$\sigma_{rb} = (1 - 2\nu)\frac{3F}{2\pi ab}\frac{\gamma}{\delta^2}(1 - \frac{\gamma}{\delta}\arctan\frac{\delta}{\gamma}) \tag{10}$$

If no stress are allowed to exceed the breaking stress σ_B, the maximum value of the tensile stresses σ_{ra} and σ_{rb} have to be smaller than the breaking stress, $max(\sigma_{ra}, \sigma_{rb}) < \sigma_B$. The tensile stress σ_{ra} and σ_{rb} can be written dimensionless as $\sigma_r{}^*$, which is illustrated in 5. This figure shows that $\sigma_{ra} > \sigma_{rb}$ for the contact angles $30° < \psi < 90°$, this means that σ_{ra} is the critical stress. If σ_{ra} equals σ_B the breaking force F_B can be calculated for the fibres, at a certain contact angle ψ. The equation for σ_{ra} (9) and the expressions for the half axis a and b, (5) and (6) results in the contact force F below:

$$F = \{\frac{3R(1 - \nu^2)}{2E}\}^2\{\frac{2\pi\sigma_r)(\alpha(\psi)^2 - \beta(\psi)^2)}{3(1 - \nu)(\frac{1}{\sqrt{1-\gamma^2}}\text{arctanh}\sqrt{1 - \gamma^2} - 1)}\}^3 \tag{11}$$

By inserting σ_B in the place of σ_r in equation (11), the breaking force F_B is determined.

4. RESULTS AND DISCUSSION

E-glass fibres exhibits almost isotropic mechanical properties /6/. Data for E-glass fibres from /7/ and /6/ are inserted into equation (11), i.e. assuming isotropic properties with $R = 10 \ \mu m$, $E = 60 \ GPa$ and $\nu = 0.22$. The contact angle ψ vary from 30° to 90° and calculations are made for four different breaking stress levels σ_B. The breaking compressional forces F_B are illustrated in 3 as a function of the contact angle.

Nicalon or alumina fibres exhibit not very pronounced anisotropy, the fibres are thus assumed isotropic with $R = 6 \ \mu m$, $E = 200 \ GPa$ and $\nu = 0.3$ /8/. The contact angle ψ vary from 30° to 90° and calculations are made for four different breaking stress levels σ_B. The breaking compressional forces F_B are illustrated in 6 as a function of the contact angle.

Due to the inverse square dependence of the elastic modulus in equation (11), stiffer ceramic fibres are more sensitive for damage. The combinations of bending, compressional and tensile loading in weaving and the compressional loads used in conventional short fibre preforming and squeeze casting results in damage on the reinforcement fibres.

5. ACKNOWLEDGEMENTS

The financial support from the Swedish Work Environment Fund is greatly acknowledged.

6. REFERENCES

1. Timoshenko S. and Goodier J., Theory of elasticity, Mc Graw Hill, New York 1951.
2. Ko F., Ceramic Bulletin 68 (1989) 401-414.
3. Andersson C-H., Mechanical Properties of Fibres and Mechanical Models for Preforming, Proc. Verbundwerk'90, Wiesbaden 1990, pp 28.1-14.
4. Månsson O., Karlsson M., Andersson C-H., A Geometrical Treatment of Contact Pressure and Tensile Stress Distribution in Crossing Fibres, Dept. of Production and Materials Engineering, Lund Inst. of Technology, Lund 1993.
5. Pennander L. and Andersson C-H., Vibrational assisted low pressure casting, Proc ECCM 5, Bordeaux 1992, pp 472-476.
6. Kawabata S., J.Text. Inst. 81 (1990) 432-447.
7. Gupta P., Glass fibres for composite materials, in Bunsell A.R. (ed), Fibre reinforcement for composite materials, Elsevier, Amsterdam 1988.
8. Bunsell A., Simon G., Abe Y. and Akiyama M., Ceramic fibres, in Bunsell A.R. (ed) Fibre reinforcement for composite materials, Elsevier, Amsterdam 1988.
9. Christensson B., et.al. ECCM 6, Bordeaux 1993.
10. Hertz H. Z., Math Physik 28, 1883.

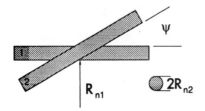

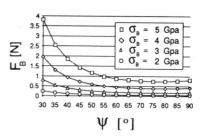

Figure 1: Fibres in contact, R_{n1} is the curvature radius and R_{n2} is the cross sectional radius, n is the fibre number.

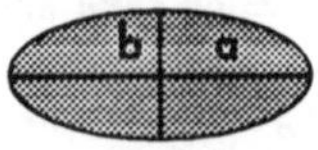

Figure 4: The contact zone, a and b are the half axis of the elliptical zone.

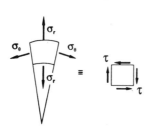

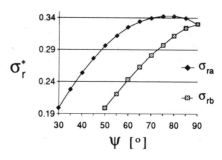

Figure 2: A finite element at the elliptical boundary.

Figure 5: Normalised stress for σ_{ra} and σ_{rb}, $\sigma_r^* = \frac{2\pi\sigma_r}{3(1-2\nu)FK^2}$ and $K = \left(\frac{3\pi F(k_1+k_2)}{4(A+B)}\right)^{\frac{1}{3}}$.

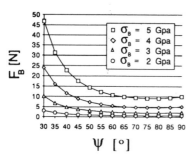

Figure 3: Breaking force F_B for E-glass fibres, $R = 10\ \mu m$.

Figure 6: Breaking force F_B for Nicalon fibres, $R = 6\ \mu m$.

MODELLING ELLIPSOIDAL FLAWS IN MULTI-LAYERED COMPOSITE MATERIAL - A CHARACTERIZATION OF THE DELAMINATION PROBLEM

S. SALDANHA, C. TRIAY, J. XING, D. JOUAN

Bureau Veritas - Research and Development Centre - Immeuble Apollo - 10 Rue Jacques Daguerre - 92565 Rueil-Malmaison - France

ABSTRACT

Based upon a methodology developed by the Research and Development Centre of BUREAU VERITAS, a structural analysis is performed on a complex structural part made of composites, in terms of its mechanical strength when discontinuities such as holes or ellipsoïdal voids are introduced into its basic material. The whole structure is analyzed for a designed loading case and a critical region is selected, taking into account the Tsai-Wu failure criterion over all integration points of the finite element model. A delamination flaw is then represented on this area, by using substructuring techniques, primarily to refine the finite element mesh of shell elements and then for the modelling of a three dimensional flaw, using continuum elements. Fracture mechanics fundamentals are considered for the analysis and J-integral calculations are performed with the ABAQUS FE code.

1 - Introduction

The FE model corresponds to a multi-layered composite structure which is submitted to a static load, applied as concentrated forces on four nodes of the model.

The basic mechanical structure is a prototype of a suspending arm for a lift cabin. It is shown in figure 1. Symmetry has been taken into consideration, and the upper head of the fork has not been represented, as it is pointed out in figure 2.

The anisotropy of the materials associated with the complex geometry of the structure have encouraged the use of specific tools offered by the ABAQUS code, such as the user's subroutine ORIENT, to define material local-axes .

Five basic materials are used to describe the behavior of almost thirty laminates and isotropic stiffeners, which together account for the modelling of the half of the mechanical structure. The main properties of these materials are listed under table 1.

The FE model is made of 266 elements, including shell elements S4R and STRI3, and beam elements of the ABAQUS library, with a total of 312 nodes. The total number of integration points all over the model is 4740 .

The RUINA "Post-Processing" program of The Bureau Veritas (1) has been used to check the Tsai-Wu failure criterion for all integration points of the model and permitted the selection of "critical elements". For these regions delamination flaws are automatically generated by running the FM4P program, an updated release of the GAMMA program.

Three categories of FE models are included in the methodology, one dealing with shell elements only, and the others based upon continuum three-dimensional elements. They are identified by FFMSE, FFMC3D and SFMC3D, respectively, corresponding to : "The First-Family Model of Shell Elements" ; "The First-Family Model of Continuum Three-Dimensional Elements" and "The Second-Family Model of Continuum Three-Dimensional Elements".

Delamination flaws are represented inside of composites by modelling ellipsoidal voids based on the three major axes "a", "b" and "c" of the ellipsoid, which one is located between two layers of the material. The plane containing the "a" and "b" major axes corresponds to the frontier plane separating the two selected layers of the laminate.

In the scope of the present paper, element number 469 has been chosen as the critical one, based on the Tsai-Wu failure criterion.

The methodology is applied at this region for a delamination flaw defined by the major axes "a" = 2000. μ, "b" = 1800. μ and "c" = 100μ, which is located between the first and the second layers.

The three categories of FE models are described, and finally a brief discussion on preliminary results is carried out, followed by the presentation of conclusions.

2 - An overview of the FFMSE, FFMC3D and SFMC3D

Figure 3 shows a central view of the FFMSE, which contains a total of 220 S4R and STRI3 shell elements, and 153 nodes. The mesh generation inside of the primitive rectangular area, the p.r.a., is carried out with quadrilateral elements only, with a total of 81 elements. The dimensions of the rectangle evolving the p.r.a. are directly related to the dimensions of the primitive parallelepipedic cell, the p.p.c., which one confines the material region corresponding to the FFMC3D. This FE model is illustrated in figure 4, emphasizing the delamination void located between the first and the second layers. The FFMC3D is entirely built with C3D15 quadratic elements of ABAQUS, totalizing 1920 elements, and 7610 nodes. The mesh volume generated around the ellipsoidal void is not enough accurate to show stress and strain singularities for points located nearby the "crack tip". The total number of elements and nodes of this family of FE model is depending on the number of layers of the laminate. For each layer there is one range of 3-D continuum elements.

The SFMC3D is constructed taking into account nodes that belong to the closest layers surrounding the delamination flaw, and that are located at external surfaces of these two layers. Figure 5 shows a cut view of the SFMC3D, pointing out the mesh generation around the crack tip. This FE model contains a total of 1408 3-D continuum elements, types C3D27 and C3D15V, and a total of 12418 nodes. The region around the crack tip is now well represented and stress and strain singularities can be visualized. The total number of nodes and elements of the model depends on the number of contours required for the J-integral calculations.

3 - Preliminary results

The present version of the FM4P program allows an automatic generation of the three FE models. They are compatible with the ABAQUS code and include all basic data required to describe the behavior of all constitutive materials of the laminate. These data are recovered directly from the basic mechanical structure.

Several structural analyses have been carried out for the FFMSE. The equilibrium conditions obtained for the nodes of the "critical element", translated in terms of displacements and rotations, have been considered as the local dynamic boundary conditions to be applied to the model. Local load fields obtained by running ABAQUS for this family of models are in good agreement with their corresponding fields in the critical element of the basic structure.

Displacements and rotations calculated for nodes that belong to the perimeter of the p.r.a. have been duly converted to displacements, only, to calculate the equilibrium values to be transmitted to their equivalent nodes located out of the neutral fiber. The shell elements theory is used to perform these conversions.

These values of displacements have been used as the loading equilibrium conditions in the FFMC3D during structural analyses with ABAQUS. As nodes that belong to the outside surfaces of the p.p.c. that are perpendicular to the critical-element normal direction, are left free, they can move across these surfaces to find their equilibrium positions. These final nodal positions do not conform very well with the results obtained from analyses with the FFMSE.

An upgraded release of the FM4P program will correct it, so that to apply dynamic boundary conditions to all nodes of outside surfaces of the p.p.c. .

Running tests with ABAQUS for the SFMC3D have shown that stress and strain singularities are duly represented in case of an arbitrary loading case. Within next few weeks structural analyses with this FE model will be performed taking into account appropriate local boundary conditions to apply to all nodes at the outward bounds of the mesh volume .

4 - Conclusions

Three families of FE models compatible with the ABAQUS code are automatically generated for a "critical element" of a structural mechanical part.

Preliminary results obtained for the FE model using shell elements are in good agreement with those related to the basic structural part.

An upgraded release of the FM4P program will ensure a better translation of the local load conditions to be applied to the FE models using 3-D continuum elements.

REFERENCES

1 - S. SALDANHA, C. TRIAY - "Failure Criteria for Composite Materials via an ABAQUS Post Processing Program ". ABAQUS User's Conference Proceedings, Oxford England, September 11-13, 1991.

2 - S. SALDANHA - " A Pre-Processor to ABAQUS in the Field of Fracture Mechanics ". ABAQUS User's Conference Proceeding, Aahen - Germany, to be held during June 23-25, 1993.

3 - HIBBITT, KARLSSON & SORENSON - " ABAQUS Guide on Fracture Mechanics " - USA, 1991.

4 - HIBBITT, KARLSSON & SORENSON - "ABAQUS User's Manual " - USA (version 5.2)

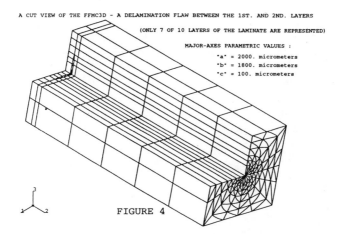

A CUT VIEW OF THE FFMC3D - A DELAMINATION FLAW BETWEEN THE 1ST. AND 2ND. LAYERS

(ONLY 7 OF 10 LAYERS OF THE LAMINATE ARE REPRESENTED)

MAJOR-AXES PARAMETRIC VALUES :

"a" = 2000. micrometers
"b" = 1800. micrometers
"c" = 100. micrometers

FIGURE 4

A CUT VIEW OF THE SFMC3D - THE MESH GENERATION NEAR THE CRACK TIP

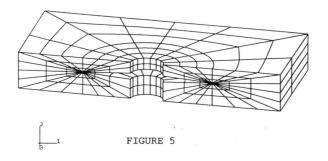

FIGURE 5

Material description	Young's Moduli (MPa)		Poisson's Ratio (1-2)	Shear's Moduli G12=G13=G23(MPa)
	E1	E2		
M1 - honey comb	10.E-12		0.30	G13=G23=22
M2 : fibers of carbon in a resinous matrix (45 % of resin)	51 160	47 277	0.39	2 971
M3 : fibers of carbon in a resinous matrix (56 % of resin)	40 400	37 300	0.39	2 117
M4 : fibers of carbon in a resinous matrix (45 % of resin)	116 477	24 705	0.39	3 740
M5 (isotropic)	37 300		0.39	

Table 1

589

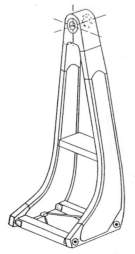

Fig.1 – a sketch of the structure

THE BASIC STRUCTURAL PART - AN OVERVIEW OF THE FE MODEL

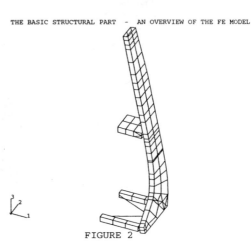

FIGURE 2

A CENTRAL VIEW OF THE FFMSE - THE MESH GENERATION INSIDE OF THIS AREA

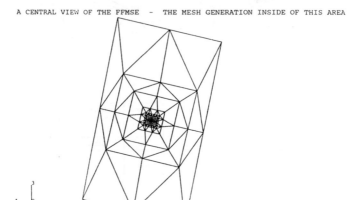

FIGURE 3

ANALYTICAL MODEL OF A WOVEN LAMINATE - SUPERPOSITION EFFECT OF TWO PLIES

Z. ABOURA, C.S. CHOUCHAOUI, M.L. BENZEGGAGH

Université de Technologie de Compiègne - LG2mS URA CNRS 1505
Groupe Polymère et Composite - BP 649
60206 Compiègne Cedex - France

I.INTRODUCTION

The woven fabric composites modelling remain complex and no much prevalent. The few existed models, in more simple case considers the woven fabric composite as an assemblage of pieces of cross-ply laminates [1] while in the more complex one basis on a three dimensional finite element analysis using an asymptotic homogenization schemes [2].

This paper propose an analytic model for the investigation of the stiffness of balanced woven fabric composites. Based in Isakawa & Chou work's [3,4], this model predict the elastic properties in each points of the structure. The effect of the ply superposition in the elastic behavior is investigated.

Such approach finds its interest in the delamination study of woven fabric composites where the front side of the crack would have to met different properties [5]. This in the correlation of the fractographic observations allow to explain the evolution of rupture energy in phase of propagation.

II.BASIC PRINCIPLE OF THE MODEL

The evolution of the warp and fill threads in the balanced woven fabric composites must be understood. The maximum thickness of the warp threads is equal to $H_t/2$ in the I-plan ($y=0$) (Fig.1). There evolution in the x direction is sinusoidal. In the case of y different of zero this thickness decrease in a sinusoidal form and vanish for $y=a/2$. Then, the best way to describe the real configuration, the fill thread evolution is refereed to the neuter fiber equation of the fill thread in "X" and "Y" directions.

These equation have the following form:

$$H_m(x)= \frac{-H_t}{4} \cos \frac{\pi x}{a} \qquad E(y)= \frac{-H_t}{4} \cos \frac{\pi y}{a}$$

The upper boundary position of the warp thread is given by:

$$W(x,y) = -H_m(x) + E(y)$$

At y=0 and x=0, the warp thread thickness value is $H_t/2$. This value is zero at y=a/2 and x=a/2.

The principle of the model is based in the classical laminated plate theory. Then, the local matrix A, B, D obtained for Δx and Δy are given by:

$$(A,B,D)ij = \int_{-H_m+E}^{H_t/2} Q(\theta)_{ij}^M (1,z,z^2)dz + \int_{H_m-E}^{H_m+E} Q(\theta)_{ij}^F (1,z,z^2)dz$$

$$+ \int_{H_m+E}^{-H_m+E} Q(\theta)_{ij}^w (1,z,z^2)dz + \int_{-Ht/2}^{H_m-E} Q(\theta)_{ij}^M (1,z,z^2)dz$$

Q^M : rigidity matrix of the resin.

Q^W : warp thread stiffness matrix.

$Q^F(\theta)$: rigidity matrix of the fill thread dependent on the fiber undulation.

In this case, the mechanical properties are determined out of the material's orthotropic axis. θ is defined as a X-function by:

$$tg\ \theta(x) = \frac{dH_m(x)}{dx}$$

The global matrix A, B, D are obtained by summation of the local matrix in the quarter of the structure.

$$(A,B,D)(y) = \frac{1}{a/2} \int_0^{a/2} A(x,y),B(x,y),D(x,y)dx$$

$$(A,B,D)_{global} = \frac{1}{a/2} \int_0^{a/2} A(y),B(y),D(y)dy$$

Seeing the complexity of the problem, numerical integrations are used.

III.VALIDITY OF THE MODEL

The results of this investigation are compared to those obtained by experimental work and by finite element analysis [Hassim et Co.] [2].

Two woven fabric composites glass/epoxy and glass/vinylester are tested. The material properties and geometrical properties are listed in Table I.

Table II and III show a good correlation between experimental results and the present model. The differences which can subsist notably in the poisson's ratio are probably du to the precision of estimation of the geometrical parameters. Moreover, in both analysis, the basic cell is considered as representative of all the structure. This modelling neglect the geometrical imperfection in the structure.

IV. EFFECT OF SUPERPOSITION OF TWO LAYERS

The fractographic observation show an aleatory disposition of the layers. Therefor, it is erroneous to consider the woven fabric composites as a perfect stacking of elementary woven layer [0/90].

Even if the mono-layer leads to well approach the global mechanical properties of the structure, the local phenomena's differ from one configuration of stacking to another one. This study considers the stacking of two layers in three specific scenarios:

IV-1. Case of stacking $[0/90]_s$ (Fig1):

The fill thread thickness is maximal in y=0.

-The superior longitudinal neuter fiber of the bundle describe the function:-Hm(x)+ Ht/2

-The inferior longitudinal neuter fiber of the bundle describe the function: Hm(x) - Ht/2

The A, B, D matrix are obtained by:

$$(A,B,D)_{ij}= \int_{-H}^{\alpha(x,y)} Q_{ij}^M (1,z,z^2)dz + \int_{\alpha(x,y)}^{-\beta(x,y)} Q(\theta)_{ij}^F (1,z,z^2)dz + \int_{-\beta(x,y)}^{-\gamma(x,y)} Q_{ij}^W (1,z,z^2)dz$$

$$+ \int_{-\gamma(x,y)}^{\gamma(x,y)} Q_{ij}^M (1,z,z^2)dz + \int_{\gamma(x,y)}^{\beta(x,y)} Q_{ij}^W (1,z,z^2)dz + \int_{\beta(x,y)}^{\delta(x,y)} Q(\theta)_{ij}^F (1,z,z^2)dz + \int_{\delta(x,y)}^{H} Q_{ij}^M (1,z,z^2)dz$$

$\alpha (x,y) = Hm(x)-Ht/2-E(y)$ $\qquad$ $\gamma (x,y)= Hm(x)+Ht/2-E(y)$

$\beta (x,y) = -Hm(x)+Ht/2-E(y)$ $\qquad$ $\delta(x,y)= -Hm(x)+Ht/2+E(y)$

The integral calculus gives a zero matrix B. This results are conformable with the selected stacking.

IV-2. Case of "Optimization 1" (Fig2):

A relative difference of y=a/2 between the two stacking layers is carried out. Then, in y=0, the inferior layer has a maximal thickness of the fill thread when the upper layer contains only transverse fibers and matrix.

At y=a/4, the two layers are identic with an average thickness of fill threads.

At y=a/2 the opposite situation of these of y=0 is produced. The fill thread thickness is maximal on superior the layer and zero on the inferior layer. This configuration is called "Optimization 1".

The A, B, D matrix are given by:

$$(A,B,D)_{ij}= \int_{-H/2-\Delta}^{\xi 1(x,y)} Q_{ij}^M (1,z,z^2)dz \int_{\xi 1(x,y)}^{\gamma 1(x,y)} Q(\theta)_{ij}^F (1,z,z^2)dz + \int_{\gamma 1(x,y)}^{\varepsilon 1(x,y)} Q_{ij}^W (1,z,z^2)dz + \int_{\varepsilon 1(x,y)}^{\delta 1(x,y)} Q_{ij}^M (1,z,z^2)dz$$

$$+ \int_{\delta 1(x,y)}^{\beta 1(x,y)} Q_{ij}^W (1,z,z^2)dz + \int_{\beta 1(x,y)}^{\alpha 1(x,y)} Q(\theta)_{ij}^F (1,z,z^2)dz + \int_{\alpha 1(x,y)}^{H/2+\Delta} Q_{ij}^M (1,z,z^2)dz$$

$\xi 1(x,y) = -Hm(x)-\Delta-E(y)$ $\quad$ $\gamma 1(x,y) = Hm(x)-\Delta+E(y)$ $\quad$ $\varepsilon 1(x,y) = -Hm(x)-\Delta+E(y)$

$\delta 1(x,y) = Hm(x)+\Delta-|E(y+a/2)|$ $\qquad$ $\beta 1(x,y) = -Hm(x)+\Delta-|E(y+a/2)|$

$\alpha 1(x,y) = -Hm(x)+\Delta+|E(y+a/2)|$ $\qquad$ $\Delta = \dfrac{1}{2}\left(\dfrac{H}{4}\sqrt{2} + \dfrac{H}{2}\right)$

IV-3. Case of "Optimization 2" (Fig3):

A relative difference of y=a/2 between the two stacking layers on x=a/2 and y=a/2 is carried out. This configuration leads to fill the matrix trapped in the interweaving. This configuration is called "Optimization 2".

The A, B, D matrix are given by:

$$(A,B,D)ij= \int_{-H/2-\Delta}^{\xi2(x,y)} Q_{ij}^{M}(1,z,z^2)dz \quad \int_{\xi2(x,y)}^{\gamma2(x,y)} Q(\theta)_{ij}^{F}(1,z,z^2)dz + \int_{\gamma2(x,y)}^{\epsilon2(x,y)} Q_{ij}^{W}(1,z,z^2)dz + \int_{\epsilon2(x,y)}^{\delta2(x,y)} Q_{ij}^{M}(1,z,z^2)dz$$

$$+ \int_{\delta2(x,y)}^{\beta2(x,y)} Q_{ij}^{W}(1,z,z^2)dz + \int_{\beta2(x,y)}^{\alpha2(x,y)} Q(\theta)_{ij}^{F}(1,z,z^2)dz + \int_{\alpha2(x,y)}^{H/2+\Delta} Q_{ij}^{M}(1,z,z^2)dz$$

$\xi_{2(x,y)} = Hm(x)-\Delta-E(y) \qquad \gamma_{2(x,y)} = Hm(x)-\Delta+E(y) \qquad \epsilon_{2(x,y)} =- Hm(x)-\Delta+E(y)$

$\delta_{2(x,y)} = -Hm(x+a/2)+\Delta-|E(y+a/2)| \qquad \beta_{2(x,y)} = Hm(x+a/2)+\Delta-|E(y+a/2)|$

$\alpha_{2(x,y)} = -Hm(x+a/2)+\Delta+|E(y+a/2)|$

With $\Delta = \dfrac{\Delta z + \dfrac{H}{2}}{2}$ et $\Delta z = \dfrac{H}{2}(\sqrt{2}-1)$

RESULTS:

The model proposed in this study leads to predict the elastic properties in each points of the structure, in "X" and "Y" directions. Firstly, the averaged Ex evolution in "X" as function of "Y" is presented. Figure 4 show an abrupt drop of modulus. This is due to the geometry of the structure. Effectively, at y=11, the structure contains only transverse fibers and matrix, even when y=1 we just find longitudinal fibers.

In the other configurations the averaged Ex as a function of "Y" presents a parabolic curve (Fig.5). Comparatively to the first configuration the drop of rigidity is particularly estimated. So we observe a trivial decrease of the modulus for "optimization 2". But no large differences in the E_{moy} evolution as function of y can be seen. Therefore if we plot Ex as function of "x" for y given, the evolution is different (Fig.6).

The drop of rigidity for the first configuration [0/90]$_s$ between the two extreme rigidity (x=1, y=1 and x=11, y=11) is in the order of 88%. For the "optimization 1" this value is in the order of 30% (calculated between x=1, y=1 and x=11, y=11) and 20% for "optimization 2" (calculated between x=6, y=6 and x=11, y=11).

CONCLUSION

This study was led to establish a well simple model to evaluate the elastic properties of balanced woven fabric composite. This model estimate the local properties in each points of the structure.

The superposition effect of two layers was been landed. Three configurations was proposed to describe the extreme positions which two stacking layers can be occupied. The large dispersion of the rigidity leads to a strong heterogeneity subsisted in the woven fabric composites. That explain the complexity confronted by the extremity of the crack during a delamination. For example, in the case of mode I cracking, the crack propagation is quite fast with a blockage in certain moments.

This phenomena can be explained by the knowledge of the local properties meted by the crack.

Current works interest to the correlation of the energy of rupture, the fractographic observation and the local properties of a balanced woven fabric composites as well as satin.

	E_1 [GPa]	E_2 [GPa]	ν_{12}	G_{12}[GPa]	a/2 [mm]	Ht [mm]
470.36/VT	57	11,42	0,28	8,37	0,175	0,145
Epoxy/VT	59,9	24,6	0,21	9,149	0,9	0,205
Mat 470.36	3,5	----	0,3	----	----	----
Mat.Epoxy	3,13	----	0,34	----	----	----

Table I : Mechanicals properties of UD glass/vinylester, UD glass/Epoxy and matrix. "a" and "Ht" are geometrical properties of Woven.

	E_x [GPa]	E_y [GPa]	ν_{xy}	G_{xy} [GPa]
Experience	24	24	6,5	0,12
Model	23,49	23,53	5,82	0,10

Table II : Comparison between model's results and experimental's results for the woven glass/vinylester.

	E_x [GPa]	E_y[GPa]	G_{xy} [GPa]	ν_{xy}
ISMCM[1]	24,9	24,9	5,9	0,275
Besançon[2]	27,3	27,3	7,8	0,15
E.F [3]	25,2	25,2	5,23	0,14
Model[4]	28,4	28,7	6,26	0,13

(1):Test realized in ISMCM
(2):Test realized by the laboratory of applied Mechanic in Besançon
(3): Finite Element's results obtained by Paumelle's model
(4):Results obtained by this study model

Table III : Comparison between model's results, finite element's results and experimental's results for the woven glass/epoxy

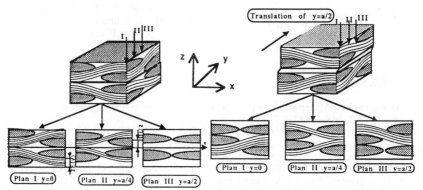

Figure1 : Two layers stacking[0/90]s

Figure2 : Two layers stacking. Optimization1

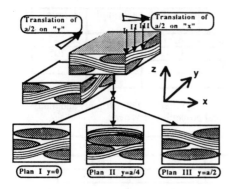

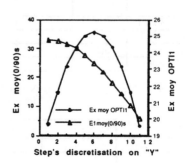

Figure3 : Two layers stacking.
Optimization2

Figure 4 : Evolution of Ex versus "y"
for [0/90]s and "opti 1".

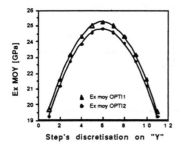

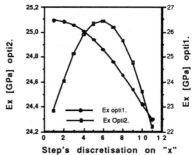

Figure 5 : Evolution of Ex versus "y"
for "opti1" and "opti 2".

Figure 6 : Evolution of Ex versus "x"
for "opti1" and "opti 2"

REFERENCES

1.**J.M Berthelot** "Matériaux composites. Comportement mécanique et analyse des structutres" Ed. Masson (1992)
2. **P. Paumelle, A.Hassim and F. Léné** " Composites with woven renforcement: calculation and parametric analysis of properties of homogeneous equivalent" Recherche Aérospatiale (1990) pp 1-12.
3. **T.Ishikawa and T.W.Chou** " Stiffness and strength behavior of woven fabrics composites" Journal of materials sciences 17 (1982) pp 3211-3220.
4. **T.Ishikawa and T.W.Chou** " Nonlinear behavior of woven fabrics composites" Journal of composites materials. Vol 17 (1983)
5 **M.L Benzeggagh and Z.Aboura** " Délaminage mode I et II de composites à renfort tissu à faibles et grandes vitesses" Journal Physique III, Decembre (1991) pp 1927-1951

RESIN TRANSFER MOULDING, THEORY AND PRACTICE

K. POTTER

Sintef Production Engineering - Rich Berkelands Vei 2B
7034 Trondheim - Norway

ABSTRACT.

RTM technology has been attracting increasing interest in recent years, with a great deal of effort being expended on the modelling of resin flow in moulds and other aspects of the process. This paper will demonstrate how a few, simple concepts can serve to illustrate the major features of the process. It will also consider how some of the features of the process that have not been extensively modelled might affect moulding problems and quality and comment on changes that need to be made in our approach to design in order to make the most cost-effective use of the technology.

INTRODUCTION.

RTM covers a very wide range of technical options, in the context of a short paper such as this we can only consider a small sub-set of these. The process that will be considered here is the manufacture of high quality, high volume fraction articles using well aligned reinforcements (most usually cloths), by the injection of high performance resins into fixed mould cavities formed by rigid tooling. The interest in RTM for such applications seems to stem from the mid to late 1970s, refs 1&2. An increasing dissatisfaction with the cost and geometrical limitations of "traditional" moulding processes led to other development and applications work in the early 1980s, ref 3. Towards the end of the 1980s the process was attracting a lot of attention from the manufacturers of aerospace components, generally using carbon/epoxy materials. Ref 4 describes the development of a complex aircraft component by the RTM route. Another driving force towards RTM, especially in the USA, was the increased effort in the manufacture of complex fibre architectures by 3D weaving and braiding technologies that cannot easily be impregnated by other technologies than RTM.

For the last ten years the interest in RTM has steadily grown, with reinforcements, matrices, reinforcement manipulation, process modelling etc all being developed and reported. The current status is that many applications of the technology are in operation or development and material systems are being developed and improved, although current RTM systems are relatively lacking in damage tolerance compared to the toughest prepreg systems. The reasons for the upsurge in interest are directly related to the advantages of RTM, these have been extensively quoted elsewhere, refs 5 & 6. This paper will first review the processes that go on when resin is injected into and cures within an RTM mould and some of the theoretical considerations relating to flow. The behaviour of reinforcements will be addressed in so far as these have practical influence on the process. The requirements for RTM processable resins will be reviewed. The technology of handling and preforming flat reinforcements to build up complex geometries will be briefly considered, and the effects of the RTM process capabilities and preforming technology on the design process will be discussed.

BASIC THEORY OF RTM.

The theory of resin flow in RTM has been extensively studied by many groups, refs 7 & 8, this section will only consider the basic aspects and concentrate on flow in woven cloths as these are the most widely used materials in "advanced" RTM.

When we inject an epoxy into a bed of cloth, two processes control how fast the resin front will expand and the quality of the resulting laminate. These are the bulk flow of resin and the wetout of individual fibres. Bulk flow rate is directly proportional to the imposed pressure and inversely proportional to the resin's viscosity and the length of already impregnated fibres. Flow rate is also proportional to two geometrical terms related to the sort of gating used (and how the length of the flow front varies with position, ie an expanding or contracting flow front) and the permeability of the reinforcement. The permeability is related to the structure of the reinforcement, eg UD knitted or woven, and the porosity (1-Vf). From a measured permeability we can use analytical relationships to predict flow rate. To predict permeability from the structure and porosity of the cloth is much more difficult and it is uncertain to what extent current models can provide accurate predictions. However, permeability is easy enough to measure so the theoretical difficulties do not restrain the practical uses of RTM.

Woven cloths, and most other reinforcing forms, can be characterised as having high volume fractions within the tow and regions of low or zero volume fraction between the tows. The bulk flow of resin occurs down the low Vf/low resistance paths between the tows rather than through the fibre tows, which makes for the modelling difficulties. Control of the size and shape of the intertow voids has allowed weavers and knitters to generate cloth forms of higher permeability for a given porosity. Impregnation of tows and fibres takes place by wicking up of resin into the tows by surface tension, from the resin that is carried between the tows. This process is not governed by the same factors as the bulk flow and so the spatial relationship between the bulk flow front and the full wetout point will change as injection rate changes.

Once the fibres in a tow are wet out, the resin in the tow is essentially static thereafter, although some mixing will inevitably occur at the interface between tow and intertow void. That this is so can be verified by observation of the flow front in a glass topped mould, or by injecting a clear resin followed by a dyed resin; the majority of the dyed resin will be seen between the tows after cure. This observation also means that we should not base our estimates of permeability under injection conditions on measurements of steady state pressure drop at a fixed flow rate in completely filled moulds. If we inject too fast into a non-evacuated mould we can "overrun" the wetout and entrap air in the tow (especially if the resin's wetout behaviour is poor). If we inject too slowly we can entrap air in the gaps between the tows, as the wetout flow front "overruns" the bulk flow front. A similar effect can be seen if we cease injecting before the mould is full, as resin will be wicked out of the intertow gaps to fill the tows. Equally, the edges of tools tend to present easy flow regions distorting the shape of the flow front, and ply drop-off regions can have similar effects. For these reasons the use of evacuated moulds tends to improve the overall quality of mouldings.

When the mould is full and we increase the temperature of the mould to cure the resin, the volume of the resin will increase. If the mould vents are not sealed, resin will be driven from the mould by this expansion. As the resin cures it will shrink somewhat and, once more, intertow voidage can be caused if the ejected resin cannot find its way back into the mould cavity to replace the volume loss due to the shrinkage. Defects such as these will not occur with all resin/fibre/process combinations and can be controlled by good tool design, the use of dry reinforcements, evacuation of mould cavities and/or sealing the tool cavity during cure.

To conclude, the basic theory of RTM is fairly simple, a few concepts serve to indicate the major effects and sources of defects. To produce a theory capable of quantitative results is much more difficult. From a quality viewpoint the local flow is as important as the bulk flow, but tends to be missing from the theoretical models, this is reasonable as this is a bounding process limit rather than a control parameter, but for practical purposes it should not be neglected. A successful model would have to predict permeability from the reinforcement structure and porosity, be able to account for edge effects and other easy paths, and to deal with the complex shapes of some RTM tools. In many cases for high performance RTM what we need to know is not so much the fill time, but the optimum positions to place inlet gates, the best shape for these gates and any internal gating, and the best positions for outlet gates. In such cases the use of software developed for injection moulding may give an acceptable approximation to the flow front positions (at least for reinforcements with isotropic permeabilities), ref 9. For simpler geometries the use of entirely manual methods, drafting out the most likely shape of the flow front as injection progresses, have been used with good results to predict the optimum positions for in and out gates. Such methods have no theoretical basis but are very quick, and are adequate in many cases, they are also a useful way of developing an understanding of basic mould design principles.

REINFORCEMENTS FOR RTM

To a first approximation all fibre types and reinforcement styles available in unprepregged form can be handled by RTM. Some problems with fibre movement have been reported with mats, but for high performance RTM these materials are unlikely to be used. In addition at reasonably high Vf the clamping pressures required to generate the high Vf will tend to prevent gross fibre movement. For reasons of space I do not intend to discuss these materials in any detail, except to say that reinforcements specially prepared for RTM (ie having a high permeability for their porosity) are now available in both woven and insertion knitted forms, and that many material forms that are hard or impossible to prepreg can be handled by RTM. These include shape woven and knitted items, woven sections and woven sandwich panels, 3D weaves and braids, and fabrics with a pile raised on one surface.

One critical factor for RTM in rigid moulds of fixed cavity is the relationship between the applied pressure and the reinforcement's fibre volume fraction, and it is to this that we will now turn. All reinforcements have a characteristic pressure/volume fraction relationship. These are of the form of a minimum Vf at zero pressure, a rapid rise in Vf as pressure rises, and a gradual tailing off in the curve to a plateau at high pressure. The volume fraction will be proportional to the log of pressure over a wide range in pressures. Typical values for a cloth material might be a zero pressure Vf of 45%, a pressure of 1 bar at 55% and 100 bar at 70%. For unidirectional materials the zero pressure Vf will be higher and the slope of the curve with respect to pressure will be shallower, on the other hand random mats will have a lower initial Vf and a higher slope.

The practical significance of this can easily be seen. If we wish to make a laminate of 4 plies of cloth in 1mm at a Vf of 55% on a projected area of $0.5m^2$, we need a closing pressure of about 0.1MPa, or a total thrust of 5000kg. If, by error, 5 plies of material had been inserted into the mould the effective volume fraction would be 68.8% and a pressure of close to 10MPa would be needed, a total thrust of 500,000kg. Only about 1% of the total tool area would have to be affected by the increased volume fraction for the mould closing force to be doubled. Whilst the accidental insertion of an extra ply might be considered a remote possibility, we may well have areas of the laminate with 4 plies and others with 5, 6 or 7 plies. If these plies extend beyond their desired position the mould will simply not shut to stops. We can deal with this problem by taking all our ply cutting and positioning tolerances on the negative side to avoid entrapped fibres, but we do this at the risk of creating resin rich zones in the laminate.

Similar problems can occur from changes in thickness caused by preforming of reinforcements and these will be considered later. Lastly, if we force a reinforcement around an external single curvature radius we will generate an additional compaction pressure, leading to a somewhat higher local Vf. If the mould cavity is of constant section around the radius we will tend to produce a region of lower Vf or a resin rich zone on the outside of the radius, which may also affect the shape of the flow front.

In conclusion, RTM in rigid tooling is very intolerant of local variations in fibre packing:- if the baseline Vf is high and the moulding is thin. Semi-rigid or flexible mould faces can be used to offset this effect, at the cost of a loss of tolerance control. The key to the production of good components lies in the concurrent design of layup, tooling and process to take into account all the properties of the materials and process.

RESINS FOR RTM.

The basic requirements for an RTM resin are the process related ones of an adequate pot life at the injection temperature and low enough viscosity to ensure that mould filling can occur within the pot life. For most purposes about 30 minutes and 1Pas would be reasonable target figures. For high speed RTM it is usual to inject at a temperature that will cure the resin and thus modelling of the cure kinetics, viscosity changes etc is needed. Most high performance RTM is carried out by injecting the resin at a temperature below the cure temperature, greatly reducing the modelling requirements. A very wide range of resins is now available for RTM including epoxies and BMIs, refs 10 & 11. There is no space here to discuss any of these resin systems in any detail. Volatiles in resins cannot escape from a closed RTM cavity so resins containing or evolving volatiles are non-preferred, especially if high vacuum levels are used in moulds of the most complex geometries. Other important factors might be low shrinkage, especially in the liquid state for reasons noted earlier, low CTE and a cured Tg above the cure temperature to minimise cure stresses in cured components, high toughness, ease of producing a uniformly mixed resin/hardener combination etc.

Resins meeting these various requirements are now available from a number of the major resin suppliers. The most difficult requirements to meet are probably to give high levels of toughness without losing the RTM processability.

REINFORCEMENT PREFORMING.

This is a large area which can only be outlined here. The process to be considered is that of applying a small percentage of (generally thermoplastic) binder material to the reinforcement, stacking up and hot pressing a series of reinforcement plies, and hot fonning this stock material in cold, matched dies. The binders used can be in a variety of forms, the most common of which is a coarse powder of particle size about the same as the reinforcement's pressed thickness. The function of the binder is to form links between the plies that can be softened by the application of heat, followed by forming of the reinforcement and cooling to lock the preform into the formed shape.

A component may consist of one or many of these semi-rigid preforms, each of which is formed in separate tooling that can be considered as a series of geometrical offsets from the main tool faces. When cloth is preformed into double curvatures there is a realignment between warp and weft fibres away from the original +/-45° to some new angle, dictated by the formed shape. This change of angle is always associated with a change in thickness of the reinforcement, ref 12, such that preforming of cloth can lead to problems of high local Vf, as noted earlier. For all reinforcements there are limiting deformations that can be made before wrinkles and folds occur in the cloth, these are related to the shape to be formed and the orientation of the reinforcement with respect to that shape. In addition any forming operation must be associated with inter-ply slip and shear if out of plane wrinkling is to be avoided, this requirement often sets the preforming temperature requirements.

EFFECTS ON COMPONENT DESIGN AND CONCLUSIONS.

The use of preform technology presents us with an opportunity to assemble complex shapes from preformed flat sheet in a rapid and process, with opportunities for automation and low labour cots. What is created, however, is a series of ply blocks within the component that do not taper off gently as recommended by the normal practice for composites design. This will lead to a series of internal resin rich zones that may be considered as potential sites for damage or delamination growth. Care should be taken to place such ply block boundaries in areas remote from peak stresses as far as possible. For ply blocks terminating on the surface of mouldings it is best if the mould geometry replicates the ply drop to avoid the generation of resin rich zones.

In conclusion the RTM process offers many advantages over competing processes, in terms of accuracy of geometry, flexibility with respect to reinforcement choice, cost of manufacture etc. These advantages are not purchased without a price, and great attention to detail is required if we wish to make complex, high volume fraction mouldings of high quality. To maximise the advantages requires that some form of preform technology is required, this may lead us away from some of the traditional features associated with quality design in composites. The differences between RTM and other composites processing techniques are such that in order to obtain the maximum benefits from RTM a design approach taking into account the specific capabilities of the process is mandatory.

REFERENCES.

1. M.C. Cray. Proc 3rd Int Conf on Electromagnetic Windows. Paris, Sept 1975.
2. W.R. Jones & J.W. Johnson. In, Fabrication techniques for advanced reinforced plastics. IPC 1980. 40-47.
3. P. Medlicott and K.D. Potter. In, High-tech the way into the 90s. Elsevier, 1986. 29-42.
4. D. Morgan. Proc 34th Int SAMPE symposium. 1989. 2358-2364.
5. K.D. Potter. In, Progress in advanced matls and processes. Elsevier 1985. 247-254.
6. W. Becker & M. Wadsworth. In, RTM for the aerospace industry. SME, 1990. 1-33.
7. J. Molnar. L. Trevino & L.J. Lee. Modern Plastics. Sept 1989. 120-126.
8. Zhong Cai. J, Comp Mat. vol26, no 9, 1992. 1310-1338.
9. Y.M. Lee, J.M. Castro, G. Tomlinson, E. Strauss. Proc Annual Tech Conf. SPE. 1990. 994-1001.
10. E. Stark. 35th Int SAMPE symposium. 1990. 782-794.
11. K.D. Potter & F. Robertson. 32nd Int SAMPE symposium. 1987. 1-12.
12. K.D. Potter. Proc ICCM3. AMAC/Pergamon, 1980. 1564-1579.

FORCED CHEMICAL VAPOUR INFILTRATION : MODELLING DENSIFICATION OF A 3D ORTHOGONAL WOVEN FIBRE STRUCTURE

C. STEIJSIGER, A.M. LANKHORST, Y.G. ROMAN

*TNO Institute of Applied Physics (TNO-TPD) - Stieltjesweg 1
PO Box 155 - 2600 AD Delft - The Netherlands*

ABSTRACT

A quasi-stationary mathematical model has been developed for the prediction of the laminar gas flow, transport phenomena and chemical reactions in a Forced Chemical Vapor Infiltration (FCVI) reactor. The densification process of a C/SiC ceramic matrix composite (CMC) is studied by looking at the mass transport of the reactants, such as methyltrichlorosilane (CH_3SiCl_3, MTS) and hydrogen (H_2), by convection, and the subsequent deposition of the solid products on the pore surfaces. The simulations with the CFD-model have resulted in a prediction of optimal process conditions (reactor pressure, pressure gradient and flow rate) leading to a CMC with a low density gradient and low residual porosity in combination with a short total infiltration time.

1. INTRODUCTION

Fabrication of ceramic composites can be done by forced chemical vapour infiltration (FCVI), for which gaseous reactants flow and diffuse into a fibrous structure (preform) and react to deposit a solid matrix surrounding the fibers. The FCVI process depends on three different phenomena which are closely coupled during the process: (i) the heat and mass transfer, (ii) the dynamical evolution of the porous structure and (iii) the chemical kinetics. By using a physical/chemical model of the FCVI process, better understanding can be achieved and optimal process conditions be found.

2. MATHEMATICAL MODEL
2.1. Fluid flow and heat transfer

The flow and mass transfer in an isothermal porous medium are determined by the continuity equation, the momentum balance equation and the concentration balance equations. As growth rates are small compared to gas velocities, a quasi-stationary approach can be adopted. This implies that it is not necessary to take time-dependent terms in the partial differential equations into account.

Continuity equation:

$$\frac{d}{dx}(\rho\, u_D) = -\mathcal{S}\, m(\text{SiC}) \sum_{s=1}^{S} \mathcal{R}_s \,\backslash_{\text{SiC},s} \tag{1}$$

Momentum balance equation (extended Darcy law):

$$\frac{\rho}{\varepsilon^2}\left(u_D\, \frac{d\, u_D}{d\, x}\right) = -\frac{d\, P}{d\, x} - \rho\, g - \mu\, \frac{u_D}{K} - \rho\, \frac{F}{\sqrt{K}}|u_D|\, u_D \tag{2}$$

Concentration equation:

$$\frac{d}{dx}\{\rho\, u_D\, \omega_j\} = \frac{d}{dx}\left\{\rho\, \mathcal{D}\, \frac{d\omega_j}{dx}\right\} + \mathcal{S}\, m_j \sum_{s=1}^{S} \mathcal{R}_s \,\backslash_{js} + $$

$$\varepsilon\, m_j \sum_{k=1}^{K} \nu_{jk}\, (\mathcal{R}_k - \mathcal{R}_{-k}) \tag{3}$$

where ρ is the density of the gas mixture (kg m^{-3}), u_D the superficial mass averaged Darcy velocity in an N component gas mixture (m s^{-1}), $\mathcal{S}$ the specific surface area (m^2/m^3), $m(\text{SiC})$ the mole mass of SiC (kg mole^{-1}), $\mathcal{R}_s$ the growth rate of the heterogeneous reaction s (mole m^{-2} s^{-1}), $\backslash_{\text{SiC},s}$ the stoichiometric coefficient for SiC in reaction s, P the local pore pressure (Pa), ε the porosity, g the gravitational constant (m s^{-2}), μ the molecular dynamic viscosity of the gas mixture (kg m^{-1} s^{-1}), K the permeability (m^2), F the Forchheimer number according to the Ergun eq. /2/, ω_j the mass fraction of species j, m_j the mole mass of species j (kg mole^{-1}), $\mathcal{D}$ the diffusion coefficient (m^2 s^{-1}), ν_{jk} the stoichiometric coefficient for species j in the kth gas-phase reaction.

The gas mixture is considered to behave as an ideal gas and its properties are a function of temperature, pressure and composition of the gas mixture /1/. We assume that K reversible chemical reactions take place in the gas-phase, with a forward reaction rate $\mathcal{R}_k$ ($k = 1,K$) (mole m^{-3} s^{-1}) and a reverse reaction rate $\mathcal{R}_{-k}$. We further assume that S heterogeneous reactions take place.

The diffusion term in the stream-wise direction, the first term on the right-hand side of eq. (3), has been neglected for the time being. Due to the complexity of the chemical kinetics also the homogeneous reactions in eq.(3) are neglected for the time being.

The boundary conditions for the transport equations (1)-(3) are:

$$u_D(\text{x} = 0) = u_{\text{inlet}}\ ;\ \left[\frac{d\, u_D}{d\, x}\right]_{\text{x=L}} = 0\ ;\ P(\text{x} = \text{L}) = P_{\text{outlet}}\ ;\ \omega_j(\text{x} = 0) = \omega_{\text{inlet}}$$

2.2. Preform description

No general relationship exists between effective porosity and permeability. The few empirical, semi-empirical, and first principle-based correlations all have to be used within the restrictions for which they have been developed.

The 3D porous structure of the preform, under consideration, is rather complex. Each yarn consists of approximately 3000 fibers. A cross-section of the initial preform was photographed under an optical microscope. A cross-section of the yarn itself is not cylindrical but elliptically shaped. A yarn width and yarn height of respectively 2000 μm and 110 μm were measured/7/.

We have chosen to describe the permeability by a Carman-Kozeny relationship:

$$K = K_{\text{fit}} * \frac{\varepsilon^3}{180(1 - \varepsilon)^2} d^2 \tag{4}$$

where d is an effective cylindrical yarn diameter, based on the number of fibers in one yarn. The variable K_{fit} has been experimentally determined by measuring the pressure drop over a preform in an Argon flow at the CTK-TNO reactor /7/.
For the geometrical characterization of the porous structure a cubic model, first introduced by Scherer /8/ for describing viscous sintering, has been used. We assume that on each edge of the cubic cell a yarn is positioned with length l and radius a. The volume fraction solid phase (or fractional density) in a unit cell is then given by the following equation:

$$f = 1 - \varepsilon = \left(\frac{c_1}{c_0}\right) x^2 - \left(\frac{c_2}{c_0}\right) x^3 \qquad \text{where}: \quad x \equiv \frac{a}{l} \tag{5}$$

Equation (5) can be inverted to obtain x as a function of f /8/:

$$x = \frac{d_{\text{yarn}}}{2l} = \frac{c_1}{3c_2}\left[1 + 2\cos\left(\frac{4\pi}{3} + \theta\right)\right] \tag{6}$$

$$\text{where} \quad \theta = \frac{1}{3}\cos^{-1}\left[1 - \left(\frac{27c_0c_2^2}{2c_1^3}\right)f\right] \tag{7}$$

For the cubic model the following constants apply: $c_0 = 1$; $c_1 = 3\pi$; $c_2 = 8\sqrt{2}$
The surface area per unit volume of packing for the cubic model can be given by the following equation:

$$S = \frac{1}{c_0 l}\{2c_1 x - 3c_2 x^2\} \tag{8}$$

2.3. Chemical kinetics

A lot of different relationships have been published in the literature regarding the deposition of SiC based on MTS and H_2 for CVD type reactors. The system Si-C-H-Cl seems to be very complex as is shown for instance by the study of Langlais c.s./3/, based on the Langmuir-Hinshelwood mechanism. The overall reaction of the SiC reaction is:

$$\textbf{CH}_3\textbf{SiCl}_3\,(\textbf{g}) + \textbf{H}_2\,(\textbf{g}) \longrightarrow \textbf{SiC}\,(\textbf{s}) + \textbf{3\,HCl}(\textbf{g}) + \textbf{H}_2\,(\textbf{g}) \tag{9}$$

The reaction rate $\mathcal{R}_1^S$ (in mole m^{-2} s^{-1}) is experimentally determined as/5,9/:

$$\mathcal{R}_1^S = K_{0,1}\,[MTS]^{0.6}\,\exp\left\{\frac{-E_{\text{act},1}}{RT}\right\} \tag{10}$$

$$K_{0,1} = 2.487 \cdot 10^{13}\ \text{mole}^{0.4}/\text{m}^{0.2}\text{s} \ ; \ E_{\text{act},1} = 400.0 \cdot 10^3\ \text{J/mole}$$

where $[MTS]$ is the MTS concentration (mole m^{-3}), R the universal gas constant (J mole^{-1} K^{-1}) and T the preform temperature (K).

3. NUMERICAL SIMULATIONS

A Computational Fluid Dynamics (CFD) model, based on the finite volume discretization method /4/, has been developed. In the model the porous preform is divided into a large number of volumes in the axial direction. All variables are considered constant within a unit cell. At the beginning of the process the porosity, specific surface area and permeability are set to the initial value. The densification in each unit cell is calculated as follows:

1. During the time step Δt we calculate the deposited matrix.
2. This gives a new value for the solid volume fraction in a unit cell.
3. Based on the porous structure model we calculate a new value for the yarn radius, porosity and the fractional density.
4. These new characteristic values are used to calculate the new specific surface area.
5. Properties of the gas mixture are updated.
6. The conservation, momentum balance and the concentration equations are solved with the new updated values and the next time step is taken.

This whole procedure is repeated until the pressure drop over the preform is larger than the pressure drop termination criterium (ΔP).

We simulate an FCVI process for infiltration of a disc-shaped substrate with a diameter of 80 mm and thickness of appr. 3 mm. This consists of Torayca T-300 fibers (appr. 52 vol %) woven into a preform with fibers in a 3D configuration. The objective of the simulations with the CFD-model is to seek optimal process conditions (reactor pressure, pressure gradient, flow rate and mixture composition) to obtain a CMC with a low density gradient and low residual porosity in combination with a short total infiltration time.

4. RESULTS

The influence of flow rate (Q_{inl}), preform temperature $(T_{preform})$, reactor pressure (P) and pressure drop termination criterium on the infiltration process has been investigated. Table 1 shows the characteristic FCVI process conditions under consideration.

The simulation of the FCVI process with the process conditions as given in table 1 results in an infiltration time of 8.65 hrs. The predicted average fractional density is 0.813 and the predicted non-uniformity of the density is 12.6 %. Fig. 1 shows the fractional density as a function of process time on different locations in the preform.

When increasing the preform temperature from 975 °C to 1050 °C , fig. 2 shows that the pressure drop across the preform rises to its final termination value of 70 Torr in an increasingly shorter time: the infiltration time is appr. halved by raising the temperature only 25 °C. However, the non-uniformity also increases for increasing temperature. Due to rapid heterogeneous reactions and subsequent depletion of MTS the preform becomes more and more coated instead of infiltrated. This is shown in fig. 3, where both infiltration time and non-uniformity are shown as a function of preform temperature for two different flow rates. The density gradient across the preform for a flow rate of 0.5 slm (standard litre per minute) and a large value of the preform temperature (e.g. 1050 °C) is 30.5 %. From fig. 3 it is also found that the infiltration time decreases when increasing the flow rate. This corresponds with experimental observations/6/. According to Darcy's law a larger flow rate leads to a larger pressure drop. Therefore, in order to obtain the same final average fractional density, the pressure drop criterium should also be increased for increasing flow rates.

When the simulations are repeated with the adjusted termination criterium, namely that the final average fractional density is the same for both cases, the infiltration time for the lower value of the flow rate is still larger than for the higher flow rate (fig. 4). If an infiltration time of less than 10 hrs. is required with a non-uniformity of less than 15 %, the choice of the preform temperature is less critical at larger values of the flow rate. The working zone for the higher flow rate is almost twice as large.

The reactor pressure itself is also an important variable for the FCVI process. Varying the pressure has shown that for this particular gas mixture around 100 Torr a local minimum in the density gradient can be found.

5. CONCLUSIONS

- The isothermal FCVI process is extremely temperature dependent. A lower preform temperature results in a more uniformly densified CMC, but the infiltration time is considerably longer. A decrease of 25 °C relative to 1000 °C results in a more than twice as long infiltration time.
- The pressure drop criterium should be set to a higher value when increasing the flow rate if the same final relative density is required.
- Increasing the reactor pressure decreases infiltration time and increases the average fractional density. For the non-uniformity an optimum around 100 Torr was found.

Acknowledgements

This work was supported by EC, programme Brite Euram, project CT91-0447.

REFERENCES

1. Hirschfelder, J.O., Curtiss, C.F. & Bird, R.B., "Molecular Theory of gases and Liquids", John Wiley and Sons Inc., New York, USA (1967)
2. Kaviany, M., "Principles of Heat Transfer in Porous Media", Springer-Verlag (1991)
3. Langlais, F., Prebende, C., Tarride, B. & Naslain, R., J. Phys. Colloq., 50 (1989), 93-103
4. Patankar, S.V., "Numerical Heat Transfer and Fluid Flow", Hemisphere Publishing Corporation (1980)
5. Roman, Y.G., "Silicon carbide deposition from methyltrichlorosilane, Part 1: Kinetics of the decomposition of MTS: a review", TPD-CTK-RPT-93-039 (1993)
6. Roman, Y.G., Steijsiger, C., Gerretsen, J. & Metselaar, R., in proceedings 17th annual conf. on composites and advanced materials (1993), January 10-15, Cocoa Beach, USA
7. Roman, Y.G., private communications
8. Scherer, G.W., J. Am. Ceram. Soc. 74(1991), 1523-1531
9. Steijsiger, C., Lankhorst, A.M. & Roman, Y.R., in Proceedings Numerical methods in Engineering '92, (Ch. Hirsch et al., ed.), 7-11 September, Brussels

Number of fibers per yarn (N_{fiber})	3000
Fiber diameter (d_{fiber})	7 μm
Width of the elliptic yarn (a_{yarn})	1000 μm
Height of the elliptic yarn (b_{yarn})	55 μm
Preform thickness ($L_{preform}$)	2.95 mm
Preform diameter ($D_{preform}$)	80 mm
initial fractional density (f_0)	0.517
initial specific surface (S_0)	3850 m^2/m^3
initial permeability (K_0)	2.05 10^{-11} m^2
Flow rate (Q_{inl})	0.504 slm
Outlet pressure (P_{outlet})	50 Torr
Preform temperature ($T_{preform}$)	1000 °C
Mole fraction MTS ($f_{in}(MTS)$)	0.128
Mole fraction Ar ($f_{in}(Ar)$)	0.076
Mole fraction H_2 ($f_{in}(H_2)$)	0.796
Pressure drop termination criterium (ΔP)	70 Torr

Table 1: Typical operating conditions of the FCVI infiltration process carried out at the TNO reactor in Eindhoven.

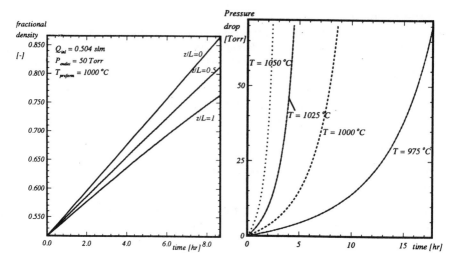

Fig. 1 Fractional density on different positions in the preform as function of time.

Fig. 2 Pressure drop over the preform for different values of the preform temperature.

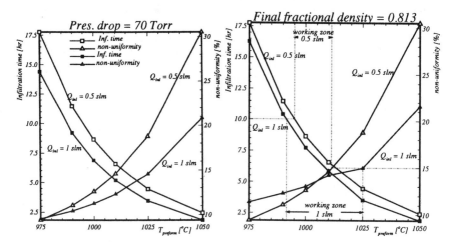

Fig. 3 Influence of the flow rate on the infiltration time and non-uniformity. The termination criterium is a pressure drop of 70 Torr

Fig. 4 Influence of the flow rate on infiltration time and non-uniformity. The termination criterium is an average fractional density of $f_{avg}=0.813$.

608

INFORMATION SOURCES FOR COMPOSITE MATERIALS

J. FELDT, W. JACKSON

Materials Information - Institute of Materials - 1 Carlton House Terrace SW1Y 5DB London - UK

Abstract

The amount of available data on the properties, fabrication, and performance of composite materials continues to increase as a result of the rapid progress in the development and applications of these materials. In this paper, some of the many information sources for composite materials will be reviewed, including the published literature and computerised bibliographic and numeric database systems. Problems involved in accessing and utilising these wide-ranging information sources will be discussed, and some solutions presented which might lead to better and more efficient dissemination of information about current developments in composite materials.

1. INTRODUCTION

The increasing importance of composite materials for engineering applications is reflected in the growth of research and development activities in many different countries. The results of these activities may be made available in a number of different ways, but their relevance to others working in the field will depend on a number of factors, including accessibility, quality, completeness and presentation. In the following, these factors will be discussed in relation to the types of information source available.

2. INFORMATION SOURCES

2.1 Primary Literature Sources

A number of journals and newsletters deal exclusively with composite materials (Table 1); there are, however, many materials-related periodicals which provide at least some coverage of the subject, and other important information sources include conference proceedings, reports, theses, patents and standards. Some indication of the size of the information pool was obtained by analysing the documents covered by one information service over a one-year period (1992) as shown in Figure 1. Of the more than 7,500 documents covering composite materials processed during the year, 46% were journal articles, 33% conference papers and 13% patents, the rest being made up of dissertations, government reports and books.

Those needing to keep abreast of current developments in composite materials face a number of problems when dealing with the primary literature. These may include: lack of awareness of published documents; the number and size of different published information sources, which no individual can hope to monitor and scan; and problems of accessibility associated with difficult-to-obtain material and documents published in a foreign language.

2.2 Secondary Information Sources

A traditional solution to some of these problems has been the development of abstracts journals which bring together, in a single publication, summaries of documents from a wide range of primary sources. Most abstracts journals have a subject index (as well as author and corporate author indexes) so it becomes relatively easy to locate those papers of relevance to a particular aspect of the properties, processing and/or applications of a given type or class of material. Most abstracts journals cover more than one class of material, and some are devoted to multi-disciplinary coverage of particular information sources such as patents or translations.

Abstracts journals are also available as computerised bibliographic databases which may be searched online (by connecting to a remote host computer) or locally (using, for example, compact discs supplied by the computer host and/or the database producer). The time required to extract information from computerised systems is very short, and a database containing many hundreds of thousands of records may be searched in a matter of minutes. The larger host systems include databases from a number of suppliers, so that cross-searching of different databases becomes possible.

Some of the more important abstracts journals/databases are listed in Table 2.

2.3 Numerical Databases

The wider application of composite materials will require that designers and engineers have ready access to reliable data on the properties, processing and performance of these materials. A few computerised property database systems containing such information are available, and a number of others are under development.

There are some important issues which need to be resolved before these systems become of general applicability. In particular, the lack of standardisation in materials descriptions, test methods, test data reporting, and data presentation mean that data from different systems may not usefully be compared or integrated. There is also a need for a "neutral interface format" so that data can be imported/exported between different systems.

These issues are currently being addressed by a number of groups, including committees of the American Society for Testing and Materials (in particular ASTM Committee E49 on the Computerization and Networking of Materials Property Data), and standards on "Identification of Composite Materials in Computerized Material Property Databases" (Ref 1) and "Identification of Fibers, Fillers and Core Materials in Computerized Material Property Databases" (Ref 2) have been published. In the composites testing area, the Versaille Project on Advanced Materials and Standards (VAMAS) has established a Technical Working Area to evaluate the mechanical properties of composite materials and to promote standardization of test methods (Ref 3)

2.4 Directories

In order to make full use of the available information sources, it is first of all necessary to know what sources exist and how they may be accessed. Fortunately a number of directories are published which provide listings of the many and varied sources of information about materials, including composites. Most of these are regularly updated, and they provide essential guides to what is currently available. Examples include *Source Journals in Metals and Materials*, a listing of current journal titles (Ref 4); and the *Gale Directory of Databases*, which gives details of databases available on-line and in CD ROM, diskette and magnetic tape formats (Ref 5). For these engaged in building property databases for composite materials, a useful directory is the *International Register of Materials*

Database Managers published by CODATA (Ref 6)

3. THE COSTS OF INFORMATION SUPPLY AND DEMAND

It is clearly of benefit to be able to keep abreast of current developments in the exploitation of composite materials, to monitor competitor activities and to prevent duplication of effort. For large organizations with extensive library and information facilities this may not be too much of a problem. For small and medium size enterprises with limited information gathering and processing resources the situation becomes more difficult.

Many information providers do have the capability, however, to tailor their information products specifically to suit the requirements of smaller companies. Computerised systems enable very specific information (about certain properties of one particular grade of material, for example) to be quickly retrieved and presented to the customer in whatever medium is required (as a printed document or on computer disc). Such information can be provided on a "one-off" basis, for example if someone needs to know about developments in a particular field over the past five years, or as a regular updating service, where abstracts of newly published information are supplied on a monthly or quarterly basis. The cost of such services is not high (perhaps of the order of £200-300 per year in a typical case) and can be further reduced if demand is sufficient, for example by distributing the information to a number of companies which are members of the same research or trade association.

4 CONCLUSIONS

Whenever there is a problem concerning the production, processing, performance and application of composite materials, the speed with which relevant information can be obtained is second in importance only to knowing what information is available. With the growth in research and development activities, the total knowledge base continues to expand, and efficient use of the information sources representing the knowledge base assumes increasing importance.

References

1. Identification of Composite Materials in Computerized Materials Property   Databases, ASTM E1309, ASTM, Philadelphia, USA

2. Identification of Fibres, Fillers and Core Materials in Computerized Materials Property Databases, ASTM E1471, ASTM, Philadelphia, USA

3. C. Bathias, VAMAS Bulletin, 15 (1992), 19

4. Source Journal in Metals and Materials, 6th Edition (1993), Materials Information, The Institute of Materials, London, England

5. Gale Directory of Databases, (1993), Gale Research Inc., Washington, USA

6. International Register of Materials Database Managers, 2nd Edition (1993), CODATA Special Report, CODATA, Paris, France

Advanced Composites (USA)
Advanced Composites Bulletin (UK)
Advanced Composite Materials (JAPAN)
Annals des Composites (FRANCE)
ASTI Journal of Composites Technology & Research (USA)
Capitol Composite (USA)
Cement and Concrete Composites (UK)
CI on Composites (USA)
Composites and Adhesives Newsletter (USA)
Composites in Manufacturing (USA)
Composites Manufacturing (UK)
Composite Polymers (UK)
Composites Science and Technology (UK)
Composite Structures (UK)
Composites (UK)
Composites (France)
Highlights (USA)
IRPI International Reinforced Plastics Industry (UK)
Itogi Nauki i Tekhniki, Kompozitsionnye Materialy (RUSSIA)
Journal of the Japan Society for Composite Materials (JAPAN)
Journal of Reinforced Plastics and Composites (USA)
Journal of Thermoplastic Composite Materials (USA)
Kompozitsionnye Polimernye Materialy (UKRAINE)
Mekhanika Kompozitnykh Materialov (LATVIA)
Plastics, Rubber and Composites Processing and Applications (UK)
Poliplasti e Plastici Rinforzati (ITALY)
Polymer Composites (USA)
Reinforcement Digest (USA)
Reinforced Plastics (UK)
Science and Engineering of Composite Materials (UK)

Table 1: Some Composites Materials Journals/Newsletters

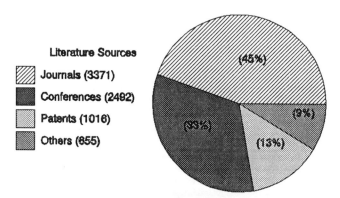

Literature Sources

Journals (3371)

Conferences (2492)

Patents (1016)

Others (855)

(45%)

(9%)

(33%)

(13%)

Figure 1. Coverage of Composites by Type of Source Literature
(Based on records added to the Engineered Materials
Abstracts Database in 1992)

Abstracts Journal/Database	Coverage
ALUMINIUM INDUSTRY ABSTRACTS	Aluminium
CERAMIC ABSTRACTS (CERAB)	Ceramics
CHEMICAL ABSTRACTS	Chemistry
COMPENDEX PLUS	Engineering
ENGINEERED MATERIALS ABSTRACTS (EMA)	Polymers, ceramics, composite materials
INSPEC	Physics, electronic engineering; coverage of relevant materials
KKF	Chemical technology - plastics, rubber, fibres
MATERIALS BUSINESS FILE	Technocommercial aspects of materials
METADEX	Metals science and technology
MIRA	Automotive industry
NTIS	US government reports
PASCAL	Multidisciplinary
PLASPEC DAILY NEWS (PLASNEWS)	Plastics
RAPRA	Rubber and Plastics
SYSTEM FOR INFORMATION ON GREY LITERATURE IN EUROPE (SIGLE)	Multidisciplinary
SILICA	Ceramics, glass, composite materials
SOCIETY OF AUTOMOTIVE ENGINEERS	Automotive
TRIBOLOGY INDEX	Tribology
WELDASEARCH	Welding
WORLD CERAMICS ABSTRACTS	Ceramics
WORLD PATENTS INDEX	Patents
WORLD TRANSINDEX	Translations

Table 2. Abstracts Journals/Databases with Coverage of Composite Materials

APPLICATION OF X RAY TOMOGRAPHY TO THE NON DESTRUCTIVE TESTING OF HIGH PERFORMANCE POLYMER COMPOSITE

C. BATHIAS, C. LE NINIVEN*, D. WU*

CNAM - Dept of Mechanic Engineering - 292 Rue Saint Martin
75141 Paris cedex 03 - France
*CNAM - Dept of Mechanic Engineering - 2 Rue Conté
75003 Paris - France

ABSTRACT

It is well known that many defects can appear inside composite materials without deterioration of the surface. In the same way the damage mechanism involves more the core than the surface of composite materials. In this respect, the application of the medical scanner for observation of composite materials and for NDT looks a very good solution. Several examples given in this paper show the performance and the limitation of the method.

INTRODUCTION

Born in the medical field, X ray tomography is gradually operating in industry. First, United States then Europe and Japan have developed different industrial instruments since the end of the eighty years. X ray tomography supplants the ultra sound by its spatial resolution and by its attenuation measure which are well adapted for the material study. We can discern four application fields of X ray tomography.
1 - Non destructive testing of big components
2 - Non destructive of small pieces
3 - Materials study
4 - Micro tomography
In the last case, resolution reaches a micron's fraction but usually the method is not applicable at the non-destructive testing and is used more than transmission electronic microscopy. In this paper we have choosen the second and the third points from which objectifs are hereafter:
- observation of internal geometric defects
- display homogeneous variations
- disclose texture, chemical structure, concentration differences
- observation of the evolution of the mechanical, physical, chemical damage

- in any case, localization in the geometrical space and quantitative calibration of the phenomena.

Thus our laboratory has set up a CGR ND 8000 scanner which is appropriated to reach the objectives. We can notice that this medical scanner is able to accept samples with a maximal diameter of 250 mm. The spatial resolution is better than 250μ. The mass density resolution (Hounsfield density) is as low as 0,2% that is to say unnecessary for classical non destructive testing.

This instrument has been equiped by a monotonic compression tension machine enabled of 50KN which permits observations under mecanical charge.

I- PRINCIPLE OF X RAY TOMOGRAPHY

The medical scanner is able to measure the X ray attenuation in Hounsfield unit or tomography density even if it doesn't give local mass density. Thus, in order to obtain it, it is necessary to make a calibration versus a reference which is water. The tomography density (DT) is a relative measure of the attenuation coefficient with regard to the water. For this reason we need to know the value given to water in the tomography scale in Hounsfield units (H). Basicaly with a constant X photon energy, the DT depends of the attenuation coefficient by a theorical relation where K is a constante equal at 1000.

$$D = \frac{\mu - \mu_w}{\mu_w}$$

This relation of conversion between the attenuation coefficient and the tomographic density is based on water attenuation coefficient value, which is 1,8 cm-1 at 73 Kev, corresponding to density zero in the hounsfield's scale of tomography.

This conversion is done in two times:
- On one hand, an intermediate scale is choosen. In this new scale, for the water, there is a relation between K and C:

$$K = C \ \mu_w$$

and the attenuation coefficient for a given material is given by:

$$\mu_{int} = \frac{K}{\mu_w} \mu$$

The tomographic density is linear with a regard to the intermediate attenuation μ_{int} and the water density must be equal a zero.

- On the other hand, the process is to make a translation in order to fill condition so that:

$$DT = \mu_{int} - K$$

so

$$DT = \frac{K}{\mu_w} \mu - K = \frac{\mu - \mu_w}{\mu_w} K$$

Calibration has been done with several materials in order to have an empirical relation between average DT of materials and respective attenuation coefficient. The table n°1 presentes some results.

II- PRESENTATION OF MAIN RESULTS

In order to promote the X ray scanner performance, and to show the large investigation field of the method when tomodensitometry is associated at tomography, few typical examples have been choosen.

2.1. Non Destructive Testing And Quality Control

First example is on a adhesive joint in woven glass fiber. The specimen is a beam, the dimensions of which are: L=800 mm   W=40 mm   l=50 mm
The shape of joint looks like a stair step. The goal of our project is to reveale the adhesive joint and to detecte the defects in the joint and in the composite materials. Using the ND 8000 scanner the beam has been examined all along the adhesive joint. The results are obtained to the 3mm joined slides: the scanner axe and the beam axe are the same.

It is very easy to show that the two parts of the beam are different. A measure of the density confirm a difference of 150 Hounsfield units between the two parts (850 and 700). Whithout any defects consideration, it may be explained by the fiber volume fraction or resin contain. The adhesive joint is excellent. A density histogram with Hounsfield density, gives 546 for the mean value and a standard deviation of 89.
Nethertheless, at one adhesive joint extremity, there is a porosity zone. Figure n°1 shows defects and the associate histogram with a density ranging between -28 to 974 with a mean value of 472.
This example shows that the method is convenient to control an adhesive joint in glass fiber composite materials. The porosity is shown both by the image and by the opacity measurement. In fact Hounsfiled density keeps far away, the most reliable indication for quality control.
The second example is more orientated to the control of manufacturing process.
Using ND 8000 medical scanner we have studied metal matrix composites proceded by squeeze casting. The first step has been to test the alumina preforms which are disks of 85 mm diameter and 15 mm thickness. In some disks, spherical defects were introduced during the processing. The figure 2 and 3 present the results of tomographic examination. The average Hounsfield density of the ceramic preform is about - 215 . However there is a large difference of X ray attenuation between the center of the specimen and the edge where the density is maximum - 118. In fact, the density is lower in the center of the specimen because the volume fraction of alumina is smaller than near the edge. Thus, it is possible to cheek the processing of the preform with tomographic observation as well as the repartition of alumina platelets which is very difficult using regular experimental methods. Of course, it is very easy to detect defects in ceramic preforms as it is shown in figure 3.

2.2. Tomodensitometry for chemical analysis

In order to put in light the possibility of chemical analysis of tomodensitometry, we have choosen to present a study of natural rubbers. Axisymetric test specimen have been studied with and without deformation. According to the chemical rubber composition, we find Hounsfield density quite different: 165 H for one rubber, 710 H for the other one. This difference is related to the chemical composition of the rubbers (fig 4 to 5).
On the other hand, when the rubber is submitted to an elongation in tension

the Hounsfield density increases of about 10 units due to the order of the micro molecules.
Thus the X ray scanner is able to give many informations about chemical composition and the physical structure of polymer materials.

CONCLUSIONS

In conclusion, three points have to be underlined:
- First, X ray tomography is a powerful tool for non destructive testing of polymer, ceramic and metal matrix composite materials
- It is also a good way for quality control and processing control; for example to follow the volume fraction of fibers or particules
- Finaly it is a possible solution to recognize chemical composition or to locate different chemical phases inside materials.

Table 1: Attenuation coefficients and tomographic densites of some materials got with the medical scanner

Matériau	μ (cm-1)	DT(H)
Polyéthylène	0,172	-72
Eau	0,191	-0,4
Nylon	0,210	90
Polyester	0,217	139
Araldite	0,219	147
Elastomère 18160/52	0,224	165
Delrin	0,262	349
Ebonite	0,288	434
Elastomère 19199/48	0,340	729
Téflon	0,340	729

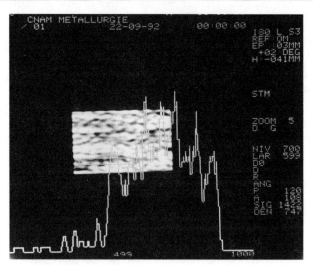

Fig 1: Cross section of the beam and density histogram showing a porus area. The adhesive joint is in the middle.

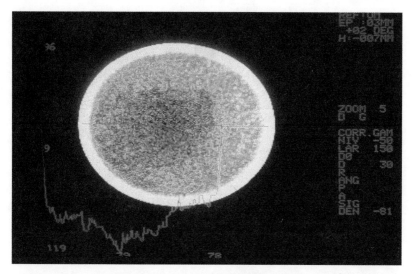

Fig 2: Ceramic preform showing a bad repartition of alumina platelets

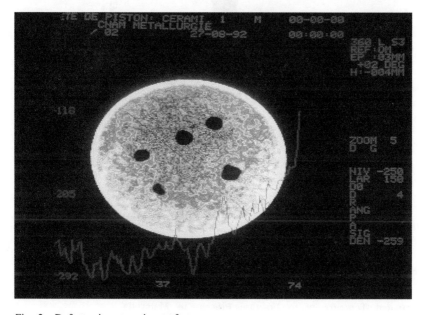

Fig 3: Defects in ceramic preform

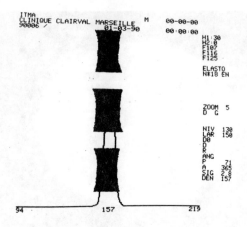

Fig 4: Cross section of a rubber specimen

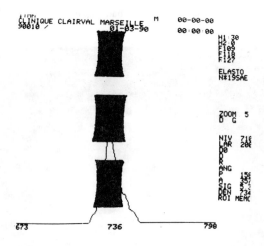

Fig 5: Cross section of a rubber specimen

REFERENCES

1. S.D Antolovich and al: "Application of Quantative fractography and Computed Tomography to Fracture Process in Materials"
- STP 1085 - ASTM PP 3-25 - (1990)
2. C. Bathias and A. Cagnasso: "Application of X Ray Tomography to the Non Destructive Testing of high Performance Polymer Composites"
- STP 1128 - ASTM - pp 35-54 - (1992)

DESIGN AND NUMERICAL MODELLING

NON LINEAR EFFECTS IN COMPOSITE MATERIALS

G. KANIADAKIS, P. DELSANTO, M. SCALERANDI

Dipartimento Di Fisica - Politecnico Di Torino
Corso Duca Degli Abruzzi 24 - 10129 Torino - Italy

ABSTRACT

Slit porosity, which is present e.g. in polygranular graphite composites, greatly affects the diffusion of impurities. In fact, in porous materials, due to saturation occurring in intergranular sites, the diffusion coefficient becomes dependent on the impurities distribution, thus contributing a non linear term to the diffusion equation. In the present paper we compare the kinetics of impurities in porous, semiporous and compact materials, using a local interaction simulation approach for the solution of both linear and nonlinear differential equations.

INTRODUCTION

An organic matrix composite exposed to humid air absorbs, besides moisture, air molecules, such as O_2, etc./1,2/. We shall refer to H_2O and all such molecules as impurities. The presence of impurities in a material affects considerably its performance. In fact dimensional and thermal variations may result, inducing stress gradients within the material and changes in the physico-chemical properties. Therefore a reliable characterization requires a detailed knowledge of the spatial concentration of the impurities and its time evolution. In the one dimensional case this involves the solution of the diffusion equation:

$$\frac{\partial p}{\partial t} = \frac{\partial}{\partial x}\left(D\frac{\partial p}{\partial x}\right), \tag{1}$$

where p is the impurities density and D the diffusion coefficient.
In elementary cases, such as in compact homogeneous materials, D may be constant in time and throughout the material. In more complex situations, hovever,

D may depend on the position x and the local density p /3/. E.g. a relevant feature in polygranular graphite composites is slit porosity, which is formed by irreversible shrinkage during graphitization and by microcrystalline anisotropic reversible shrinkage during cooling from the graphitization temperature /4/. As a result /3,5/, saturation occurs first in intergranular sites and D will have a linear dependence on p:

$$D = D_1 p \quad . \tag{2}$$

If, hovever, the amount of porosity is low, among the pore sites, where D is given by Eq.(2), there are compact regions with constant D. Thus, throughout the material

$$D = D_0 + D_1 p \quad . \tag{3}$$

A great deal of theoretical attention /3,6-8/ has been recently devoted to nonlinear kinetics problems, in which D is a given function of p, such as in Eq.(2). In the present contribution we discuss a method of solution, based on the local interaction simulation approach (LISA) /9/, which is briefly reviewed in the next section.

THEORY

We assume that the kinetic behaviour of the impurities in the matrix of the composite material satisfies Eq.(1) with a diffusion coefficient $D(p)$ given by Eq.(3). We also assume that the density distribution $p(x,t)$ at the time $t = 0$ is assigned:

$$p(x,0) = p_0(x) \quad , \tag{4}$$

and that the material specimen is finite (with proper boundary conditions at the extremes).
Under these conditions an exact solution of the problem is extremely difficult /10,11/. We therefore adopt the simulation approach LISA /9/, which is based on a space and time discretization. Calling ε the elementary "cell" size of the lattice and δ the time unit and writing

$$D(p) = k\frac{\partial[p\alpha(p)]}{\partial p} \quad , \tag{5}$$

where

$$k = \frac{\varepsilon^2}{\delta} \quad , \tag{6}$$

and

$$\alpha = \alpha_0 + \alpha_1 p \quad . \tag{7}$$

Eq.(1) becomes, for each lattice nodepoint i,

$$p_{i,t+1} = (\alpha_0 + \alpha_1 p_{i+1,t})p_{i+1,t} + (\alpha_0 + \alpha_1 p_{i-1,t})p_{i-1,t} + [1 - 2(\alpha_0 + \alpha_1 p_{i,t})]p_{i,t} \quad (8)$$

Eq.(8) is an iteration equation which allows to compute for each site i and time $t+1$ the impurities density $p_{i,t+1}$ as a function of the densities in i and neighbouring sites $i \pm 1$ at the previous time t. Since all the $p_{i,t+1}$ can be computed simultaneously and independently, Eq.(8) is ideally suitable for parallel processing. In the limit $\varepsilon \to 0$ and $\delta \to 0$, Eq.(8) leads back to Eq.(1).

RESULTS AND DISCUSSION

To start, we consider the diffusion of a single impurity located at the center of a specimen, represented by a lattice $i = 1, 100$. Fig.1 shows the time evolution of the distribution density (probability) in the cases of a compact material (dotted line) and of a semiporous material (solid line). In the former case the diffusion coefficient is constant, while in the latter case (nonlinear problem) it is given by Eq.(3). Both cases have the same amount of "compact" diffusion ($\alpha_0 = .01$), but the additional "porous" contributions ($\alpha_1 = .05$) speeds up considerably the diffusion process in the second case, thus lowering the peak of the gaussian distribution. The effect is less conspicuous at larger times (e.g. $t = 35000\ \delta$) since, after so many time steps, most of the diffusion has already taken place, albeit at a slower pace in the first case.

Fig.2 shows the effect of a purely porous diffusion ($\alpha_0 = 0$: solid line), compared with the "semiporous" case already discussed. The striking feature here is that the probability goes abruptly to zero, instead of declining gently as in the gaussian distribution. In fact, beyond a certain range (say $p_0 \pm \Delta p$) there is no impurity distribution to support any diffusion. Thus a front (or interface) is created which travels at a finite velocity, rather than at an infinite velocity, as in the semiporous or compact cases. Under special initial conditions it can be proved that the front does not move immediately at $t = 0$, but "waits" for a finite time before starting to move /12-14/.

Fig.3 shows the impurities distribution at $t = 2000\ \delta$ for various values of the nonlinear parameter α_1. The "compact" diffusion parameter is kept constant: $\alpha_0 = .01$, as in Figures 1 and 2 (dotted lines). Again we see that the effect of the nonlinear term is to speed up the diffusion process.

Finally, in Fig.4 we consider the case of an initial distribution with two impurities (at $i = 30$ and 70). The dotted lines refer to the linear (compact) case: $\alpha = \alpha_0 = .01$. The solid lines refer to the nonlinear (semiporous) case: $\alpha = .01 + .05\ p_i$. Due to the parallel processing kind of approach adopted, any initial distribution can be treated with the same ease.

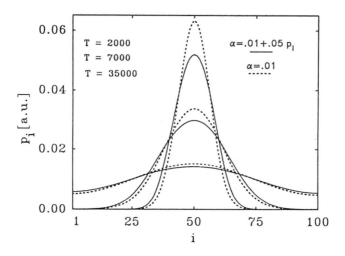

Figure 1: Impurities distributions at various times, from a single impurity located in the center of the specimen at $t = 0$. Comparison between the linear (compact) and the nonlinear (semiporous) case.

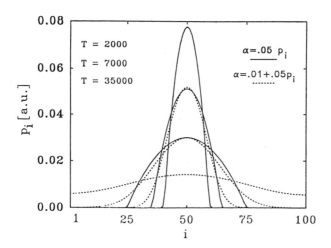

Figure 2: Impurities distributions at various times, from a single impurity in the center of the specimen at $t = 0$. Comparison between the porous and the semiporous case.

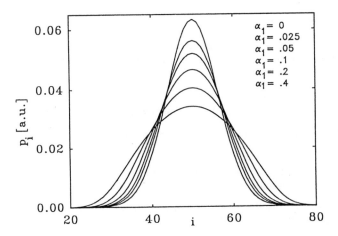

Figure 3: Distribution densities at $t = 2000\ \delta$ for different values of the nonlinear parameter α_1; $\alpha_0 = .01$; from a single inpurity in the center of the specimen at $t = 0$

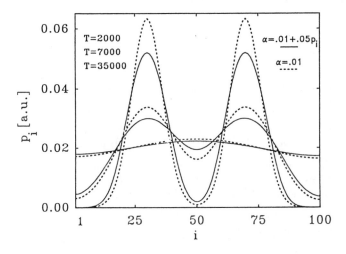

Figure 4: Impurities distributions at various times, from two impurities initially located in $i = 30$ and $i = 70$. Linear and nonlinear cases.

References

[1] E. Fitzer: *Carbon Fibres - the Miracle Material for Temperatures Between 5 and 3000 K, High Temperatures - High Pressures*, 1986,V.18,p.479-508 (10 ETPC Proceedings)

[2] G.S. Springer: *Environmental Effects*, in Composite design, sec. 16, edited by S.W. Tsai and T.N. Massard (Ippei Susuki Publ., Dayton, Ohio 1988)

[3] W. Kath: *Waiting and Propagating Fronts in Nonlinear Diffusion*, Physica D12 (1984) p.375

[4] E. Fitzer, M. Heine: in *Fibres for Composite Materials*, ed. A.R. Bunsell (Amsterdam:Elsevier 1988)

[5] P.Y. Polubarinova-Kochina: *Theory of Ground Water Movement* (Princeton Univ. Press, Princeton, 1962)

[6] J. Herrera, A. Minzoni, R. Ondarza: *Reaction-diffusion Equations in One Dimension: Particular Solutions and Relaxation*, Physica D57 (1992) p.249

[7] H. Wilhelmsson: *Explosive Instabilities of Reaction-diffusion Equations*, Phys.Rev.A, V.36, n.2 (1987) p.965

[8] J. Powell, A. Newell, C. Jones: *Competition Between Generic and Nongeneric Fronts in Envelope Equations*, Phys.Rev.A, V.44, n.6 (1991) p.249

[9] G.Kaniadakis, P.P. Delsanto, C.A. Condat: *A Local Interaction Simulation Approach to the Solution of Diffusion Problems*, Math. Comp. Modelling, in press

[10] M.J. Baines, N.R.C. Birkett, P.K. Sweby: *Non-linear Diffusion in Process Modelling*, Int.J.of Num. Modelling, V.3 (1990) p.79

[11] J. King, C.P. Please: *Diffusion of Dopants in Crystalline Silicon*, IMA J.App.Maths., 37 (1986) p.185

[12] D.G. Aronson: *Regularity Properties of Flows Through Porous Media: a Counter Example*, SIAM J.Appl.Math., 19 (1970) p.299

[13] S. Komin: *Continuous Groups of Transformation of Differential Equations; Applications to Free Boundary Problems*, in: Free Boundary Problems, (E. Magenes ed., Rome, 1980)

[14] B.F. Knerr, *The Porous Medium Equation in One Dimension*, Trans.Amer.Math.Soc., 234 (1977) p.381

TAYLORING THE THERMAL CONDUCTIVITY OF A CHOPPED CARBON FIBER/POLYMER COMPOSITE

A. DEMAIN, J.-P. ISSI

Unité de Physico-Chimie et de Physique des Matériaux
Université Catholique de Louvain - Place Croix du Sud 1
1348 Louvain-la-Neuve - Belgium

ABSTRACT

Thermal conductivity of different chopped carbon fibers / polycarbonate composites is studied as a function of concentration and orientation of the fibers. Effect of fiber length and thermal conductivty is also discussed. We show that, even with chopped fibers, at room temperature, thermal conductivity comparable to that of stainless steel can be attained.

1 INTRODUCTION

The growing needs for materials dedicated to thermal management applications such as heat removal from a thermally sensitive device or heat exchanger has led to design and investigate new composite materials.

Polymers exhibit outstanding mechanical and chemical properties. Unfortunately, their poor thermal conductivity confined them to the field of thermally insulative materials. In the past ten years, we have shown that there exist several ways to enhance the thermal conductivity of a polymeric material. One way is to act at the molecular scale by orienting the chains (1). Another solution is to add thermally conductive entities to a polymeric matrix, thus forming a composite material.

Up to now, most of the studies were mainly devoted to composites filled either with particles whose aspect ratios were ranging from 1 to 5 (2-6) or with continuous fibers (7). On the other hand, very little attention has been paid to polymers loaded with fillers whose aspect ratios were ranging from 5 to 100 such as chopped fibers.

The work we report here is part of a wider on-going systematic study aimed at determining the parameters that influence the thermal and electrical conductivities of polymer / chopped fiber composites.

The plates of composites investigated in the present work were amorphous thermoplastic matrices (polycarbonate) filled with short pitch-based carbon fibers. We mean by "short" that the length of the fiber is much smaller than the smallest dimension of the sample.
The concentration and the average orientation of the fibers were varied. The room temperature thermal conductivity was measured in directions parallel and perpendicular to the plates

2 MATERIALS

The thermoplastic polymer used in this study was the polycarbonate PC Lexan 145 from General Electric, which has a room temperature thermal conductivity of 0.20 W m^{-1} K^{-1} and a specific gravity of 1.2 kg/dm^3 (8). An amorphous polymer was chosen in order to avoid a matrix with varying crystallinity ratios since the crystallinity ratio may have a strong influence on the thermal conductivity of a polymer (9) (10).
The fibers were P55 pitch-based carbon fibers from AMOCO, which have a room temperature thermal conductivity of 100 W m^{-1} K^{-1} (11), a room temperature electrical resistivity of 10^{-5} Ω m, a diameter of 10 μm, a specific gravity of 2.0 kg/dm^3 and a Young modulus of 380 GPa.

The fibers were incorporated in the polycarbonate by means of a compounding process described elsewhere (12). The resulting material was hot-pressed into plaques of 50X50X3 mm^3.
Fiber length measurements were performed on fibers extracted from the matrix after dissolution. These revealed that the fiber lengths were ranging from a few microns to 1 mm with an average fiber length around 500 μm. We also observed a sligth decrease in the average fiber length with the increase of the fiber concentration.
SEM examination of different polished sections showed that the fibers were evenly dispersed in planes roughly parallel to the main faces of the plaques. This was ascribed to the effect of the hot-pressing.

3 EXPERIMENTAL

All the experimental devices we have used were specially designed and built in order to adapt the standard thermal conductivity methods to the specific needs of the present investigation. Indeed, in order to determine the influence of the orientation of the fibers on the thermal conductivity of a composite, measurements have to be made in two perpendicular directions. The first direction was parallel to the plane (in-plane direction), which was also the direction to which the fibers are almost parallel. The second direction was perpendicular to the plates (out-of-plane direction) and, therefore, perpendicular to the average orientation of the fibers.
The different techniques of measurement are fully described elsewhere (12).
The principle used for the in-plane thermal conductivity measurement was based on the four probe technique (figure 1a). However, in order to minimize the strong effect of the radiative losses, a heated guard surrounding the sample was added. The samples were cut out from the plates in the form of 32X5X3 mm^3 parallelepipedic bars.
For the transverse thermal conductivity measurements, a modified version of the guarded hot plate method was used (two probe technique). The samples were in the form of a 3 mm thickness and 32 mm diameter discs (figure 1b).
The precision of the measured values was estimated to be around 10%.

4 RESULTS

4.1 In-plane thermal conductivity

The in-plane thermal conductivity dependence on the fiber volume concentration is shown in figure 2. If we except the apparent saturation of the thermal conductivity above 20%, the dependence seems to be linear.

It is worth noting that a thermal conductivity close to 10 W m^{-1} K^{-1} can be attained, that is 60 times the thermal conductivity of the neat polycarbonate. Furthermore, the values obtained here are close to the ones that we might expect for continuous fibers evenly oriented in-plane.

4.2 Out-of-plane thermal conductivity

The out-of-plane thermal conductivity dependence on the fiber volume concentration is reported on in figure 3. Unlike in-plane thermal conductivity, only a maximum fourfold increase with respect to the neat polycarbonate is observed. Furthermore, the dependence on fiber concentration seems to be linear throughout the whole range of concentration.

5 DISCUSSION

Thermal conductivity close to 10 W m^{-1} K^{-1} can be attained, what is 60 times the thermal conductivity of the neat polycarbonate. This value is close to the ones that we might expect for continuous fibers evenly oriented in-plane and is comparable to that of stainless steel. Such high thermal conductivities can only be achieved thanks to the use of highly thermally conductive fibers. Therefore, we suggest that still higher values could be obtained with more conductive fibers. This assumption is under investigation at the moment.

The rapid increase of the in-plane thermal conductivity from 0.2 W m^{-1} K^{-1} to 10 W m^{-1} K^{-1} within a 20% concentration range shows the dominent effect of this parameter.

Furthermore, a comparison between in- and out-of-plane thermal conductivities clearly emphasize the effect of fiber orientation.

Indeed, because of their "quasi in-plane" orientation, average projection of the fibers in the out-of-plane direction is close to the fiber diameter while it is proportional to the fiber length in the in-plane directions. Therefore, when the fiber axis is roughly perpendicular to the heat flux direction, the heat can not benefit from the elongated shape of the fibers since it only "sees" the diameter of the fiber instead of the length.

In that case, the number of thermal resistances (consisting in layers of polymer separating fibers) that have to be crossed by the heat flux per unit length increases, thus reducing the thermal conductivity in that direction.

The saturation observed in the in-plane thermal conductivity above 20% can be ascribed to a geometrical effect. Indeed, there exists a maximum for the packing of fibers corresponding to each combination of length and orientation of the fibers in the composite which can be evaluated by computerized simulations (13). Any attempt to exceed that maximum by incorporating more fibers during the processing of the composite will automatically lead either to the breakage of the fibers or to a more ordered distribution of the fibers in the material. According to these studies, that maximum lies around 20 % by volume in our composites. As a better alignment seems to be very difficult to realize during the preparation of the composite, we can assume that a few fibers break when the fiber concentration is raised above 20%. Therefore, in the range of 20% to 30% of volume fiber concentration, we might observe the competition between, in the one hand, the possible increase of the number of defects in the polymer and the breakage of fibers that should decrease the thermal conductivity of the composite and, in the other hand, the increase of the in-plane thermal conductivity owing to the increase in fiber concentration.

6 CONCLUSION

The process used in elaborating composites leads to a high orientation of the fibers which is directly reflected, as expected, in the high anisotropy of the thermal conductivity. It is interesting to note for a practical purpose that a thermal conductivity parallel to the plane of the plaques as high as that of the common metallic alloys such as stainless steel may be attained. By contrast, only a small increase of thermal conductivity compared to the neat polymer was observed in the transverse direction.
We also suggested that any further increase of the fibers thermal conductivity should improve the in-plane thermal conductivity of the composite.
Finally, it is shown that the geometrical effect due to the existence of a maximum for the packing of fibers results in the occurence of a maximum in the longitudinal thermal conductivities.

REFERENCES

1. Issi J-P, Nysten B., Jonas A., Demain A., Piraux L. and Poulaert B., October 15-18, 1989, in Thermal Conductivity XXI, Proceeding XXIst International Conference on Thermal Conductivity, Lexington, Kentucky, H.A. Fine and C.J. Cremers, ed., Plenum Press, New-York.
2. D.M. Bigg, June, 1986, Polymer Composites, Vol.7, N° 3, pp. 125-140.
3 D.P.H. Hasselman and L.F. Johnson, june, 1987, Journal of Composite materials, vol.21, pp 508-515.
4 R.C. Progelhof, J.L. Throne and R.R. Ruetsch, September, 1976, Polymer Engineering and Science, Vol. 16, N°9, pp. 615-624.
5. L. Nielsen, 1973 , J. Appl. Polym. Sci., 17, 3819.
6. L. Nielsen, 1974 , Ind. Eng. Chem. Fundam., 13, 17.
7. J. Ashton, J. Halpin and P. Petit, "Primer on composite materials: Analysis", Stamford, Conn., Technomic Pub.Co..
8. M. and I. Ash., 1982, "Encyclopedia of plastics, polymers and resins", vol.II, Chemical Publishing Co., New-York.
9. D. Hansen and G.A. Bernier, May 1972, Polymer Engineering and Science, Vol. 12, N° 3, pp 204-208.
10. C.L. Choy and K Young, Aug. 1977, Polymer, vol.18, pp. 769-776.
11. Nysten B., Piraux L. and Issi J.P., 1985, Thermal Conductivity 19, Proceedings of the 19th ITCC, Cookeville, Tennessee, D.W. Yarborough (Plenum Press, New-York, 1987), 341.
12. Demain A. and Issi J.P., Journal of Composite Materials, 8, august 1993, in press.
13. K.E. Evans and M.D. Ferrar, 1989, J. Phys. D: Appl.Phys. 22, pp. 354-360.

LIST OF FIGURES

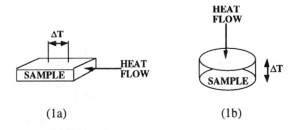

(1a)   (1b)

Figure 1 :   Principle of the techniques used for the different measurements.
              In-plane thermal conductivity is measured by a four probe technique
              (1a), while out-of-plane thermal conductivity is measured by a
              two-probe technique (1b).

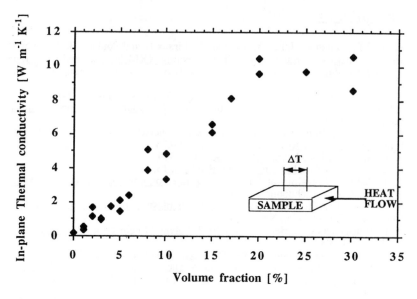

Figure 2 : In-plane thermal conductivity dependence on the fiber volume concentration

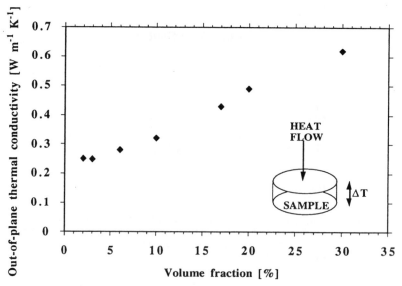

Figure 3 : Out-of-plane thermal conductivity dependence on fiber volume concentration.

KNITTED CARBON FIBERS REINFORCED BIOCOMPATIBLE THERMOPLASTICS, MECHANICAL PROPERTIES AND STRUCTURE MODELLING

S.-W. HA, J. MAYER, J. DE HAAN, M. PETITMERMET, E. WINTERMANTEL

Chair of Biocompatible Materials Science and Engineering
Federal Institute of Technology ETH Zürich - Wagistrasse 13
8952 Schlieren - Switzerland

ABSTRACT

The knit structure of the carbon fibers as reinforcement in thermoplastic composites allows an adjustable anisotropy of the mechanical composite-properties due to its drawability. It has been shown that the rate of drawing and the mechanical anisotropy have a linear correlation. Specific weakening of the fiber-matrix interface revealed a dependence of fiber-matrix adhesion on the failure behaviour. These observations were confirmed by FEM calculations.

1. INTRODUCTION

Knitted carbon fibers as reinforcement of thermoplastics are interesting for loaded parts, due to their drapability and the possibility of waste free near net shape production methods /1,2,5/. The coherence of the knit structure and the effect of strain stiffening and strain hardening during controlled drawing offer properties which are unknown from angle-ply or from woven reinforcements. Compared to common composite manufacturing processes this may lower costs. Their application as load bearing structure in medical implants is of special interest due to their mechanical properties close to cortical bone properties and the possibility to adapt the composite implant to individual bone shapes /4/.

2. MATERIALS AND METHODS

To study the influences of knit structure, knit drawing, fiber matrix adhesion and matrix properties on the mechanical properties, different knitted carbon fiber-matrix systems, i.e. T300-polyethylenemethacrylate (PEMA), T300-polyamide 12 (PA12) and AS4-polyetheretherketone (PEEK), were investigated. The influence of the fiber-matrix-adhesion was investigated with the T300-PA12-composite specimens. Samples were kept in 1 molar sodium hydroxide during 5000 hours at 37°C, and their mechanical properties and failure behavior were compared with untreated specimens. The selected test directions were either the wale or the course direction (fig. 1). Referring to DIN 29971 stress/ strain behavior has been analyzed during tensile and 4-point-bending loading; interlaminar shear strength was determined in some cases. Measurements were performed in a universal testing machine (Zwick 1474) with a testing speed of 0.5 mm/min.

The weft knit structure can be described by the definition of a geometrical unit cell (fig. 1). Drawing of the knit structure change the fiber orientation distribution which can be determined by analysing ground section specimens (spatial distribution) and x-ray radiographs (plain distribution) through image analysis /6/.

Finite element models (MARC) were used to estimate stress-strain behaviour in a single stitch which corresponds to the smallest functional unit of the knit structure. According to the rule of mixture the Young's modulus of the fiber bundels were calculated based on the fiber volume fraction. A linear elastic material behaviour and perfect adhesion were assumed.

3. RESULTS

Undeformed weft knit structures show a smoother anisotropy than unidirectional, woven and 1x1 plain weave reinforced composites (fig. 4). As shown in figure 2 drawing induces a higher anisotropy in the fiber orientation distribution. The out of plain orientation of fibers (elevation) of undrawn specimens were measured to be 15° due to the knit structure. Specimens of the AS4-PEEK-composites showed that stiffness and strength reached or even exceeded the properties of 1x1 plain weaves due to drawing in the deformation direction (fig. 5). Figure 3 displays that the rate of drawing can directly be correlated to the mechanical anisotropy. The measured energy release rate during crack growth in K_I case (DCB test, area method) was twice as high for knits (T300-PEEK, undrawn: $G_{Ic}=4.5$ kJ/m^2) than for unidirectional reinforced composites (T300-PEEK, 0°: $G_{Ic}=2.0$ kJ/m^2).

SEM-analysis of aged T300-PA12-composite specimens revealed a strong influence of the adhesion on the failure behaviour (fig. 7). Increase in fiber pull-out, roughness of fracture surface and location of primary failure in the stitch binding zones are characteristical features for low adhesion failure.

Assuming total fiber-matrix adhesion the calculations with FEM showed that the maximum appeared in the straight part of the fibers oriented in the direction of the applied force. In comparison to other models /3,5/ the FEM calculations gave lower values of Young's modulus in both longitudinal and transversal directions. The values were found to be close to the experimental values of the PEMA-composite. Figure 8 shows the stress distribution in the model.

4. DISCUSSION

Through drawing of the textile knit-structure a higher portion of fibers in the direction of the load can be achieved. A nearly linear correlation of the rate of drawing and mechanical properties has been observed in experimental investigations.

The reinforcement of the thermoplasic matrix with knitted carbon fibers is more effective in the AS4-PEEK- and T300-PA12- than in the T300-PEMA-composites. SEM-analysis reveals that the AS4-PEEK- and T300-PA12-composite show higher fiber-matrix-adhesion than the T300-PEMA-composite leading to a more effective load transmission between fiber and matrix. These observations have been confirmed by the mechanical results of the chemical ageing experiments. After the treatment of the T300-PA12-composite the interphase between fiber and matrix is already partly destroyed. Ageing leads to a decrease of the tensile strength, whereas the tensile stiffness still remains unchanged after 5000 hours of exposure.

5. REFERENCES

/1/   J. Mayer, E. Wintermantel, F. De Angelis ., M. Niedermeier, A. Buck,
      M. Flemming. Euromat Cambridge UK, 1991, in: Advanced structural Materials,
      Ed. T.W.Clyne, The Institute of Materials, Cambridge UK, 1991, pp. 18-26
/2/   R. Ruffieux, M. Hintermann, J. Mayer, E. Wintermantel, VTT Congress, Textiles
      and composites '92, Tampere, Finland, 15-18 June, 1992, p. 326-332.
/3/   C. C. Rudd, M. J. Owen, V. Middleton,, Comp. Sci. Techn., 39, 1990, p. 261-277.
/4/   J. Mayer, K. Ruffieux, B. Koch, E. Wintermantel, T. Schulten, A. Hatebur, 1992
      International Symposium Biomedical Engineering in th 21st Century, Taipei,
      Sept. 23-26, 1992
/5/   S.-W. Ha, J. de Haan J., J. Mayer, E. Wintermantel, Jahrestagung der
      Polymergruppe Schweiz, PGS, Nov. 1992
/6/   J. Mayer, P. Lüscher, E. Wintermantel, VTT Congress, Textiles and composites
      '92, Tampere, Finland, 15-18 June, 1992, p. 315-521

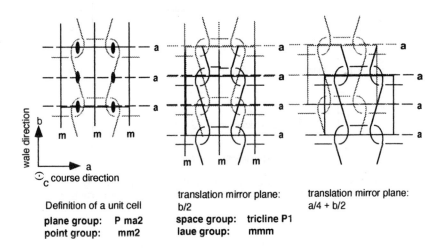

Figure 1.: Definition of a weft knit and description of the geometrical unit cell of a single layer and of the spatial textile knit-structure /5/

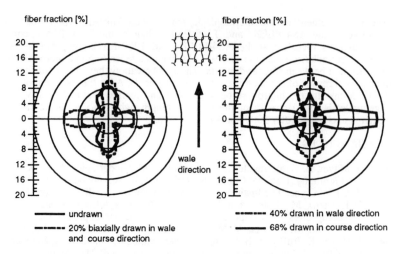

Figure 2: Effect of the mode of drawing on stitch-shape and the correlating fiber orientation distribution, through image analysis /6/. Drawing induces a higher anisotropy in the fiber orientation distribution and the corresponding mechanical properties.

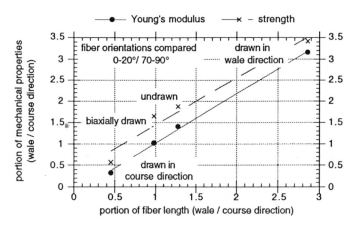

Figure 3: Linear correlation between the rate of drawing expressed by the portion of fiber length and the mechanical anisotropy shown by its portion

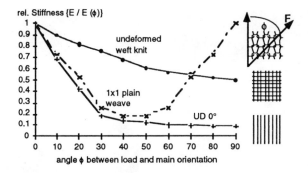

Figure 4: Comparison of the anisotropy between unidirectional, woven and undrawn weft knit reinforced composites

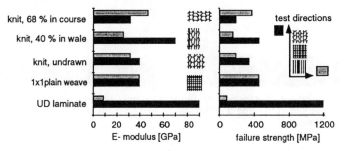

Figure 5: Comparison of the mechanical properties of 0°UD, 1x1 plain weave and weft knit. The material is PEEK with a AS4 fiber and all properties are reduced corresponding the fiber volume content of the knitting tested (40 vol.%).

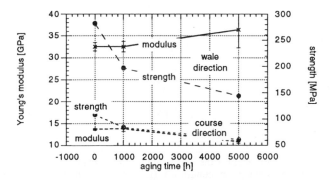

Figure 6: Reduction of fiber- matrix adhesion by aging in NaOH (1 mol/l, 37°C) over 5000 hours. Adhesion has a significant influence on strength and failure behaviour. The material is PA12 with T300 fibers.

Figure 7: Influence of adhesion on the microscopic failure behaviour comparing a unaged (left) and a 1000 hours aged tensile specimen, tested in wale direction.

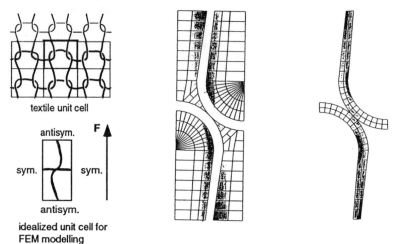

Figure 8: Plain FEM model for the knit structure;(left): Idealization of the textile unit cell and its symmetry conditions; (midle): Stress distribution (von Mises) in the matrix; (right): Stress distribution (von Mises) in the fiber indicating maximum stress in the straight part of the fiber oriented in the direction of force applied.

MACHINING, BONDING, AND REPAIR

EFFECTS OF RESIDUAL STRESSES, LOADING GEOMETRY AND LOADING RATE ON THE DEBONDING OF GLASS FIBRES FROM POLYPROPYLENE

C.-H. ANDERSSON, J. WAHLBERG*, E. MÄDER**

Dept of Production and Materials Engineering
Lund Institute of Technology - Lund - Sweden
*Swedish Institute for Textile Research - Göteborg - Sweden
**Institute for Polymer Research - Dresden - Germany

ABSTRACT

The debonding of E-glass fibres in polypropylene has been studied by micto-droplet and pull-out testing on sized and non sized fibres. The testing methods are compared. The time and temperature of processing, non uniform distribution of sizing and contamination in the interface have influence on the debonding behaviour like geometry and strain rate for the application of load.

BACKGROUND AND INTRODUCTION

The processes of wetting and bonding of a fibre to a matrix material depends on temperature, time, geometrical features and applied pressure. In some way, the relevant conditions of processing and geometrical features have to be simulated in order to make the measurments techically relevant.

Single fibre testing for characterization of the fibre/matrix interfacial bonding properties is frequently done by several different methods, according to the therminology of Piggott: i) pull-out, ii) microtension, iii) microcompression and iv) fragmentation [1-6]. All these methods however have drawbacks in the skill and care needed to perform the experimental work.

These methods are designed from physical first principles. Of fundamental physical reasons however, conflicting results have been obtained in many systems, as reviwed in refs [1,2].

Most fibres are surface treated in order to improve handleability and compatibility to matrix materials. Non uniform distribution of surface coatings, sizings, is however a common problem, and thus also a probable source of scatter in results.

In this work methodology for the preparation of specimens and application of load for microtensional debonding tests on droplets will be compared with pull out testing on low pressure moulded specimens.

The influence of defomation rate on the debonding shear stress and the work of debonding is studied by the use of different loading rates in tests on microtension droplet specimens.

THEORY

The most simple analysis of the microdebonding of fibres in polymeric materials is based on the shear lag assumptions. The interface yields at some shear stress, T_{iy}. A constant shear stress along the interface is thus assumed. Using a simple force equilibrium for the applied load, F_A, and the length of contact, l, can be related to the interfacial shear stress, T_{iy}, by the relation:

$$F_A = \Pi dl T_{iy}$$

where d is the diameter of the fibre.

The limitations of the shear lag analysis are thoroughly discussed elsewhere, for example by Piggott or by Day and Marquez /1,2,9/.

The fundamental merits of ring loading for the application of the load have been demonstrated analytically by Piggott and by FEM simulations by Wu and Claypol /2,10/.

EXPERIMENTAL

The materials studied are of industrial qualities used for co-knitting for hot stamping.

Glass fibres: Scandinavian Glass Fibre AB, E-glass $\phi=20$ μm, silane sized and E-glass non sized, IPF Dresden

Textile PP fibres $\phi=20$ μm

- Preparation of specimens:

i) Micro tension specimens - Droplets and single fibre mouldings are made by methods simulating the geometries of co-knitting. Thermoplastic fibres of matrix material are knotted around the load bearing fibres, melted to circular symmetrical shape at lowest possible temperature and subsequently heat treated in a closed tubular furnace. The droplets however have to be individually characterized after processing with respect to dimensions and shape before testing, an obvious drawback.

ii) Pull out specimens - The conditions of low pressue moulding are simulated in open isothermal tools, where the ends of fibres are embedded to a fixed length in the molten polymer under a microscope. The kinetics of the developing contact angle for example, can thus be measured in situ. The temperature for this processing was 210°C.

- Geometry for the application of load:

i) The specimens are mounted in capillary tubes in a self ajusting rig in an Instron 1122 machine according to Ref/4/, Fig1.

The contact edges taking the load are melt formed in order to blunt protruding edges, to make fibre mounting easy and the load bearing holes of diameters close to that of the fibres, typically 50-100 μm. The shape of the droplets and the blunting of edges however give a compressional radial force component in the testing.

ii) The tests on moulded specimens are performed in a micro tensile testing machine with the samples mounted in the tools used for the moulding. The distance to the bonding walls is thus big compared with the diameter of the fibre. The fibres will therefore be subjected to a radial tensile force component.

- The influence of defomation rate on the debonding shear stress and the work of debonding is measured by the use of different loading rates in tests on microtension droplet specimens.

The gauge length of the fibres is kept constant L=100 mm, in order to keep the elastic energy constant.

- Debonded fibres are characterized by SEM.

RESULT AND DISCUSSION

The typical three stage behaviour of the micro-droplet tension testing with i) debonding, ii) rebonding and iii) sliding is illustrated by Fig.2. The results are presented in Fig 3-5.

The pull-out testing gives significantly lower figures for debonding, typically 6-10 MPa, than the micro tension. An obvious possible source of this apparent discrepancy however lies in the higher processing temperature.

The conditions of heat treatment are of vital importance . Oxidation of the polypropylene is a possible source of the thermally activated decrease in the interfacial strength and transitions from adhesive to cohesive debonding.

The the high level of rebonding of the micro-droplets is activated by applied radial compression force component.

Beside the differencies processing temperature and loading geometry are non relaxed radial tensile thermal shrinkage stresses in the pull-out specimens.

Increasing loading rates also give increased debonding strength. Analysis of data using viscoelastic models and SEM-characterization of fractured surfaces is going on.

Most reinforcement fibres are modified with lubricating surface treatments, sizings, primarily in order to improve handleability and minimize the risk of fibre damage and dust formation in processing. Some important common requirements on sizing performance are:

-Lubricated friction
-Ductile surface behaviour
-Coupling to matrix materials
-Barrier for chemical attacks

The handling properties and the interfacial performance of fibres in composites are however physically closely related, as illustrated by the apparently contradictory requirements for boundary lubrication: i) good bonding i.e low surface energy or reactive bonding, ii) high compression strength, iii) very low shear strength.

Sizing systems for E-glass are usually silane based and thus chemically coupled to the fibres /7/.It thus very likely that fracture will occur in the boundary phase or in the matrix. Debonding and fragmentation of sizings on E-glass due to shear combined with very high levels of tensile contact stresses in fibre handling over cylindrical surfaces have however been demonstrated /8/.

Of vital practical importance for the performance of composites is the transition from the adhesive to cohesive fracture modes. Contamination of glass fibres with dust has however also been found to initiate early cohesive fracture /4/. This kind of fracture mode transition belongs to the material properties and to the realities to be dealt with in virtually any technical application, but is hard to separate from scatter due to the inperfections in experimental procedure. As expected from the geometry of load transfer, inperfections in droplet geometry can give very bad measured figures.

REFERENCES

1. M.R.Piggott : Debonding and friction at fibre-polymer interfaces. I:Criteria for failure and sliding. Compos.Sci.Techn. 30(1988) 295-306
2. M.R.Piggott : Micromechanics of fibre polymer interfaces.
Proc. IPCM'91, Leuven, Belgium September 1991, pp 3-8
3. B.Miller, P.Muri, L.Rebenfeld : A Microbond Method for Determination of the Shear Strength of a Fiber/Resin Interface
Comp.Sci.Techn. 28(1987) 17-32
4. C-H Andersson, T.Dartman, J.Wahlberg and C.Klasson : Wetting and adhesion of some thermoplastic materials to fibres.
Proc. IPCM'91, Leuven, Belgium September 1991, pp 61-64
5. E.Mäder, K-H Freitag : Interface properties and their influence on short fibre composites. Composites 21(1990) 397-402.
6. A.N.Gent, Chi Wang : What happens after a fibre breaks, pull-out or resin cracking. J.Mater.Sci. 28(1993) 2494-2500
7. D.Wang, F.R.Jones : Surface analytical study of the interaction between γ-amino propyl triethoxysilane and E-glass surface. J.Mater.Sci. 28(1993) 2481-2488
8. B.Christensson et.al.: Friction and damage accumulation, formation and morphology of dust from fibre handling. ECCM 6, Bordeaux 1993.
9. R.J.Day and M.Marquez : The micromecanics of the microbond test
Proc. IPCM'91, Leuven, Belgium September 1991, pp 65-68
10.H.F.Wu and C.M.Claypool : An analytical approach of the microbond test method used in characterizing the fibre-matrix interface. J.Mater.Sci.Lett. 10(1991) 260-262

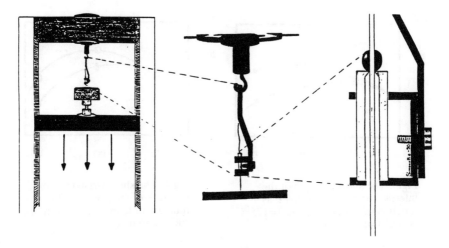

Fig.1. Set up for micro tension testing on droplets according to /4/, i) the tensile testing machine with the micro-tension clamp and a standard clamp, ii) the micro-tension clamp and iii) a droplet supported by a capillary tube, note the clearance to the fibre.

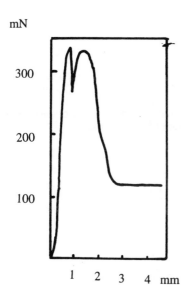

Fig.2. Load displacement curve of micro droplet tensile testing, sized E-glass / PP illustrating the three stages, i) debonding, ii) rebonding and iii) frictional sliding.

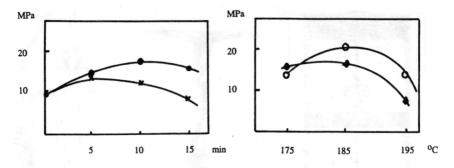

Fig.3. The shear strength of the interface of sized E-glass/PP as a function of annealing time T=175°C (●) and T=195°C (×).

Fig.4. The shear strength of the interface of sized E-glass/PP as a function of temperature, t=5min (o) and t=15min (●).

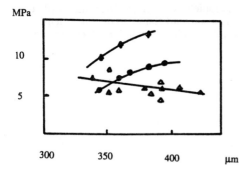

Fig.5. Illustrates the influence of surface contamination and geometry of the droplets, (●) good, (●) contaminated and (▲) geometrically non perfect droplets.

JOINING ALUMINIUM-BASED COMPOSITES USING ALLOY INTERLAYERS

R. BUSHBY, V. SCOTT, R. TRUMPER*

School of Materials Science - University of Bath - Claverton Down
BA1 7AY Bath Avon - UK
*DRA - Molton Heath - Dorset - UK

ABSTRACT

The joining of particulate-reinforced and fibre-reinforced aluminium using copper-silver alloy interlayers has been investigated. Joint strengths have been measured using a shear jig designed to avoid the peel stresses associated with the more conventional lap-shear test. Microstructural characterisation of the bonds, carried out using electron microscopy and electron-probe microanalysis, showed that for both types of composite the formation of a liquid phase was a crucial factor in developing the joint. For the fibre-reinforced composite, shear testing resulted in failure away from the joint whilst, for the particulate-reinforced composite, failure took place mostly at the bond interface.

1. INTRODUCTION

Diffusion bonding is a joining technique normally used for metals and involves applying pressure to a work-piece held at high temperature. The method works well when bonding titanium because surface oxides and contaminants are dissolvable in the metal at bonding temperatures of ~900°C, but with aluminium the tenacious oxide forms a barrier to bonding. Fragmentation of the oxide film by deformation provides a solution, but this method is not suitable for composite material as the reinforcement may be damaged. Another approach uses an interlayer which reacts with the aluminium to promote diffusion but brittle second phases may then be formed which are deleterious to joint strength. If, however, a combination of materials are chosen so as to form a low melting point liquid, the reaction product may be ejected from the joint, a process usually termed diffusion brazing, transient- or liquid-phase bonding.

The present paper is concerned with the diffusion bonding in air of fibre-reinforced aluminium and particulate-reinforced Al-Cu-Mg (2124) alloy, using a copper-silver alloy interlayer in each case. The integrity of the bonds are assessed by shear tests, and related to microstructural and fractographic observations.

651

2. EXPERIMENTAL PROCEDURE

2.1. Materials

The fibre-reinforced composite consisted of commercial-purity (Aluminium Association No. 150.0) containing Nicalon fibres (a SiC-based fibre, 15 μm diameter) as reinforcement. Composite sheet (Al/Nic), 3 mm thick, was produced by liquid metal infiltration of a woven fibre preform using an applied pressure of >7 MPa and a melt temperature of ~750°C; the fibres were aligned in both the 0° and 90° directions with a nominal fibre volume fraction of ~0.4.

The particulate-reinforced composite consisted of Al-4.2Cu-1.5Mg-0.6Mn (2124) alloy and silicon carbide particulate for the reinforcement. The composite material (2124/P) was manufactured using the powder metallurgy/extrusion route, and supplied as 25 mm diameter bar. The SiC reinforcement had a nominal size of 3 μm and the volume fraction was 0.35.

A copper-silver alloy of eutectic composition (~29 wt% copper) was used for the interlayer, which was 50 μm thick.

2.2. Diffusion bonding

Specimens were cut from composite plate and the surfaces to be bonded were ground, using progressively finer silicon carbide paper, down to 1200 grit. They were then cleaned with Teepol and degreased in acetone. The copper-silver alloy foil was cleaned in 10% vol H_2SO_4 followed by degreasing. Diffusion bonding was carried out using a modified Instron testing machine fitted with a furnace held at a temperature of 510 ± 5°C; an applied pressure of 20 MPa was used for the fibre composite and 8 MPa for the particulate composite.

2.3. Microstructural analysis.

Metallographic sections through the bond region were prepared by grinding on successively finer grades of diamond compound down to 1 μm abrasive, with a final polish using colloidal silica. Light microscopy was usually followed by examination in the scanning electron microscope (SEM), composition analysis of microstructural features being performed using an energy-dispersive spectrometer (EDS) attached to the SEM. Quantitative analysis was carried out using a wavelength-dispersive spectrometer (WDS) system fitted to an electron-probe microanalyser (EPMA).

Thin (~few 100 nm thickness) sections suitable for transmission electron microscopy (TEM) were prepared from the bond regions by cutting 3 mm diameter discs and mechanically grinding them to a thickness of ~300 μm. The discs were dimpled on both sides using a VCR model 500 and then thinned by argon bombardment in a Gatan Duomill. TEM examination was carried out in a 200 kV electron microscope fitted with an EDS system.

Differential scanning calorimetry (DSC) using a DuPont 9900 was carried out on the liquid metal bead expelled from the joint during the bonding operation.

2.4. Shear tests and fractography

The shear strength of a bond was determined by carrying out shear tests on a series of coupons cut from the original joint and then averaging the individual data. Tests were carried out using a 100 kN Instron machine fitted with the jig shown diagrammatically in Fig. 1. The shear strength was calculated from measurement of the failure load and the overlap area of each tested coupon. The bonded area of the total joint was established from a study of the fractured surfaces, bonded areas being identified by the presence of ductile dimples and/or metal cleavage facets and non-bonded areas by a planar surface covered with oxide. A value for the specific shear strength for each tested coupon was calculated using the relevant failure load and bonded area. These were averaged to give the specific shear strength of the whole bonded specimen.

3. RESULTS

3.1. Commercial-purity aluminium reinforced with Nicalon fibre (Al/Nic)

The as-received composite had satisfactory infiltration of fibres by the metal to give a fibre volume fraction of ~0.4, Fig. 2. The dark phases present in the aluminium matrix were found to contain iron and silicon. TEM studies of the fibre/matrix interface revealed the presence of needle-like crystals which EDS showed contained aluminium and carbon. These crystals, probably Al_4C_3, are believed to have formed during composite manufacture by reaction of the liquid aluminium with the free carbon originating from the Nicalon fibre, as reported in a microstructural study of a similar composite material /1/. With regard to the shear strength of the composite, no tests were performed since the as-received plate was too thin (3 mm) for use in the shear-jig.

Bonding for 30 minutes at 510°C with a copper-silver alloy interlayer using an applied pressure of 20 MPa gave a sample deformation of ~3% and resulted in the expulsion of a liquid bead, the solidified microstructure of which is shown in Fig. 3a. DSC analysis, on heating, indicated endothermic reactions at ~495°C and ~505°C, Fig. 3b. A micrograph of the joint, Fig. 3c, shows a reaction zone which extended up to ~100 μm into the composite. The phases present in both the bead and joint were identified using EPMA as aluminium, $CuAl_2$ and Ag_2Al along with an intermetallic containing iron and silicon. Within the reaction zone extensive fibre breakage was observed, with solidified eutectic phases present between fibre ends, Fig. 3d. This indicated that fibre fracture was not produced as a result of prior surface preparation but had occurred as a consequence of the bonding operation. The aluminium matrix adjacent to the reaction zone contained copper and silver, TEM studies revealing the presence of oriented particles which EDS and diffraction studies showed were γ' (Ag_2Al). The shear strength of the joint was 54 ± 16 MPa, failure occurring mainly through the composite well away from the bonded interface. The bonded area was 90 ± 22 % to give a specific strength of 61 ± 10 MPa. Unbonded regions at the periphery of the joint were associated with oxidised copper.

3.2. Aluminium alloy (2124) reinforced with SiC particulate (2124/P)

Examination of the as-received composite revealed silicon carbide particulate in a range of sizes from 5 μm to less than 1 μm and these were distributed fairly evenly, although due to the nature of the manufacturing route some matrix-rich regions were present, Fig. 4a. Thin foil TEM studies showed that the silicon carbide had not reacted with the aluminium as evidenced by the 'clean' particle/matrix interfaces. TEM also revealed the presence of θ, S and Al-Cu-Mn-Fe phases in the alloy as well as aligned precipitates of θ' and S' although they were not distributed evenly throughout the matrix. Shear tests on specimens cut from as-received composite gave a strength of 288 ± 6 MPa.

Bonding for 30 minutes at 510°C with a copper-silver alloy interlayer using an applied pressure of 8 MPa resulted in a permanent specimen deformation of ~6%. No liquid was expelled from the joint, although the presence of SiC particles in the joint area, Fig. 4b, and the copper- and silver-rich phases at the bond line indicated that a liquid had formed. The liquid is believed to have been 'absorbed' by the composite, as evidenced by results from EPMA which found significant amounts of copper and silver up to a distance of ~150 μm away from the bond interface. The phases present in the joint region were too small for quantitative EPMA analysis, although EDS did show that they contained copper, aluminium and silver. These phases were probably $CuAl_2$ and Ag_2Al, similar to those formed in bonding experiments on the Al/Nic composite using a copper-silver interlayer. The shear strength of the joint was 193 ± 91 MPa with failure taking place mainly at the bonded interface, although the roughened appearance of the fracture surfaces indicated that some failure had taken place also through the composite. The bonded area accounted for some 80% of the original mating surfaces; in peripheral regions of the joint oxidation of the interlayer had taken place and inhibited bonding.

4. DISCUSSION

4.1. Bonding commercial-purity aluminium reinforced with Nicalon fibre (Al/Nic)

The bonding of Al/Nic composite with a copper-silver alloy interlayer at 510°C using a pressure of 20 MPa resulted in the formation of a liquid phase, some of which was expelled from the joint. However, full coverage was not achieved because oxidation of outer regions of the interlayer acted as a barrier to metal diffusion. Previous work /2/ has shown that the development of a joint is aided by the presence of a liquid phase as this disrupts the thin residual oxide layer present on the aluminium to allow more intimate contact between the mating surfaces. Analysis of the solidified structure of the expelled bead revealed aluminium, $CuAl_2$, Ag_2Al phases as well as an intermetallic which contained iron and silicon. The silicon is considered to have originated from impurities in the aluminium; there was no evidence to indicate that it was introduced as a result of a fibre reaction. DSC analysis of the bead indicated melting at ~495°C and ~505°C. The reaction at 505°C is associated with the formation of Al-Cu-Ag eutectic liquid; melting at the lower temperature is attributed to the modification of this reaction by silicon, although the reaction is limited due to the low level of silicon impurity in the composite matrix. It would be interesting to substitute the composite matrix with an Al-Si alloy so that full melting could be achieved when bonding at the lower temperature.

It was encouraging to observe that where the joint was established (~90%) failure occurred away from the joint interface, the enhanced strength of the composite in the joint region being due to (a) the 'crack-resistant' nature of the solidified eutectic and, (b) precipitation hardening of the adjacent aluminium by γ' phase. With regard to the effect of the reinforcement, although the fibres restricted the initial stages of bonding, they do offer easy diffusion paths, as evidenced by the depth of penetration of eutectic liquid, Fig. 3c.

4.2. Bonding aluminium alloy (2124) reinforced with SiC particulate (2124/P)

The bonding of 2124/P composite with a copper-silver alloy interlayer at 510°C using a pressure of 8 MPa resulted in the formation of a liquid phase, although none was expelled from the joint; the liquid was instead absorbed by the composite, the process being aided by its rapid diffusion at particulate/matrix interfaces. This resulted in the disruption of the planar bond interface to give keying with the SiC particulate within the joint region and this allowed the full strength of the joint to be realised (~190 MPa), a value comparable with (~170 MPa) that reported by Partridge and Dunford /3/ for liquid-phase diffusion bonded 8090/17% SiC$_p$ composite.

5. CONCLUSIONS

Although both types of composite were readily bonded using a copper-silver interlayer at 510°C the joints exhibited different strengths and fracture behaviour. The bonded Al/Nic composite was weaker than the joint, resulting in failure away from the bonded region (~60 MPa). The higher strength of the particulate composite allowed the full strength of the joint to be realised and failure took place at the bond interface (~190 MPa).

ACKNOWLEDGEMENTS

To the SERC and DRA for support. British Crown Copyright 1993/DRA. Published with permission of the Controller of Her Britannic Majesty's Stationery Office.

6. REFERENCES

1. V.D. Scott, S.M. Bleay, A.R. Chapman and G. Love, 'Interface compatibility in Nicalon-fibre reinforced metal and ceramic composites' *J Microscopy* **169**, 2 (1993) 119-129.

2. R.S. Bushby and V.D. Scott, 'Liquid-phase bonding of aluminium and aluminium/Nicalon composite using copper interlayers' *Mater Sci and Tech, in press.*

3. P.G. Partridge and D.V. Dunford 'The role of interlayers in diffusion bonded joints in metal-matrix composites' *J Mater Sci* **26** (1991) 2255-2258.

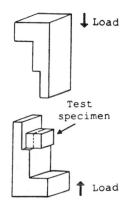

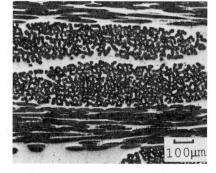

Fig. 1. Schematic diagram of the shear jig.   Fig. 2. As-received Al/Nic composite.

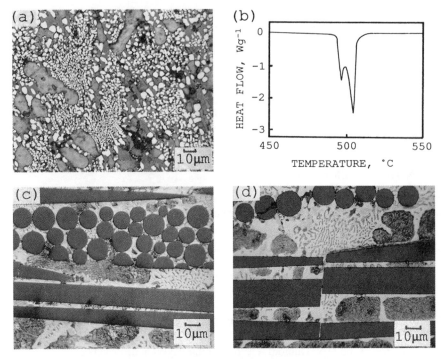

Fig. 3. Al/Nic composite bonded with a copper-silver
alloy interlayer, 20 MPa, 30 mins, 510°C.
a. Microstructure of expelled bead.
b. DSC curve of expelled bead.
c. Section of joint.
d. Broken fibres in bond region.

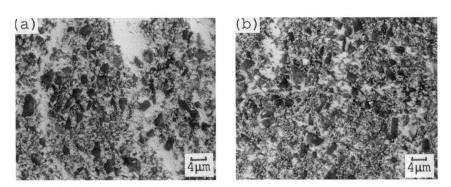

Fig. 4. 2124/35P composite.
a. As-received.
b. Bonded with a copper-silver alloy
interlayer, 8 MPa, 30 mins, 510°C.

PHYSICAL (NON-MECHANICAL) PROPERTIES

STRETCHED EXPONENTIAL MODEL
OF THE DESORPTION IN COMPOSITE MATERIALS

G. KANIADAKIS, E. MIRALDI, M. SCALERANDI, P. DELSANTO

Dipartimento Di Fisica - Politecnico Di Torino
Corso Duca Degli Abruzzi 24 - 10129 Torino - Italy

ABSTRACT

In order to develop an understanding of the desorption mechanism and be able to predict the temporal behaviour of the value of the so called moisture expansion coefficient, we study the dynamics of the migration of impurities from the bulk towards the specimen surface and their consequent loss into the environment. This can be done using the LISA approach which allows to predict, in a very efficient way, the time evolution of the impurities distribution inside the specimen and on its surface. From the knowledge of the impurities desorption, it is possible to evaluate the rate of mass and volume variation of the specimen.

INTRODUCTION

Carbon-fibers-reinforced composites (CFRC's) combine a very high amount of strength and stiffness with low thermal expansion in the fibers direction. These properties contribute to make CFRC's particularly appealing as structural materials for the aerospace industry [1,2]. In fact, space structural materials must keep a very high level of performance in an extremely wide range of temperatures: from very high during the crossing of the earth atmosphere to extremely low, in the interplanetary space [3-5].
The CFRC matrix is most frequently made of epoxy resin, such as Bisglycidyl-bisphenol A-resin with Diethylentriamin as curing agent or Tetraglycidylmethylene-dianiline with Diaminodiphenylsulfon as curing agent. For a high performance a good adhesion of the carbon fibers with the resin is essential. This can be obtained through a surface treatment with formation of oxides on the carbon surface. These oxides may react with polar groups in the resin, such as water molecules or other impurities. Thus the presence, absorption and desorption of moisture and/or other

impurities affect the physical properties and the mechanical behaviour of CFRC's and may lead to a gradual deterioration of the material.

A detailed knowledge /6,7/ of the moisture expansion coefficient of a composite material, defined as the ratio between the relative length increment and the corresponding percentage of mass variation, is very useful for the design of space structures. For instance, the moisture expansion coefficient of CFRC's is one or two orders of magnitude larger than the thermal expansion coefficient. Its effects must be considered in order to design a frame, tailored to have zero thermal expansion when exposed to almost zero pressure.

The most convenient way to measure the moisture expansion coefficient consist in the recording at various times of the weight and length of a specimen kept in a suitable environment. If the surrounding atmosphere is saturated with gas that can be absorbed by the sample, its weight and volume grow and reach an asymptotic value. If, instead, the specimen is kept in the vacuum, within an isothermal shield, a nearly complete lack of moisture is reached by outgassing. The desorption of impurities induces microscopical structural changes in the plastic material (the primary component of the sample), that give raise to a macroscopic shrinkage of the specimen. The recorded dimensional changes are assumed to be proportional to the desorbed vapour mass.

In the present contribution, we study the mechanism of isothermal desorption of impurities through the external surface of the specimen kept in vacuum. We analyze the migration of impurities in a plane normal to the fibers considering first a single "unit region", which we assume to be rectangular, with the cross section of the fiber in its center and isotropic resin all around. Then it is possible to consider the case of many unit regions joined together to represent a specimen.

THEORY

We divide the unit region R, defined above, into $M \times N$ "cells". At each nodepoint (i, j) of this lattice and time t, the impurity distribution density $p_{i,j}(t)$ satisfies the master equation /8/:

$$\dot{p}_{ij} = \sum_{kl} \left(c_{klij} p_{kl} - c_{ijkl} p_{ij} \right) - \Gamma_{ij} p_{ij} \ , \tag{1}$$

where c_{ijkl} represents the transition rate from the node (ij) to (kl) and Γ_{ij} is the transition rate from (ij) to the region outside of R, which we shall call Limbo (L). If we discretize the time and assume that transitions are possible only to the immediate neighbours, we obtain the iteration equation

$$p_{i,j}(t+1) = \sum_{s=\pm 1} \left[\alpha^{(s)}_{i-s,j} \ p_{i-s,j}(t) + \beta^{(s)}_{i,j-s} \ p_{i,j-s}(t) \right] +$$
$$+ \left[1 - \sum_{s=\pm 1} \left[\alpha^{(s)}_{i,j} + \beta^{(s)}_{i,j} \right] - \gamma_{i,j} \right] p_{i,j}(t) \ , \tag{2}$$

where

$$\alpha^{(s)}_{i,j} = \tau c_{i,j,i+s,j} \ , \tag{3}$$

$$\beta^{(s)}_{i,j} = \tau c_{i,j,i,j+s} \ , \tag{4}$$

$$\gamma_{i,j} = \tau \Gamma_{i,j} \ , \tag{5}$$

and τ is the elementary time unit for the time discretization.

Eq.(2) with properly defined boundary conditions is ideally suitable for parallel processing, which becomes necessary when the lattice size $M \times N$ is very large or many regions like R are considered simultaneously.

In Fig.1 the unit region R is schematically represented as a 7×7 lattice. A full square, empty square, full circles and empty circles represent the fiber core, fiber surface cells, internal resin cells and surface resin cells, respectively. Correspondingly, of all the possible $\alpha^{(s)}_{ij}$ and $\beta^{(s)}_{ij}$, we distinguish the following seven kinds of transition rates: k_{rr} , k_{rs} , k_{sr} , k_{ss} , k_{rf} , k_{fr} , k_{ff} , where r, s and f denote an inside resin cell, surface resine cell and surface fiber cell, resp.

We assume that transitions to and from the core fiber cell are forbidden and that transitions to the Limbo are possible only from the surface resin cells, with a transition rate γ. It follows

$$\gamma_{ij} = \gamma(\delta_{i1} + \delta_{iN} + \delta_{1j} + \delta_{Mj}) \ , \tag{6}$$

with a double transition rate from the four cells at the corners of R. We next define the amount of impurities escaped from R to the Limbo up to the time t

$$L(t) = \sum_{t'=0}^{t-1} \Delta L(t') \ , \tag{7}$$

where

$$\Delta L(t') = \gamma \left[\sum_i \left[p_{i1}(t') + p_{iN}(t') \right] + \sum_j \left[p_{1j}(t') + p_{Mj}(t') \right] \right] \ , \tag{8}$$

represents the number of impurities escaping from R in the unit time interval $(t', t'+1)$.

A knowledge of $L(t)$ is essential for the understanding of the desorption mechanism and, consequently, of the dimensional variation of the specimen. The experimental trends of the moisture expansion coefficient show /10/ a "stretched exponential" behaviour

$$L(t) = L(\infty) \left[1 - exp \left[- (t/T)^{\beta} \right] \right] \ , \tag{9}$$

where the asymptotic value $L(\infty)$ is also equal to the initial number of impurities in R. For the determination of the parameters β and T from the experimental curve $L(t)$ it is convenient to introduce the function

$$\Phi(t) = log \left[- log \left[1 - L(t)/L(\infty) \right] \right] \ . \tag{10}$$

From Eq.(9) it follows

$$\Phi(t) = \beta \left(\log\, t - \log\, T \right) \tag{11}$$

From the linear plot $\Phi(t)$ vs. $\log t$ it is then easy to determine β an T.

RESULTS AND DISCUSSION

We wish to show in this Section that it is possible to reproduce with an elementary model the stretched exponential behaviour of $L(t)$, with $\beta \neq 1$. For simplicity we assume $k_{rr} = k_{ff} = k_{rf} = k_{fr} = k_{ss} = k_{rs} = k$ and $k_{sr} = 0$, i.e. the diffusion within R is isotropic everywhere except from and to the fiber core, where it vanishes, and at the surface, from where no return into the interior is allowed. Thus we have two "free" parameters: γ and k.

Fig.2 shows the temporal evolution of $L(t)$ for two sets of parameters: $\gamma = k = 0.1$ (curve A) and $k = 0.1$, $\gamma = 0.02$ (curve B). The corresponding curves in the linear plot defined by Eq.(11) are shown in Fig.3. From this figure one obtains $\beta = 1$ and $T = 14.7\ \tau$ for the curve A and $\beta = 1.25$, $T = 20.9\ \tau$ for the curve B. The larger value of T, i.e. a slower desorption for the curve B is an obvious consequence of the lower value of γ. It is also easy to explain the value $\beta = 1$ for the curve A: in fact, being $\gamma = k$ there is no real separation between R and the Limbo. More interesting is the value $\beta = 1.25$ for the curve B, which shows that we truly have a stretched exponential behaviour for the desorption, due to the slower transition rate γ from the surface of R, which controls the entire process, slowing it down.

In Fig.s 4 and 5 we consider a case in which $\gamma \gg k$: $\gamma = 0.2$, $k = 0.02$. As a result the process of internal diffusion is very slow, while the escape from the surface of R is very fast. Therefore we can distinguish three phases in the desorption process. The first one, which consists almost exclusively of a depletion of the surface impurities, is represented in Fig.4: it is much faster ($T = 27.7\ \tau$) and is characterized by a remarkable level of "stretching" ($\beta = 0.5$). The second phase, shown in Fig.5, is controlled both by surface desorption and internal diffusion. In this phase we observe a transition from a stretched behaviour ($\beta = 0.5$) to a Debye kind of relaxation ($\beta = 1$), which characterizes the third phase of the process. In that last phase the process is exclusively controlled by internal diffusion, which is very slow ($T = 79.0\ \tau$).

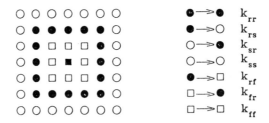

Figure 1: Schematic representation of the unit region and of the transition rates. A full square, empty squares, full circles and empty circles represent the fiber core, fiber surface cells, internal resin cells and surface resin cells. resp.

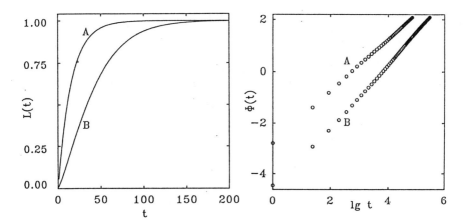

Figure 2: Time evolution of the Limbo accumulation for two sets of parameters: $\gamma = k = 0.1$ (curve A); $\gamma = 0.02$, $k = 0.1$ (curve B)

Figure 3: Time evolution of the Limbo accumulation as represented by the linear function Φ vs. *log t* for the same sets of parameters as in Fig. 2.

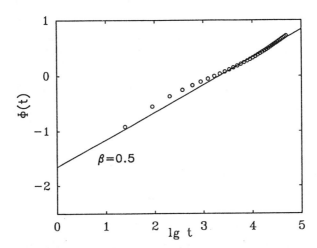

Figure 4: First phase of the time evolution of the Limbo accumulation for the case $\gamma = 0.2$, $k = 0.02$.

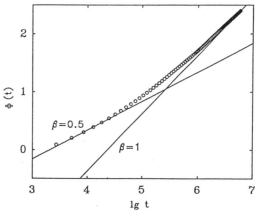

Figure 5: Second and third phases of the time evolution of the Limbo accumulation for the case $\gamma = 0.2$, $k = 0.02$.

References

[1] E.Fitzer: *Carbon Fibres - the Miracle Material for Temperatures Between 5 and 3000 K, High Temperatures - High Pressures*, 1986,V.18,p.479-508 (10 ETPC Proceedings)

[2] E.Fitzer,M.Heine: in *Fibres for Composite Materials*, ed. A.R. Bunsell (Amsterdam:Elsevier 1988)

[3] G.Romeo,E.Miraldi,M.Gazzi: *Thermal and Moisture Expansion Properties of Advanced Composite Materials* , in Advanced Structural Materials: Design for Space Applications, proc. ESA,WPP-004, (1981) p.67

[4] V.Baudinaud,M.Canevet:*High Temperatures-High Pressures*,V.19(1987)p.411

[5] E.Miraldi,G.Romeo,G.Ruscica,F.Bertoglio,G.Ruvinetti, proc. Int. Conf.on Spacecraft Structures and Mechanical Testing , Estec Noordwijk, The Netherlands (1991)

[6] G.S.Springer: *Environmental Effects*, in Composite design, sec. 16, edited by S.W.Tsai,T.N.Massard (Ippei Susuki Publ., Dayton, Ohio 1988)

[7] G.Kaniadakis,E.Miraldi,G.Frulla: *Stretched Exponential Structural Relaxation of Composite Materials Induced by Impurities Desorption in Vacuum*, Mod.Phys.Lett.B, V.5 (1991) p.1911

[8] G.Kaniadakis,P.P.Delsanto,C.A.Condat: *A Local Interaction Simulation Approach to the Solution of Diffusion Problems*, Math. Comp. Modell.,in press

[9] P.J.Denning,W.F.Ticky: *Highly Parallel Computation*, Science V.250 (1990) p.1222

[10] G.Williams,D.C.Watts, Trans. Faraday Soc., V.66 (1970) p.80

A NEW WAY OF PROGNOSIS OF BUILDING CERAMIC DURABILITY UNDER FROST ACTION

M. ZYGADLO

*Kielce University of Technology - Al 1000
Lecia P.P. 7 - 25-314 Kielce - Poland*

ABSTRACT

Presented method is a new approach concerning frost durability testing of building ceramics based on the results of statistical analysis of experimental data. It has been confirmed that easily measurable physical properties like: specific gravity, absorbability, capillary absorption, compressive strength, altogether are suitable for the prediction of building ceramics frost resistance. The result of the statistical analysis is the classification rule. This rule enables us to classify materials into two groups: frost resistant and non-frost resistant.

INTRODUCTION

The traditional way of assessment in building ceramic frost resistance consisted in the freeze/thaw cycling of materials in a freezing chamber. This test required a long term investigation extending to a period of several weeks. But in spite of this it is generally recommended in current standards due to the absence of a better one.

In the paper a new test is described which enables us to receive the results of the frost durability classification after two days.

The common opinion of many authors /1-3/ is that the pore size distribution in a material is responsible for its destruction under frost action.

It is known that the higher level temperature of the clay ceramic production process involves a complete porosity decrease, whereas medium pore size increases /4-6/. Increasing of the medium pore size leads to the higher frost resistance. But manufacturing temperature of

clay products is limited considering other features like melting point, deformation, thermal insulation.
It is known also that building materials of higher total absorbability possess lower frost resistance. But this criterion is not confirmed in each case /7, 8/. This fact shows the significance of other parameters influencing porous material behaviour under frost action.

1.LABORATORY TEST

Three kinds of building materials from many plants have been considered: clay bricks, hollow clay blocks and calcareous - sand blocks.
15 batches from clay brick factories, 19 batches from hollow clay block factories and 14 batches from - calcareous sand block factories were investigated. Each batch consisted of at least 10-15 units chosen from the production line as a representative group for the bigger part (according to standard).
The proper physical features have been included based on the authors own previous experencies described in /9, 10, 11/.

2.STATISTICAL APPROACH

The parameters to be determined as most suitable for prediction of a given assortment of frost resistance have been selected using a discriminant analysis. The selected parameters were chosen from among the many physical properties determined investigating the materials.
This conclusion was drawn: which parameters ensure the best classification of each batch in comparing with the direct method using a laboratory chamber.
A linear function for materials classification according to their frost resistance has been derived. In order to select a material the total form is:

$$y = A \cdot R_c + B \cdot T_g - C \cdot A_c - D \cdot H_t - E$$

where:

R_c - compressive strength [MPa], adopted as a average value for considered batch,

T_g - specific gravity [g/cm^3],

A_c - absorbability in cold water [%],

H_t - capillary absorption after t hours being in contact with water [g/cm^2],

y - results of batch classification. If it occures for 10 statistically chosen objects:

$y > 0$, the batch is frost resistant
$y < 0$, the batch is frost nonresistant

A, B, C, D, E - constants.

The above rule possesses good classification ability of investigated objects for two considered groups. The results of classification of a particular kind of investigated assortments of building materials are presented on Fig.1 on a canonical axis.

Prognosis of building material behaviour under frost action has an essential practical value because it enables to obtain assessment of a material's frost resistance in a short time (not longer that 2 days).

3.CONCLUSIONS

The method has an essential practical value because one can obtain assessment of a material's frost resistance in a short time, no more then two days.

Assessment can be realized using ordinary available laboratory equipment.

The verification process of presented method carried out by laboratory personell in industrial conditions brought enough satisfactory results. It became the basis for a project and later was published as a new polish standard PN-92/B-12017.

REFERENCES:

1. G. Fagerlund, Proc.Inter.Symp.RILEM, Part IV, Prague,(1973), pp. 56-78.
2. A. Ravaglioli, Ceram.Inf., vol.12, no 136,(1977), pp.543-547.
3. M. Maage, Mat.Struct., vol.17, no 101,(1984) pp.345-350.
4. G. Piltz, Ziegelindustrie, no 15/16,(1970) pp.327-336.
5. A.J. Rijken, Ziegelindustrie, no 2,(1971) pp.78-79.
6. O.I. Whittemore, Powder Techn., vol.29,(1981), pp.167-175.
7. A. Ravaglioli, G. Vecchi, Ceramurgia, vol.4, no 2,(1974) pp.107-117.
8. P. Vinzenzini, Ceramurgia, vol.4, no 3,(1974) pp.176-188.
9. M. Żygadło, Z. Piasta, Industrial Ceramics, vol.8, nr 4.(1988), pp.129-133
10.M. Żygadło, Z. Piasta, A. Lenarcik, Proc. Intern. Conf. Brittle Matrix Comp. Warszawa,(1991) pp.377-385
11.Z. Piasta, M. Żygadło, Proc. Int.Conf.Composite Mat., Honolulu, 15-19.July(1991)

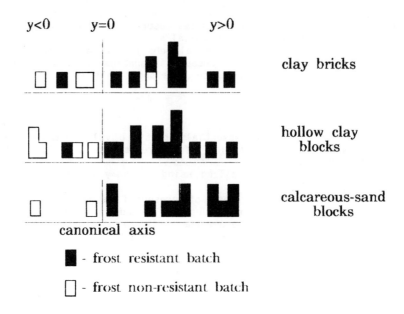

Fig.1. Results of classification of different kind of assortments of building materials using classification formulae.

VISCO-PLASTIC PROPERTIES OF MULTICOMPONENT COMPOSITE MATERIALS

L. SARAEV, J. MAKAROVA

Samara State Technical University - 23-7 Dimitrov St.
443095 Samara - Russia

ABSTRACT

In the present paper it is proposed a variant of the averaging method of the accidental fields theory. The application of this method permits to construct the models of behaviour for visco-plastic microheterogeneous media with the various power of the composite components connection. The macroscopic constitutive equations describing behaviour of the composite are constructed. There are given some numerical results.

Let incompressible visco-plastic medium is formed by n different components and occupies a volume V, which is bounded by a surface S. All components are coupled with an ideal adhesion. Let mechanical properties of this composite are described by means of non-linear visco-plastic law of flow /1/:

$$s_{ij} = k_s \frac{e_{ij}}{\sqrt{e_{kl}e_{kl}}} + \mu_s(\sqrt{e_{kl}e_{kl}}) \, e_{ij} \, , \quad (s=1,2,\ldots,n). \quad (1)$$

Here $s_{ij} = \sigma_{ij} - \frac{1}{3}\delta_{ij}\sigma_{pp}$; σ_{ij} is the stress tensor; k_s is the yield points, $\mu_s(\sqrt{e_{kl}e_{kl}})$ is the non-linear viscosity coefficient of the s'th component; e_{ij} is the strain rates tensor, it satisfies to the incompressibility condition: $e_{pp}=0$.

The structure of this composite may be described through the indicator function of co-ordinates $\mathscr{æ}_1(\mathbf{r})$, $\mathscr{æ}_2(\mathbf{r})$,...., $\mathscr{æ}_n(\mathbf{r})$, where $\mathscr{æ}_s(\mathbf{r})$ equals to 1 on the multitude of s'th component material points V_s and 0 beyond this multitude. Now the Eq.(1) may be written in the form:

$$s_{ij}(\mathbf{r})=\sum_{s=1}^{n}\left[k_s\frac{e_{ij}(\mathbf{r})}{\sqrt{e_{kl}(\mathbf{r})e_{kl}(\mathbf{r})}} + \mu_s(\sqrt{e_{kl}e_{kl}})e_{ij}(\mathbf{r})\right]\mathscr{æ}_s(\mathbf{r}) \qquad (2)$$

The indicator functions $\mathscr{æ}_s(\mathbf{r})$, stresses and strain rates and values k_s, μ_s are supposed to be statistically homogeneous and ergodical fieldes. Hence, their expectations may be replaced by the average values within corresponding volumes:

$$\langle f\rangle= \frac{1}{V} \int_V f(\mathbf{r})d\mathbf{r} , \qquad \langle f\rangle_s= \frac{1}{V_s} \int_{V_s} f(\mathbf{r})d\mathbf{r} .$$

To use well-known methods of the elasticity theory for constructing of the effective relations let us suggest that within composite volume the values $e_{kl}(\mathbf{r})e_{kl}(\mathbf{r})$ are homogeneous and may be replaced by their expectations: $\Lambda =$ $=\sqrt{\langle e_{kl}(\mathbf{r})e_{kl}(\mathbf{r})\rangle}$. In this case the Eq.(2) can be written in the form:

$$s_{ij}(\mathbf{r})= \lambda(\mathbf{r}) \, e_{ij}(\mathbf{r}) , \qquad (3)$$

$$\lambda(\mathbf{r})= \sum_{s=1}^{n} \lambda_s \mathscr{æ}_s(\mathbf{r}) , \qquad \lambda_s= \frac{k_s}{\Lambda} + \mu_s(\Lambda).$$

Let us add to Eq.(3) the equilibrium equations and the strain-displacement relations writing in rates:

$$\sigma_{ij,j}(\mathbf{r})= 0; \quad 2e_{ij}(\mathbf{r})=v_{i,j}(\mathbf{r}) + v_{j,i}(\mathbf{r}); \quad v_{p,p}= 0 \qquad (4)$$

Here v_i are the displacement rates vector components. The boundary conditions for the closed system of the Eq.(3),(4) are the equality to 0 any fluctuations of all values on the bound of the surface S:

$$f(\mathbf{r})= \langle f\rangle, \quad \mathbf{r}{\in}S \qquad (5)$$

Following to the method of the generalized singular approximations used in the theory of accidental fields, let us introduce the indeterminate value λ:

$$\min_{s}\{\lambda_s\} \le \lambda \le \max_{s}\{\lambda_s\}$$

and write the system of the Eq.(3),(4) in terms of displacement rates:

$$\lambda\, v'_{1,pp} + 2\left[\sum_{s=1}^{n} [\lambda_s]\text{\ae}_s e_{1p}\right]'_{,p} + 2\,\sigma'_{,1} = 0 \qquad (6)$$

Here $[\lambda_s]=\lambda_s-\lambda$, $\sigma = \frac{1}{3}\sigma_{pp}$, the stroke symbol denotes a fluctuations of values within the volume $V{:}f'(\mathbf{r})=f(\mathbf{r})-\langle f\rangle$. By using the Green's tensor:

$$G_{1k}(\mathbf{r})= \frac{1}{8\pi}\left[\delta_{1k}r_{,pp} - r_{,1k}\right], \qquad r=|\mathbf{r}|$$

the system of the Eq.(6) together with the boundary conditions (5) can be written in the form:

$$e'_{1j}(\mathbf{r})= \int_{V} G_{1k,1j}(\mathbf{r}-\mathbf{r}_1)\tau'_{kl}(\mathbf{r}_1)d\mathbf{r}_1 \ , \qquad (7)$$

$$\tau_{kl}(\mathbf{r})= -2\sum_{s=1}^{n} \frac{[\lambda_s]}{\lambda}\,\text{\ae}_s(\mathbf{r})e_{kl}(\mathbf{r}).$$

To define the composite effective constants let us statistically average the Eq.(3) within full volume V:

$$\langle s_{1j}\rangle = \sum_{s=1}^{n}\lambda_s c_s \langle e_{1j}\rangle_s \qquad (8)$$

Here $c_s=V_s/V$ is the volume content of the s'th component. The values $\langle e_{1j}\rangle_s$ can be found from the known relation /2/

$$\langle e_{1j}\rangle_s = \langle e_{1j}\rangle + c_s^{-1}\langle \text{\ae}'_s e'_{1j}\rangle.$$

To find the value $\langle \text{\ae}'_s e'_{1j}\rangle$ let us multiply the Eq.(7) through by $\text{\ae}'_s(\mathbf{r})$ and then average the result within the volume V:

$$\langle \text{\ae}'_s e'_{1j}\rangle = \int_{V} G_{1k,1j}(\mathbf{r}_1)\langle \text{\ae}'_s(\mathbf{r})\tau'_{kl}(\mathbf{r}+\mathbf{r}_1)\rangle d\mathbf{r}_1 \ .$$

To calculate the right part integral let us limit oneself to the singular approximation without the formal components of the Green's tensor second-order derivatives /3/:

$$\langle \alpha'_s e'_{ij} \rangle = \frac{1}{5} \langle \alpha'_s \tau'_{ij} \rangle.$$

Substituting the expression for the τ_{ij} into the last equation and taking into account the relations /2/:

$$\langle \alpha'_q \alpha'_s f \rangle = \begin{cases} c_q c_s (\langle f \rangle - \langle f \rangle_q - \langle f \rangle_s); & q \neq s, \\ c_s (c_s \langle f \rangle + (1 - 2c_s)\langle f \rangle_s); & q = s, \end{cases}$$

let us find

$$\langle e_{ij} \rangle_s = \frac{2\langle s_{ij} \rangle + 3\lambda \langle e_{ij} \rangle}{5\lambda + 2[\lambda_s]} \tag{9}$$

Substituting the Eq.(9) into the Eq.(8), let us receive:

$$\langle s_{ij} \rangle = \lambda \frac{3\xi}{5 - 2\xi} \langle e_{ij} \rangle, \quad \xi = \sum_{s=1}^{n} \frac{c_s \lambda_s}{\lambda + 0.4[\lambda_s]} \tag{10}$$

The strain rates averaging within the components volumes V_s are expressed by relation:

$$\langle e_{ij} \rangle_s = \frac{3}{5 - 2\xi} \cdot \frac{\lambda}{\lambda + 0.4[\lambda_s]} \langle e_{ij} \rangle$$

The value Λ is found by using the integral equilibrium equation which can be written in the form /2/:

$$\int_V s'_{ij}(\mathbf{r})e'_{ij}(\mathbf{r})d\mathbf{r} = 0$$

From this equation the next relation follow:

$$\lambda \langle e'_{ij} e'_{ij} \rangle + \sum_{s=1}^{n} [\lambda_s]\langle \alpha_s e'_{ij} e'_{ij} \rangle = 0.$$

Excepting from this relation the fluctuations of strain rates and taking into account the singular approximation we can obtain the equation:

$$\Lambda = \frac{3\lambda}{5 - 2\xi} \sqrt{\sum_{s=1}^{n} \frac{c_s}{(\lambda + 0.4[\lambda_s])^2}} \; e \tag{11}$$

Here $e = \sqrt{e_{kl} e_{kl}}$.

The Eq.(11) may be solved numerically when the form of functions $\mu_s(\Lambda)$ are defined. The system of the Eq.(10),(11) describes macroscopic properties of considered composite material.

Selecting the indeterminate value λ we may construct various models of multicomponent composites. For example, if $\lambda = \lambda_1 = \dfrac{k_1}{\Lambda} + \mu_1(\Lambda)$ then we have the model of composite in which the first component plays a role of adhesive matrix and all others are separate inclusions. If $\lambda = \{\lambda\} = \sum\limits_{s=1}^{m} c_s \lambda_s$ ($m \leq n$) we have another model of composite the first m interpenetrated components of which form a matrix and the others (n-m) components play a role of separate inclusions. In particular, if $k_s = 0$ ($s = 1, 2, \ldots, n$) then the Eq.(10),(11) describe the flow of non-linear viscous medium.

As an example of proposed method let us use the Eq.(10), (11) for description of two-component mixture melt flow by the temperature 463 $^\circ$K. The mixture consists of two polyethylenes of low (PELD) and high (PEHD) density. As the mixture is the interpenetrated frames then $n = m = 2$, $\lambda = \mu = \langle \mu \rangle = c_1\mu_1 + c_2\mu_2$. The sections of flow diagram of pure polyethylenes may be approximated by power law of flow:

$$\mu_s(\Lambda) = 2\eta_s (\sqrt{2}\,\Lambda)^{\alpha_s - 1} \tag{12}$$

Here, η_s is the coefficient of consistency and α_s is the non-linearity index of s'th component.

Let us plot the shear diagram of considered composite. The Eq.(10) can be written as

$$s_{12} = \lambda \frac{3\xi}{5 - 2\xi} e_{12} \tag{13}$$

The Eq.(11)-(13) were solved numerically by the successive approximation method. Figure shows the experimental data (dots) /4/ compared with theoretical shear diagram of mixture flow (full line) calculating from the Eq. (11)-(13). The dotted lines are the flow diagrams of component melts (1 - PELD, 2 - PEHD).

The calculated values are:

$$c_1 = 0.1, \quad \eta_1 = 2.437 \text{ KPa} \cdot e^{\alpha_1}, \quad \alpha_1 = 0.615$$
$$c_2 = 0.9, \quad \eta_2 = 14.642 \text{ KPa} \cdot e^{\alpha_2}, \quad \alpha_2 = 0.515$$

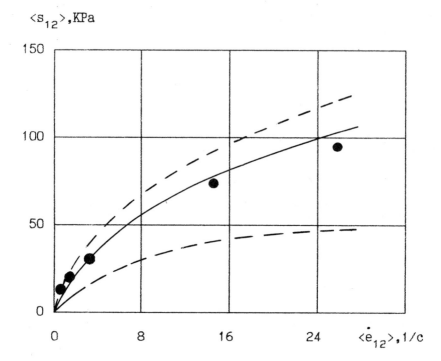

Figure. Comparison theoretical shear diagram of the mixture flow with experimental data.

REFERENCES

1. L.A.Sarajev, J.of Appl. Mech. and Theoret.Phys., 4(1988) 124-130.

2. L.A.Sarajev, J.of Appl. Mech. and Theoret.Phys., 6(1984) 157-161.

3. L.A.Sarajev and T.D.Shermergor, Applied Mechanics, V.21, 5(1985) 92-97.

4. G.V.Beregnaja and al, in "Rheology of Polymers and Disperse Systems and Rheophysics" Part 1, (1975) 108-112, Inst. of Theoret.Mech., Minsk.

PAPERS RECEIVED AFTER THE DEADLINE

INTERFACIAL CHARACTERIZATION OF VAPOUR-GROWN CARBON FIBRE COMPOSITES USING THE SINGLE FIBRE COMPOSITE FRAGMENTATION TEST

P.J. NOVOA, M.T. SOUSA, J.L. FIGUEIREDO, A. TORRES MARQUES

Department Mechanical Engineering - Applied Mechanics - Faculty of Engineering - Rua dos Bragas - 4099 Porto codex - Portugal

ABSTRACT

In the last few years there has been a growing interest in the application of vapour-grown carbon fibres in composites. While a few results on the mechanical properties of vapour-grown carbon fibres have already been reported, there is yet no known study on the interfacial properties of systems involving these fibres. To address this question we have studied a vapour-grown carbon fibre/epoxy matrix system using the single fibre composite fragmentation test. Preliminary results are presented and discussed along with results for conventional unsized, untreated carbon fibres.

1. INTRODUCTION

One of the reasons for the growing interest in the use of carbon fibres produced by chemical vapour deposition (CVD) is the greater degree of graphitic perfection they can exhibit when compared with polyacrylonitrile (PAN)- or pitch-based carbon fibres graphitized at similar temperatures [1]. Another main reason for this is the potential for obtaining vapour-grown carbon fibres (VGCF's) at a low price using a cheap raw-material, like natural gas.

While there is some literature concerning the mechanical performances of CVD produced carbon fibres [1-2], there seems to be no record whatsoever of a study on the interfacial properties of VGCF/matrix systems. This fact is surprising if it is intended that these fibres are to be used as reinforcement in composite materials, as then, the interfacial properties are of the outmost importance. The interface between matrix and reinforcing fibre has a major role in the mechanical properties of the composite.

The authors have been involved in the production of CVD carbon fibres and their surface characterization /3/, as well as in the measurement of their mechanical properties /4/. A study to evaluate the interfacial properties of VGCF/matrix systems was thus initiated in order to make a first comparison with existing results for commercially available carbon fibres in similar matrixes.

This paper presents preliminary results for the interfacial evaluation of a VGCF/epoxy matrix system, measured as the interfacial shear strength (ISS), using the single fibre composite fragmentation test.

2. THE FRAGMENTATION TEST

This technique has its origin in a study described by Kelly /5/ where a multiple fracture phenomenon was observed upon tensile loading of a system consisting of a low concentration of tungsten wires axially aligned and embedded in a copper matrix. This is in fact expected to occur at wire loadings below a certain critical value /5, 6/, and constitutes the basis for the interfacial evaluation procedure which we are about to describe.

In this method a single fibre is embedded in a matrix material whose strain to failure is at least three times higher than that of the fibre. As the system is loaded in the direction of the fibre, the stress is transferred from the matrix to the fibre at the interface through shear forces. The fibre axial stress that is introduced by the interfacial shear stress builds up in a roughly linear way from the ends until the fibre ultimate strength (σ_f) is reached /7/. At this point the fibre breaks at some location along its middle section where the tensile stress is constant - fracture should occur near the middle point if the fibre is homogeneous. As we continue to increase the applied load this fragmentation process repeats itself over and over again and the fibre fragments become smaller each turn until, at some point, the maximum shear stress (τ) which the interface (or matrix) can sustain will be reached. If we assume that the interfacial region fails in shear by yelding and flowing plastically, the shear stresses at the interface can be considered to be constant along the short fibre fragments now present. Thus, the average shear stress is given by a simple force balance resulting in:

$$\tau = \frac{\sigma_f(l_c)d}{2l_c} \qquad \text{(eq. 1)}$$

where l_c is the critical length over which stress builds up, counting from both ends of each fragment length, $\sigma(l_c)$ is the fibre tensile strength at a gauge length equal to l_c, and d is the diameter for a fibre with constant circular cross-section.

If we consider the definition of critical length, the analysis of the fragment length distribution of such an experiment is expected to reveal lengths between $l_c/2$ and l_c. The longest fibres capable of avoiding fragmentation would have a length just under l_c, and the shortest fragments would be those resulting from the fracture of fibres with lengths just above l_c, with fracture expected to occur near the middle point. This was experimentally verified by Kelly /5/ in his, above cited, investigation.

This has led some authors /8/ to accept that the lengths will be uniformly distributed over this range, and so, that $0.75l_c$ would be a reasonable estimate for the mean fragment length($\bar{l}$). Taking this into account leads to a modified form of eq. 1:

$$\tau = \frac{\sigma_f(l_c)d}{2\bar{l}}K \qquad \text{(eq. 2)}$$

where $K=0.75$.

However, in fragmentation experiments of glass fibres in thermoplastics /6/ and carbon fibres in an epoxy resin /9/ it was found that the fragment length distributions were non-symmetrical and extended well beyond the 2:1 range expected from the critical fibre length concept. This was recognized as being due to the statistical nature of fibre strength. These fibres possess flaws randomly distributed which control the fragmentation process.

Several approaches have been used which take these aspects into consideration. Using of a Monte-Carlo simulation of a Weibull/Poisson model /10/ for fibre strength and flaw occurrence, Netravali *et al.* found that a value of 0.889 for K in eq. 2 would be better fit for most brittle fibres. This leads to values of ISS which are 19% higher then those that obtained when the K value for a flawless material is used. We thus feel that the mean fragment length extracted from the asymmetric fragment length distributions which result from the random nature of brittle fibre fracture will differ from $0.75l_c$, and furthermore accept the suggestion of Netravali *et al.* ($\bar{l} = 0.889l_c$) as a better choice.

3. EXPERIMENTAL PROCEDURE

Carbon fibres were grown in a mixture of 30% methane in hydrogen by a CVD process using iron as catalyst. The experimental conditions were those described by Gadelle and co-workers /11/. A support of grafoil® (a registered trademark of Union Carbide) is vaporized with a solution of $Fe(NO_3)_3.9H_2O$ in ethanol and then submitted to a temperature program. This consists of heating up to 950 °C at 30 °C/min. and then at 7 °C/min. up to 1150 °C, keeping this temperature constant for 100 min. The fibres lengthen during the heating and thicken during the dwell at 1150 °C. The average diameter of these fibres is 5.5 ± 1.4 μm /4/.

The matrix material was a diglycidyl ether of bisphenol-A (DGEBA) based liquid epoxy (LY554, Ciba-Geigy) cured with a diamine hardener (HY956, Ciba-Geigy) in a 5:1 weight ratio as recommended by the manufacturer. This epoxy resin has a yeld strain of about 6%, high transparency and good photo-elastic behaviour.

The test specimens were dog-bone shaped and had a 10 mm gauge length. They were manufactured by axially aligning a single fibre in the center of a silicone rubber mould, filling the mould with a degassed mixture of liquid epoxy and hardener in the recommended proportions, and then oven curing the assembly at 70 °C for two hours.

Before testing, fibre diameters were measured using an optical microscope fitted with a calibrated eye-piece. Specimens were strained on an Instron machine at a cross-head speed of 1 mm/min. to approximately their yeld elongation and then analyzed between crossed polarizing filters under a light microscope. Fibre fragments were measured observing photo-elastic patterns near fibre breaks.

4. RESULTS AND DISCUSSION

In order to provide a statistically significant number of fibre fragments for data analysis, fibre measurements from several separate samples were aggregated. The fragments were normalized by the average diameter for the set of specimens and the mean fragment aspect ratio $(\bar{l}/d)$ and fragment aspect ratio (l/d) distributions evaluated. In Fig. 1 a relative frequency distribution is shown. It can be seen from this aspect ratio histogram that the distribution is skewed and extends beyond the 2:1 range expected from the simple critical length concept. This is characteristic of fibres exhibiting flaws, as stated earlier, and comes as no surprise. This dispersion is related to the degree of fibre damage, *i.e.*, the flaw distribution /6/, and can also be used as a measure of fibre size efficiency in the protection against damaging (*e.g.*, by handling). Of interest to us, is that the actual range for our system is close to 4:1, this falling within the values usually found for commercial fibres. Our VGCF's had no size so this simply tells us that they exhibit flaw distributions similar to those of commercial fibres.

For the evaluation of the interfacial shear shear strength an estimate of $\sigma(l_c)$ was needed. Unfortunately, there are few results for tensile strength of these VGCF's, and only for a gauge length of 6 mm. Therefore, the use of Weibull statistics is not yet possible. An average of 0.80 GPa was used, bearing in mind it may be a low estimate for the tensile strength at l_c.

In Table 1, our results along with results from fragmentation tests performed by Jacques and Favre /8/ for unsized, untreated ex-PAN commercial carbon fibres and the respective DGEBA resins used, are listed. The cited results were originally calculated with the assumption that $\bar{l} = 0.75l_c$, so we performed their recalculation considering $\bar{l} = 0.889l_c$ as suggested above.

Before proceeding to the discussion of the listed results we should state that some difference between the ISS of our system and the other two may arise from differences in the matrix and curing cycle (cf. Table 1). On inspection of the ISS values we verify that the VGCF/DGEBA system has the lowest value. However, the value we used as $\sigma(l_c)$ in eq. 2 may have been itself a low estimate so, it seems fair to say that this may be the main cause for the low ISS. Attention must be drawn to the fact that this value is greatly dependent on the low critical aspect ratio of this system. Taking this into consideration, and also the possible underestimation of $\sigma(l_c)$, we may conclude that the ISS for our system could be greater than the one reported in Table 1.

As stated above, our system produced a low critical aspect ratio although similar values have been reported /12/. From the values listed we find that our distribution of fibre fragments produces a much shorter critical length. This absolute value is normally related to fibre/matrix interactions and an immediate conclusion would be that our system exhibits better efficiency in shear stress transfer at the fibre/matrix interface. If we accept that the matrix differences between the two systems are negligible we can ascribe this to differences in fibre surface. However, due to the nature of VGCF's surface functionality is not the cause. We believe it has to do with fibre surface irregularity. Our microscopic observations have shown that their cross-section is quite irregular. This gives rise to stress concentrating points which may lead to the extensive fibre breakage, thus being the major cause for the observed low absolute value of the fibre fragments. This may constitute a limitation to the simple interpretation of the fragmentation phenomenon upon which we based our results.

5. CONCLUSIONS

We have addressed the interfacial mechanical properties of a VGCF/DGEBA resin system using the single fibre composite test. We have concluded that:

• the fibres investigated exhibit flaw distributions similar to those of commercial fibres;

• in spite of the preliminary nature of the results obtained for fibre tensile strength and fragmentation testing, the value of the interfacial shear strength seems to be reasonable possibly due to the small critical aspect ratio determined;

• the irregularity of these fibres may be the reason for the small critical aspect ratio and, if so, be a limitation to the simple model used in the evaluation of the interfacial shear strength.

REFERENCES

1. G. G. Tibbets and C. P. Beetz, Jr, J.Phys. D: Appl. Phys. 20 (1987) 292-297

2. Jianchui Chen, Yan Lu, D. B. Church and D. Patel, Appl. Phys. Lett. 60 (1992) 2347-2349

3. M. Teresa Sousa and J. L. Figueiredo in "The Interfacial Interactions in Polymeric Composites" (G. Akovadi, ed.) (1993) 449-450, D. Reidel Publishing Company, Dordrecht

4. M. Conceição Paiva and C. Bernardo, unpublished work

5. A. Kelly, Proc. Roy. Soc. (London) A 288 (1964) 63-79

6. W. A. Fraser, F. H. Ancker and A. T. DiBenedetto, 30th Ann. Tech. Conf. Reinf. Plastics/Composites Inst., Soc. Plast. Industry, Sec. 22-A (1975) 1-13

7. P. J. Herrera-Franco and L. T. Drzal, Composites 23 (1992) 2-27

8. D. Jacques and J. P. Favre *in* ICCM VI & ECCM 2, vol. 5, (F. L. Mathews, N. C. R. Buskell, J. M. Hodgkinson and J. Morton, eds.) (1987) 471-480, Elsevier Applied Science, London

9. N. J. Wadsworth and I. Spilling, Brit. J. Appl. Phys. (J. Phys. D), Ser. 2, 1 (1968) 1049-1058

10. A. N. Netravali, L. T. T. Topoleski, W. A. Sachse and S. L. Phoenix, Comp. Sci. Tech. 35 (1989) 13-29

11. F. Benissad, P. Gadelle, M. Coulon and L. Bonnetain Carbon 26 (1988) 61-69

12. W. D. Bascom, R. M. Jensen and L. W. Cordner *in* ICCM VI & ECCM 2, vol. 5 (F. L. Mathews, N. C. R. Buskell, J. M. Hodgkinson and J. Morton, eds) (1987) 425-438, Elsevier Applied Science, London

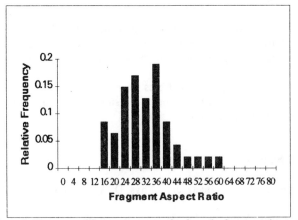

Figure 1. Histogram of fragment aspect ratios from the fragmentation tests.

TABLE 1

Interfacial shear strength and critical aspect ratios
for several CF/DGEBA-amine systems

Fibre Type	Epoxide (DGEBA)	Curing Agent	Curing Cycle	d (μm)	lc/d ($l_c = \bar{l}/0.889$)	$\sigma_f(l_c)$ (GPa)	τ (MPa)
Vapour-grown	LY554	HY956	2h/70°C	5.0	36 (177 μm)	0.8 *	11.3
Grafil HT (COURTAULDS)	LY556	HT972	2h/140°C	7.62	171 (1301μm)	5.18	15.2
T 300 (TORAYKA)				6.90	77 (532 μm)	5.70	37.0

* This is a value of σ_f for a gauge length of 6 mm. See text.

COMPARISON WITH EXPERIENCE AND NUMERICAL MODELLING OF FATIGUE DAMAGE IN COMPOSITE TUBE IN TORSION-TORSION

B. MACQUAIRE, J. RENARD, A. THIONNET

Ecole Nationale Supérieure des Mines de Paris - Centre des Matriaux P.M. Fourt - BP 87 - 91003 Evry Cedex - France

ABSTRACT

Fatigue behaviour studies on composite structures are often conducted on flat specimens using an uniaxial test. In order to model a real stress field we have developped a test on tubes which avoids the free-edge effects and allows modifications of the loading mode : tension, torsion, compression. We developped a non destructive method to characterize the crack density in each ply without removing the specimen.

Concerning the theoritical modelling, assuming as a first approximation, that the cracks appear physically and geometrically in the same way under quasi-static and cyclic loading. We use a formulation already developed for quasi-static loading in the cyclic case. We use experimental data to describe the cumulative damage process. Then we write a damage law. We identify the constants of the model and we numerically simulate two level fatigue tests .

INTRODUCTION

Damage tolerance of cross-ply laminates is of great importance for fatigue properties of a composite. Most studies have been performed on flat specimens, such as $(0°, 90°)_s$. Two main disadvantages of this sample type are the edge effects and the impossibility to follow the propagation of the damage, in situ, when increasing loading. The non destructive test most widely used to follow crack propagation are replicas or dry-penetrant and X-ray analysis. Tubes were used for this study rendering these two methods inappropriate. We developed a new non-destructive test to characterize the damage state. As the tubes were wound to nearly reproduce the stacking sequence $(+45°, -45°)_{4s}$, the damage state can be defined by two types :
- the transverse matrix cracking;
- delamination.

Fibre breaks were generally not observed. However when fibre failure was detected it was not taken into account. The $(\pm 45°)$ layers were used in order to achieve several goals :
- a pure shear test in tension with equal crack densities (Fig.1a).
- a multicracking state in the $+ 45°$ angle plies of the tube in torsion (for a non symmetric loading). In this case, the fibres and the principal axis are in the same directions (Fig. 1b).
- a generalized asymmetric multicracking state in $+ 45°$ and $- 45°$ directions using tension and torsion. In this specific case, more transverse cracks are formed in one direction due to the superposition of the tensile and torsion loadings (Fig. 1c).

The first aim of this work is to measure the crack density, which appears in each ply during a test (static or fatigue) without removing the specimen.

The second goal is to model the cracking phenomenon, and to predict the damage evolution. In this initial approach to transverse cracking under fatigue loading, the cracks are assumed to have the same geometric characteristics as those occurring under quasi-static loading. In particular, it is assume that they go through the thickness of the plies and that their propagation is quasi-instantaneous. This hypothesis allows us to extend the methodology developed in the quasi-static case by Renard and Thionnet [1] to the fatigue case. Tests we made and encountered in the literature then guide us in the choice of the accumulation process.

EXPERIMENTAL PROCEDURES

Tubes were made of E-glass and epoxy resin LY 556/HY 932. They were 230 mm long, external diameter 18.2 mm, 1.6 mm thick with 8 layers. The fibre content is 0.54.

Optic fibres were used to light the tube from the inside. Then, no temperature rise can warm the inner surface of the tube. On the outside, a video-camera and a camera were fitted on a binocular to record the damage as the load increases (Fig. 2). The light is strong enough to see the formation of the cracks in the outer plies (may be 2 or 3). Cracks appear as dark parallel lines. It is important to note, that we measure a surface damage state for given a layer, and not the superposition of several damage states for each ply as the case with the X-ray technique.

Because the specimen is a wound tube, the outer ply may be either at $+ 45°$, or at $- 45°$ (Fig. 3). The video and photographs show that the cracks with different lengths seemed to appear throughout the sample. Crack density in a flat specimen is calculated as a function of the non destructive control test used, the stacking sequence and the material. If X-ray analysis is used on a $(0°, 90°)_s$ specimen, crack density is calculated as a number of cracks per unit length per damaged ply. For a more complex stacking sequence involving $\pm \theta°, 0°, 90°$ plies, the use of replicas is very useful. The crack density for the $\pm \theta°$ layers is calculated as a number of cracks on the edge per unit length per the sinus of the angle of the ply. In our case, the crack density is defined as the total length of cracks per unit area .The width of the surface is chosen so as not to exceed 5 mm to avoid the spherical distorsion. From the different photographs, crack densities were calculated for each step of the loading. We first used a classical method using tracing paper. All the calculations were made at the same time. Then, we were sure not to be influenced by the final result. To avoid this subjective part of the technique, we are developing a computer image analysis.

To identify the damage accumulation type and the parameters of the model :
- first, static tests are performed;
- second, tests were made at different values of maximum torque for different number of cycles and given frequency.

DESCRIPTION OF THE CONVENTIONAL FATIGUE EQUATIONS

Generally, a fatigue evolution equation can be expressed as follows :
$$\frac{dD}{dN} = f(D, V_E, p(N))$$
where :
- D is the damage variable (which is a state variable);
- V_E denotes a set of state variables;
- $p(N)$ denotes a set of parameters depending on the number of cycles and describing the environment in which the structure is located. Certain of these parameters characterize the mechanical environment, such as R, the loading ratio, and the frequency F.

The temperature, humidity, chemical activity can be used to describe the external environment. Now that the parameters of the equation have been describes, it is necessary to take their dependences into account, i.e. how they are involved in the damage accumulation process. For instance, in the case where one of parameters p(N) is the maximum stress σ_{max} how does damage accumulation proceed when a test is conducted on two loading levels ? Linearly or not ? Actually, the answer results from the form adapte for function f, which is itself based on experimental observations. Let us take the case where the damage evolution equation depends on only one parameter, denoted P_0. It is shown that when f has separable variables, the accumulation is necessarily linear. It can simply be stated that an evolution equation which may or may not be linear with respect to the damage is linear with respect to the accumulation if the damage variable is in a one-to-one relation with the percentage of life elapsed, i.e. the ratio between N and the number of cycles to failure (in our case, the failure is the saturation state). It is necessary to carry out fatigue tests on several levels on a representative volume element of the material to be modeled in order to see what type of equation should be adopted.

CHOICE OF THE ACCUMULATION RULE

The tests we conducted (Fig. 4) and tests encoutered in the literature [2, 3] in the case of $(0°, 90°)_s$ are highly significant from the standpoint of the fatigue behavior of long fiber composite materials. We therefore used them to decide on the type of accumulation to be used. For each level, the number of cracks versus the number of cycles is recorded (Fig. 4). It can easily be seen that the loading levels have a large effect on the damage kinetics. In effect, plotting the network of curves for the number of cracks versus N/N_s where N_s is the number of cycles at which the damage reaches saturation state, it can be seen that this network cannot be reduced to a single curve. As the relation between the damage and N/N_s is not a one-to-one relation, the damage accumulation cannot be linear. A formulation in which the damage and number of cycles variables are inseparable is necessary. In addition, it is assumed that the damage at a given time has no memory of the path followed to reach it. This means that to determine the damage evolution, it is necessary only to know the initial value of this damage in addition to the loading parameters.

In the case of two-level loading, it can be shown how the damage is accumulated through three possible cases. We denote as C_1 and C_2 the two loading levels applied consecutively for N_1 and N_2 cycles. We denote as D_1^0 and D_2^0 the damage that would have been reached if C_1 and C_2 had been applied quasi-statically. It can be seen that D_1^0 and D_2^0 had are the values reached by the damage during the first cycle if the tests are conducted on a single loading level with C_1 and C_2 respectively. This is why it is assumed in all cases that, depending on the applied loading intensity, static damage can occur during the first loading, i.e. the first cycle :

- case a : low level C_1, then high level C_2 such that the damage D_1 reached at the end of C_1 is less than the damage D_2^0 at the start of static loading with C_2. When the level is changed, there is first an increase in the damage from D_1 to D_2^0 due to the new static loading, followed by fatigue damage during cycling at C_2;
- case b : low level C_1, then high level C_2 such that damage D_1 reached at the end of C_1 is higher than damage D_2^0 at the start of static loading with C_2. When the level is changed, there is no longer an increase in the damage due to the new static loading but fatigue damage immediately during cycling at C_2;
- case c : high level C_1 then low level C_2. When the level is changed, the damage increase is due immediately to the fatigue phenomenon during cycling at C_2.

Now that we have chosen the accumulation process, we must model it. First, we recall how damage under static loading is modeled, since this model was used as inspiration to write our equation for damage evolution under fatigue loading. It is possible to use a model written for static loading to develop a mode lunder fatigue loading only because it is assumed that the defects generated in both cases are physically and geometrically the same.

QUASI-STATIC LOADING DAMAGE MODEL [1, 4]

We consider a laminate structure assumpting plane stresses. Transverse cracking is modeled on the ply scale and is describe by a single variable denoted α_f, defined from the the physical quantities of this defect $\alpha_f = e_f /L$ where e_f and L represent the thickness of the cracked ply and the intercrack spacing respectively (Fig. 5). We place ourselves in the local orthotropic reference system of the ply where x_1 is the fiber axis and x_1 is the direction perpendicular to its plane. The first step in modeling is to compute the stiffnesses of the homogeneous ply equivalent to the cracked ply in the sense of damage mechanics, as a function of α_f by a homogenization method. Curve fitting of these quantities is carried out with the form :

$$R_{\gamma\delta}(\alpha_f) = R_{\gamma\delta}^0 e^{-k_{\gamma\delta}\alpha_f} \quad (\gamma, \delta = 1, 2, 6)$$

where $R_{\gamma\delta}^0$ and $k_{\gamma\delta}$ are the stiffnesses of the undamaged material and the intrinsic constants of the material determined by curve fitting respectively. Two sets of constants $k_{\gamma\delta}$ are identified one when the cracked ply is inside the lamination and the other when it is on the outside. The behavior of the ply assumed to be under plane stresses and expressed in its local reference system is denoted $\sigma_\alpha = R_{\alpha\beta}^0 \varepsilon_\beta$ where σ_α and ε_β denote the ply stresses and strains respectively. The second step is to find an evolution equation for α_f when the structure containing the damaged ply(ies) is subjected to static loading. To do so, we first construct a thermodynamic potential ψ, then we define variable A_f, which is the dual variable of α_f, by analogy with the constitutive equation :

$$\psi = \psi(\varepsilon, \alpha_f) = \frac{1}{2} R_{\gamma\delta}(\alpha_f)\varepsilon_\gamma\varepsilon_\delta \Rightarrow A_f(\varepsilon, \alpha_f) = \frac{\partial\psi(\varepsilon, \alpha_f)}{\partial\alpha_f} = \frac{1}{2}\frac{\partial R_{\gamma\delta}(\alpha_f)}{\partial\alpha_f}\varepsilon_\gamma\varepsilon_\delta,$$

then we write an initiation criterion :

$$f = A_f^c(\alpha_f) - A_f(\epsilon, \alpha_f) \leq 0$$

from which we determine the desired evolution equation assuming consistency (f = 0 during the damage phase). The damage growth is then expressed as a function of the growth in the local strains of the cracked ply :

$$d\alpha_f = -\frac{\dfrac{\partial f}{\partial A_f}\dfrac{\partial^2\psi}{\partial\alpha_f\,\partial\epsilon_\gamma}d\epsilon_\gamma}{\dfrac{\partial f}{\partial A_f}\dfrac{\partial^2\psi}{\partial\alpha_f^2}+\dfrac{\partial f}{\partial\alpha_f}} = \frac{-\dfrac{\partial^2\psi}{\partial\alpha_f\,\partial\epsilon_\gamma}d\epsilon_\gamma}{\dfrac{\partial^2\psi}{\partial\alpha_f^2}-A_f^{c'}(\alpha_f)} = \frac{-R_{\gamma\delta}(\alpha_f)\epsilon_\gamma d\epsilon_\delta}{\frac{1}{2}R_{\gamma\delta}^{''}(\alpha_f)\epsilon_\gamma\epsilon_\delta - A_f^{c'}(\alpha_f)}.$$

The third step consists of identifying the critical damage threshold $A_f^c(\alpha_f)$. This identification is performed using the experimental results obtained on a stacking of type $(0°_n, 90°_m)_s$, this lamination makes it possible to correctly identify and observe the cracking phenomenon. $A_f^c(\alpha_f)$ is identified using, in a particular algorithm [5], $A_f(\epsilon, \alpha_f)$ and the experimental data giving the damage (crack density) versus the applied stress. The form of the calculation point suggests taking $A_f^c(\alpha_f)$ with the following form :

$$A_f^c(\alpha_f) = -be^{\dfrac{\ln\{-\ln(1-\frac{\alpha_f}{c})\}}{a}}$$

Coefficients a, b and c are determined by curve fitting.

SETTING OF THE FATIGUE EQUATION

The criticism that can be made of the conventional expression for the fatigue equations is that they include loading explicity through the mean or maximum stress, the amplitude, etc ... This remark applies both to metals and to composites. For instance, the evolution equation proposed by Beaumont [6] uses the amplitude of the external loading in addition to R and F. Another equation, written by Ye [7], uses the maximum longitudinal load applied to the edges of the structure. Such evolution equations are therefore very difficult to apply in the case of complex structures, when the loadings vary over time both in intensity and in point of application. In effect, if the behavior of the volume element depends on external factors, the equations do not really have an intrinsic character. Also, at present, most of the cases are based on uniaxial tests providing a uniaxial model whose extension to the multiaxial case is not necessarily justified. A more logical approach is to formulate a multiaxial model developed and identified on tests which could be uniaxial.

We therefore directly construct an evolution equation from an energy formulation making it unnecessary to have a 3D generalization which is difficult to achieve. It should be noted that parameters R and F characteristics of the external loading can be included in the evolution equation since these values are preserved throughout the structure subject to working in the framework of linear elasticity (damageable or not). In addition, since the aim is to account for the loading induced by fatigue effects as well as possibly by static effects, we attempt to write an evolution equation as follows : $d\alpha_f = g(\epsilon, \alpha_f, R, F)\, d\epsilon + h(\epsilon, \alpha_f, R, F)\, dN$, where ϵ denotes the strain state in the volume element when the loading reaches maximum value. The part related to the increases in strain takes static damage into account whereas that related to the increases in the number of cycles is a pure fatigue effect. Our idea was to obtain this formulation by extending the critical initiation threshold developed for the static case to the case of fatigue. We therefore decided to make it depend not only on α_f, as in the static case, but also on the number of cycles and the loading parameters (R and F for example). It is therefore written $A_f^c(\epsilon, \alpha_f, R, F)$, in which it is obvious that the static threshold equation should be found when N = 1. At a given time, there is therefore a unique threshold which acts as damage state memory. We therefore chose $A_f^c(\alpha_f)$ with the same form as for the static case, the difference being that coefficient a, b, c are functions of N and the following parameters :

- static threshold : $A_f^c(\alpha_f) = -be^{\dfrac{\ln\{-\ln(1-\frac{\alpha_f}{c})\}}{a}}$;

- fatigue threshold : $A_f^c(\alpha_f, R, F, N) = -b(R, F, N)e^{\dfrac{\ln\{-\ln(1-\frac{\alpha_f}{c(R, F, N)})\}}{a(R, F, N)}}$

It can then be noted that :

- the fact that variables α_f and N are inseparable appears here;

- the influence of parameters (R, F, etc ...) of the model is included in coefficients a, b and c of the critical threshold A_f and only there. Since the identification of A_f requires experimental tests, it is sufficient to conduct those showing the effect of a given parameter in order to take it into account in curve fitting of coefficients a, b and c;
- the equation for the criterion remains unchanged with respect to the static approach :

$$f = A_f^c(\alpha_f, R, F, N) - A_f(\epsilon, \alpha_f) \leq 0.$$

As only tests conducted with R and F fixes are currently available, we place ourselves in this context until the end of work. However, it is important to note that this does not detract in any way from the generality of the methodology insofar as the establishment of the evolution equation and the identification of A_f^c remain completely unchanged. The evolution equation is determined by writing, again as for the static case, that we are and remain on the threshold during the damage phase (consistency hypothesis), and by eliminating N from its expression because this quantity is not a state variable :

$$d\alpha_f = \frac{-\dfrac{\partial^2 \psi}{\partial \alpha_f \, \partial \epsilon_\gamma}}{-\dfrac{\partial^2 \psi}{\partial \alpha_f^2} + \dfrac{\partial A_f^c}{\partial \alpha_f}(\alpha_f, g(A_f, \alpha_f))} \, d\epsilon_\gamma + \frac{\dfrac{\partial A_f^c}{\partial N}(\alpha_f, g(A_f, \alpha_f))}{-\dfrac{\partial^2 \psi}{\partial \alpha_f^2} + \dfrac{\partial A_f^c}{\partial \alpha_f}(\alpha_f, g(A_f, \alpha_f))} \, dN.$$

The process for the fatigue case is fully comparable to the quasi-static case, since we also use A_f (ϵ, α_f) to calculate critical threshold A_{fc} in a particular algorithm [5]. The difference is that there are several threshold to be identified (as many as there are cycles in the absolute) and the calculation is therefore conducted for each curve of the available experimental network.

The experimental data give the damage (the crack density) versus the number of cycles for each loading level with R and F fixed. If other parameters are inclued in the model such as (R, F, etc..) curve fitting has to take into account the possible effect of each of these parameters through coefficients a, b and c which are then written a (R, F, N,...) b (R, F, N,..) and c (R, F, N...).

NUMERICAL SIMULATIONS

We compare now the response of the model with two levels loadings tests : first a weak loading followed by a strong loading, second a strong loading followed by weak loading. Both cases are plotted on figure 6.

CONCLUSION

The experimental device permits to follow of the damage with a microscopic approach using an optical method which characterize the crack density. The specimen is not removed from the machine, so that no more damage is added during the different manipulations.

For the model of fatigue in laminated composites whose plies have unidirectional fibers, we arrived at a new form of the transverse cracking damage evolution equation. The originality of the work resides in three points :
- first, in the evolution equation per se, which can take possible static loadings into account. This makes it very flexible for simulations as regards the diversity of possible applied loadings;
- next, in the process by which the equation was established. The assumption that the defects created in a structure are similar for quasi-static loading and for cyclic loading makes it natural to extend the quasi-static model to the fatigue model;
- finally, contrary to what is frequently the case, the parameters of the equation do not explicitly include the external loading applied to the structure. This makes the model intrinsic to the material and therefore applies to all types of loadings and structure.

It should also be noted that a parameter (R, F, etc ...) can easily be added to the model without any change in the methodology. In effect, this addition is made in the critical threshold A_f^c. It is sufficient to conduct the required experimental tests taking introduction of the new parameter into account. We described the different cases of damage accumulation, in particular the increase due to a major change in the level of the applied stress. We also showed the influence that the order of application of the loads had for tests on two levels.

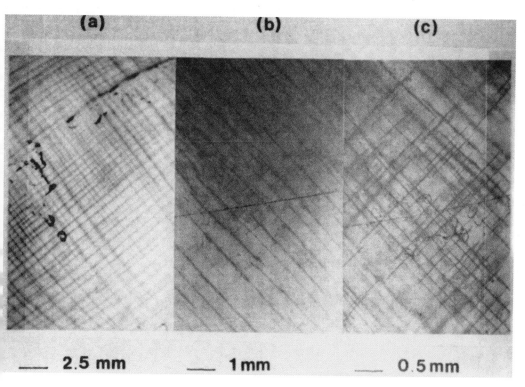

Fig. 1 - Views of matrix cracking for different loadings
(a) tension; (b) torsion; (c) tension-torsion

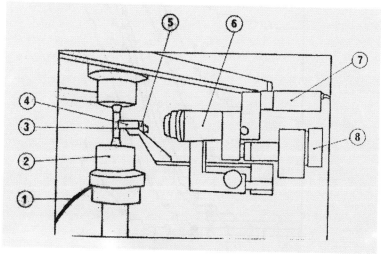

Fig. 2 - Schematic of the experimental equipment
1 : optic fibers; 2 : tubes fixtures; 3 : lighted specimen; 4 : optical system with mirrors;
5 : extensometer; 6 : binocular; 7 : video camera; 8 : camera

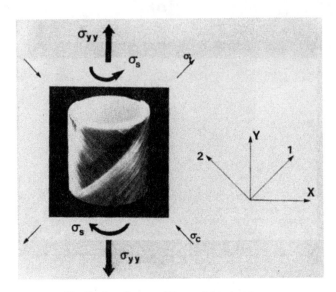

Fig. 3 - Loading conditions of the structure.
View of a pyrolysed tube showing the ±45° directions of the tows

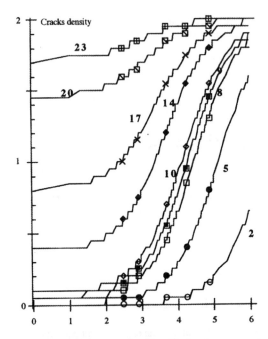

Fig. 4 - Experimental results used to identify the damage evolution law
(The bolded numbers indicates the maximum applied torque)

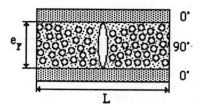

Fig. 5 - Quantities characterizing the defect

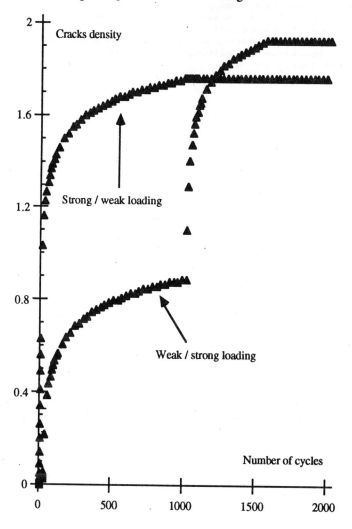

Fig. 6 - simulation of two levels loading tests

REFERENCES

[1]   RENARD J. and THIONNET A. - Une loi d'évolution de la fissuration transverse dans les composites stratifiés soumis à un chargement quasi-statique. Revue des composites et des matériaux nouveaux, Vol. 1, n° 1-1991

[2]   DANIEL I.M. and CHAREWICZ A. - Fatigue Damage Mechanisms and residual Properties of Graphite/Epoxy Laminates. Engineering Fracture Mechanics, vol. 25, 5/6, (1986).

[3]   LESAGE P., J. RENARD and THIONNET A. - Damage model for composites materials submitted to quasi-static and cyclic loadings based on meso-macro formula. MECAMAT 93, International seminar on micromechanics of materials, Moret-sur-Loing, France

[4]   THIONNET A. and RENARD J. - Meso-macro approach to transverse cracking in laminated composites using Talreja's model. Composites Engineering (à paraître).

[5]   THIONNET A. and RENARD J. - An evolution equation for transverse cracking in laminated composites under fatigue loading. La Recherche Aérospatiale, n°1992-1

[6]   BEAUMONT P.W.M. - The Failure of Composites an Overview. Journal of Strain Analysis, vol. 24, 4, (1989).

[7]   YE L. - On the Damage Accumulation and Material Degradation in Composite Materials. Composites Sciences and Technology, 36, (1989)

KINETIC OF TRANSVERSE MATRIX CRACKING IN (0n, 90m)s COMPOSITES UNDER TENSION OR SHEAR STATIC LOADING

B. MACQUAIRE, J. RENARD

Ecole Nationale Supérieure des Mines de Paris - Centre des Matériaux P.M. Fourt - BP 87 - 91003 Evry Cedex - France

ABSTRACT: The present paper deals with the continuum damage mecanism in cross-ply laminates such as [0,90]s. This approach includes three steps which are : 1- the threshold of the first crack in the damaged ply, 2- the determination of the kinetics of the progressive damage, 3- the expression of the saturation state, (CDS). The proposed analysis can be transposed to others off-axis specimen where the opening crack mode is mixed. The cumulative law is based on a Fermi-Dirac distribution for which the parameters used have a greater physical significance.

1. INTRODUCTION

Damage occuring in cross-ply laminates are generaly transverse matrix cracks. This damage is of great importance for static tests for Glass/fibre specimen. Most of the non-linearity of the behaviour curve can be attributed to these defects. To modelize the behaviour curve of different stacking sequences, it is necessary to know the intrinsic transverse stiffness loss. Therefore, it is easy to recover the main behaviour with several methods such as the Laminate Plate Theory. We used a shear-lag model [1] which impose the shear strain in half of the 90 ply. The obtained curves for the different materials were fit with decreasing exponentials involving the thickness of the ply, the crack density and a parameter k22, [9]. This parameter takes into account the global behaviour of the matrix and the interface. For each resin system, a value k22 will be associated. The knowlage of this parameter is enough to calculate the CDS of every cross-ply laminate. A second parameter which is the threshold of the first crack will be sufficient to determine the cumulative damage as a fonction of the applied stress.

I - Theory
I-1 Stiffness reduction in the damaged ply

The stiffness loss are calculated with the Swanson model [1]. This 1D shear-lag analysis considers that the shear strain between two neighboured layers only takes place in the weakest ply. In the case of a [0,90]s laminate, this assumption is reasonnable.

As shown in figure n° **1**, the equilibrium of each cell gives the relations between the normal and shear stresses for an inner 90 ply :

$$t_0 \frac{\partial \sigma_0}{\partial x} - \tau_{xz} = 0 \tag{1}$$

$$t_{90} \frac{\partial \sigma_{90}}{\partial x} - \tau_{90} = 0 \tag{2}$$

in terms of displacements these equations lead to :

$$\alpha_1 \frac{\partial^2 U_{90}(x)}{\partial x^2} - U_{90}(x) + U_0(x) = 0 \tag{3}$$

$$\alpha_2 \frac{\partial^2 U_0(x)}{\partial x^2} - U_0(x) + U_{90}(x) = 0 \tag{4}$$

and the transverse stiffness loss is given :

$$\frac{\tilde{E}_{90}}{E_{90}} = \frac{\overline{U}_{90}(L)}{U_{90}(L)} = \frac{1 - \dfrac{\tanh(\beta L)}{\beta L}}{1 + C_2 \dfrac{\tanh(\beta L)}{\beta L}} \tag{5}$$

$\overline{U}_{90}(L) = \overline{\varepsilon}L$: the average damaged displacement for the length L
$U_{90}(L)$: the uniform displacement at x=L

for an inner 90 ply:

$$\beta^2 = t_{90} G_{23} \left(\frac{1}{E_{90}t_{90}} + \frac{1}{E_0 t_0} \right) = \frac{1}{\alpha_1} + \frac{1}{\alpha_2}; \qquad C_2 = \frac{E_{90}t_{90}}{E_0 t_0} = \frac{\alpha_1}{\alpha_2}$$

t_{90}: half thickness of the whole 90° ply
t_0: half thickness of the outer 0° ply

for an outer 90 ply :

$$\beta^{2'} = \beta^2 \text{ and } C^{2'} = C^2 \text{ with :}$$

t_{90} : thickness of the outer 90° ply
t_0 : half thickness of the inner 0° ply

in both case, the crack density is $d = \dfrac{1}{2L}$ (6)

694

The stiffness loss can be approximated by this function [9] :

$$\tilde{Q}_{22} = Q_{22} \exp(-k_{22} \, t_{90} \, d) \tag{7}$$

where $\tilde{Q}_{22}$: damaged transverse stiffness

                t_{90} : the whole thickness of the 90 ply, (mm)

                d : the crack density, (1/mm)

                k_{22} : a constant

In that case, we approximated :

$$\tilde{E}_{90} = \tilde{Q}_{90}. \tag{8}$$

Hence, we defined the k_{22} (outside) and k_{22} (inside) for a $[0,90_3]s$ and a $[90_3,0]s$ in glass/epoxy. Stiffness parameters were taken after reference [4].

Figure n°2 shows the transverse reduction as a function of the damage parameter ($t_{90} \, d$) for both stacking sequences mentioned above.

I-2 Threshold of the first transverse crack ($\sigma_{90}{}^{f}$) in the 90° ply

We did not take into account the residual curing stresses which have a very large effect on this threshold. We identified an average value from results found in the litterature,[5][6][7][8]. This threshold is around 36 MPa and corresponds to the maximum stress obtained for a $[90]_n$ laminate. In the figure n°3, the different thresholds after[8] are plotted with the number of 90 plies. Because no previous damage appeared before (except for the curing stresses), these points are perfectly aligned. With a simple elastic calculation and assuming that the crack corresponds to the matrix failure in mode I :

$$\frac{\sigma_0}{Q_0} = \frac{\sigma_{90}}{Q_{90}} = \varepsilon_{app} \tag{9}$$

$$\sigma_{app} = v_0 \sigma_0 + v_{90} \sigma_{90} \tag{10}$$

the threshold will be obtained for the critical value in the 90° ply

$$\sigma_{fpf} = \left(v_0 \frac{Q_0}{Q_{90}} + v_{90} \right) \sigma_{90}{}^{f} \tag{11}$$

$$\sigma_{90}{}^{f} = v_f Q_f \varepsilon_f + v_m \sigma_m{}^{f} \cong v_m \sigma_m{}^{f} \tag{12}$$

where σ_{fpf} : the first ply failure

            σ_m^{f} : the maximum stress of the epoxy resin in tension

            v_0, v_{90} : ratio of the two types of layers

            v_m, v_f : matrix and fibre volume content

I-3 The characteristic damage state, (CDS)

As soon as no delamination exists in the laminate, and the ultimate strain of the 0° plies is not obtained, a CDS in the damaged ply can be reached. To calculate this CDS, we defined an equivalent damaged elastic material where the stiffness values are lowed in regard with the number of transverse cracks, see equation (7). From equation (2), we calculated the shear stress at the crack tip :

$$\tau_{xz}(0) = \frac{\partial \sigma_{90}}{\partial x}\bigg|_{x=0} = t_{90}\sigma_{90}\beta\tanh(\beta L) \tag{13}$$

because L is very small, i.e. the CDS is reached , we approximated :

$$\tanh(\beta L) \cong \beta L \text{ , so that :}$$
$$\tau_{xz}(0) = t_{90}\sigma_{90}\beta^2 L \tag{14}$$

At saturation, the shear stress is a linear function of L. This evolution is drawn on figure n°4. This schema represents a 90 ply with two cracks at the edges. Part n°1 is an equivalent elastic damaged material with a reduced transverse stiffness. Part n°2 is the half of the real cracked material. The left tip is the proper undamaged material and the right edge is a crack. For an applied strain ε, the normal damaged stress in the 90 ply is :

$$\tilde{\sigma}_{90} = \tilde{Q}_{90}\varepsilon \tag{15}$$
$$\tilde{\sigma}_{90} = \sigma_{90}\exp(-k_{22}\ t_{90}\ d) \qquad \text{with (7)} \tag{16}$$

The equilibrium of the boudary between the cells n°1 and n°2 gives (for an inner ply)

$$t_{90}\tilde{\sigma}_{90} = t_{90}\sigma_{90} - 2\bar{\tau}_{xz}L. \tag{18}$$

where $\bar{\tau}_{xz}$ is the average shear stress; finally with (6),(15),(16) we have :

$$F(d) = d\left(1 - \exp(-k_{22}t_{90}d)\right) = \frac{1}{t_{90}}\left(\frac{\bar{\tau}_{xz}}{\sigma_{90}}\right) \tag{19}$$

When saturation is reached, the load transfert is maximum. So, for a small variation of d, the ratio $\dfrac{\Delta\bar{\tau}_{xz}}{\Delta\sigma_{90}}$ must be as big as possible. In terms of derivative :

$$F'(d) \text{ maximum}$$
$$F''(d) = 0 \tag{20}$$

$$d_s = \frac{2}{k_{22}\ t_{90}} \tag{21}$$

This model can be easily extended to the ±45 plies. It just requires the values of k_{66} (outside) and k_{66} (inside). The function F'(d) versus d is presented in figure n°5.

I-4 Cumulative damage law

Most of the studies dealing with matrix cracking refer to a Weibull distribution law. More precisely, the Weibull parameters (σ_u , m) have no accurate physical meaning, so that, we proposed a looked like Fermi-Dirac distribution :

$$N(d) = \cfrac{1}{1 + \exp\left\{\cfrac{-A\,d_s\sigma_{app} - \sigma_{lam}^f + \sigma_{90}^f}{4\,\sigma_{90}^f}\right\}} \qquad (22)$$

with

A	:	a material identified parameter, (2 for CFRP and 4 for GFRP)
σ_{app}	:	the macroscopic applied stress
σ_{lam}^f	:	the ultimate failure stress of the laminate
σ_{90}^f	:	the threshold of the first transverse crack in the 90 ply
ds	:	the crack density at saturation, (CDS)

the crack density during the loading sequence is :

$$d = d_s \; N(d) \qquad (23)$$

II - Conclusion

The threshold of the first transverse crack is obtained with a deterministic criteria and an elastic analysis. Because on figure n°3 the experimental thresholds are perfectly aligned, we can conclude that this approximation is good. The behaviour of a damaged ply, totaly or half constrained is strongly modified. From figure n° 2 we can see that the ratio between the two k_{22} values is not 2 as we could have expected. Nevertheless, the approximation with the decreasing exponential is not very accurate. Because these different k_{22} values, a CDS for each type of laminate can be calculated with equation (21). The agreement with the results from [3] is very good, figure n°6. The model predicts that the CDS is inverse proportional to the ply thickness.

If the proposed damaged law, eq. (22), fits quite the experiment for a glass/epoxy specimen, the evolution in the outer plies still causes some difficulties. In fact, several damage modes are involved for [90,0]s type specimen such as delamination along the crack tip. This is particulary amplified during fatigue tests. Those delaminations share a amount of the released energy and slow down the kinetics of the transverse cracking mecanism. The results are reproductible when we compare the curves obtained by Laws & al and Highsmith & al for a

$[0,90_3]$. We identified the A parameter which should be proportional to a length in the case of a dimensionless equation. Figure n° 7 shows the comparison between the experimental results and the model for a carbon-epoxy system. The general shape of the curve is recovered, but the fitting is high dependant of the parameters. To fit the tests from [8], we choose respectively 2.4; 3 and 4.2 mm for the thickness of the 90° plies. It can be noticed that the thickness of the ajacent plies does not change the damage kinetics of the inner 90 ply.

We proposed a simple analytical model to predict the damage in a cross-ply laminate in mode I. This analysis can be extended to a specimen loaded in mode II. This model correctly describes the experimental behaviour in the case where only matrix cracking occurs in the damaged ply : delamination, splitting, fibre break and residual curing stresses are not taken into account.

III - References

[1] SWANSON S.R., "On the Mechanics of Microcracking in Fibre Composite Laminates under Combined Stress", Journal of Engineering Materials and Technology, Vol. 111, April 1989, pp 145-149.

[2] HIGHSMITH, A.L. and REIFSNIDER, K.L., "Stiffness Reduction Mechanisms in Composite Laminate ", ASTM STP 775, (1982), pp 103-117

[3] LAWS, N. and DVORAK, G.J.,"Progressive Transverse Cracking in Composite Laminates", Journal of Composite Materials, Vol. 22, October 1988, pp 900-916

[4] PETITPAS, E. & al , "Comportement en fatigue de stratifiés carbone-résine à base de fibre T300 ou T400", proc. of JNC6, Eds. J.P. Favre et D. Valentin, Paris, (1988), pp 621-633

[5] HULL, D., An Introduction to Composite Materials",Cambridge Solid State Science Series, (1981), p 171

[6] BENZEGGA, M.L.,"Matériaux Composites", Tome 2,automne 1991, cours de l' Université de Technologie de Compiègne

[7] LACAZE, S., ANQUEZ, L., "Modeeling of transverse crack growth and saturation in cross- ply laminates", Journal of Materials Science 27, (1992) , pp 5892-5988

[8] WANG, A.S.D.,"Fracture Mechanics of sublaminate Cracks in Composite Materials", Composites Technology Review, Vol. n°2, Summer 1984, pp 45-62

[9].RENARD, J., FAVRE J.P. and JEGGY, T.,"Influence of Transverse cracking on ply behaviour : Introduction of a Characteristic Damage Variable, Composites Science and Tehnology, 46, (1993), pp 29-37.

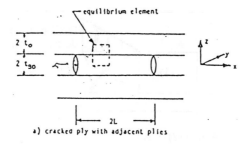

figure n°1 : schema of the cross-ply laminate

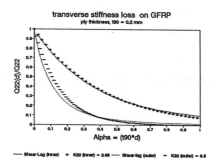

figure n°2 : transverse stiffness loss for a GFRP as a function of the thickness ply and the crack density.

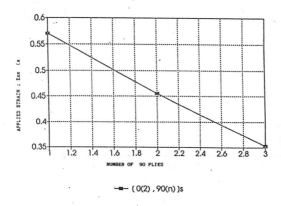

figure n° 3 : experimental thresholds [8] as a function of the number of 90 plies for a CFRP

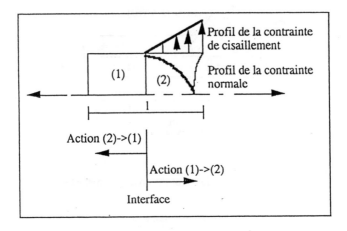

figure n°4 : schema of the two half damaged cells corresponding to a cracked ply

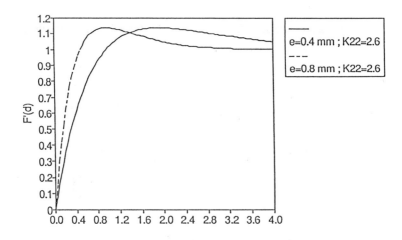

figure n° 5 : F'(d) as a function of the crack density

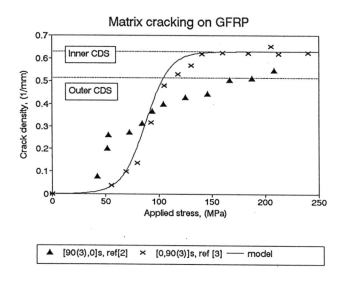

figure n° 6 : crack density versus the applied stress for GFRP specimen

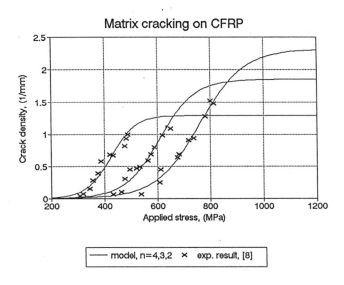

figure n° 7 : crack density for $[0,90_n,0]_t$ CFRP versus the applied stress, from [8]

EACM STANDING COMMITTEE

The role of the European Association for Composite Materials is to further contacts between scientists and engineers working in the field of composites and to encourage the development of industries using high quality composite materials. Towards this goal, the European Association for Composite Materials is directed by a standing committee consisting of :

A.R. BUNSELL
Chairman
ENSMP
Paris - France
Tel. (33) 60 76 30 15
Fax (33) 60 76 31 50

A. MASSIAH
General Secretary
Comité d'Expansion Aquitaine
Bordeaux - France
Tel. (33) 56 01 50 20
Fax (33) 56 01 50 05

A. KELLY
University of Surrey
Great Britain
Tel. (44) (0) 483 57 12 81
Fax (44) (0) 483 57 24 80

H. LILHOLT
Risø National Laboratory
Denmark
Tel. (45) (42) 37 12 12
Fax (45) (42) 35 11 73

I. CRIVELLI VISCONTI
Universita di Napoli
Italy
Tel. (39) 81 76 82 366
Fax (39) 81 76 82 362

K. SCHULTE
Technical University of Hamburg-Harburg
Germany
Tel. (49) 40 771 83138
Fax (49) 40 771 82002

MEMBERSHIP

Whether you are in Europe, the United States, Japan or elsewhere, if you have a need to know what it is happening in all the areas of the composite materials field, become a member of EACM.

Advantages of EACM membership

• Free subscription to EUROCOMPOSITE LETTER (6 issues per year)

• Reduced registration fees for ECCM conferences

• Provide a phone-in information service

Other advantages

Being a member of EACM will allow you to receive 12 issues per year of Advanced Composite Bulletin, a publication of ELSEVIER ADVANCED TECHNOLOGY at EACM preferential rate for 1993 (20% discount) : £ 192.

For any further details, please contact :
EACM - 2 Place de la Bourse - 33076 Bordeaux Cedex - France
Tel. (33) 56 01 50 20 - Fax (33) 56 01 50 05.

MEMBERSHIP APPLICATION

To be returned to :

EACM
2 Place de la Bourse
33076 Bordeaux Cedex
France
Tel. (33) 56 01 50 20 - Fax (33) 56 01 50 05

NAME .... FORENAME ....

ORGANISATION/COMPANY ....

ADDRESS ....

....

POST CODE .... CITY ....

COUNTRY ....

TEL .... FAX ....

• Check type of membership for which you qualify :

• COMPANY MEMBERSHIP /__/

    - large sized firms .... 10 000 FF
    - small and medium sized firms .... 1 500 FF

• RESEARCH LABORATORY MEMBERSHIP /__/ .... 1 000 FF

• INDIVIDUAL MEMBERSHIP /__/ .... 500 FF

Payment is enclosed .... /__/ (make out your cheque to EACM)

Please invoice me .... / _ _ /

Please charge my credit card .... /__/ Visa/Eurocard/Mastercard
                                        /__/ American Express

Card : .... N° : ....

Expiry date : .... Signature : ....

The proceedings have been printed from texts written by the authors. The editors and the organisers are not responsible for the contents of the papers and for the possible mistakes make in the authors' texts.

LA NEF
I M P R I M E U R   C O N S E I L
22, RUE DU PEUGUE
33000 BORDEAUX

Dépôt légal septembre 1993, N° imprimeur 40544

Imprimé en France